Introduction to Engineering Design:
Modelling, Synthesis and Problem Solving Strategies

To Eva and Cathy

Introduction to Engineering Design: Modelling, Synthesis and Problem Solving Strategies

Andrew Samuel

John Weir

OXFORD AUCKLAND BOSTON JOHANNESBURG MELBOURNE NEW DELHI

Butterworth-Heinemann Ltd
Linacre House. Jordan Hill. Oxford OX2 8DP
225 Wildwood Avenue, Woburn, MA 01801-2041
A division of Reed Educational and Professional Publishing Ltd

A member of the Reed Elsevier plc group

First published 1999

British Library Cataloguing in Publication Data
A catalogue record for this book is available from the British Library

Library of Congress Cataloguing in Publication Data
A catalogue record for this book is available from the Library of Congress

ISBN 07506 4282 3

Typeset by EI&Associates Pty Ltd. Melbourne
Printed and bound in Great Britain by Antony Rowe Ltd, Chippenham & Reading

Special Acknowledgment

We make consistent reference throughout this text to engineering judgement and wisdom. Some of this wisdom is acquired through years of practice in the design profession, but much of it is gained through close association with inspired and creative colleagues. Engineering design is an emerging discipline. Most learning institutions offering courses in engineering struggle with the nature of engineering design experience to be offered to students. It is a labour intensive activity. Our experience and wisdom in this emerging field has been nurtured and developed by our continued, close association with William (Bill) Lewis. It is now recognised that major technical change can occur through a series of paradigm shifts. In engineering education at Melbourne, a paradigm shift took place when engineering design became a core part of the engineering curriculum. This change can be, almost entirely, attributed to the inspired course management and planning of Bill Lewis. Bill gained his Ph. D. in 1974 in a then almost unheard of field of research in engineering courses, engineering creativity. Bill was the first to obtain a doctorate at Melbourne in engineering design, and may well have been the first in the world to do so. Even before that achievement, in 1967, Bill was the first to identify and publish a reasoned evaluation of formal educational objectives in engineering design courses. These days engineering courses around the world offer problem-based and project based learning experiences to undergraduates. With Bill's guidance and inspiration we have been offering such learning experiences for more than 25 years. This book is a distillation of these experiences and Bill's contribution is gratefully acknowledged.

Acknowledgment

Writing a text on engineering design, based on an on-going course programme, makes wide use of ideas gleaned from both students and colleagues. Since its serious introduction to the undergraduate programme at Melbourne in the early 60s, the design programme has been an organic component of the engineering course. It has grown and has been honed and nurtured by ideas and contributions from many sources. Some of these ideas were developed in formal planning sessions, but many are the result of informal coffee-table discussions. For all of these, and the many opportunities to work with insightful colleagues and inspiring students, we are grateful. Special thanks are due to Colin Burvill, Bill Charters, Bruce Field, Jamil Ghojel, Errol Hoffmann, Barnaby Hume, Janusz Krodkiewski, Wayne Lee, Stuart Lucas, Jonathan McKinlay, Peter McGowan, Peter Milner, Alan Smith, Craig Tischler, and the many hundreds of undergraduates who continue to keep us on our mettle.

Andrew Samuel
John Weir
Melbourne, 1999

CONTENTS

Acknowledgement

Contents

Commonly used symbols and design terminology

Preface : The need for this book

PART 1 MODELLING AND SYNTHESIS

1 Introducing modelling and synthesis for structural integrity 3

1.1 Structural integrity and the nature of failure 5
1.1.1 Some spectacular structural failures 5
1.1.2 Some spectacular mechanical failures 8
1.1.3 Summarising our review of failures 13
1.2 Units, estimation and things we expect you to know 13
1.2.1 Units 14
1.2.2 Estimation 15
1.2.3 What we expect you to know 16
Case example 1.1: Dolphin enclosure 20
Case example 1.2: A Christmas tale
1.3 Structural distillation 21
1.3.1 Simple engineering components 23
1.3.2 Simple beam behaviour 25
Case example 1.3: Pliers 27
Case example 1.4: Multigrips
Case example 1.5: Paper clip 28
Case example 1.6: Scissors
1.4 Assignment on Structural distillation
1.5 Solutions to exercises 29
1.6 Assignment in modelling: Rural power distribution system 32
1.6.1 Instructions
1.6.2 Some relevant results from engineering science 33

2 Design against failure 49

2.1 Introducing engineering materials and modes of failure 50
2.1.1 Engineering materials 52

2.1.2 Corrosion and oxidation of metals 57
2.1.3 Plastics and composites 59
2.1.4 Wood and concrete 60
2.2 Material selection
2.3 Design for static loading, introducing failure predictors and factors of safety 62
2.3.1 Maximum principal stress failure predictor (*MPFP)* 66
2.3.2 Maximum shear stress failure predictor (*MSFP*) 67
2.3.3 Distortion energy failure predictor *(DEFP)* 69
2.3.4 Comparison of the three failure predictors 70
2.3.5 Factors of safety 72
2.4 Assignment in design for structural integrity :Airline service table 74
2.4.1 Background
2.4.2 The design problem
2.5 Design for dynamic loading (fatigue) 82
2.5.1 Factors influencing the onset of fatigue failure 83
2.5.2 A simplified design procedure for uni-axial fatigue 90
2.6 Assignment on design to resist fatigue failure 94
2.6.1 Background
2.6.2 The design problem 95
2.6.3 Data
2.7 Design for contact loading 101
2.7.1 The nature of contact between elastic bodies 103
2.7.2 Some applications 108
2.8 Chapter Summary 110

3 Design synthesis of some generic engineering components 111

3.1 Shafts: bending and torsion 112
3.1.1 Historical notes
3.1.2 Factors affecting shaft design 114
3.1.3 Modes of failure for shafts 118
3.1.4 Design of shafts to resist yielding and fracture
3.1.5 Principal stresses in shafts 119
3.1.6 Failure prediction for dynamic loading 120

	3.1.7	Failure by excessive lateral deflection	121
	3.1.8	Failure by excessive torsional deflection (excessive twist)	122
	3.1.9	Summary of design rules for shafts	
3.2	Assignment on shaft design		124
	3.2..1	Aims	
	3.2.2	Background	
	3.2.3	The design problem	
3.3	Mechanical springs: bending and torsion		131
	3.3.1	Historical note	
	3.3.2	The nature of mechanical springs	132
	3.2.3	Linear compression springs, static loading	136
	3.3.4	Stresses in linear compression springs, dynamic loading	138
	3.3.5	Other practical issues for compression springs	140
	3.3.6	Tension springs	
	3.3.7	Torsion springs	
	3.3.8	Properties of materials used for springs	142
	3.3.9	Vibrations and natural frequency	
	3.3.10	Summary of design rules for springs	143
3.4	Columns: axial loading and static instability		144
	3.4.1	Identifying columns	
	3.4.2	Design against failure in tension - an example of a *General Method* of design against structural failure	146
	3.4.3	Design against failure in compression	147
	3.4.4	Modes of Failure for columns	
	3.4.5	Buckling of Long Columns	
	3.4.6	Some key terminology	148
	3.4.7	Effective length	149
	3.4.8	Radius of Gyration	150
	3.4.9	Behaviour of short and intermediate columns	151
	3.4.10	Design Method	152
	3.4.11	Design Algorithms for Columns	
	3.4.12	Eccentric Loading	153

	3.4.13	Further extensions: beam columns	155
	3.4.14	Summary of design rules for axially loaded components	
3.5	Assignment on column design		155
	3.5.1	Background	
	3.5.2	The design problem	
	3.5.3	Additional information	156
3.6	Pressure vessels: internal pressure		164
	3.6.1	Failure of pressure vessels	
	3.6.2	Identifying vessels	165
	3.6.3.	Thin-walled cylindrical pressure vessels	167
	3.6.4	Thick-walled cylindrical pressure vessels	172
	3.6.5	Deformations in pressure vessels	174
	3.6.6	Stresses due to external bending	175
	3.6.7	Further complexities	177
	3.6.8	Summary of design rules for cylindrical vessels, internal pressure	179
3.7	Assignment on pressure vessel design		
	3.7.1	Background	
	3.7.2	The design problem	180

4 Design of mechanical connections 187

4.1	The nature of mechanical connections		
	4.1.1	Mechanical versus integral connections	
	4.1.2	Classes of mechanical connections	188
	4.1.3	Assignment on abstract thinking about fasteners	194
4.2	Bolted joints and screws: design for clamping contact		195
	4.2.1	Function of a bolted (clamped) joint	196
	4.2.2	Bolted joint failure modes	198
	4.2.3	A simple mathematical model for bolted (clamped) joints	
	4.2.4	Bolted joint design equations	203
	4.2.5	Fatigue of bolted joints	208
	4.2.6	Threaded fasteners for use in bolted joints	209
	4.2.7	Further details regarding threaded connections	212
	4.2.8	Summary of design rules for bolted (clamped) joints	215

4.3 Pinned joints and shear connectors
4.3.1 Typical pinned joints 216
4.3.2 Modes of failure 217
4.3.3 Simple design inequalities 218
4.3.4 Summary of design rules for pinned joints and shear connectors 221
4.4 Welded joints
4.4.1 Types of welded joint 223
4.4.2 Design of welded joints for strength 224
4.4.3 Welded fillet joints under parallel loading 225
4.4.4 Welded fillet joints under transverse loading 226
4.4.5 Welded joints subject to oblique loads 228
4.4.6 Transverse welds subject to net bending moments
4.4.7 Fillet welds subject to combined stresses 229
4.4.8 Summary of design rules for welded joints 230
4.5 Assignment on bolted and welded joints
4.5.1 Background
4.5.2 The design problem 231

5. Review: Structural integrity of engineering systems 239

5.1 Designing engineering systems: the 'big picture'
5.1.1 The structural *audit* 241
5.1.2 Some simple insights on structural analysis 245
5.2 Some 'radical new theories' in engineering design 248
5.3 Review questions 250
5.3.1 Engineering estimation
5.3.2 Failure
5.3.3 Shafts 251
5.3.4 Mechanical springs 252
5.3.5 Columns
5.3.6 Pressure vessels
5.3.7 Bolted joints
5.3.8 Welded joints 253
5.4 Design assignments

5.4.1 Optimum design of tripod
5.4.2 Optimum design of octahedral platform 255
5.4.3 Determination of factors of safety
5.4.4 Assignments on fatigue 257
5.4.5 Assignment on columns 261
5.4.6 Assignment on bolted joints
5.5 Some cautionary notes 264
5.5.1 A cautionary note about sign conventions
5.5.2 A cautionary note about material selection 265

PART 2 PROBLEM SOLVING STRATEGIES: AN ENGINEERING CULTURE

6. The evolution of problems 269
6.1 Cultural development of the designer 270
6.1.1 Engineering design as wisdom
6.1.2 Engineering design as discipline 271
6.2 Design problem solving 272
6.2.1 The designer as a person 273
6.2.2 What constitutes design? 274
6.3 Evolution and enformulation of design problems 278
6.3.1 The nature of information and problems
6.3.2 Problem evolution 279
6.3.3 Problem enformulation 284
Case example 6.1: Level crossing problem 285
Case example 6.2: Portable water heater 288
6.3.4 Problem intensity parameters 290
6.4 The design process 292
6.4.1 The components of the process
6.4.2 The Initial Appreciation 294
6.5 Generating ideas (solutions) 295
6.5.1 Mind games 297
6.5.2 Deferred judgement 299

6.5.3 Flexibility and fluency 301
6.6 Generic barriers to idea generation 302
6.6.1 Perceptual barriers
6.6.2 Cultural and environmental barriers to idea generation 305
6.6.3 Emotional barriers 306
6.6.4 Intellectual and expressive barriers 307
6.6.5 Answers to problems on conceptual barriers to creative problem solving
6.7 Evaluating designs and decision support systems 323
Case example 6.3: Site for a power station
6.8 Decision making strategies 325
6.8.1 Input/output analysis
Case example 6.4: Design of a washing machine.
6.8.2 Benefit-Cost-analysis 326
6.8.3 Morphological analysis
6.8.4 Fishbone diagrams 327
6.8.5 The design tree
6.8.6 Decision tables and other scaled check lists 328
Case example 6.5: proposal for a new processing plant
Case example 6.6: Automobile horn 329
6.8.7 Mathematical modelling 330
Case example 6.7: Electric power distribution
Case example 6.8: Wind load on chimney 331
Case example 6.9: Heat-exchanger design 332
6.8.8 A comparative evaluation of case examples 333
6.9 Assignment on problem evolution and enformulation: Bicycle Security 335
6.10 A brief note about technical reporting 337
7. Economic, social and environmental issues 339
7.1 Economic imperatives in design
7.1.1 The components of projects cost
7.1.2 Opportunity loss: a simple approach to evaluating project costs 340
Case example 7.1: Proposed widget plant 341
7.1.3 The annual cost equivalent 342

Case example 7.2 : Re-lamping a factory
7.2 Project evaluation and cost-benefit analysis 344
Case example 7.3: The Ford Pinto case 345
Case example 7.4: Injectable contrast agents 347
7.3 Design for human users: introductory ergonomics 348
7.3.1 Anthropometry 349
7.3.2 People in engineering systems 354
Case example 7.5: A cement mixer 357
7.4 Managing risk and hazards: fault, failure and hazard analysis in engineering systems 359
7.4.1 Causal networks 360
Case example 7.6: Failure of main bearing on tunnelling machine
7.4.2 Fault Tree Analysis 361
7.4.3 Event Tree Analysis 363
References 365
APPENDICES
Appendix A: Conversion tables
Appendix B: Standard sizes and preferred number series
Appendix C: Properties of sections
Appendix D: Beam formulae
Author Index
Subject Index

PREFACE: THE NEED FOR THIS BOOK

"...Engineers are proud of their profession, anxious to sing its praises. But they cannot seem to get beyond perfunctory and non personal expressions of the satisfactions they derive from their work. Many of them are 'turned on' by what they do. But they are unwilling or unable to reveal their inner emotions to an audience" Samuel C. Florman - The existential pleasures of engineering

Theodore von Karman, that polymath of the early 20th century, defined the difference between scientists and engineers thus:
"Scientists look at things that are and ask 'why'; engineers dream of things that never were and ask 'why not'"[1]
Engineering design is all around us every day of our lives. Almost every artefact we touch, chairs, doors, cutlery, light-globes, the list is endless, and every system we use, transport, mail, supermarket, banking, health care, power distribution and so on, has an engineering design component. Engineers were involved in the design of these artefacts and systems. How did this happen? What is involved in designing something? Even more importantly, how does one become a designer?

This book is addressed to young engineers on the threshold of entering the profession. The objective is to give experience in successfully designing simple engineering artefacts. These artefacts are the essential building blocks of larger more complex systems yet to be experienced. Yet the design elements introduced here encapsulate the richness of the full design experience of much larger and more complex engineering systems. The design steps involved here mimic the experiences to be had in the design of much more complex systems, but the examples offered provide easily evaluated success with the much simpler systems addressed in this book.

The activity of designing artefacts offers many challenges. The process is essentially synthetic rather than analytic in substance. Identifying the *real* substance of the design problem is probably the greatest challenge encountered by designers. In the examples offered in this text we have diluted this challenge by clearly identifying the building block to which each specific design problem belongs. Nevertheless we will still experience the challenge of estimating and, occasionally, making arbitrary design choices, tempered by insight into the problem. Yet another challenge is to find the right modelling method for analysing our chosen design. Here too we offer guidance, but the ultimate decision is left up to the designer. We do not prescribe how to solve problems. On the contrary, we only ask that the design challenges offered are solved by rational argument. Occasionally we ask that the rationale include sufficient information to allow a certificating authority to evaluate the solution. In these cases cogent argument will be needed to support the soundness of design choices. Apart from these minor limitations the field is there to be explored. It is not intended that anyone memorise any formulae or codified information offered in the text. We sincerely hope to engage the interest of young engineers and to develop their understanding of the material presented. Moreover, we hope to stimulate further enquiry or even challenges to our approaches, where it is felt we have skimmed over important issues. We act only as guides in the journey through this book, pointing out the significant features of the *design scenery*. However, the journey is there for the reader to enjoy!

The book is subdivided into two major parts. *Part 1: Modelling and synthesis*, including Chapters 1 to 5, deals with some focused issues of designing engineering components for structural integrity. *Part 2: Problem-solving strategies: An engineering culture*, including Chapters 6 and 7, introduces some broader issues in design.

Chapter 1 introduces structural integrity and the nature of failure, units and the notion of *structural distillation*. Chapter 2 offers a brief introduction to engineering materials, failure modes and

1 A paraphrase was used by John F. Kennedy in a political speech

introduces *failure predictors* (theories of failure), based on engineering mechanics. Chapters 3 and 4 deal with the design synthesis of specific engineering components, with extensively explored case examples. Chapter 5 reviews the work of Part 1, and offers case examples in the design of engineering systems. We conclude this part of the text with some explorations of engineering misconceptions and some experiences of engineering design judgement under severe time constraint.

In Chapter 6 we introduce the operational model of the design process and its various stages, through the evolution of design problems. Chapter 6 identifies and explores some broad, nontechnical issues that significantly influence the successful delivery of design project objectives. Finally, in Chapter 7 we briefly introduce economic, social, and environmental influences on design.

Most introductory texts in design explore either Part 1 or Part 2 of this text, but it is rare to find them together in a single book. We suggest that this separation is the direct result of Western European, mainly German, influence on design thinking. In the European school of thought, design for structural integrity has been identified with focused analytical objectives of deriving some specific details of component geometry. Additionally, in the European school, that type of *design* took place after the gross structural or embodiment decisions had been made by some broader creative process. In that school of *serial design*, allowing the design analyst to influence the embodiment decisions would be seen as opening the *floodgates* to a whole range of iterative redesign issues, that might severely impact on *concept-to-launch* times.

Current thinking in design has more-or-less rejected these limitations on the total design programme. We subscribe to the notion that successful design must take place in a creative environment, where all influences are considered, as far as possible, concurrently. So, why do we deal with particular design experiences in structural integrity, before considering the broader issues in design? The answer to this question lies in pedagogy. This book is addressed to both teachers and students of design. Our experience with design education has led us to the conclusion that some early success with component design, as offered in Part 1, is far more motivating and accessible for students than the broader issues of design dealt with in Part 2 of this book. Nevertheless, we do not prescribe how others should use the text. Both parts are self contained and may be taught or read separately in any order. We hasten to point out, however, that even in Part 1 we offer design experiences with opportunities for creative expression in design choices. We also conjecture that the particular design experiences offered in Part 1 may be clearly described by the general design procedures introduced in Part 2. We leave the support or refutation of this conjecture to readers.

Andrew Samuel
John Weir
Melbourne, 1999

COMMONLY USED SYMBOLS AND DESIGN TERMINOLOGY

A area, cross-section area

C,c most commonly a parametric constant of proportionality

c corrosion allowance

D,d diameter

e base of Naperian logarithm, eccentricity

E modulus of elasticity (*Young's modulus*)

f stress intensity, function

F force

F_a actual factor of safety

F_d design factor of safety

g acceleration due to gravity

G shear modulus of elasticity

I_{xx}, I_p second moment of area about x-axis;polar second moment of area

k,K spring constant , stiffness, kinetic energy

K_C fracture toughness coefficient

K_p, K_f modifying factors for endurance limit

l, L length measure

m mass, end-fixity ratio

p pressure

q load/unit length

r,R radius

r *real* interest rate (lending rate - inflation rate)

S'_e, S_e material endurance limit(material property); component endurance limit

S_U ultimate tensile stress (material property)

S_y yield stress (material property)

t,T material thickness, time, temperature

u strain energy/unit volume

v,V velocity, potential energy, shear force

W transverse load

P axial load

x, X axis direction

y, Y axis direction

z, Z axis direction

α thermal coefficient, angle

β hollowness ratio

δ small change in quantity, extension/compression

ε strain

η efficiency, welded joint efficiency

θ,ϕ angle, twist per unit length (torsion)

μ coefficient of friction, Poisson's ratio, dynamic viscosity

ν kinematic viscosity (μ/ρ)

π ratio of circumference of circle to diameter

ρ density

σ direct-stress (general)

τ shear-stress (general)

Φ diameter ratio

Allowable value: bound or limit on some specified output variable (e.g. $\tau_a \leq 96$ MPa)

Compromise: arbitration between conflicting objectives (e.g. accepting an increase in mass for a reduction in stress level: this is also called a *trade-off*)

*Conflicting objectives (*also, occasionally, *competing objectives)*: design objectives that negatively influence each other (e.g. low mass coupled with high strength, large volume coupled with small surface area; high reliability coupled with low cost)

Constraint: mandatory requirement to be fulfilled by the design (e.g. must be manufactured in Australia; must meet Federal Drug Administration Authority requirements)

Criteria: scales on which we measure the relative level of achievement of design objectives (e.g. the criterion for a *"low mass"* objective is mass in kg). See also *performance variable*

Design goal: the primary functional objective of a design (usually expressed in terms of a *design need*)

Design need: expression of a means for solving a problem, withot reference to embodiment (e.g. *"a means of removing dirt from clothing"*, instead of *"a washing machine"*)

Design variable: variables in the control of the designer (e.g. material choice; length; diameter; number of spokes)

Design parameter: combination of performance variables (or *criteria*) whose improvement usually contributes to success in achieving the design objectives (e.g. strength to weight ratio for mass limited design; strength to stiffness ratio for deflection limited design)

Design Audit (sometimes *Design Review*): identification of the *design goal* of a given embodiment and the *objectives*, *criteria*, *requirements* and *constraints* (the *design boundaries*) that lead to the current embodiment of a design. While the *Design Audit* does not seek to modify the design, it seeks to evcaluate how well the design embodiment meets its *design goal within the identified design boundaries*

Effectiveness: capacity of component or system to perform as required (e.g. an automobile is an effective personal transport device, but it is an inefficient high-volume people mover)

Efficiency: ratio of output to input (e.g. transmission efficiency of a gear box is the ratio output power/input power)

Endurance limit: limiting stress level for ferrous materials: the material will withstand infinite numbers of cyclic load applications if stresses are kept below the endurance limit (S_e')

Failure predictor: combinations of multi-axial stresses that predict component failure (e.g. maximum shear stress failure predictor, *MSFP*, $\tau_{allowable} < \tau_{fail}(=S_y/2)$

Goal: the ultimate purpose of the design (e.g. to reduce accidents at level crossings; to provide safe and effective storage for liquid fuel under pressure; to provide a simple mechanical cleaning system for clothes)

Governing requirement: the most demanding constraint or requirement on a system facing multiple requirements or modes of failure; the design requirement that imposes the most severe value on a particular design variable (e.g. for a shaft required to resist both fracture and excessive deflection, the governing requirement is that which necessitates the larger shaft diameter)

Iteration: systematic, goal-directed, trial and error solution to a problem

Mode of failure: technically precise description of the failure (e.g. yielding, rupture, fatigue, buckling, excessive deflection)

Objectives: the desired features, or characteristics of the design, that determine its ultimate effectiveness or suitability for a given task (e.g. the driver's seat must be comfortable; the can opener must be safe, cheap and portable)

Objective function: single unit mathematical combination of design objectives (e.g. total cost = some function of component costs; or, total energy use = some function of the energy used by the several design components)

Performance variable: output variables (not directly under the control of the designer, but indirectly determined by the values of the design variables) that describe the performance of a system in fulfilling design objectives (e.g. strength, cost, pump flowrate, engine power, mass). See also *Criteria*

Optimisation: maximising or minimising an objective function so as to achieve the *best* combination of design objectives (e.g. minimising a cost function, or maximising a benefit function)

Requirement: non-mandatory, *flexible*, design limit (e.g. *"the cost must be less than some specified value"*)

Rule-of-thumb (heuristic): informal decision making procedure based on experience and wisdom (e.g. the fundamental frequency of vibration for most mechanical systems is less than 1 kHz)

Stress: force per unit area

Trade-off: arbitration procedure that allows devaluing one objective in favour of improving the value of some other objective (e.g. accepting a loss in efficiency for an improvement in reliability)

Worst credible accident: limiting design condition that assigns the maximum loads to be sustained by a component (e.g. for selecting the loads on standard office chair, we consider the heaviest person – 97.5th percentile male – balancing the chair on two of its legs; considering wind loading on electric power distribution cables, we work with winds of the highest velocity observed in any 50 year period. Such winds are identified by the bureau of meteorology as 50 year return winds)

Yielding: condition of failure observed in metals, generally taken to be the limit of linear elastic behaviour (e.g. in CS 1040 steel yielding occurs at approximately 210 MPa)

PART 1

MODELLING AND SYNTHESIS

1

INTRODUCING MODELLING AND SYNTHESIS FOR STRUCTURAL INTEGRITY

With them the seeds of Wisdom did I sow,
And with mine own hands wrought to make it grow; Edward Fitzgerald, The Rubaiyat of Omar Khayyám, 1889

This is a book about the basic elements of engineering design. We begin with an emphasis on structural integrity, and later broaden our concern to various philosophical and practical issues. The book is addressed to engineering undergraduates who are being exposed to engineering design thinking and reasoning for the first time in their courses. Many texts on engineering design of machine elements treat structural integrity from a purely analytical point of view. Readers of such texts are rarely offered opportunities to apply the process of design synthesis to *real-life* engineering problems. Engineering texts present encapsulated experiences with design problems pared down to a manageable scale. *Real-life* engineering is full of uncertainties and risks, impossible to replicate effectively in the formalised medium of a textbook. Nevertheless, first-time students of engineering need to be introduced to the myriad choices and decisions that designers must face.

Our focus in this first part of the text is on design for structural integrity. For the purposes of our study, we define structural integrity as the capacity of engineering components to withstand service loads, effectively and efficiently, during their service life. This definition involves several concepts that need further elucidation.

Engineering component in this context means any engineering structure, which may be constructed from several interconnected elements into a *single entity*. Examples are pressure vessels, bicycle frames, flywheels, springs, shafts of electric motors, airframes and the frames of motor vehicles or buildings;

Effectively withstanding loads is defined as the capacity to accept service loads without exceeding either the specified maximum stress, specified maximum deflection, or both of these specifications. Examples are: 140 MPa maximum shear stress allowed in certain qualified welding materials, or maximum deflection of 1/300 the span in architectural beams;

Service loads are those loads, specified or unspecified, that the designer considers as credible to be imposed on the component during its service life. In the context of structural integrity, a *service load* will usually be specified as a *force* (N), a *moment* (Nm), or a *pressure* (Pa). More generally, *design loads* can be thought of as *thermal* (kW/m^2), *electrical* (A/m^2; kW), *information-processing* (decisions/hr; bits), or even in terms of such concepts as *traffic intensity* (e.g. passengers/hr).

The mature description of service loads is probably the aspect of component design requiring the greatest creative input by the designer, in terms of engineering wisdom and judgement. Often the service loads are unknown and need to be estimated before design can begin. In the case of critical design, such as airframes, considerable in-service test data is used to enable suitable service load estimates. For critical components, and those subject to fatigue failure, (e.g. airframes and motor vehicle components) these estimates are almost invariably followed up by field testing of prototypes. We explore service loads a little further in section 1.3, *Structural distillation*.

Efficiently, in this context, means either *at least cost* or *at least mass*. Cost-limited designs, the most common case, must consider all aspects of cost, including material, manufacture, maintenance, and most importantly, design costs. In general, for relatively simple components, the cost of manufacture is of the same order of magnitude as the material cost. In contrast to this, design time is expensive and needs to be spent wisely. Hence, the wise, or experienced designer will not spend expensive design time to eliminate a few extra grams of material from a relatively simple and cheap component. Quite often, for simple components, *one that does the job* is likely to be most efficient. Mass-limited designs require considerable analysis. With complex shapes, finite element analysis (a computer-based stress analysis procedure) offers opportunities for mass optimisation[1.1].

For the purposes of design, once the service loads have been established, the mechanical analysis of a component requires mathematical modelling of the component structure. The model allows us to translate loads into stresses or deflections. Naturally, the more closely the mathematical model represents reality, the more accurately we can identify stresses or deflections. As a precursor to analysis, we must make absolutely clear that the results of our analyses reflect only the behaviour of our model, and not the behaviour of the *real* component. Even experienced designers sometimes overlook this consideration. The spectacular failure of the *Tacoma Narrows* suspension bridge is an example of the complacency of structural engineers with models. As Professor Henry Petroski notes (Petroski, 1994):

"In the half-century following the accomplishment of the Brooklyn Bridge, suspension bridge design evolved in a climate of success and selective historical amnesia... to the Tacoma Narrows Bridge. Now, a half-century after that bridge's colossal collapse, there is reason to be concerned that the design of newer bridge types will be carried out without regard to apparent cycles of success and failure."

The Tacoma Narrows Bridge was of lighter design than previously built bridges, since it was designed to carry vehicular, rather than rail traffic. Wind loads on previously designed suspension bridges were modelled as causing static lateral deflections only. The lighter deck structure of the Tacoma bridge was susceptible to severe torsional vibrations, a loading clearly overlooked in the design, and this eventually caused its failure (Farquharson, 1949)[1.2].

Another famous case of improper modelling is the recorded tragic disasters suffered by the Comet aircraft of the DeHavilland company in the early 1950s. Although controversial at the time, it is now generally agreed that the aircraft failures were due to stress concentrations near the sharp corners of some rectangular windows in the fuselage. It seems clear that the Comet designers were unaware of the significant variability of fatigue data, and failed to take this into account in their design (Hewat and Waterton, 1956).

On a grand scale, we accepted Sir Isaac Newton's models of the physics of motion universally for nearly 400 years. In fact, they were, and still are, referred to as *Newton's Laws*. Einstein's hypotheses on relativity forced a paradigm shift in physics thinking. We no longer regard the *laws* of physics as absolute truths, but merely as conjectures, or working hypotheses, to be used until refuted by physical experiment (Popper, 1972).

In Section 1.1 we identify some characteristic features of engineering failures. We start by offering four propositions about failures, and we describe some simple engineering tools to be used in design for structural integrity (*failure avoidance*). We also offer some spectacular examples of failures, characterised by their social, economic or environmental impact.

The key elements of design for structural integrity are:

- service load estimation; and
- structural modelling.

Both of these elements require sound engineering judgement. We need to establish a base reference frame for our capability in making such judgements. In Section 1.2 we explore the notion of *engineering estimation*. We also set the reference frame for what we expect first-time users of this text to know, in order to make effective use of the remaining chapters.

Material properties and material selection are closely interwoven with design for structural integrity. Chapter 2 introduces material properties, formal definitions of failure and failure predictors to be used in design for structural integrity.

In Chapters 3 and 4 we offer simple mathematical models for several generic engineering components: *columns*, *shafts*, *pressure vessels*, *springs* , as well as *bolted joints*, *pinned connections* and *welded joints*. These components, and their respective mathematical models, represent the basic building blocks of our approach to modelling the failure–behaviour of more complex engineering structures.

1.1 Minimum mass with imposed stress or deflection constraints.

1.2 Quoted in Petroski (1994)

We propose that most engineering structures may be treated as a combination of several cooperating, much simpler, engineering structural elements. Moreover, if we understand how these simpler elements behave under loads, and we can recognise the underlying, simpler engineering elements which constitute a given *practical* structure, we are exercising *structural distillation* and *design thinking*.

This simplistic approach to structural integrity is clearly applicable to practical design situations where a relatively conservative or *safe* design is required. For designs of complex structures, exposed to limit loads (or beyond in some cases, such as racing cars, or racing yachts), we advocate the application of more sophisticated techniques of structural analysis. Nevertheless, even the more subtle problems become easier to understand and to explore, once we have a good basis for understanding how simpler structures behave. In Section 1.3 we introduce *structural distillation*, and we investigate its application to a wide range of common engineering components.

1.1 Structural integrity and the nature of failure

Proposition 1: Engineers are inherently concerned with failure and our vision of success is to develop modelling tools to avoid it. Moreover, by studying failures we develop clear ideas about causal relationships in complex *real-life* engineering situations, often too difficult to model completely realistically for structural analysis.

Proposition 2: Engineering failures may be categorised as technical, operational or unpredictable:

- technical failures are most commonly due to insufficient information about the nature of the structure, its material, its loading, or its operating conditions;
- operational failures are most commonly due to improper operating practices or conditions;
- unpredictable failures are most commonly the result of special circumstance or *acts of God*.

Proposition 3: Incentives to avoid engineering failures are related to *failure intensity* – the degree of seriousness of the failure. *Failure intensity* is measured by the economic, environmental or social impact caused by the failure.

Proposition 4: Failure is essentially related to risk: given extreme conditions, all structures or systems can fail. The engineer's task is to:

- generate design specifications that best meet the required operating conditions of the structure or system within acceptable levels of risk;
- identify the limits of *know–how* associated with the structure or system and assign *factors of ignorance* (commonly referred to as *factors of safety*) to cope with these limits, also within acceptable levels of risk — this brings with it the notion of *worst credible accident*.

1.1.1 Some spectacular structural failures

In what follows we present some spectacular structural failures, and examine the causes and outcomes associated with them.

(a) Oil tank collapse in the field

Figure 1.1 shows two of six collapsed oil tanks at a West African installation. The tanks were 15.5m diameter and 16.5m in height. At the time of erection, the tops were being assembled when a severe storm struck. The wind blowing over the tops exerted both direct pressure on the outside of the tanks as well as causing negative pressure inside the tank on the walls facing into the wind. The loss incurred was more than $US 1 million.

According to the investigators, there were several contributing factors to this loss. All six tanks had been erected to their full height before construction of the top covers commenced, and the tanks had not been reinforced internally to support them in a storm. Furthermore, the construction took place at a time known to be the worst storm season in that part of the world.

While no lives were lost nor any human injuries incurred, the financial loss was substantial.

(b) Collapsed hoarding in Dubai

Figure 1.2 shows a collapsed hoarding around a clock-tower being built in Dubai. The hoarding was erected to shield the site from view during construction. In a storm, resulting in maximum wind speeds of 62 km/h, the hoarding collapsed

Figure 1.1 Collapsed oil tanks in West Africa (2 of 6 similarly damaged tanks shown): After Schaden Spiegel 2/94[1.3]

Figure 1.2 Collapsed hoarding in Dubai: After Schaden Spiegel 2/94

1.3 Schaden Spiegel (*'reflections of failure'*) is a publication of Munich Reinsurance Company, Munich, Germany

and caused damage to the building and a site crane. The total damage was $US 400,000.

Investigators found that the hoarding was allegedly designed to withstand wind speeds of up to 115 km/h. However a design error resulted in the failure. The contractor had to bear the cost of the damage, since design errors were excluded from the insurance policy covering the works.

(c) Terrorist bomb attack on city

On April 24 1993, a terrorist bomb was used to devastate a whole section of Bishopsgate, a London suburb. One person was killed, and 40 were injured. The major loss in this event was due to the resulting fires. The damage bill came to about $US 1 billion (Figure 1.3).

While we needn't speculate on the cause, it is surprising how little damage the larger structures suffered. The anti-terror experts claim that, due to its unpredictability, this type of accident can not be designed against. However, substantial loss minimisation is possible.

Some suggested measures are:

- reduction in size of fire areas;
- installing supply facilities and safety apparatus in different fire areas; e.g. emergency power generation on roofs;
- installing safety shutoffs in electrical equipment.

(d) Kobe-Osaka highway collapse, 1995

On January 17 1995, the Kobe–Osaka highway suffered extensive damage during an earthquake. The quake lasted 40 seconds and was measured at Richter 7.2. Experts claim that the highway failure was initiated by shear failures of the piers. The total loss is estimated in billions of dollars (Figure 1.4).

Figure 1.3 Bomb devastated city, Bishopsgate, London: After Schaden Spiegel 2/94

Clearly, this disaster was an unexpected *freak event*, and the highway designers could not regard such an event as a *credible accident* for design purposes. However, special design measures can be, and are, taken in earthquake-prone locations.

1.1.2 Some spectacular mechanical failures

This is a convenient point in our discussions of failures to make a general observation, based on the data from Munich Reinsurance, one of the largest and most experienced industrial insurers in the world today.

In the parlance of insurance assessment, disasters caused by unpredictable natural events are referred to as *acts of God*. Until 1993, the greatest loss in any one disaster was that due to hurricane Andrew, affecting the east coast of USA and the Bahamas. Wind speeds reached 280 km/hr and 44 deaths were recorded with an estimated property loss of $US 30 billion. The following is a more recent report: (from *The Age*, Melbourne, 10 February 1999):

"Hurricanes, floods and other natural disasters cost more than £53 billion ($A135 billion) in damage worldwide last year, an increase of almost 50% on 1997, according to figures from the world's largest insurance company. The rise in the scale and scope of natural disasters, linked to global warming and reflected in a report produced by Munich Re, has alarmed underwriters so much that some are considering making parts of the world uninsurable."

Of all structural or engineering disasters around the world, those due to earthquake and windstorm represent 96% of total losses. In general, a very large proportion of industrial loss is incurred due to fire damage. Consequently, while mechanical failures may be embarrassing, occasionally very costly, and in some cases result in tragedy (see for example Eddy Potter and Page, 1976), on the scales of insured industrial loss, they are a *rare event*. This doesn't mean to imply that they happen rarely. On the contrary, engineering components fail with

Figure 1.4 Collapsed Kobe-Osaka highway

monotonous regularity. However, in most cases, designers are aware of the limitations of engineering components and systems. Failures elicit planned responses, and damage and injury are minimised. In some critical designs, and in design against fatigue, mechanical failures are often used as a design tool to establish component life.

Some very special engineering components are *designed to fail* in service. On critical pressure vessels a *bursting-disc* is used to limit the effects of overpressure to a specially designed part of the system. In some machinery, expensive parts are protected by specially designed, cheap, easily replaced, sacrificial components. These types of *mechanical fuses* minimise damage and injury. What engineering designers fear most is unexpected and un-planned-for failures. Some are described in the next section. In general, if properly investigated and documented, they represent valuable case studies for future designs.

(a) Failure of 600 MW turboset in a steam power generating plant

Turbogenerators are common types of engineering systems used to convert thermal energy, usually from steam or gas, into mechanical and thence electrical energy. The subject of this failure example was steam driven by a two stage turbine set.

Following a major overhaul, carried out after 25,000 successful service hours, the 600 MW turbogenerator set was being checked for overspeed protection. After successfully checking the electrical overspeed protection, at 108% of rated speed, the machine speed was reduced and then increased to check the mechanical overspeed protection, set at 110% of the rated speed. Just prior to reaching this overspeed, the whole set *exploded* causing 85% destruction of the turbine and generator. The loss was estimated at $US 40 million.

The generator shaft was ruptured over half its length (seen in Figure 1.5), and the entire train of

Figure 1.5 Failure of the main shaft of a 600 MW turbogenerator set: After Schaden Spiegel 1/82

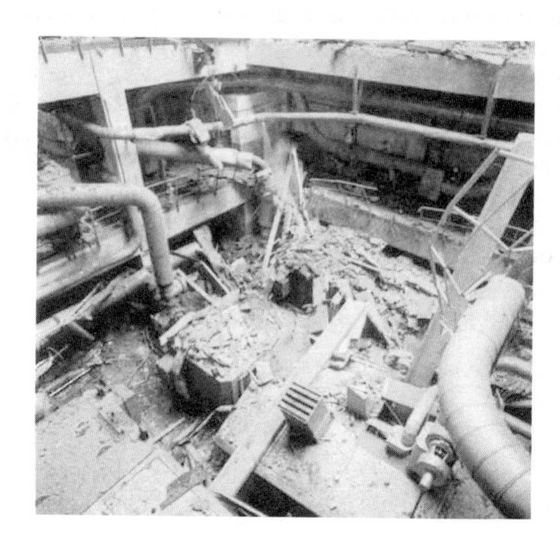

Figure 1.6 Burst high pressure preheater and devastated plant room: After Schaden Spiegel 1/82

shafts was broken in 11 places. One rotating part of the generating equipment, weighing about three tonne, was hurled into the machine hall and hit the girder of an overhead travelling crane.

The insurer's experts speculated on the cause, but they could not find any substantive reason for the failure. No material defects were found and all shaft failures were attributed to exceeding the permissible bending stresses. Subsequent speculation by rotodynamics experts conjectured that a bearing or rotor instability might have caused this type of failure.

(b) Explosion of high pressure preheater

High pressure feedwater preheaters are commonly used in steam turbine plants as a means of usefully employing exhaust steam, from the high pressure stage of the turbine. The subject of this failure was a 2m diameter, 10m high welded steel tank, with a 22mm outer jacket, operating at 300°C and 33 bars pressure (3.34MPa). Due to an emergency shutdown, the pressure in the preheater rose to 37.5 bars (3.80MPa) and the outer jacket burst, resulting in the explosion of the tank and severe damage to the plant room. The loss incurred was approximately $US 3 million.

The burst vessel and the devastated plant room are shown in Figure 1.6. Initial investigation of this failure focused on the vessel material. There was some concern that other similar vessels in this plant could suffer similar failures from what appeared

Figure 1.7 Collapsed piledriver: After Schaden Spiegel 2/96

to be a relatively minor pressure rise, well below the safe design limit. Ultimately, however, the failure was attributed to faulty welding of this particular vessel. The faulty weld was a longitudinal weld, with a manufacturing fault, that remained undetected during X-ray inspection. During 11,500 hours of operation the faults in the weld developed further, requiring only the small pressure rise to initiate the failure.

(c) Piledriver collapse

Piledrivers are essentially large mechanical hammers, used to drive large pylons (piles) into the ground during the preparatory stages of major construction works. In some pile drivers the hammer has a centrifugal out-of-balance mass driven to provide a vibrating load for hammering. The angle of the hammer guides needs to be adjustable to allow shifting of the centre of gravity of the unit. This adjustment is particularly important when operating on uneven ground, where a very small *lean* (out of vertical) of the hammer guides can result in substantial overturning moments acting on the unit. A 30m high, 140 tonne, piledriver with a vibrating hammer was being used for earthworks near the edge of a river, when the unit, driven on a crawler tractor, slowly toppled over into the river (Figure 1.7). The recovery and repair of the pile driver was estimated at approximately $US 1 million.

Investigation of this failure found that a jammed hydraulic valve prevented successful hydraulic control of the 30m high hammer guides. Once the unit commenced to lean over in the uneven soil of the river bank, the hydraulics could not correct the lean by appropriate shift of the centre of gravity and the unit toppled over into the river.

(d) Broken Liberty ships

In September 1941, the *Patrick Henry*, a rugged 10,490 tonne merchant ship, was launched in the US. This was the first of 2,210 Ec-2 class ships that became known as the *Liberty* ships. War time pressures of production schedules demanded that these ships should be produced in record time. One such Liberty ship was delivered fourteen days after laying the keel in the shipyard.

These ships were of low alloy steel, fully welded, construction. In earlier, riveted hull ships, structural cracks would be halted by the presence of rivet holes in the material.

Figure 1.8 Broken Liberty ship

In a welded construction, any crack, once initiated will continue to propagate through the structure.

Some of the Liberty ships were scheduled to operate in very cold climates. The combination of low temperature embrittlement (lowered fracture toughness in low temperatures) and the unimpeded propagations of cracks through the structure resulted in some spectacular failures in these ships (Figure 1.8).

(e) Turbine blade failure

Figure 1.9 shows a type of turbine blade connection between the disc and blading, commonly referred to by the geometry-inspired title of *fir-tree* connection. A photo elastic model of a similar *fir-tree* connector under load is also shown in the figure. Photo-elastic models are made of a photo-polymer, which can cause light to split into two components when under stress (double refracting). When polarised light is passed through a photopolymer under stress, the stresses show up as interference patterns. These patterns can display the stress in the material, with closely spaced fringe lines indicating large stress gradients.

From Figure 1.9 we can infer that the stresses in the model are high in the root region of the *fir-tree* connector and also around the *teeth* of the *fir-tree*. Figure 1.10 shows a failed turbine disc, with the blades torn out of the disc. Failures of this type can result in spectacular and costly damage. In one case, reported in *Schaden Spiegel 1/82*, a blade disc failure in a 120MW turbine caused a loss of $US 5 million. Several fragments cut through the turbine casing, damaging it beyond repair. One fragment impacted a 100,000kVA transformer causing oil to leak out and ignite. The major damage was caused by the ensuing fire.

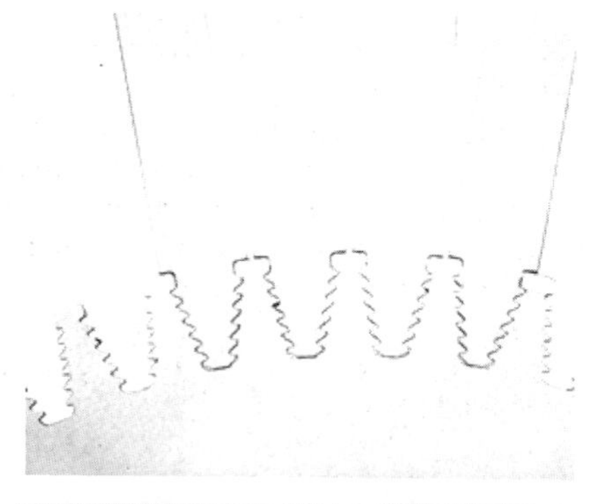

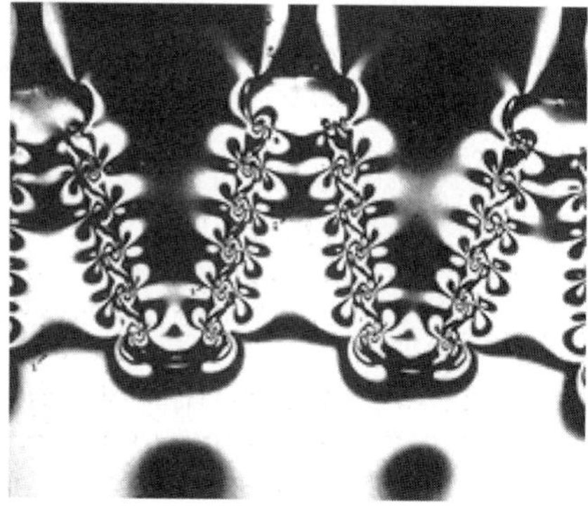

Figure 1.9 Fir-tree connection between turbine disc and blading and corresponding photoelastic model

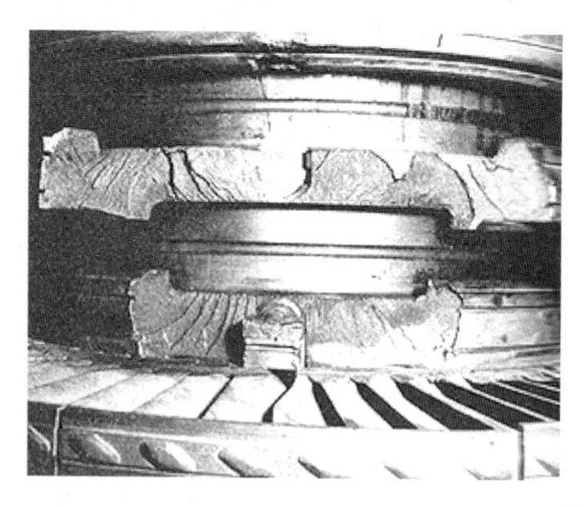

Figure 1.10 Failed turbine disc

(f) Airline propeller failure

Figure 1.11 shows a failed airscrew. These engineering components are exposed to very high time-dependent loads. Consequently, the most common mode of failure of propellers is fatigue.

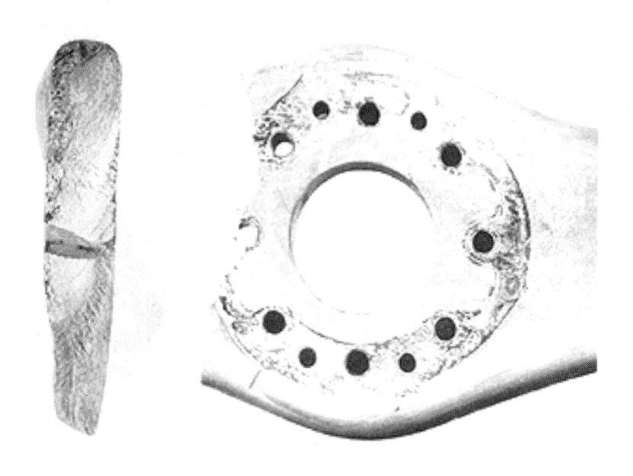

Figure 1.11 Failed aircraft propeller

In Figure 1.11, the varied discolouration on the failed section clearly indicates this type of failure. Also, the failure commenced at one of the bolt holes on the propeller, indicating that this was a region of high stress caused by the bolt holes acting as stress concentrators.

As can be expected, propeller failure will result in spectacular and occasionally tragic consequences. As a conservative safety measure, regulations require commercial aircraft to be able to fly and land with half of their normal complement of engines still operating. However, depending on the specific nature of the failure, corrective action by aircrew is not always possible.

Two cases were reported by Schaden Spiegel 1/98, in which twin engine turboprop aircraft, from the same manufacturer, both lost propellers in flight. One aircraft lost a propeller as it was about to land and eventually landed safely, incurring only mechanical damage. The other aircraft lost its propeller in mid flight, and had to make an emergency landing. Several people including the pilot were killed and the aircraft was lost. The resulting damage loss from this one incident was estimated at in excess of $US 100 million.

In both cases the failures were traced to a small surface indentation incurred during manufacture, resulting in a local stress concentration in an already highly stressed region near the root of the blade. Given the resulting high local stress, the resulting fatigue failures were predictable.

1.1.3 Summarising our review of failures

Almost all important failure cases are followed by an inquiry about the causes leading to the failure. These investigations often focus on *'who is to blame'* and *'who will bear the cost'* of the failure. Our focus is considerably more constructive. We focus on technical issues associated with failures, in order to learn from them. We hope to discover the underlying technical issues that might influence a whole class of such failures, and how to prevent this type of failure from happening again. There is an instructive story about a professional golfer, who was playing with a well known celebrity partner in a pro-celebrity golf tournament. The celebrity hit off at the first tee and landed his ball in the local car park, out of bounds. *"How the hell did I do that?"* cursed the celebrity. *"Why do you ask ?"*, questioned his professional partner, *"do you want to do it again?"*.

We have examined a series of ten failures. At the beginning of this chapter we categorised failures as *technical*, *operational* or *unpredictable*. In addition, we have identified a notion of *failure intensity* as a measure of economic, environmental and social impact caused by the failure. We should finally comment briefly on the problem of assessing *likelihood* of predictable failures.

In general, designers are aware of *normal*, or *expected*, or *most likely*, operating conditions of their engineering component or system, so that they are at least able to design for its *normal* operation. The difficulty arises when they have to plan for the *abnormal*. Ultimately, one could conceive of a situations so abnormal that it would be a *reasonable bet* that they would never arise. The implication is that design judgement is a little like gambling. Most engineers, however, are conservative, and wish to avoid the risk of gambling with financial or human loss. The most important judgement to be made early in any design, is the *worst credible accident* that is likely to befall our component or system. In many industries, codes of practice will prescribe allowable working stresses. In structural design, many codes also assign working loads. In mechanical design of engineering components, the assessment of working loads during the *worst credible accident* requires considerable experience and engineering wisdom.

Ex 1.1

(a) Categorise each of the ten failures described, as either *technical*, *operational*, or *unpredictable*.

(b) Rank order the ten failures described according to *failure intensity*.

Ex 1.2

We have identified economic, environmental, and social impact, as the three general results of failures. Are these effects independent? For example, can you identify a failure, not necessarily one of the ten listed, which might result in substantial environmental impact, without incurring proportionate economic impact?

Clearly, in most failures, the reverse is true: large economic impact with relatively minimal environmental impact.

Ex 1.3

Most of us face failure of engineering components in our daily lives. A washing machine or dishwasher fails in service, a component of our bicycle, motor car, or lawn mower, breaks, or the food processor, juicer, refrigerator, no longer perform as expected. Please identify the failure of some engineering component with which you have had direct contact. Describe the failure in terms of structural integrity as far as possible. Classify the most probable causes of failure. Use the failure categories offered above. For your component, identify the *worst credible accident* the designer should consider when designing it.

Ex 1.4

What is the *worst credible accident* for a door handle, a door hinge, the legs of an office chair, the top of an office desk, a power pole, a bar stool, a pair of scissors.

1.2 Units, estimation and things we expect you to know

In every domain of knowledge one needs to establish a frame of reference for effective communication. An amusing, albeit instructive, story is about a visit by the physicist Leo Szilard to the bacteriology laboratory of Salvador Luria, who was awarded the Nobel prize for his work on bacteriophages (viruses that attack bacteria). Szilard himself was by then an established figure in physics, partly responsible for the Manhattan project that developed the first atom bomb. Luria was initially embarrassed at showing the famous physicist through his laboratory. *"Dr. Szilard"* he asked, *"what should I assume, how much should I explain?" "You may assume"* responded Szilard, *"infinite ignorance and unlimited intelligence"* (Weber and Mendosa, 1973). In what follows we apply Szilard's mildly arrogant measure to readers of this text.

1.2.1 Units[1.4]

Whenever we are concerned with the discussion of physical quantities, for consistency and transportability of information the basic units of measure associated with each physical quantity need to be established. In this text we make exclusive use of the International System of Units (Système International d'Unités) commonly known as SI units.

The origin of SI units, based on the decimal system, dates to the aftermath of the French revolution, when in 1791 the French Academy of Sciences adopted the *metre* as the unit of length. This measure was based on 10^{-7} times the quadrant of a great circle of the Earth (meridian) passing through the poles and through Paris. It took six years to survey the arc from Barcelona to Dunkirk, resulting in a value of 39.37008 inches to the metre.

The units of *gram*, 10^{-6} times the mass of a m^3 of water at its maximum density (4°C), and *litre*, the volume of 10^{-3} m^3 of water, are both derived from the metre.

In 1875 a National Bureau of Weights and Measures was established at Sèvres, near Paris, where representatives of 40 countries meet every six years to update the international units of measurement.

For some time, the units of *metre* and *kilogram* were based on two convenient archival standards, known as *étalons*. For the metre, this was the distance between two marks on a specified bar of platinum, and for the kilogram, a carefully constructed and certified mass of platinum. At the 1983 meeting of the National Bureau of Standards, seven basic physical quantities were defined in the following way:

metre (m) = distance travelled by light in vacuum in 1/299,792,458 seconds;

kilogram (kg) = the mass of the internationally recognised archival prototype (*étalon*) at Sèvres;

second (s) = duration of 9,192,631,770 periods of the radiation corresponding to the transition between two hyperfine levels of caesium 133;

ampere (A) = the constant current which, when maintained in two straight parallel conductors 1 metre apart, generates a force of 10^{-7} newton per metre of length;

Kelvin (K) = 1/273.16 the absolute temperature of the triple-point of water (where the three phases are in equilibrium);

candela (luminous intensity) = 1/600,000 of the intensity of one m^2 of a radiating cavity at the temperature of freezing platinum (approximately the luminous intensity of one paraffin candle);

The more commonly found measure of luminous intensity is the unit of *lumen*. This is the quantity of visible light emitted per unit time per unit solid angle or steradian (recall that there are 4π steradians in a sphere). The lumen is related to the visual sensation caused by light and is dependent on the wavelength used. For example, light at a wavelength of 555 nanometres (10^{-9} metre), to which the eye is most sensitive, has a luminous intensity of 685 lumens per watt of radiant power. 1 lumen/steradian = 1 candela.

mole = the quantity of elementary substance equal to the number of atoms contained in 0.012 kilograms of carbon-12.

While we make exclusive use of SI units throughout this text, some engineering organizations still make use of the older Imperial System of units. It is expected that within the foreseeable future engineers will need to be familiar with both sets of units. A table of conversion factors is provided in the Appendix.

Ex 1.5

Convert between the following units :

(a) Fresh water has a density of 1000 $kg.m^{-3}$ (at 20°C). Express this in lbm ft^{-3} (note: 1 kg = 2.205 lbm; 1 m = 3.28 ft).

(b) Convert the thermal conductivity of copper from Imperial units to SI units (i.e. 225 Btu/hr-ft-°F to kW/m °K) (1 Btu = 1.055 kJ; 1W = 1J/s).

1.4 Source for this brief discourse on the origin of units of measurement is The Encyclopaedia Britannica, 15th edition.

1.2.2 Estimation

Estimation is a euphemism for *informed guess*. Engineering science eschews guessing or *estimating* by the very nature of its analytical content. Engineering design, however, relies on estimation for synthesis. We cannot sensibly evaluate the outcomes of all possible choices in synthesising a solution to an engineering design problem. Hence, we often rely on estimation as a *heuristic* (*rule of thumb*), for reducing the task of evaluating outcomes to only a few, informed, solution choices. Naturally, there is a risk involved in making guesses, or estimates. But then engineering design synthesis is a risky business, not for the fainthearted.

There are many industries relying on estimation for their business, and accepting risk as an integral part of that business. Perhaps the most well recognised of these risky businesses is insurance. If we wish to be insured against some loss, we estimate the value of the loss, and contract with an insurance underwriter to *carry the loss* should it actually occur. The insurer estimates the nature and extent of the risk involved in the loss event, and based on this estimation, calculates an appropriate *premium*, or annual payment, to cover the payout when, and if, the loss event is incurred. In essence, the whole process is a gamble, where the insured bets with the insurer that the loss will occur, and continues to lay an annual wager to this effect. Since, generally, the wager is only a small proportion of the eventual loss value, both parties, the insurer and insured, benefit from the process.

In the business of insuring against industrial loss, a key parameter of insurance risk is experience with the technology involved. Insurers go to inordinate length in estimating risk with new technologies. The industrial insurance of space technology involved an almost unacceptable risk at the beginning of the 1970s. Not only was the technology new, with no data to rely on for loss estimation, but also the losses that might be incurred could be very large. A single space launch, not including the huge expense involved in development, can cost between $US 500 and 700 million (based on 1992 $US).

One would be tempted to question the value of offering industrial insurance in such a risky technological environment. However, the benefits to be gained are also substantial. At the end of 1992 there were some 300, mostly geo-stationary, satellites in orbit around the Earth. As experience with the technology grows, the insurance estimation becomes less risky. Figure 1.12 is a view of the space shuttle, the current pinnacle of technological achievement and complexity in the space industry. Also shown in Figure 1.12 is the insurance loss and gain data for the years 1976 to 1992. Clearly, while the insurers made substantial early losses, the industry appears to be breaking even cumulatively. In future years the space insurance industry can look forward to strong gains from the experience built up in these early startup years.

As already noted, engineering judgement is a key determinant of design success. The underlying skill in engineering judgement is the capacity to marshal

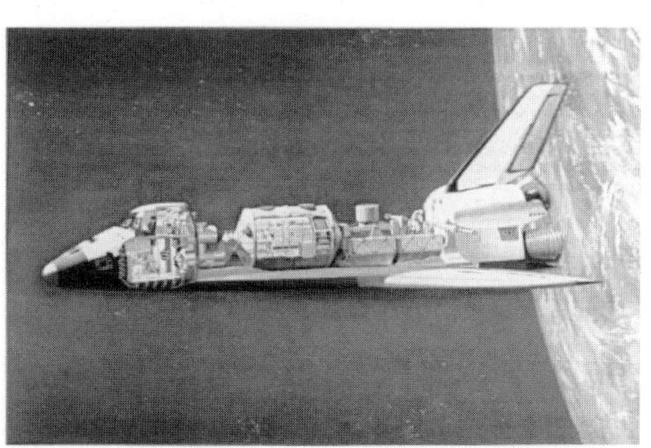

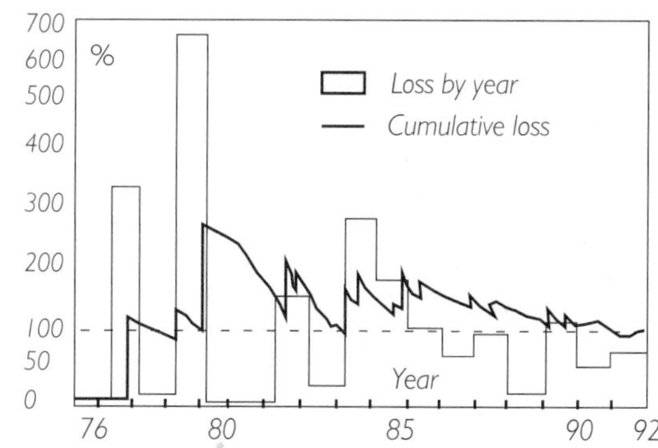

Figure 1.12 European space lab in cargo bay of the space shuttle and space insurance loss data: After Schaden Spiegel, 'Space Flight and Insurance' 1993,

all the available data and to make informed, intelligent estimates about loads and failure scenarios. Typically, we can find almost limitless examples where quite realistic population estimates may be drawn from personal experience and observation.

An example will demonstrate the process of estimation. Suppose we were interested in designing hospital equipment to support birthing in hospital maternity wards. As a matter of course, we would need to estimate the birthrate to asses the size of our market. Figure 1.13 provides the necessary data for our estimate.

Australian population $\approx$ 18 million

Victorian population $\approx$ 0.25 $\times$ 18 million = 4.5 million

$\Rightarrow$ approximately 4.5/2 = 2.25 million women.

Figure 1.13 indicates the average life expectancy of Australian females is 75 years and the birthrate has stabilised at around 2 children per female. This means that every female will , on average, give birth to 2 children in a lifetime of 75 years.

We can now estimate the total number of children born as follows :

$$N_{births} = 2.25 \times 10^6 \times (2/75) = 60,00 \text{ births.}$$

Bureau of Statistics data shows that in 1986 there were 60,650 live births to a Victorian population of 4.25 million.

Clearly, we were able to make a close estimate of birthrate from the demographic data of Figure 1.13, as well as our own wisdom about the total population of Australia, and the proportion of population living in the State of Victoria. Engineering estimation proceeds in a way not very different from the above example. Of course, for success in estimation and engineering judgement, we need to start with a list of what we expect you to know.

1.2.3 What we expect you to know

We begin with a brief, albeit random, list of *general knowledge* information. These are listed in Table 1.1 as a series of questions and we encourage first time readers to test themselves, by covering the answer column, and attempting to answer the questions posed. Where calculations are needed, these should be done mentally. Total time estimate for answering all these questions successfully is 10 minutes.

While we focus here on factual knowledge, it is useful to recall the admonishment of Henri Poincarré (1854-1912) :

"Science is built up of facts, as a house is built of stones; but an accumulation of facts is no more science than a heap of stones is a house."

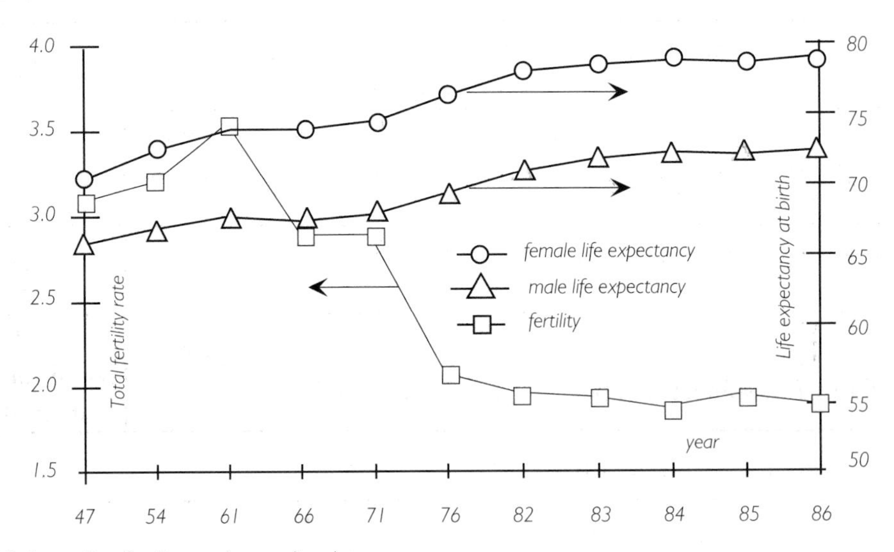

Figure 1.13 Australian fertility and mortality data
(Source: Australian Bureau of Statistics Cat. No. 3101.0, Australian Demographic Statistics)

Table 1.1 General knowledge quiz

Question	Answer
What is the speed of light in vacuum (in $m.s^{-1}$) – 2 significant figures ?	3.0×10^8
What is the 2nd moment of area of a rectangular section of width b and height d about a horizontal centroidal axis ?	$bd^3/12$
What is the 2nd moment of area of a circular section of diameter D about a transverse centroidal axis ?	$\pi D^4/64$
What is the density of steel (in $kg.m^{-3}$) ?	7,800
What is the density of aluminium (in $kg.m^{-3}$) ?	2,700
What is the speed of sound in air (in $m.s^{-1}$) ?	300
What is the value of Young's Modulus for steel ?	210 GPa
What is the value of Young's Modulus for aluminium ?	70 GPa
What does the unit prefix *giga* indicate ?	10^9
What does the unit prefix *pico* indicate ?	10^{-12}
What is the diameter of the earth ?	12,756 km (equatorial)
What is the area of a sphere, as a function of its radius ?	$4\pi R^2$
What is the volume of a sphere, as a function of its radius ?	$4\pi R^3/3$
What is the maximum bending moment in a simply supported beam of length L when loaded by a central concentrated force P ?	PL/4
What is the maximum bending moment in a simply supported beam of length L when loaded by a total force P distributed uniformly along its length ?	$PL/8$
What is the shape of the BMD for a cantilever subject to a uniformly distributed load ?	parabolic
What is the expression for the bending stress in a beam having a 2nd moment of area I_{zz}, subject to a bending moment of M ?	$\sigma = My/I_{zz}$ (y=distance from neutral axis)
What is the rotational analogue of the relationship: $F = m(d^2x/dt^2)$?	$T = J(d^2\theta/dt^2)$
What is the yield stress of mild steel ?	240 MPa
What is the current human population of the earth (to one sig. fig.) ?	6×10^9

However, attempting to work on problems of significance in engineering design, without the background knowledge of basic facts, is like trying to build a house without foundations.

One of the oldest cliché (*hackneyed, overused phrase*) in engineering is the *back of the envelope* calculation. No doubt, this phrase predates computers and electronic calculators. It was often used by practising engineers as a pseudonym for *estimation involving a few facts and numbers*. In essence, the above estimation of birthrates is an example of a *back of the envelope* calculation.

(a) An estimation quiz

The following quiz is to do with estimation. We don't need to calculate any of the results in this quiz, or, at least, if there is to be calculation, it should be of the *back of the envelope* type. In this second quiz, you should try to answer all the questions in 60 seconds.

(a) What mass can the average person lift and carry comfortably over 100 metres ?

(b) What is the height of a bull elephant ?

(c) What is the mass of a bull ant ?

(d) What is the weight of a moderate-sized apple ?

(e) What is the maximum squeezing force which could be delivered to a pair of pliers from the average hand ?

(f) How many locusts in a locust swarm ?

Approximate answers: (a) 30kg; (b) 3m; (c) 0.1gm; (d) 1N; (e) 200N; (f) 10^9.

(b) Calculator quiz

The next quiz allows the use of calculators, and in this case the aim is for accuracy rather than quick results. Correspondingly more care is needed in estimating answers than in the previous two quizzes.

(a) What is the contact pressure under a stiletto heel ?

(b) What is it under the foot of an elephant ? (average weight 5 tonnes)

(c) How many cars cross Golden Gate Bridge in San Francisco each day ?

[Hint: The population of San Francisco is approximately 800,000. We can guess that the Golden Gate Bridge, 1280m midspan, is the major thoroughfare for crossing the harbour and about $1/16^{th}$ of the population cross it daily: is this estimate reasonable ?]

(d) What is the average number of cars on the Golden Gate Bridge at one time ?

[Hint: Consider the length of time taken to cross the midspan as 4 minutes: is this reasonable ?]

(e) What is the average daily electrical energy consumption in California (in GJ/day) ?

[Hint: The power consumption of the USA is approximately 3×10^{12}kW-hour per annum, and the population of USA is 260 million and that of California 30 million]

(f) How much water is flushed down London toilets in one day ?

[Hint: The population of Greater London is approximately 7 million]

Approximate answers : (a) 10MPa; (b) 0.17MPa; (c) 50,000; (d) 280; (e) 3.4×10^6 GJ; (f) 20×10^6 litres.

(c) Mental calculations

The final quiz involves mental calculation only.

(a) How many seconds in a year ?

(b) Evaluate $\sqrt{800}$.

(c) The Empire State Building has 102 floors and a TV mast above ground. How high is it ?

(d) What volume of water is contained in an Olympic swimming pool ?

(e) How many Monday afternoons have there been since 1 A.D. ?

Answers : (a) 32 million; (b) 449 metres; (c) $20\sqrt{2}$; (d) 2,000 m^3; (e) 104,000.

We place considerable importance on the capacity to make rational estimations in engineering. This has become particularly important since the widespread use of electronic calculators. It is so easy to *slip a digit or two* when entering data in a complicated set of calculations. In general, one needs to invoke self-checking mechanisms to ensure realistic results are obtained for problems where there is little or no previous experience on the expected order of magnitude of the results. Order of magnitude, and dimensional consistency, checks must be routinely used for all calculations. Moreover, the number of significant figures available from electronic calculators bears no relation to the real precision of our results. Consequently, we never present more significant figures in our results than we can credibly justify, within the levels of approximations made along the way. Ultimately, these procedures build confidence and credibility in evaluations for both engineer and client. Our guiding principle must be that we engineers are *in the credibility business*.

(d) Rational estimations:

- focus on the essential issues behind an unfamiliar problem;
- help to gain insight into the way key controlling parameters influence a process or outcome;
- help to concentrate on primary rather than second-order effects (*see the wood for the trees*);
- help to identify major sources of uncertainty.

Ex 1.6

Calculate the number of barbers that could be fully employed within the state of California [population 30 million] (see solution at 1.5.1).

Ex 1.7

Check the following *dimensionless* numbers for dimensional consistency

(a) Reynolds Number $Re = \dfrac{\rho VD}{\mu}$

(b) Nusselt Number $Nu = \dfrac{h_x x}{k}$

(c) Grashof Number $Gr = \dfrac{g\beta(T_w - T_\infty)x^3}{\nu^2}$

[ρ = density; V = velocity; D = characteristic length; μ = dynamic viscosity; h_x = convective heat transfer coefficient at a distance x from initial contact with the surface; k = thermal conductivity; g = gravitational acceleration; β = volumetric coefficient of thermal expansion; ν = kinematic viscosity = μ/ρ, with units of $m^2.s^{-1}$]

Ex 1.8

The following formula gives the wind *drag* per unit length on a cylindrical body of diameter D exposed to a wind of velocity V [C_D is a dimensionless *drag coefficient*].

$$q = C_D\left(\frac{\rho V^2}{2}\right)D.$$

Check this formula for dimensional consistency.

(e) Note on the methodical conversion of units:

- arrange everything to be clearly in the numerator or the denominator;
- use the *brackets operator*, [...], to indicate *"the units of..."*;
- place units (in parentheses) alongside each simple quantity;
- convert each simple quantity one-at-a-time;
- cancel and calculate.

Example 1: Convert a flow rate of 1000 UK gal/hr to ℓ/s

$$Q = 1000\left(\frac{\text{UK gall}}{\text{hr}}\right) \times 4.546\left(\frac{\ell}{\text{UK gal}}\right) \times \frac{1}{3600}\left(\frac{\text{hr}}{\text{s}}\right)$$

$$= \frac{1000 \times 4.546}{3600}\left(\frac{\ell}{\text{s}}\right) = 1.26\ \ell/\text{s}$$

Example 2: Check the units of the universal gravitational constant in

$$F = \frac{GMm}{r^2}.$$

$$[G] = \left[\frac{Fr^2}{Mm}\right] = \frac{[F][r^2]}{[M][m]} = \frac{\text{N.m}^2}{(\text{kg})^2} = \frac{\text{kg.m.m}^2}{\text{s}^2(\text{kg})^2}$$

$$= \text{m}^3\ \text{s}^{-2}\ \text{kg}^{-1}$$

(f) Key elements of clear engineering calculations:

- the work is well spaced;
- assumptions and simplifications are stated unambiguously;
- the argument unfolds in a logical sequence;
- there is a good balance between algebra and calculation;
- useful intermediate values are calculated;
- important intermediate results are highlighted;
- related results are summarised (maybe tabulated);
- you *know* when you've got a ridiculous answer;
- awareness of *meaningful accuracy* (significant figures).

Ex 1.9 'Chessboard, wheat, and a constipated sparrow'

According to legend, the inventor of the game of chess asked from his grateful mogul lord only one thing in payment — all the wheat which would be needed on a 8x8 chess board if one were to place 1 grain on the first square, 2 grains on the second, 4 on the third, and so on, doubling up on each subsequent square.
Imagine that all this wheat is eaten by a sparrow. Calculate the diameter of such a sparrow, assuming the bird does not excrete any wastes until it finishes its meal. (see solution at 1.5.2)
(It is a popular belief that the inventor was executed for trying to be a bit too clever)

Ex 1.10 'Energy from a cuppa'

An urban legend has it that James Prescott Joule (1818-1889), the English physicist after whom the unit of energy is named, took several thermometers with him on his honeymoon to Niagara falls.
(a) Estimate the temperature rise due to a conversion of potential energy into heat over the 49 metre falls;
(b) Calculate the height to which a cup of tea could be lifted above sea level, if all the thermal energy in the hot fluid were converted into mechanical energy. It is an interesting *party trick* to try to estimate the height initially without calculating.
(see solution at 1.5.3)

CASE EXAMPLE 1.1: DOLPHIN ENCLOSURE

A cautionary tale about engineering calculations
Source: *Checking designs is equally important as designing itself*, James Ferry, *Engineers Australia*, Feb., 1995: 52–53

"... The case concerned the design and construction of a relatively simple and inexpensive dolphin enclosure."

"... A crucial calculation, using an equation known as Merion's equation, was made at conceptual design stage."

"... The calculation contained a simple and easily identifiable arithmetical error. However, it was never checked."

"... almost immediately after the vinyl enclosure was constructed, [it] failed as a result of the presence of long waves in the marina."

"... *[the client] ultimately claimed ... the losses it had suffered as a result of the defective design. These losses, inclusive of interest, by the end of the hearing totalled $602,845.68, together with legal and expert witness costs of some $150,000.00."* [The original firm of designers and the company responsible for checking were both sued. The judge held them both to be equally liable for the failure.]

"... The case illustrates the importance of risk management systems where all designs are reviewed and checked."

CASE EXAMPLE 1.2: A CHRISTMAS TALE

"Is there a Santa Claus?"
This question is considered in terms of *realistic* estimates of what Santa needs to do in order to deliver the *goods* on Christmas eve.

1. There are two billion children (persons under 18) in the world. Since roughly 15% of the world's population is Christian, it is at least possible that there are 300 million children who believe in Santa Claus, and who therefore qualify for a visit (let us assume that they are all *good*). At an average of 3.5 children per household there are 86 million households.

2. Allowing for time zones and the rotation of the Earth, assuming he travels east to west, Santa has 31 hours of Christmas to work with. This represents 770.6 visits per second.

In other words, for every Christian household with good children, Santa has about a millisecond to park, hop out of the sleigh and drop down the chimney *etc.*

3. Assuming the 86 million households are evenly distributed on the Earth's surface (a gross approximation for the sake of this estimate), the average distance between households is 1.7 km (The Earth's land surface area is 148 million km^2). Consequently, the distance to be covered in 31 hours, neglecting the time spent in visiting each household, is approximately 148 million km, at 1,330 km/second, or about 4,400 times the speed of sound. By contrast, the fastest man-made vehicle, the Ulysses space probe, moves at a slow 40 km/sec, the space shuttle at 7.5 km/sec while a conventional reindeer moves at, maximum, 2.5 km/hour.
4. Assuming each child gets no more than (say) a simple construction kit ($\approx$ 1 kg), neglecting reindeer and Santa (usually seen as *gravitationally challenged*), the total payload of the sleigh is 86 thousand tonnes, roughly the weight of the ship Queen Elizabeth, one of the largest passenger liners ever built.

One could continue to speculate on the energy required to stop and start this mass 86 million times, even as it is reducing at a rate of 1 kg per stop, and the energy generated by air resistance at the phenomenal speed it is travelling. However, even if the sleigh and its payload were to travel in the stratosphere, it would most likely burn up on re-entry. We can safely conclude that either Santa's existence is, at best, questionable, or that *he does it by magic*.

Ex 1.11

The following estimation exercises require only simple *back-of-the-envelope* calculations.

(a) Limb-load monitors are load cells installed in the soles of special shoes for measuring the load on each leg as a person walks. Estimate the useful range of loads for which such a monitor should be designed. Assume that loading is monitored only during walking.

(b) The oldest stone pyramid is the Djoser or *Stepped pyramid* at Saqqarah in Egypt. It has a rectangular base 108 m x 120 m, and rises to a height of 60 m. Assuming it has 10% of its volume in burial chambers and corridors, estimate the number of stones used to construct it. [Hint: It has been estimated by experts that the average size of stone used in this construction weighs approximately 2 tonne].

(c) It has been estimated that it took 20 years to construct the Djoser pyramid. Estimate the rate at which stones would need to be placed in order to complete the construction in this time span. How many labourers (stone movers) would be required to carry out the construction ?

(d) The use of rain water storage containers have been suggested as a means of providing water for household use in arid regions. The average house size in Melbourne (population 3 million) has an area of approximately 300 m^2. Estimate the rainfall needed to ensure sufficient water for the needs of the average household. [Melbourne's annual rainfall is approximately 600 mm]

1.3 Structural distillation

Little drops of water,
Little grains of sand,
Make the mighty ocean
And the beauteous land.
Julia Carney, English poet, 'Little Things', 1845

This section introduces the process of decomposing relatively complex engineering structures into elementary engineering components possessing readily available mathematical models. In engineering design, we regard this process as the very *essence* of engineering thinking. Hence our term of *structural distillation*, the process which distils the essence of a complex engineering structure into its recognisably tractable elements.

While most of our engineering education tends to focus on analysis, structural distillation requires substantially different skills. We focus on the way a structure is to be used, and the type of loading it will be subjected to, during its operational life. These operational loads determine the underlying nature of the structure. After all, it was these loads that the structure or component was designed to support without failing. A typical example is shown in Figure 1.14. The supporting legs of the *director's chair* are decomposed into a pair of *beams*. We note that legs are really *beam–columns*, due to the large axial load component, particularly in the upper half of the *scissor*. *Beam–columns* are quite tricky to model (see also section 3.2 on column design), and for the purposes of this simplified example we will consider the legs as *simple beams*.

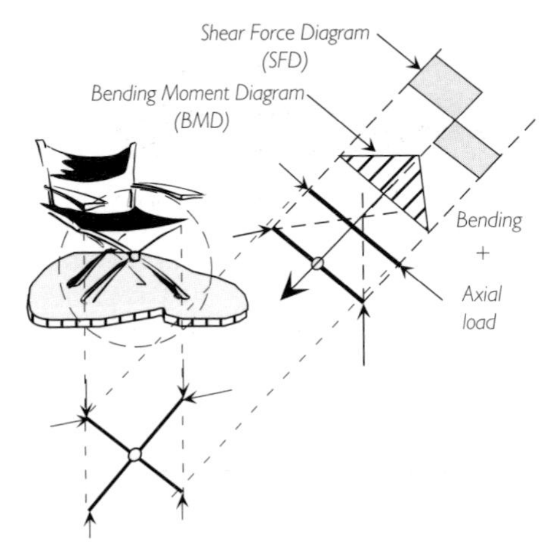

Figure 1.14 Partial structural distillation of director's chair

While there are other structural elements in the chair (the arm supports are simple cantilevers) this decomposition example only considers the front pair of legs. In this chair there are two pairs of legs, front and rear, interconnected by a central hinge element and two side *'beams'* to which the cloth seat is attached.

A more complex structure is offered by the example shown in Figure 1.15–the loaded supermarket trolley. Here we have decomposed the trolley into two structural elements, a *solid blob* for the loaded basket, and a simple portal frame, for the base and support arms.

Needless to say, there are parts of the whole structure we have not considered in this simple decomposition. The objective is to illustrate the process. Moreover, we recognise that this simple distillation process is only the first step in the design synthesis. One could go on to model the complete structure in a finite element analysis modeller to achieve mass optimisation for example. That can certainly happen, once the final embodiment of the structure is determined, but it may not be warranted for the sake of potential cost savings.

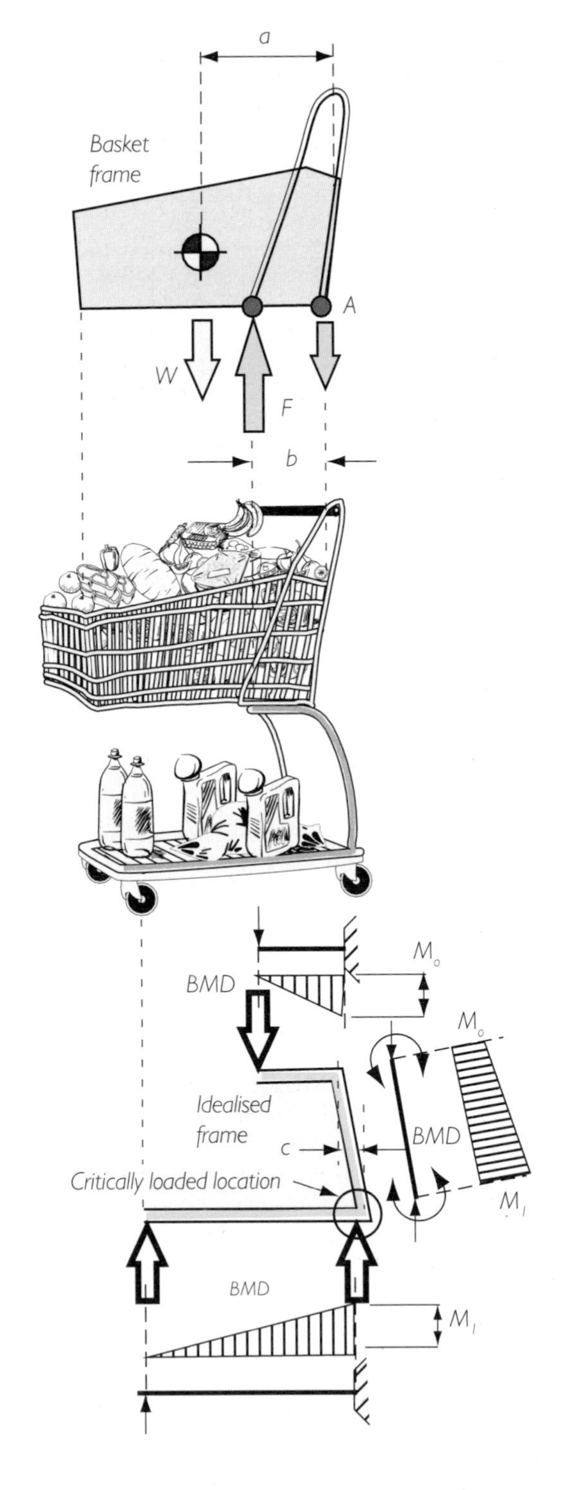

Figure 1.15 Structural distillation of supermarket trolley

Notes:

The distance a to the centre of mass of the basket is a matter of *judgement*.

The load F (= $a\ W/b$) is found by taking moments about A.

M_1 is slightly greater than M_0 (= $F\ b$), due to the lean of the rear arm of the frame, resulting in a small extra moment, Fc, acting on this part of the frame: $M_1 = F\ (b+c)$.

Structural distillation precedes the analytical evaluation process by providing early evaluation of the structure before final embodiment is decided upon. Hence, the process we are proposing here is essentially one of synthesis. While the examples we offer are those of existing embodiments, the intent is to provide design tools for the simple and credible evaluation of engineering structures as they are being created. The designer needs to be sufficiently confident to perform structural distillations of new embodiments as they occur. That, after all, is the real essence of engineering design.

1.3.1 Simple engineering components

The elementary building blocks used in structural distillation are a range of simple, generic, engineering components whose behaviours under load are relatively well understood. They are *the designer's friends*.

It is important to realise that these generic components are defined (and indeed *recognised* during the process of structural distillation) by both their *shape* and the directions and types of *load* they sustain. The designer must be able to see these components – often disguised – within larger and more complex structures.

Here we describe some of these simple components. In each case we identify the component by the type of loading it is expected to withstand, its generic name, as well as other names that are in common usage.

Figure 1.16 shows some simple axially loaded components. Clearly, these types of components are exposed to uniaxial loading only. Later on, in Chapter 2, we deal with the relationships between loading and stress conditions, and how these influence component failure. However, in structural distillation we are only concerned with the nature of loading the component is expected to carry.

The components shown in Figure 1.17 carry transverse (*beam*) or torsional (*shaft*) loads. The distinguishing feature of these components is their capacity to carry loads which result in more complicated stress fields than those in the simple axially loaded components of Figure 1.16.

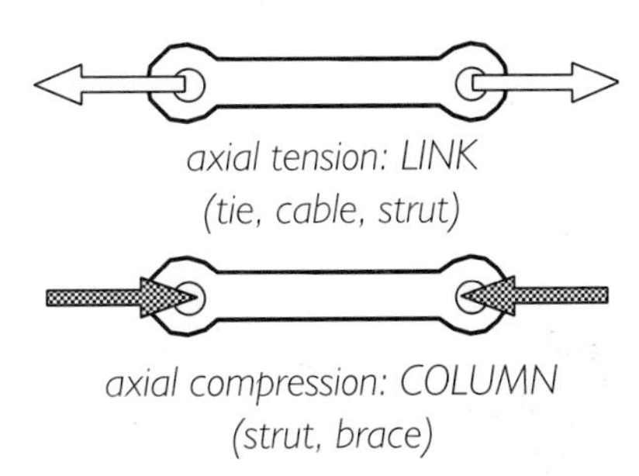

Figure 1.16 Simple axial loaded components

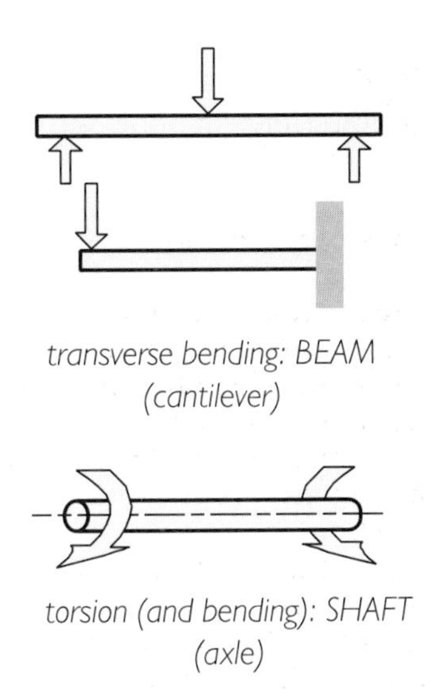

Figure 1.17 Simple components subject to transverse or torsional loading

Pressure vessels are constructed from a whole range of more complicated components. Their design is, in general, the subject of statutory codes of practice, since the stress fields generated by their loads are complex, and their failures can be devastating (refer Figure 1.6). However, for the purposes of this elementary consideration we regard the vessel as a smooth, closed, thin walled, cylindrical container, capable of carrying internal pressure.

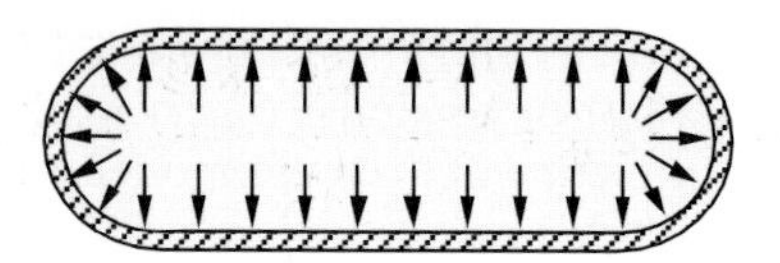

Figure 1.18 Pressure vessel

The following generic components represent loading conditions associated with special types of forces experienced in engineering structures. In our discussion of loading we need to establish when a component is subjected to:

- time varying loading (*fatigue* - Figure 1.19);
- sealing forces: bolted joints (nonlinear load behaviour- Figure 1.20);
- contact stresses: bearings (Figure 1.21);
- highly compliant behaviour: springs (Figure 1.22).

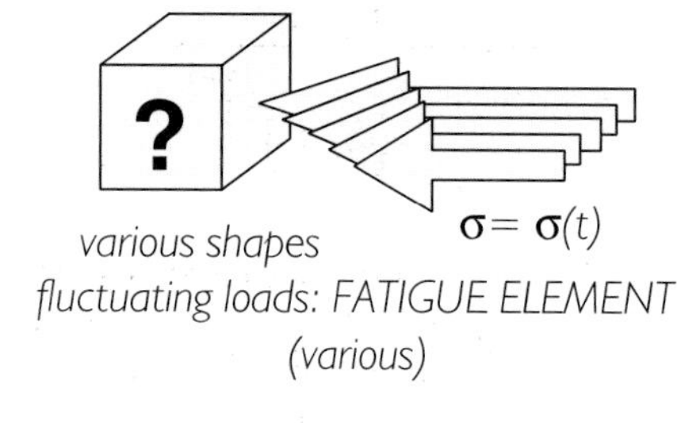

Figure 1.19 Element subject to time-varying loading: fatigue

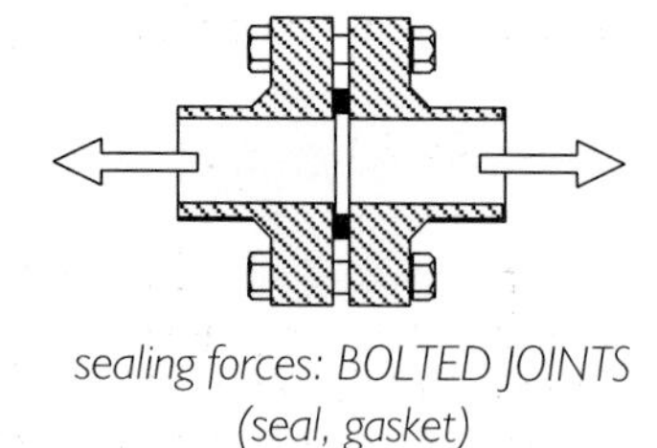

Figure 1.20 Element subject to sealing forces: bolted joint

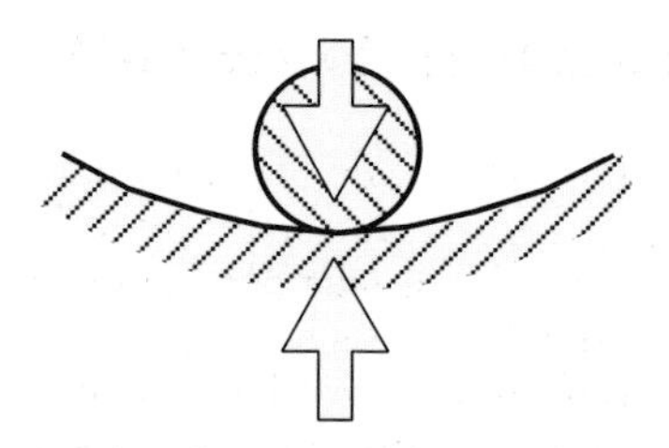

localised surface compression: BEARING
(roller, ball, cup & cone)

Figure 1.21 Element subject to contact loads: bearing

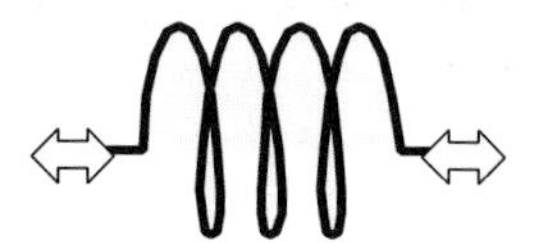

Figure 1.22 Highly compliant behaviour: springs

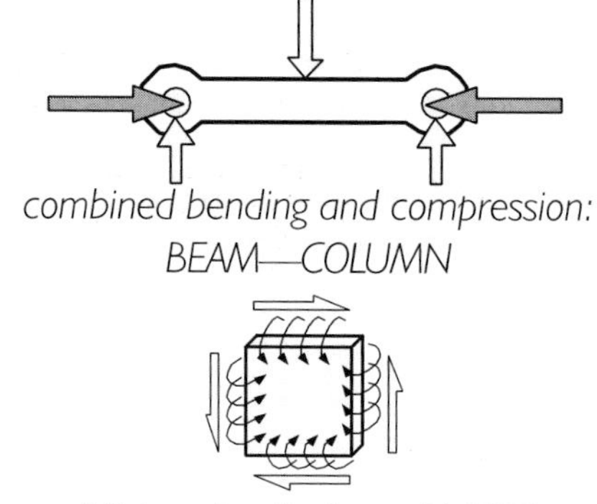

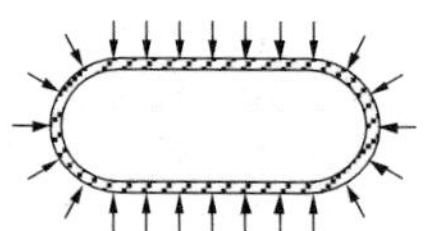

Figure 1.23 Some not-so-simple engineering elements

There are some *not-so-simple* generic engineering components we will meet in our programme of structural distillation. In general, these types of elements are beyond the cope of an introductory text. Some examples are shown in Figure 1.23.

1.3.2 Simple beam behaviour[1.5]

The following is a brief review of elementary beam formulae. We consider a beam of uniform rectangular cross section under the influence of a bending moment M. We accept the notion that the deflection of this beam is sufficiently small to allow the following simplifying assumptions to hold:

- the radius of curvature R is large compared to the span of the beam;
- plane sections remain plane during the deflection process.

Figure 1.24 shows a schematic view of a deflected section of the beam.

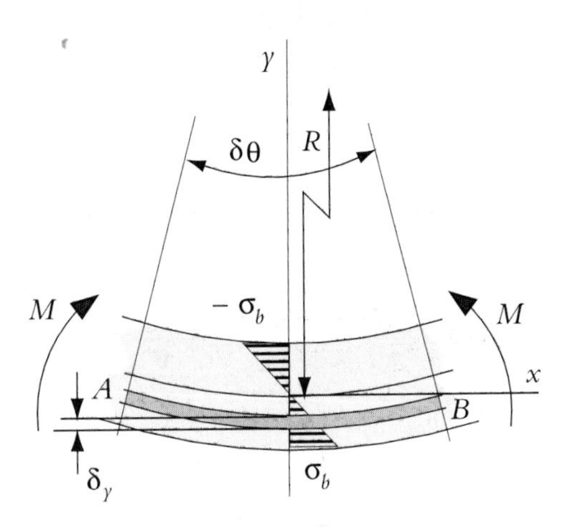

Figure 1.24 Deflected beam

Given the above assumptions, the applied moment M generates a stress field in the beam section as indicated. The radius of curvature R is taken to be that of the only undeflected *neutral layer* in the beam. The line in which any cross section of the beam intersects this *neutral layer* is called the *neutral axis*. Although some earlier conventions have measured positive distances in the beam in the direction of increasing radius of curvature, here our convention follows that adopted by more recent texts on applied mechanics (see for example Young, 1989). Also conventionally, tensile stresses are regarded positive and, correspondingly, compressive stresses are regarded as negative. Using these conventions, we can write down the extension of the small element AB at distance y from the neutral axis.

$$Strain = \varepsilon_{AB} = \frac{(R-y)\delta\theta - R\delta\theta}{R\delta\theta} = -\frac{y}{R}.$$

$$Stress = \sigma_y = E\varepsilon_{AB} = -\frac{Ey}{R}$$

The force acting on the edges of the small element AB is

$$\delta F = \sigma_y \delta A\ ,$$

where δA is δy times the z-dimension. The associated *moment of resistance* offered by the beam stress field to the applied moment is

$$\delta M = -y\delta F = \frac{E}{R}y^2\delta A$$

Integrating over the section we obtain

$$M = \frac{E}{R}\int y^2 dA = \frac{EI_{zz}}{R},$$

where I_{zz} is known as the *second moment of area* of the beam section taken about the neutral axis zz. The resulting relationship between beam section geometry, and radius of curvature is

$$\frac{M}{EI_{zz}} = \frac{1}{R} = \frac{d^2y/dx^2}{\left[1-\left(dy/dx\right)^2\right]^{3/2}}.$$

Where the beam curvature is small, and consequently $dy/dx \ll 1$, we can neglect the denominator in this equation, obtaining

$$\frac{M}{EI_{zz}} = \frac{d^2y}{dx^2},$$

or, as more commonly written, for the general differential equation of beam deflection,

$$M = EI_{zz}\frac{d^2y}{dx^2}.$$

Other conventions used in this text may be deduced from this general equation, but we note the results as applied to the generation of bending moment and shear force diagrams. Figure 1.25 shows schematically the conventions used for graphical representation of shear–force and bending–moment. (Refer to Chapter 5 for a brief discussion on sign conventions.)

1.5 A more substantial coverage of beam theory is available from several texts including Timoshenko (1955)

In general, clockwise shear force is taken as positive (although this choice inadvertently conflicts with the sign convention used for shear stress). Beams with a positive bending moment (as indicated in Figure 1.25) are referred to as *sagging* beams and beams with negative bending moment are called *hogging* beams.

Figure 1.26 shows shear-force and bending-moment diagrams plotted for a simply supported beam and a cantilever. Both point loading and distributed loading are shown.

Notably, there is a *duality* between simply supported beams and cantilevers, and this duality is drawn out in the point loaded case. Clearly, with the loading shown, the cantilever beam behaves precisely like one half of the simply supported beam. With distributed loading, the *duality* still exists, but it is a little more subtle than for the point load case. While the maximum moments are equal, the shapes of the bending-moment and shear-force diagrams are different.

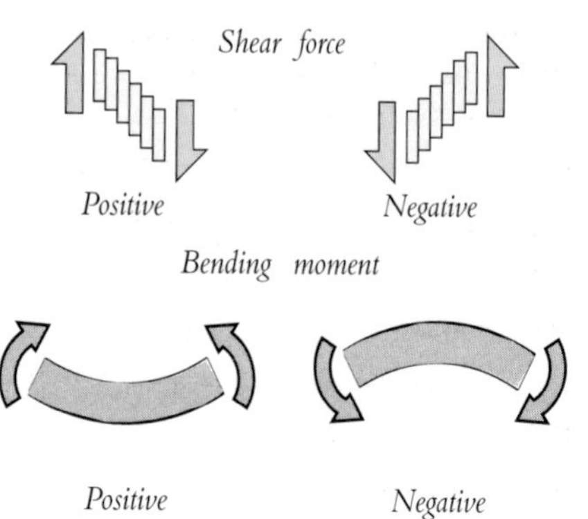

Figure 1.25 Conventions for plotting shear-force and bending-moment diagrams

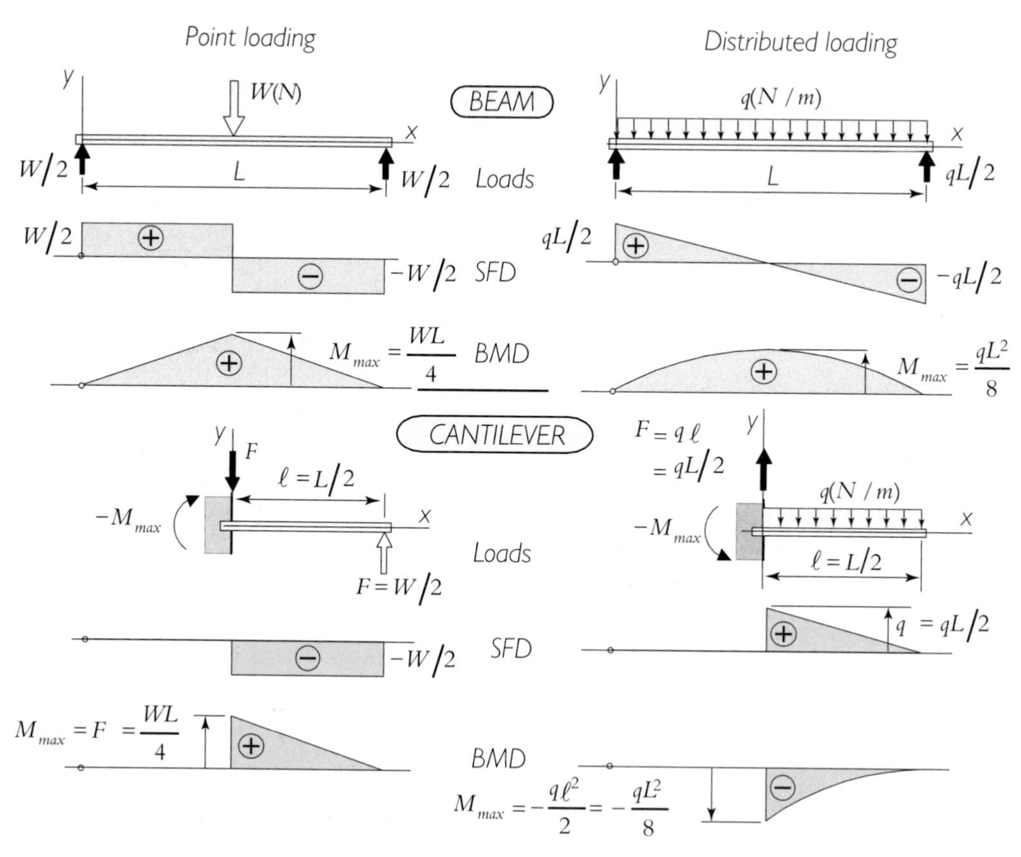

Figure 1.26 Simple beam behaviour

1.3.3 Case studies in structural distillation

We examine some simple engineering components and apply our rules of structural distillation to them. The steps involved are:

- identify applied forces, moments (i.e. *loads*);
- simplify the geometry (maybe subdivide the system);
- sketch the free-body diagram;
- show internal shear forces, bending moments, torques.

CASE EXAMPLE 1.3: PLIERS

Figure 1.27(a) shows the structural distillation of a pair of pliers. Shear-force and bending-moment diagrams are plotted, to indicate the nature of loads experienced by the component. The force F is that applied to some item held in the *jaws* and q is the distributed load applied by the human hand to the handle of the device. Clearly, the operation of each device is limited by the maximum load, q, a human is capable of applying. The loading is essentially the same as for a simple lever, and the arms of the pliers behave like simple beams.

CASE EXAMPLE 1.4: MULTIGRIPS

Figure 1.27(b) is the structural distillation for a pair of multigrips. The arms of the multigrips behave like a frame commonly referred to as a *bell–crank*. The structural behaviour of the *bell crank* is considered as two simple cantilevers at right angles to each other, joined at their *built–in* ends. *Free-body* diagrams are drawn for each of the two parts of the arm, using local coordinates, as shown in Figure 1.27(b).

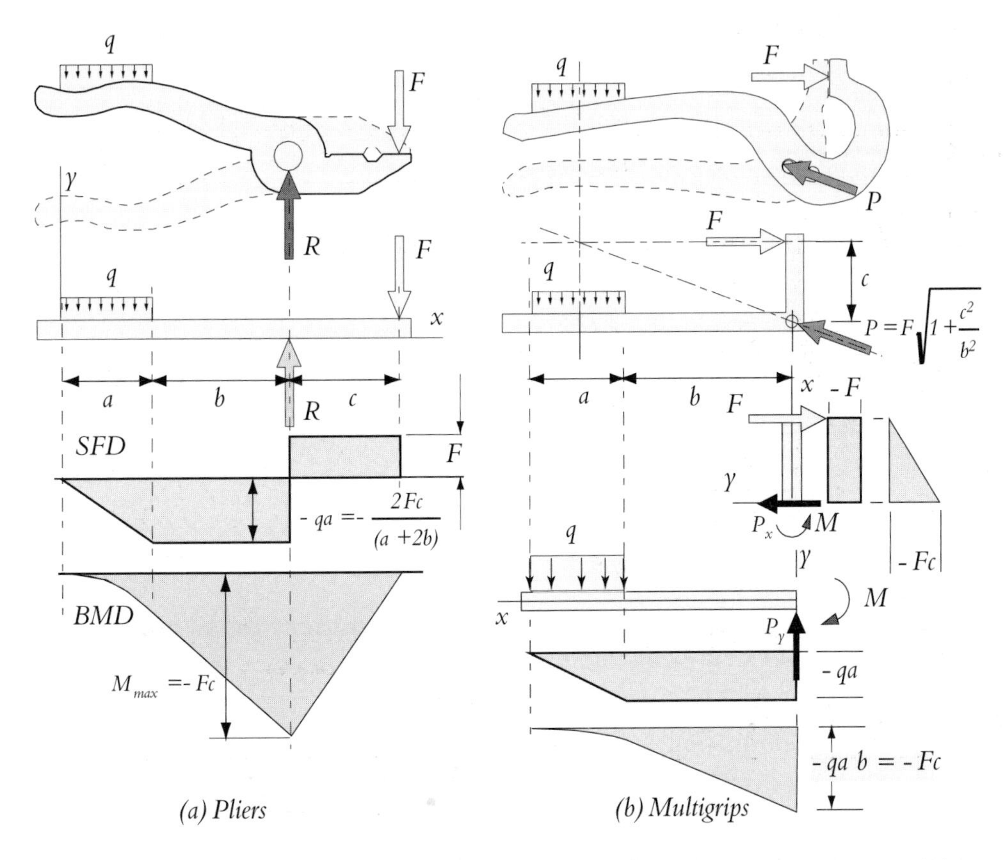

Figure 1.27 Structural distillation of a pair of pliers and a pair of multigrips
Note: all coordinate systems shown are 'local' coordinates referred to the section of the beam being modelled.

CASE EXAMPLE 1.5 PAPER CLIP

Figure 1.28 is the structural distillation of the ubiquitous paper clip, used to assemble (*clip*) several pages of paper together in a way that allows easy disassembly and reassembly. It is a ubiquitous device, since apart from its use in paper clipping it is often employed in other roles. For example, a string of paper clips, joined end-to-end, can be formed into a ring to create the office stationery equivalent of a *daisy-chain*. Some desktop computer manufacturers recommend the use of a straightened paper clip for dislodging jammed floppy disks. However, for our example we consider only the most commonly intended use of the clip, namely that of acting as a compliant connector for sheets of paper.

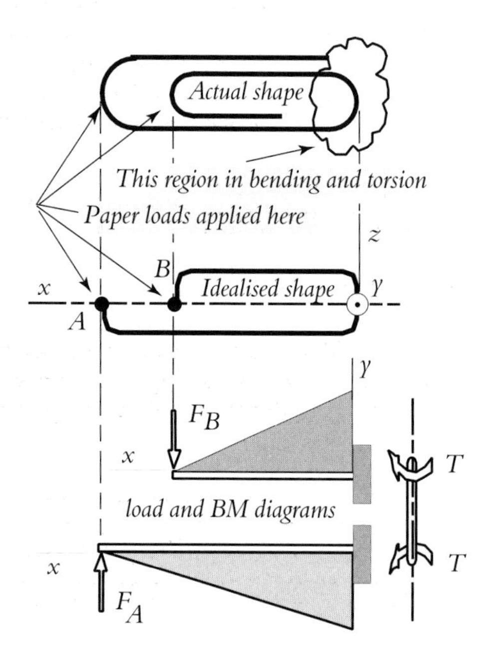

Figure 1.28 Structural distillation of a paper clip

The clip is manufactured as two planar loops of wire. In operation the two loops of wire are deformed out of their plane and the forces generated by this elastic deformation hold the sheets of paper together. Perhaps the main objective of the structural distillation process we are advocating is that it focuses on a key aspect of design, even when applied to such humble artefacts as the paper clip. *Someone had to think about the parameters of successful performance for this device.*

CASE EXAMPLE 1.6: SCISSORS

Figure 1.29 shows a pair of scissors commonly used for cutting paper or cloth. The two cutting blades need to be analysed separately, since they have slightly different shapes. However, for the purposes of this simple example we have only plotted bending-moment and shear-force diagrams for blade 'A'. The diagrams for blade 'B' will be similar, but of different signs.

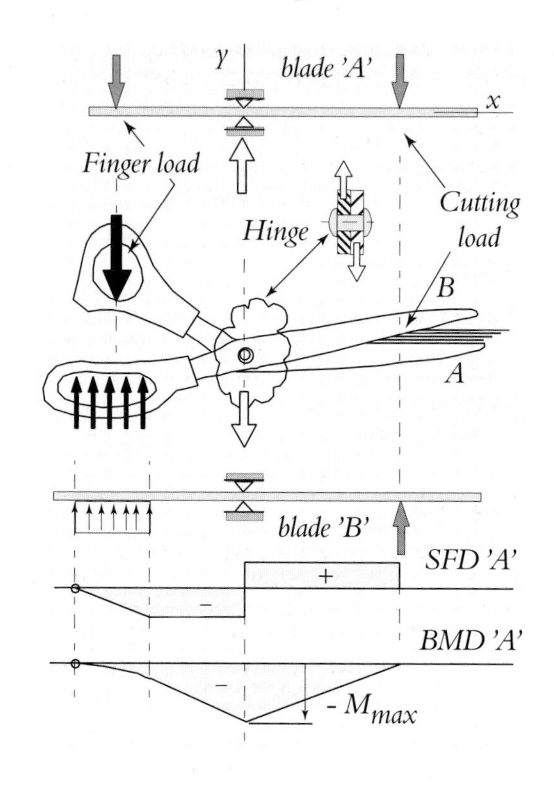

Figure 1.29 Structural distillation of a pair of scissors

1.4 Assignment on structural distillation

(a) Preamble

One of the major skills which engineers need to develop is that of *distilling the essence* of a real–life loading situation. This involves the use of insight and simple physics, to comprehend the loads which are being imposed on a device. It usually involves the ability to make simple assumptions and approximations which are conservative (that is, they tend towards the worst case), but which are not so conservative that they lead to uneconomical design. Often, the engineer will proceed by sketching a simplified diagram of the device, which reduces it to a set of simpler components, such as beams, cantilevers, shafts and columns. The aim is then, to specify the major external loads (forces and moments) acting on these components, and the internal forces and moments to which they give rise.

A relatively simple and useful way to view the devices in this assignment is to imagine they are made of some compliant material, like rubber, and imagine how they would deform under the *normally applied loads*. In essence, we are encouraging the use of a *thought–experiment* to aid thinking in these simple examples.

(b) Aims

This assignment is intended to give practice in the art of *structural distillation*, specifying loads , and in sketching simplified torque, shear-force and bending-moment diagrams for (mostly) common household items.

(c) The task

We present a set of sketches portraying 30 simple engineering devices subject to load. You may choose any set of fifteen of these for structural distillation. For your chosen set of devices, *distil the essence* of the loading situation as described in the preamble above. Specifically, you should:

- identify the major loads which will concern the designer in determining the structural integrity of the device;
- decompose the device into those basic structural components which could be used in a first-order analysis for design (don't do the analysis).

Prepare structural distillations for any set of fifteen out of 30 devices shown in Figure 1.30. In each case sketch the shear-force and bending-moment diagrams for the parts of the structure critical to the successful function of the device. Also identify the loading situation that could be considered the *worst credible accident* for the device.

The case examples, shown in Section 1.3.3, indicate the substance of what is expected in a structural distillation. One half A4 page per device should be ample, and each device should take no longer than 10 minutes to complete to the standard indicated in case examples 1.5 and 1.6.

1.5 Solutions to exercises

1.5.1 Barbers in California (Ex 1.6)

Bertrand Russell (1872-1970) formulated his famous *Russell's paradox* on set theory thus (Monk, 1996):

"In a town with only one barber, all those who do not shave themselves, are shaved by the barber. Who shaves the barber ?"

There are 30 million Californians. We assume that human hair grows, on average, at the rate of about 30 mm in 40 days (a guess that may need experimental evidence for its support). Additionally, we assume that most people feel a need to be *trimmed up* once they have sustained this growth. Hence, we obtain the annual number of Californian haircuts:

$$= 365\frac{\text{days}}{\text{year}}\times\frac{\text{haircuts}}{40\,\text{Calif.}\times\text{days}}\;30\times10^{6}\left(\text{Calif.}\right)$$

$$= 274\ \text{million}\left(\text{haircuts/year}\right).$$

If we are considering a simple hair trim, we can guess that each one takes approximately 20 minutes, and over 240 working days per year we get

$$N_{barber} = \frac{274\times10^{6}\ \text{haircuts/year}}{\left(\dfrac{240\ \text{day}}{\text{barber year}}\right)\times\dfrac{\left(8\,\text{hr/day}\right)\times\left(60\,\text{min/hr}\right)}{20\,\text{min/haircut}}}$$

$$= 47{,}500\,.$$

1.5.2 Constipated sparrow (Ex 1.9)

The number of wheat grains on the chess board are found according to the series $2^0 + 2^1 + 2^2 + \ldots + 2^{63}$, which sums to $(2^{64} - 1)$.

The volume of a grain of wheat is estimated as a cylinder of 1mm diameter and 2mm long. We also assume that during the consumption of the grains, they suffer a 10% compaction (these estimates also need corroboration by experiment).

Hence, we can calculate the total volume of the sparrow as

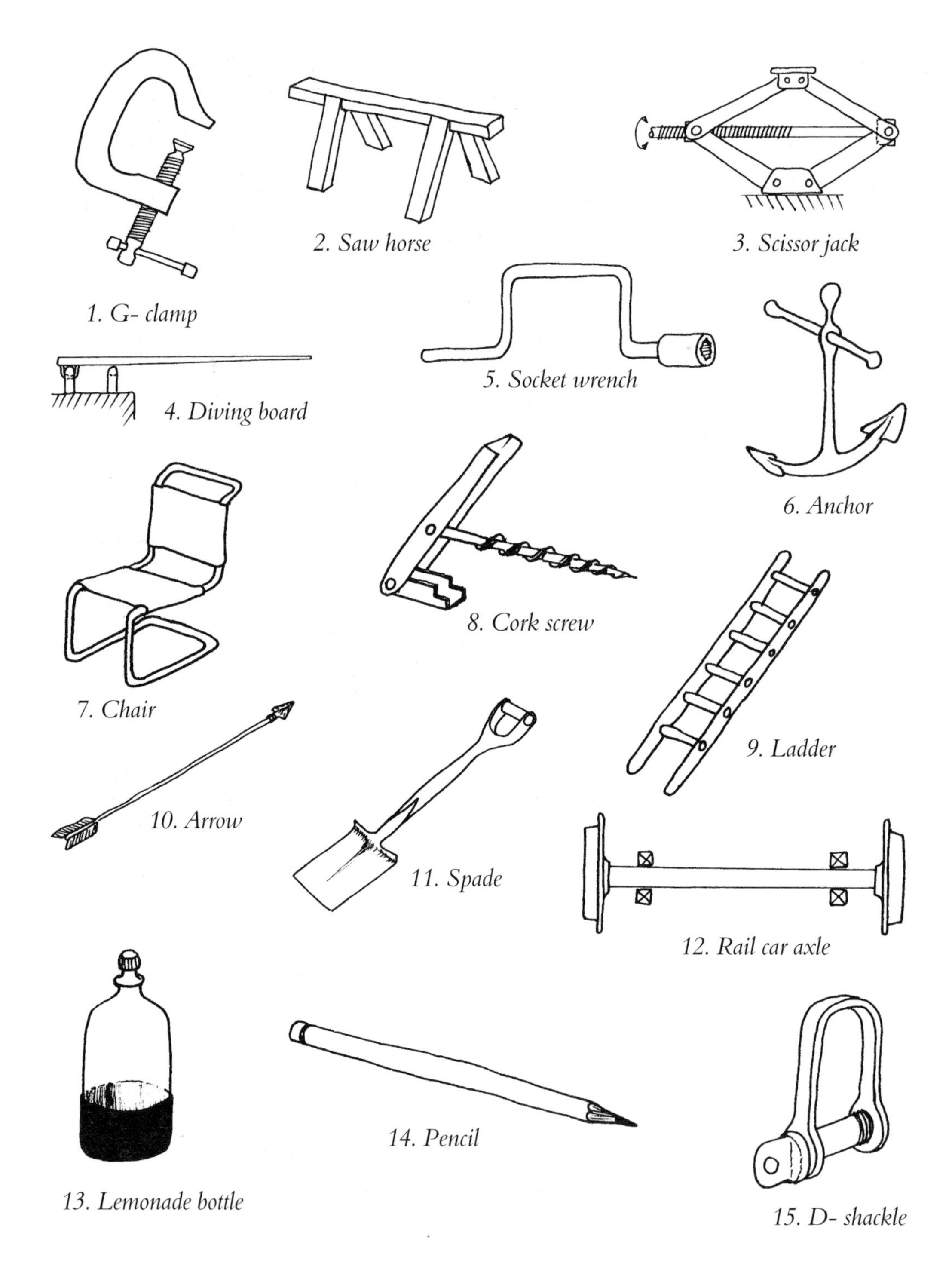

Figure 1.30(a) Sample components for structural distillation assignment (Items 1-15)

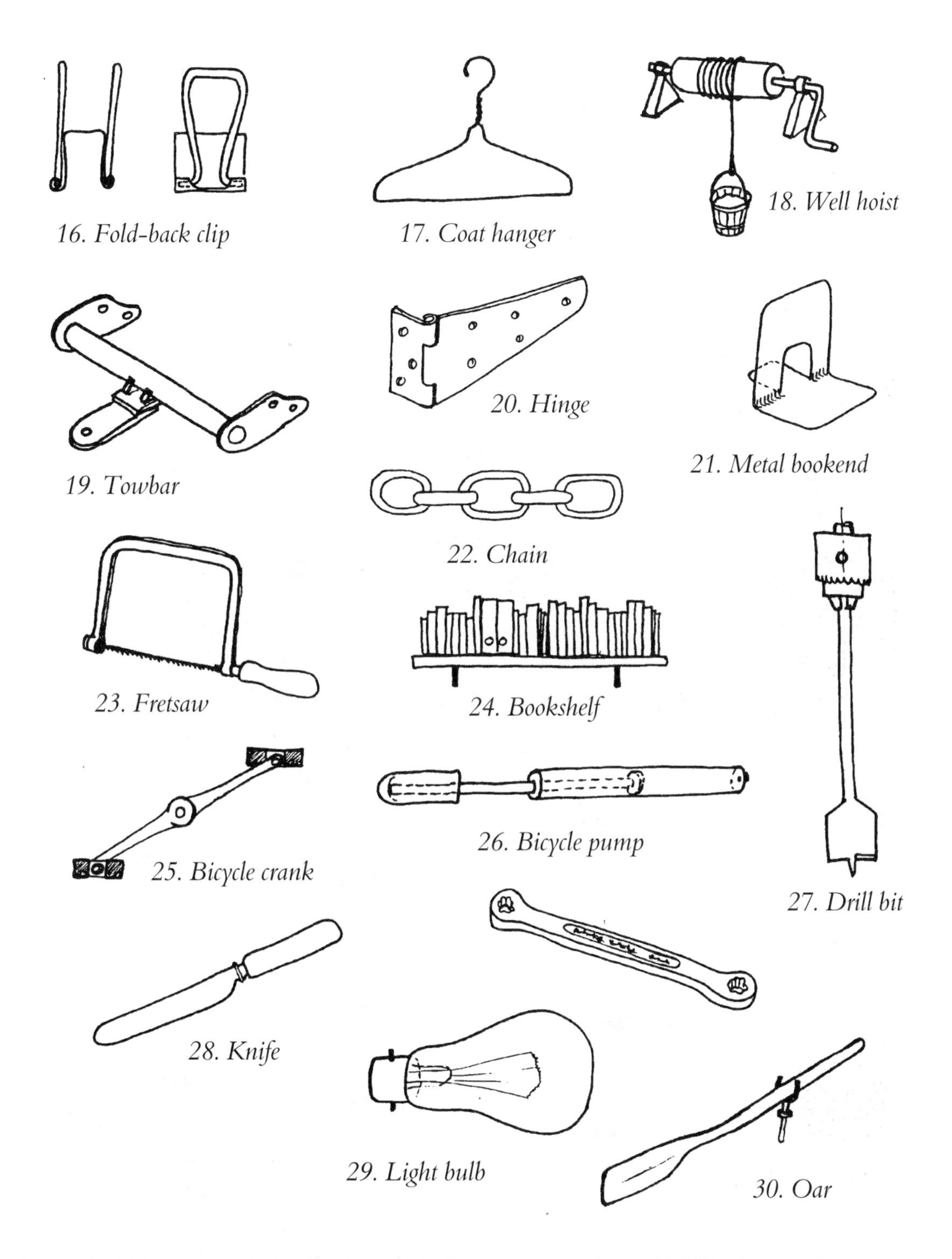

Figure 1.30(b) Sample components for structural distillation assignment (Items 16-30)

$$V_{sparrow} = 0.9\left(2^{64}-1\right)\left(\frac{2\times\pi\times1^2}{4}\right)\left(\text{m m}\times\text{m m}^2\right)$$

$$= 8.301\,\pi\times10^{18}\ \text{m m}^3$$

$$= V_{sphere} = \frac{4\pi R^3}{3} = \frac{\pi D^3_{sparrow}}{6}$$

$$D_{sparrow} = \sqrt[3]{6V_{sparrow}/\pi}$$

$$= \sqrt[3]{\frac{6\times8.301\pi\times10^{18}}{\pi}}\ \text{m m}\ \times10^{-6}\ \frac{\text{km}}{\text{m m}}$$

$$= 2.67\text{km}\,.$$

1.5.3 Energy from a 'cuppa' (Ex 1.7)

(a) Niagara falls, 49 m height; Specific heat of water, C_p = 4.18x10³ J/kg °K.

Assuming that all the potential energy of the falls is converted to heat (an optimistic assumption: it is interesting to speculate about other forms of energy conversion that might also take place), we can write

$$g\times h = C_p\times\Delta T$$

$$\Delta T = \frac{9.81\left(\frac{\text{N}}{\text{kg}}\right)\left(\frac{\text{m}}{\text{m}}\right)\times4.9(\text{m})}{4.18\times10^3\left(\frac{\text{N m}}{\text{kg °K}}\right)} = 0.12°\text{K}$$

Mr. Joule could not have detected this with ordinary thermometers.

(b) Energy from a cup of tea (Approximately 0.2kg water at 90°C). In this case we have a temperature drop (to ambient at 20°C) converted to potential energy.

$$g\times h = C_p\times\Delta T$$

$$h = \frac{4.18\times10^3\left(\frac{\text{Nm}}{\text{kg°K}}\right)\times(70)\,(°\text{K})}{9.81\left(\frac{\text{N}}{\text{kg}}\right)\frac{\text{m}}{\text{m}}\times10^3\,\frac{\text{m}}{\text{km}}} \approx 30\text{km}$$

Of course, it is impossible to convert heat to *mechanical* energy completely. But even if we applied an ideal *Carnot efficiency* of

$(T_h - T_l)/T_h$ = (363 – 293)/363[1.6] = 19%

to the conversion, the result is a modest surprise.

1.6 These temperatures are in absolute °K.

1.6 Assignment in modelling: Rural power supply

This assignment is concerned with the design of a structural system for *minimum cost*. A farm is to be connected to the electricity grid, requiring the installation of a set of low tension transmission poles and a single cable over a distance of 2.0 km from the farmhouse to the nearest grid tie-in point. Your responsibility is to design the poles and cable so that they possess adequate structural strength at the lowest possible installed cost.

The total cost to be considered is $C_{TOT} = C_{pole} + C_{cable} + C_{erect}$, where C_{pole} and C_{cable} are the cost of the pole and cable materials, respectively, and C_{erect} is the cost to erect the poles and run the cable. For simplicity, we will take C_{erect} to be equal to C_{pole}.

The poles will be made from steel and have a hollow cylindrical shape. They must be designed to withstand horizontal forces due to winds up to 100 km/hr , using an allowable stress of 85 M Pa.

The electrical cable will be copper with an allowable stress of 45 MPa, and designed to withstand yielding due to axial tension under self-weight. Electrical transmission requirements dictate that the cable must be no less than 10mm in diameter (it can be regarded as being effectively a single solid wire). For safety reasons, the cable must always be kept at least 6.0 m above the ground.

1.6.1 Instructions

(a) *Design* the system of cable and poles. Initially you should concentrate on producing a design which fulfils just the structural integrity requirements;

(b) In as systematic a fashion as possible, improve your design by reducing its cost;

(c) *Prepare* a report setting out clearly the reasoning leading to your results. In your report, use a flow chart to *show* your sequence of calculations and the major steps in your design. *Discuss* which of the variables you have treated as design variables. *List* all the assumptions used in your design calculations and critically review them. If you think that any of the assumptions lead to oversimplified calculations which could be seriously in error, make this point clear in your report.

Notes

1. The system will require one pole to be installed at the household end, one at the grid tie-in point, and several in between. At the termination points, and at any intermediate poles where the horizontal direction of the cable may change, it may be assumed that sufficient guy wires are provided to ensure that no net sideways force is imposed on the poles due to the tension in the electrical cable.
2. Although the poles experience downwards axial force due to self weight and the weight of the electrical cable, it is assumed that this effect will be of minor significance compared to the effect of lateral bending due to wind forces. It should be ignored.
3. The densities and specific costs of steel tube and copper wire are as follows:

$$\rho_{St} = 7800\,\mathrm{kg/m^3}\,; \quad \rho_{Cu} = 8860\,\mathrm{kg/m^3}\,;$$

$$c_{St} = 0.60\,\$\mathrm{Aus/kg}\,; \quad c_{Cu} = 6.50\,\$\mathrm{Aus/kg}.$$

4. The design of this simple system is an introductory exercise to illustrate features common to all design problems. The important ideas of *allowable stress*, *design variable* and *optimization* are introduced.
5. Beware of getting bogged down in algebraic manipulations. Some important background formulae and definitions are given below.

1.6.2 Some relevant results from Engineering Science

(a) Wind load on cylinders

The transverse load acting on a cylinder in an air flow of V m/s is uniformly distributed along the cylinder's axis, and can be specified as a force per unit distance q N/m along that axis.

$$q = C_D\left(\frac{\rho V^2}{2}\right)D_o,$$

where D_o is the outer diameter of the cylinder (*note*: the term in parentheses is called the dynamic pressure);

$C_D = 0.4$ — C_D is called the *drag coefficient*, and is assumed constant over the range of wind velocities relevant to this problem;

$\rho = 1.2\,\mathrm{kg/m^3}$ — for air at atmospheric temperature and pressure.

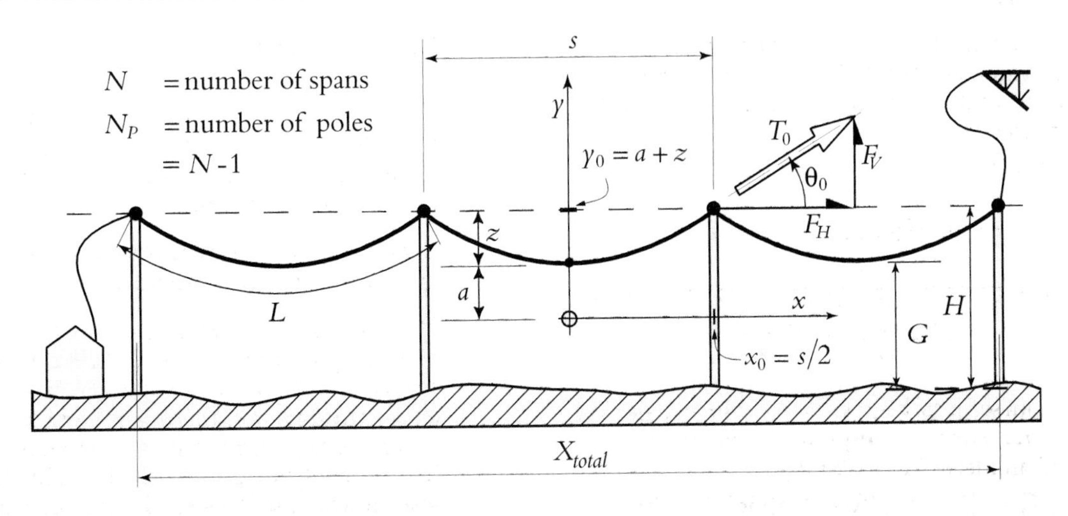

Figure 1.31 Rural power distribution system

(b) Bending of beams

A beam subject to a local bending moment M at some point along its length, and having a second moment of area I_{zz} about a neutral bending axis, will experience axially oriented bending stresses σ_B on the cross section at that point which vary linearly from zero at the neutral axis to $\sigma_{B,max}$ at the extreme fibres.

$$\sigma_{B,max} = \frac{My_{max}}{I_{zz}}$$

— where y_{max} is the furthest distance in the cross section from the neutral axis;

$$I_{zz} = \frac{\pi}{64}\left(D^4 - D_i^4\right),$$

— for hollow circular cross sections, where D and D_i are the outer and inner diameters, respectively. This expression is exact.

It is often useful to define a dimensionless shape parameter, such as the *hollowness ratio*, $\beta = (D_i/D)$, to help simplify expressions relating to hollow sections such as these.

(c) Statics and geometry of catenary cables

A uniform, inextensible, flexible cable subjected only to its own uniform linear weight, w N/m, and suspended from two horizontally separated points, will adopt a curved shape described by

$$y = a \ Cosh\left(x/a\right),$$

where $a = F_H / w$

(F_H is the horizontal force in the cable);

$$y_0 = a \ Cosh\left(x_0/a\right),$$

— x_0 and y_0 are coordinates of the support, where $y_0 = s/2$;

$$L = 2a \ Sinh\left(s/2a\right),$$

— L is the cable's length over horizontal span s;

$$y_0 = a + z,$$

— defines the relationship between z and a, where z is the mid-span sag of the cable below its supports;

$$T = w\,y$$

— T is the cable tension at a general height y above the x-axis;

$$T_0 = w\,y_0 = w\left(a + z\right),$$

— T_0 is the cable tension at the support.

The cable's curved shape is called a *catenary*, after the Latin word *catena*, meaning chain. Some useful nomenclature is summarised in Figure 1.31

(d) Some additional notes

The solution offered to this problem is by no means unique, and while it is *sufficient* it is by no means the *necessary* solution. This non-uniqueness is a natural feature of all design problems. We fully expect that the creative problem solver will find some more elegant solution. However, the one presented here is one that certainly seems to work.

Two interesting features of the problem, which are not obvious at first glance, are:

- it is not possible to obtain an explicit relationship between the sag and the span of the cable (i.e. between z and s) in

$$z = a \ Cosh\left(s/2a\right) - a,$$

so the designer must use some intermediate parameter (probably a) as a design variable, and very likely iterate for a solution;

- one apparently important design variable (the cable diameter, d) can be eliminated from consideration. This is because the shape of the cable catenary turns out to be independent of d.

One good way to start the solution is to work out an expression for the total cost of the system. This will quickly identify the major design variables.

Two important symbols which have not yet been provided are:

— the maximum allowable stress in the copper cable, $S_{A,Cu}$;

— the maximum allowable stress in the steel pole, $S_{A,St}$.

Engineering Design Associates

Project Title : **RURAL POWER DISTRIBUTION SYSTEM**

Date 02:14:98

Page 1 of 14

Ckd. mat

"Electricity Transmission Line"

Ⓐ Mathematical Background

The well-known results for the geometry & statics of a catenary are provided in the problem description. They are available also from standard texts on calculus or statics, and can be easily checked by considering the equilibrium of the "free body" cable shown below. (Note that this sketch has been non-dimensionalised using the parameter "a".)

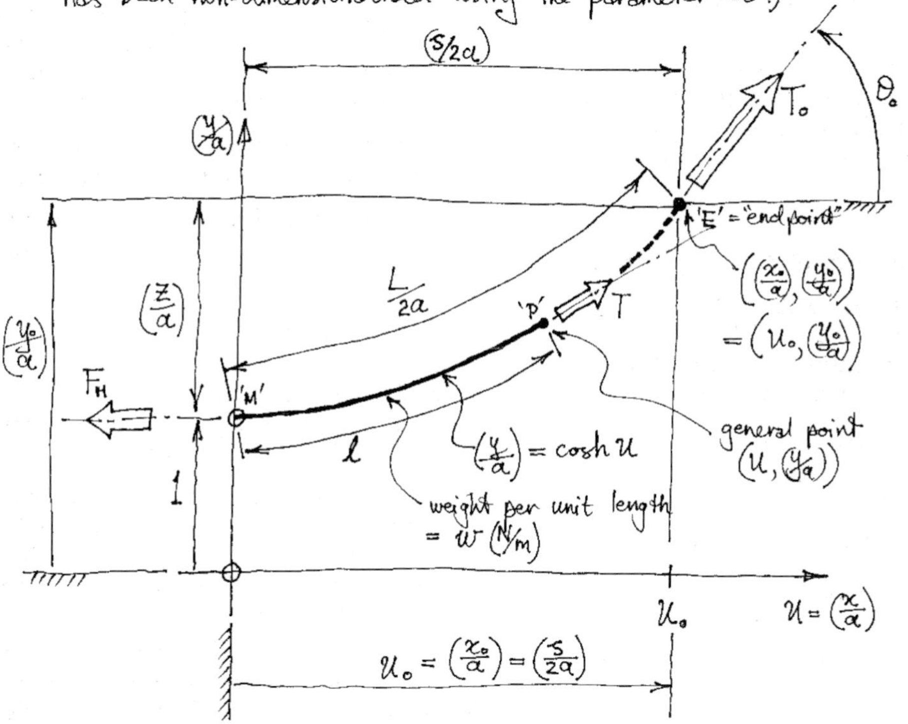

Note, if ρ_c = density of cable material (kg/m³),
A = cross-section area of cable (m²),
then $w = g.\rho_c.A$

[units: $[w] = [g\rho_c A] = (N/kg)(kg/m^3)(m^2) = (N/m)$ ✓✓]

Engineering Design Associates

Project Title : **RURAL POWER DISTRIBUTION SYSTEM**

Date 02:14:98

Page 2 of 14

Ckd. mat

(B) Application to Design of Transmission Line

In applying the mathematical model (i.e. the given equations) to this practical design problem, we must introduce a few more variables relevant to the strength of the system, as well as a convenient practical measure of the shape of the catenary. I think that the mid-span "sag", Z, is useful for the latter — it is defined in the diagrams supplied with the problem, such that

$$\boxed{Z \equiv y_0 - a}$$

Now,

$$y_0 = a \cosh\left(\frac{x_0}{a}\right) = a \cosh\left(\frac{S}{2a}\right)$$

$$\Rightarrow \quad \boxed{Z = a \cosh\left(\frac{S}{2a}\right) - a}$$

which gives us an implicit relationship for $a = a(S, Z)$, but cannot be manipulated to give an explicit expression for a in terms of S & Z. One response to this might be to use the distance a as a design parameter.

Some other useful quantities for design purposes are:

H = the height of the pole above ground

G = the clearance between the cable & ground

(note that $G \neq a$ in general, and $H \neq y_0$ in general, since the location of the ground is unrelated to the vertical location of the x-axis — another way of saying this is that a is a mathematical quantity describing the shape of the catenary curve for a particular span)

D = outer diameter of the poles,

D_i = inner " " " ,

d = diameter of the copper cable

N = the number of spans of cable,

N_p = the number of poles used $(= N+1)$.

Engineering Design Associates

Project Title : **RURAL POWER DISTRIBUTION SYSTEM**

Date 02:14:98

Page 3 of 14

Ckd. mak

Ⓒ Develop the "objective function"
(ie. develop an expression for the cost of the system)

In every optimisation problem, we need an "objective function" — in this case it is the total cost of the system.
By making some bold approximations, we can estimate :-

$$C_{TOT} = C_{pole} + C_{cable} + C_{erect}$$

$$\simeq 2\,C_{pole} + C_{cable} \qquad (\text{since } C_{erect} \approx C_{pole})$$

$$\simeq 2\left[M_{pole} \times c_{st}\right] + \left[M_{cable} \times c_{cu}\right]$$

$$\simeq 2\left[\rho_{st} \times \underbrace{H \times A_{pole} \times N_p}_{\text{total metal volume in pipes}} \times c_{st}\right] + \left[\rho_{cu} \times \underbrace{L \times A_{cable} \times N}_{\text{total metal volume in cable}} \times c_{cu}\right]$$

$$\simeq 2\rho_{st}\,c_{st}\,\tfrac{\pi}{4}\left(D^2 - D_i^2\right)H\,(N-1) + \rho_{cu}\,c_{cu}\,\tfrac{\pi}{4}\,d^2 L\,N$$

[DESIGN VARIABLES] : $\underbrace{D \;\; D_i \;\; H}_{\text{(pole)}} \quad \underbrace{N}_{\text{(system)}} \quad \underbrace{d \;\; L}_{\text{(cable)}}$

It will be much more elegant later if we characterize the pole cross-section by one dimensional variable (D), and one dimensionless shape parameter $(\beta \equiv D_i/D)$, rather than using both D & D_i. So we re-write :–

$$\boxed{C_{TOT} \simeq \left[2\rho_{st}\,c_{st}\,\tfrac{\pi}{4}\right] D^2 (1-\beta^2)\,H\,(N-1) + \left[\rho_{cu}\,c_{cu}\,\tfrac{\pi}{4}\right] d^2 L\,N} \quad -(1)$$

[DESIGN VARIABLES] : $\underbrace{D \;\; \beta \;\; H}_{\text{poles}} \quad \underbrace{N}_{\text{system}} \quad \underbrace{d \;\; L}_{\text{cable}}$

Engineering Design Associates

Project Title : **RURAL POWER DISTRIBUTION SYSTEM**

Date 02:14:98

Page 4 of 14

Ckd. mak

(D) Consideration of Pole Strength

On each pole there is assumed to act a maximum lateral wind force of: $F_W = F_{w,cable} + F_{w,pole}$

$F_{w,cable} = C_D \frac{\rho V^2}{2} L.d$ (due to wind on cable, one span)

$F_{w,pole} = C_D \frac{\rho V^2}{2} H.D$ (" " " " pole, one pole)

This produces a maximum resisting moment M at the base of the cantilever given by:

$$M = \left(F_{w,pole} \times \frac{H}{2}\right) + \left(F_{w,cable} \times H\right)$$

$$\boxed{M = C_D \frac{\rho V^2}{2} \cdot H\left(\frac{HD}{2} + Ld\right)} \quad \text{(P1)}$$

Max stress resulting from this is

$$\sigma_{B,max} = \frac{M\ D/2}{\frac{\pi D^4}{64}(1-\beta^4)}$$

$$= \frac{32\ M}{\pi D^3 (1-\beta^4)} \quad \text{(P2)}$$

Design Inequality: $\boxed{\sigma_{B,max} \leqslant S_{A,st}}$ (P3) ← (to ensure structural integrity)

$$\Rightarrow \frac{32\ C_D\left(\frac{\rho V^2}{2}\right) \cdot H\left(\frac{HD}{2} + Ld\right)}{\pi\ D^3\ (1-\beta^4)} \leqslant S_{A,st}$$

(rearrange to separate constants from des. variables)

$$\Rightarrow \boxed{\frac{32\ C_D\left(\frac{\rho V^2}{2}\right)}{\pi\ S_{A,st}} \leqslant \frac{D^3(1-\beta^4)}{H\left(\frac{HD}{2} + Ld\right)}} \quad \text{(P4)}$$

constant (dimensionless, as it happens!) — design variables

Engineering Design Associates **Project Title :** **RURAL POWER DISTRIBUTION SYSTEM**	Date 02:14:98
	Page 5 of 14
	Ckd. mat

(E) Cable Strength (& shape of catenary curve)

There are many ways to go about this part of the problem, some more elegant than others. It is important that we see the "big picture" — the cable will be fully defined when we have specified its size, and its path through space. This will indirectly determine its support tension, but it's possible to think of the tension as being dependent on the size (i.e. d) and the path or curve shape (i.e. S, Z, H) relative to the ground. The trick is that there are many potential variables describing the curve shape ($N, S, a, Z, G, H, y_o, T_o, F_V, F_H, \theta_o, L$), but many are interdependent.

* SUGGESTION

One way of starting to approach this problem is to ask what is the simplest system of poles & cables we can think of — (the answer ⇒ $N=2$ poles), & then to ask what a system might be like with $N=2$, using the lightest possible cable. Could it be safely built? If the poles can be as high as you like, what would be the limiting requirement? (answer: that the cable stress be OK) So what is the best value for the cable stress? (answer: $\sigma_{T,max} = S_{A,Cu}$)

[This is a crucial point, so spend some time on it!]

We require $\boxed{\sigma_{T,max} \leq S_{A,Cu}}$ ← (design inequality for cable) — (2)

Now, where does $\sigma_{T,max}$ occur? (answer: at the supports since $T = w \cdot y$, $T_o = w \cdot y_o$)

So put $\sigma_{T,max} = \dfrac{T_o}{\left(\frac{\pi d^2}{4}\right)}$ — (3)

$= \boxed{\dfrac{w\, y_o}{\left(\frac{\pi d^2}{4}\right)} \leq S_{A,Cu}}$ — (4)

But now note that $w = g \rho_{Cu} \left(\frac{\pi d^2}{4}\right)$ — (5)

Engineering Design Associates

Project Title : **RURAL POWER DISTRIBUTION SYSTEM**

Date 02:14:98

Page 6 of 14

Ckd. mak

So substituting (5) into (4), we get some cancellation which demonstrates that the height of the supports above the origin of the catenary curve, y_0, has a minimum value which depends only on material constants, and not on the size of the cable, i.e.

$$\frac{g\,\rho_{Cu}\left(\frac{\pi d^2}{4}\right) y_0}{\left(\frac{\pi d^2}{4}\right)} \leqslant S_{A,Cu}$$

$$\Rightarrow \boxed{y_0 \leqslant \left(\frac{S_{A,Cu}}{g\,\rho_{Cu}}\right)} \qquad (6)$$

For copper, this turns out to be

$$y_0 \leqslant \left(\frac{45\times10^6\ \text{N/m}^2}{9.81\ \frac{\text{N}}{\text{kg}} \times 8860\ \frac{\text{kg}}{\text{m}^3}}\right)$$

$$\leqslant 517.7\ \text{m} \qquad (6.a)$$

It is very useful (essential? — I'm not sure) to understand the implications of equation (6.a). Note the following:

— the result is independent of span;

— " " " " d, so that a larger diameter wire would increase the weight for a given catenary curve, and so increase the tensile force T_0, but would not alter the stress in the cable, since the cross-section area also increases to keep stress constant;

— a value of y_0 greater than 517.7 m (in this case) would cause too much stress in the cable ($\Leftrightarrow$ yielding);

— a value of y_0 less than 517.7m would lead to stress values in the cable lower than the allowable limit, so smaller y_0 values are increasingly wasteful of the inherent strength of the cable;

— the optimum value of y_0 is $\boxed{y_0 = y_0^* = 517.7\,\text{m}}$ since this ensures the cable is optimally stressed.

(F) A Possible Calculation Sequence

The observation in (6) that there is an optimum y_0^* for a given material, gives us the basis for a possible calculation sequence. There are many possible sequences — this one is not necessarily the best :—

(0) — choose $d = d_{MIN}$; also calculate $y_0^* = \left(\frac{S_{A,Cu}}{g\rho_{Cu}}\right)$

(1) — calculate w : $w = g\rho_{Cu}\frac{\pi}{4}d^2$

(2) — choose N (arbitrarily)

(3) — calculate s : $s = \frac{X_{TOT}}{N}$

iteration for a, y_0:

(4a) — choose value for a (see note overleaf)*

(4b) — calculate $y_0 = a\cosh\left(\frac{s}{2a}\right)$

(4c) — check if $y_0 > y_0^*$ (if yes, increase a, re-calculate) (if not, check if y_0 is too small...)

(4d) — check if $y_0 < y_0^* \times 95\%$ (if yes, reduce a, recalculate) (if not, value of a & y_0 OK)

(5) — calculate $Z = y_0 - a$

(6) — calculate $H = G_{MIN} + Z$

(7) — calculate $L = 2a\sinh\left(\frac{s}{2a}\right)$

(8) — arbitrarily choose value of β (say $\beta = 0{\cdot}9$)

(9) — iteratively evaluate D (from equ'n (P4) on p.4)

} DESIGN OF POLE

(10) — now we have $D\,\beta\,H\,N\,d\,L$ ⇒ evaluate C_{TOT} from equ'n (1)

(11) — choose a different value of N, & re-calculate everything to see the effect on C_{TOT}

Engineering Design Associates **Project Title :** **RURAL POWER DISTRIBUTION SYSTEM**	Date 02:14:98
	Page 8 of 14
	Ckd. mak

*NOTE — re. the iteration for a

— the graph on the next page shows a set of curves of the function $y_o = a \cosh\left(\frac{s}{2a}\right)$ with "a" as the independent variable. Each curve has a different, constant value of s, based on a different value of N (i.e. $s = \frac{X_{TOT}}{N}$).
The graph enables us to quickly find the "optimum" († see note) value of "a" (i.e. the one corresponding to $y_o = y_o^*$) for a particular value of N (and s).

— some calculators (and spreadsheets) allow the user to solve automatically for "a" in the implicit relationship of equation: $Z = a \cosh\left(\frac{s}{2a}\right)$ on p. 2.

Finally, p.10 shows some sketches of a catenary having a constant sag, z, but with various values of a.
This could be used to help people understand that the parameter "a" is able to act as a design parameter.

† note: the graphs overleaf correspond to the "deep catenary" branch of the dual solution, not to the "shallow catenary", and so are not the shortest length of cable.
Refer to the discussion of this issue on p. 11, and the "improved" optimum charts which follow on pp. 12/13.

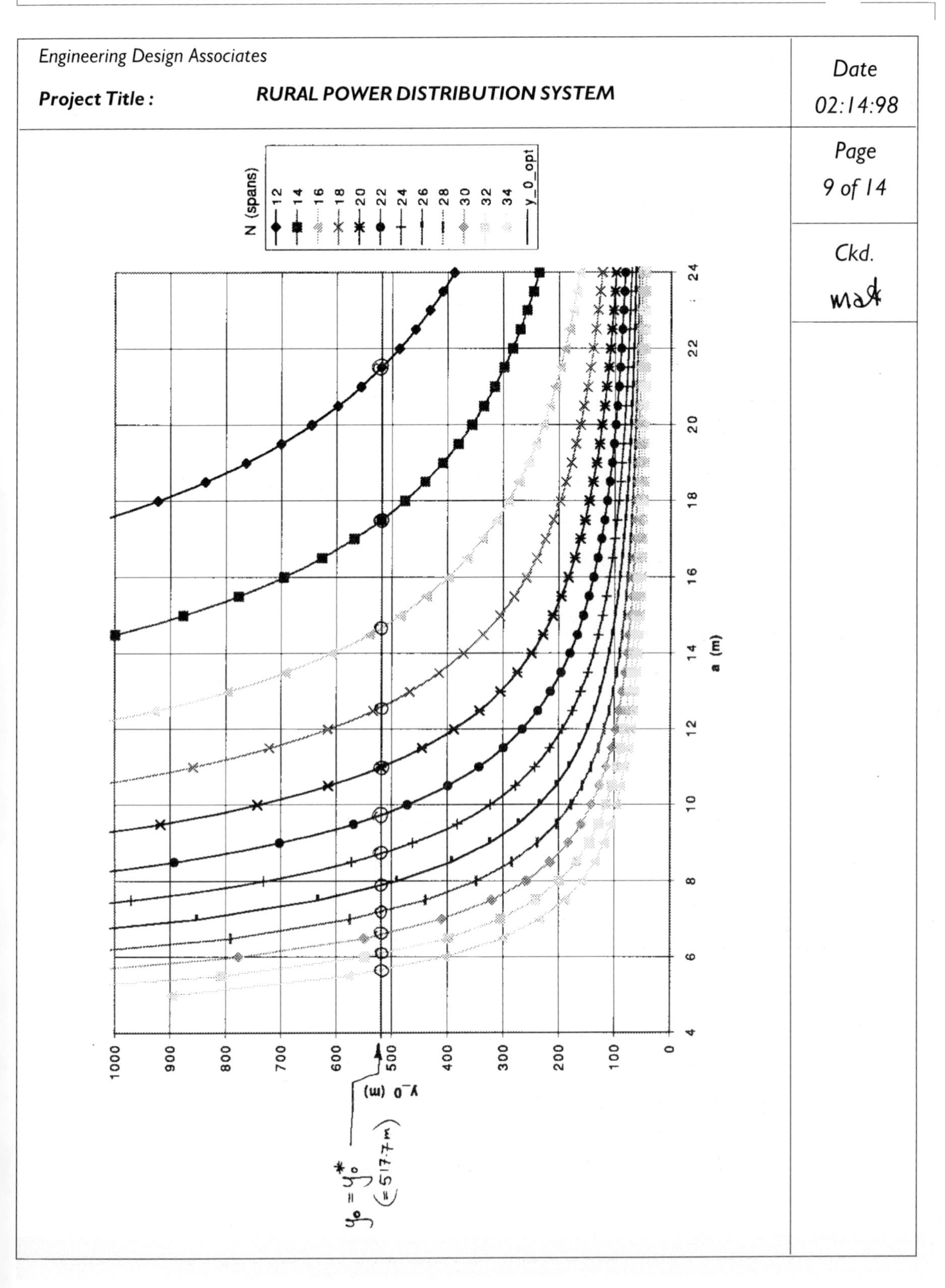
Engineering Design Associates
Project Title : RURAL POWER DISTRIBUTION SYSTEM
Date
02:14:98
Page
9 of 14
Ckd.
N (spans)
12
14
16
18
20
22
24
26
28
30
32
34
y_0_opt
a (m)
y_0 (m)
$y_0 = y_0^*$
(= 517.7 m)

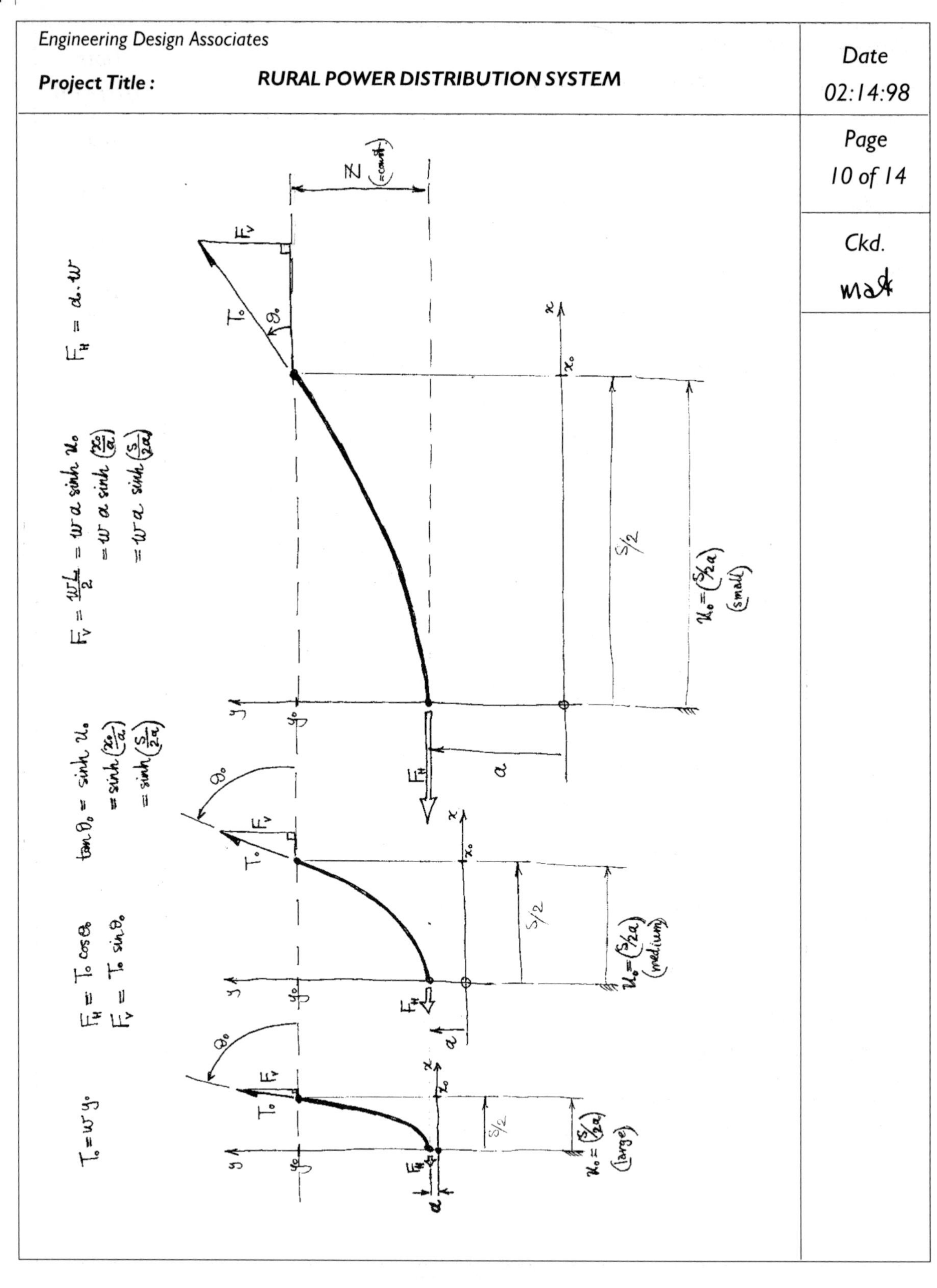
Engineering Design Associates
Project Title : RURAL POWER DISTRIBUTION SYSTEM
Date
02:14:98
Page
10 of 14
Ckd.

Engineering Design Associates

Project Title : **RURAL POWER DISTRIBUTION SYSTEM**

Date 02:14:98

Page 11 of 14

Ckd. mak

Further Notes

Catenary Problem.

This "supplementary note" is just to raise an awareness of one potentially confusing issue :

In general, the equation $y_0 = a \cosh\left(\frac{S}{2a}\right)$

(which we solve iteratively to give "a" in terms of y_0 & S) will lead to two possible values of "a" for a given pair of S and y_0 values.

Thus for a given span, S, & support coordinate, y_0, there are two possible values of "a" which will produce cable tension on the stress limit.

Clearly, we are interested only in the "shallow" catenary, since this allows shorter poles.

The graph supplied earlier (on p. 9) corresponded to to the "deep catenary" branch of solutions.
Please instead refer to the attached chart & table of "a" values corresponding to "shallow" catenaries.

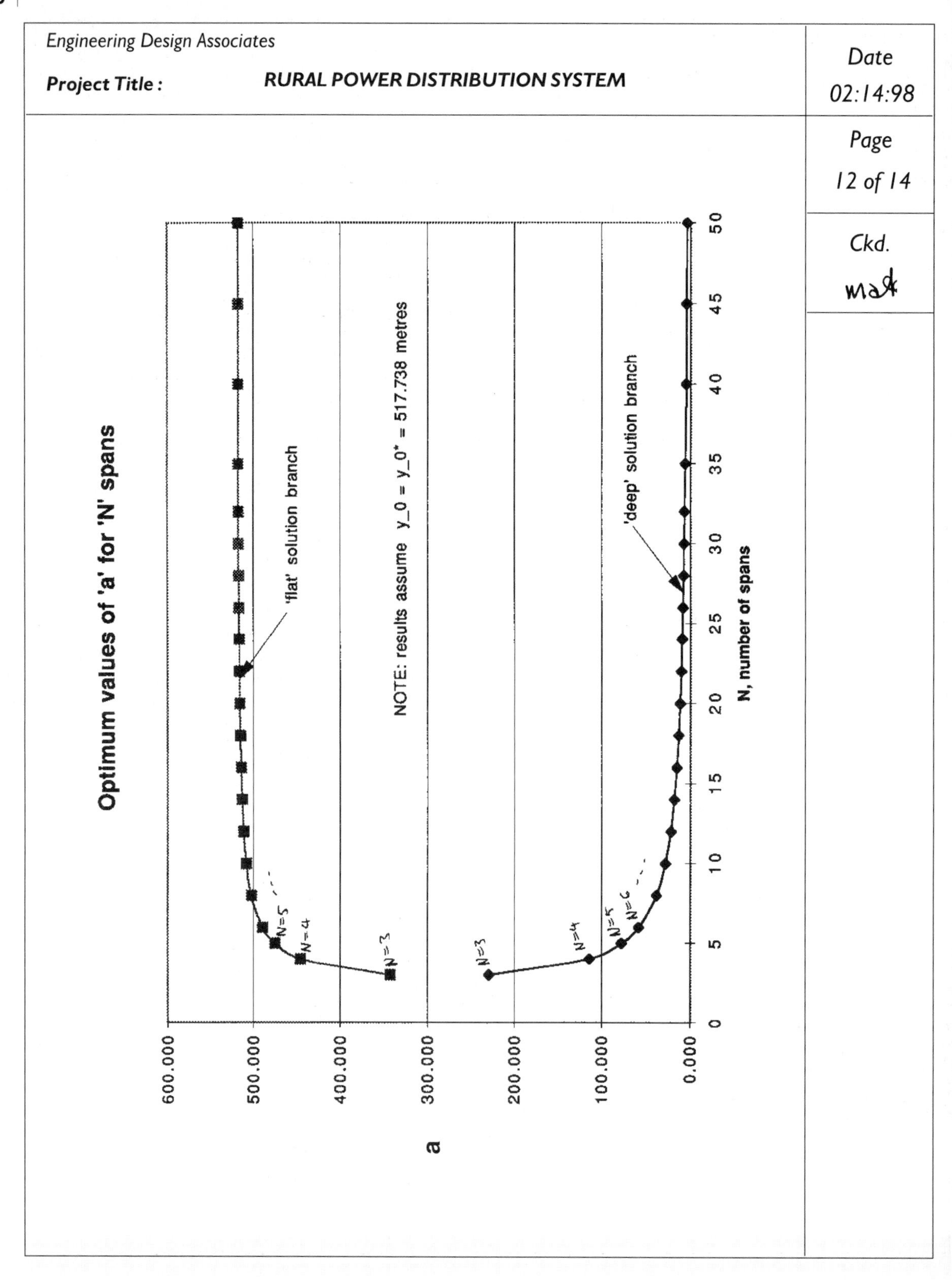
Engineering Design Associates
Project Title :
RURAL POWER DISTRIBUTION SYSTEM
Date
02:14:98
Page
12 of 14
Ckd.
Optimum values of 'a' for 'N' spans
NOTE: results assume y_0 = y_0* = 517.738 metres
'flat' solution branch
'deep' solution branch
N=3
N=4
N=5
N=6
a
N, number of spans
600.000
500.000
400.000
300.000
200.000
100.000
0.000
0
5
10
15
20
25
30
35
40
45
50

Engineering Design Associates **Project Title :** ***RURAL POWER DISTRIBUTION SYSTEM***	*Date* *02:14:98*
	Page *13 of 14*
	Ckd. mak

X_TOT =	**2000** m		g =	**9.81** N/kg
rho_Cu =	**8860** kg/m3		y_0_opt =	517.738 m
S_a_Cu =	**45** MPa		G_min =	**6.000** m

N	s	(shallow solution) a_flat	y_0	z	u_0	H
2	1000.000	416.778	754.440	337.661	1.199673	
3	666.667	**342.051**	517.738	175.687	0.974514	181.687
4	500.000	**445.779**	517.738	71.959	0.560816	77.959
5	400.000	**475.007**	517.738	42.730	0.421046	48.730
6	333.333	**489.063**	517.738	28.675	0.340788	34.675
8	250.000	**502.098**	517.738	15.640	0.248955	21.640
10	200.000	**507.861**	517.738	9.877	0.196904	15.877
12	166.667	**510.927**	517.738	6.811	0.163102	12.811
14	142.857	**512.755**	517.738	4.983	0.139304	10.983
16	125.000	**513.933**	517.738	3.805	0.121611	9.805
18	111.111	**514.737**	517.738	3.001	0.107930	9.001
20	100.000	**515.310**	517.738	2.428	0.097029	8.428
22	90.909	**515.734**	517.738	2.004	0.088136	8.004
24	83.333	**516.055**	517.738	1.683	0.080741	7.683
26	76.923	**516.305**	517.738	1.433	0.074494	7.433
28	71.429	**516.503**	517.738	1.235	0.069146	7.235
30	66.667	**516.662**	517.738	1.076	0.064517	7.076
32	62.500	**516.793**	517.738	0.945	0.060469	6.945
35	57.143	**516.948**	517.738	0.790	0.055269	6.790
40	50.000	**517.134**	517.738	0.604	0.048343	6.604
45	44.444	**517.261**	517.738	0.477	0.042961	6.477
50	40.000	**517.351**	517.738	0.387	0.038658	6.387

↑ (optimum value of α) ↑ (optimum value of sag, Z)

[these are the optimum values of a and Z for a given span S, since they correspond to cable tensile stress at the allowable limit.]

(deep solution) a_deep	y_0	z	u_0	H
416.778	754.440	337.662	1.199679	
228.769	517.738	288.970	1.457077	294.970
113.894	517.738	403.844	2.195024	409.844
77.203	517.738	440.535	2.590573	446.535
57.831	517.738	459.907	2.881955	465.907
37.766	517.738	479.972	3.309882	485.972
27.591	517.738	490.148	3.624435	496.148
21.514	517.738	496.224	3.873492	502.224
17.509	517.738	500.230	4.079642	506.230
14.687	517.738	503.051	4.255448	509.051
12.601	517.738	505.137	4.408655	511.137
11.003	517.738	506.735	4.544370	512.735
9.741	517.738	507.997	4.666149	513.997
8.723	517.738	509.015	4.776566	515.015
7.885	517.738	509.853	4.877541	515.853
7.185	517.738	510.553	4.970548	516.553
6.592	517.738	511.146	5.056740	517.146
6.083	517.738	511.655	5.137040	517.655
5.444	517.738	512.294	5.248035	518.294
4.619	517.738	513.118	5.412415	519.118
3.999	517.738	513.739	5.556472	519.739
3.518	517.738	514.220	5.684640	520.220

note: no solution is possible for N = 2, since the minimum possible value of y_0 is 754.44 m (for s = 1000 m), which is greater than $y_0^* = 517.7$ m, so the cable stress limit will be exceeded.

(ignore these "deep catenary" solutions, since they lead to very tall poles.)

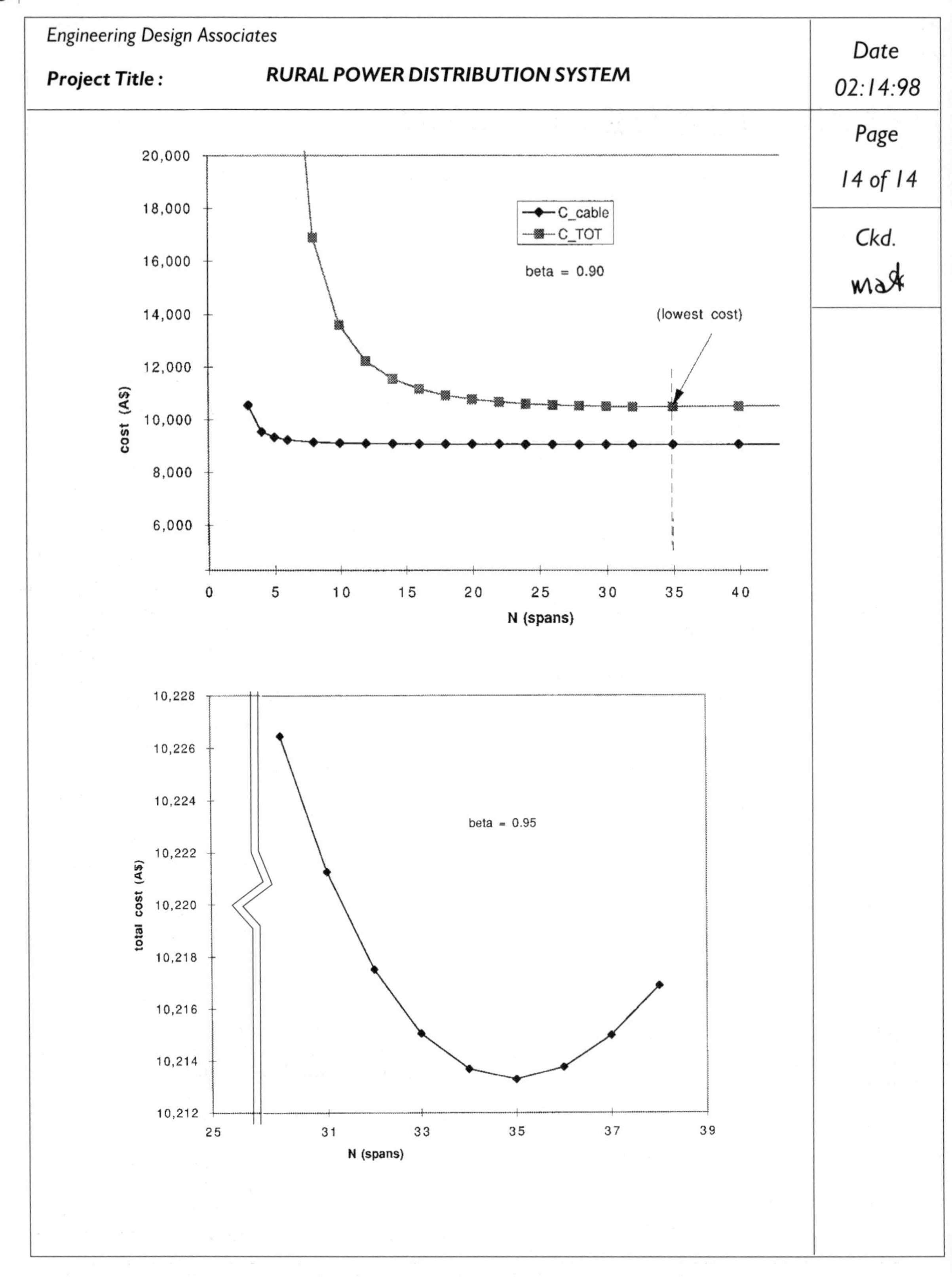

Engineering Design Associates
Project Title : RURAL POWER DISTRIBUTION SYSTEM
Date 02:14:98
Page 14 of 14
Ckd.
20,000
18,000
16,000
14,000
12,000
10,000
8,000
6,000
cost (A$)
C_cable
C_TOT
beta = 0.90
(lowest cost)
0 5 10 15 20 25 30 35 40
N (spans)
10,228
10,226
10,224
10,222
10,220
10,218
10,216
10,214
10,212
total cost (A$)
beta = 0.95
25 31 33 35 37 39
N (spans)

2

DESIGN AGAINST FAILURE

Engineers use materials, whose properties they do not properly understand, to form them into shapes, whose geometries they cannot properly analyse, to resist forces they cannot properly assess, in such a way that the public at large has no reason to suspect the extent of their ignorance. John Ure, Engineer, 1998

Design against failure might seem a thoroughly pessimistic objective when compared to the more optimistic objective of studying successful designs. The title of this chapter intends to signify that, in general, designers are a pessimistic breed of engineers.

In an unpredictable world a pessimist, designing against failure, is less often disappointed than an optimist designing for success. In the perfect world of the scientist's model of reality, stresses and loads, material properties and environmental variables all behave in some ideal, perfectly predictable way. In that *model world* manufacturing errors don't exist. After all, this is the way a *model world* is expected to be. In engineering design we also deal with idealised models, but we *know* that they are merely simplified to make them tractable for analysis. The results of our analyses need to be adjusted for the uncertain nature of the *real world*, where we can never be certain of loads or environmental variables, and where manufacturing errors abound.

Of course, we are not alone in considering failure as a critical starting point in a discourse on engineering design. Professor Henry Petroski (Petroski, 1994) in his introduction to the nature of design notes:

"The concept of failure is central to the design process, and it is by thinking in terms of obviating failure that successful designs are achieved. It has long been practically a truism among practicing engineers and designers that we learn much more from failures than from successes. Indeed, the history of engineering is full of examples of dramatic failures that were once considered confident extrapolations of successful designs; it was the failures that ultimately revealed the latent flaws in design logic that were initially masked by large factors of safety and a design conservatism that became relaxed with time."

Our discourse on failure is entirely based on two associated studies: the study of the properties and mechanics of materials and the predictive power of experimental science.

The history[2.1] of mechanics of materials dates to the construction of the earliest structures that required practical prediction of material behaviour under load. It is now generally agreed that the Djoser pyramid in Upper Egypt, known as the Great Stepped Pyramid, is the earliest important stone structure requiring engineering estimation of material performance. This extraordinary structure, built during the third Dynasty in Egypt (2650-275 BCE) has a base measuring 120 m by 108 m and rises to a height of 60 m. The temple of Karnak, also in Upper Egypt, on the east bank of the Nile, dates to the Middle Kingdom (1938-1600 BCE). The ruins of the great temple of Amon, which form part of the Karnak complex, dates to the New Kingdom (1539-1075 BCE). The fourteen most impressive columns in this temple rise 24 meters in height to massive capitals and lintels still standing today.

Clearly, the Egyptian constructors of these large and complex structures needed a considerable store of practical predictive wisdom for assessing the behaviour of their materials under load, however no written records of these predictive skills have been handed down to us.

The first comprehensive treatise on architectural design was compiled by Vitruvius (1st century BCE) in the ten books of *De Architectura (On Architecture)*, a handbook for Roman architects (Morgan, 1960).

The beginnings of the science of structural mechanics dates to George Bauer (1494 – 1555) a German scholar and scientist, who adopted the Latin name Georgius Agricola. A physician by

2.1. The sources for these brief historical notes are the *Encyclopaedia Britannica*, 15th Ed., 1991, and Timoshenko, *History of Strength of Materials*, 1953, New York:McGraw Hill

training, he is regarded as the founder of modern mineralogy, and among the first to base scientific deduction on experimental observation rather than on fruitless speculation. His major work, *De re Metallica*, dealing with mining and smelting, was published in 1530. This work was translated into English in 1912 by the mining engineer Herbert Hoover (later US president).

Together with his contributions to almost all other aspects of technology, Leonardo da Vinci (1452-1519) also recorded some early experiments on the mechanics of materials. Among several other experiments on structures, Leonardo investigated the strengths of columns and the tensile strength of wire. However, the first recorded study of the *safe* dimensions of structural elements appear in Galileo's (1564-1642) book *intorno á due nuove scienze* (Introduction to Two New Sciences), published in 1638. In this work Galileo describes studies on columns and beams (Timoshenko, 1953).

In 1795 the *Ecole Polytechnique* was established in Paris, mostly due to the inspired leadership of the French mathematician, Gaspard Monge, who is widely regarded as the father of analytical geometry. The establishment of this school by famous engineers and scientists such as Fourier, Lagrange, Monge, Prony and Poisson, and eventually Navier and Poncelet, lent great impetus to the development of the science of mechanics of materials and structures. Some of the important contributors to the early development of structural mechanics and the elastic theory of structures are:

- Robert Hooke (1635-1703), responsible for the discovery of the elastic properties of materials;
- Thomas Young, English physicist, 1807, who defined the elastic modulus of materials under direct stress (Young's modulus);
- Claude-Louis-Marie Navier whose lecture notes on strength of materials, *Résumé des Leçons de Méchanique*, were published in 1830. This work was substantially edited and republished in 1864 by Saint-Venant;
- A. J. C. Barré de Saint-Venant, 1853, French engineer, who derived the torsional or shear modulus of elasticity;
- Alberto Castigliano, Italian mathematician and engineer, 1875, developer of the notion of strain energy, based on the principle of virtual work;
- Eugenio Beltrami, 1885, the Italian mathematician, who proposed a total strain energy based theory of failure.

In what follows, we first briefly examine those properties of elastic materials that substantially influence our material choices for the design of engineering components. We then proceed to study the way in which the mechanical properties of our materials of choice allow us to predict component behaviour under complex, practical, loading conditions.

2.1 Introducing engineering materials and modes of failure

Success or failure of engineering components, devices and systems is a function of technical, environmental, economic and social issues. Our focus in this chapter is on technical issues that relate to structural integrity, and the design of engineering components to resist some form of structural failure. We begin by considering elementary failure mechanisms of some materials of construction used in engineering.

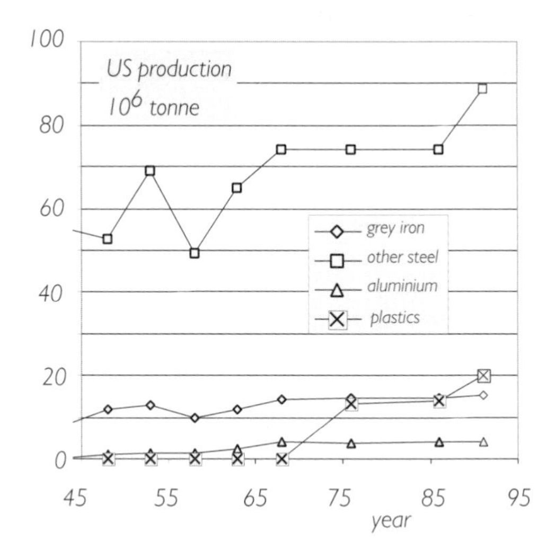

Figure 2.1 Most important materials of construction, 1945 to 1991. Source: Bureau of statistics

Although there has been a consistent, almost monotonic, increase in the use of polymers in engineering components since the turn of this century, we will concentrate on elastic materials, and principally metals, only.

Figure 2.1 indicates the major materials of construction used in the second half of this century. As these graphs indicate, there has been a steady slight (between 4% to 4.5%) annual increase in steel production. Aluminium production, by far the largest component of all nonferrous metal production, has remained moderately constant over the last 30 years. Based on 1976 data, the world steel production was 214 million tonnes and aluminium production was 12 million tonnes. In 1992 the corresponding production was 800 million tonnes for steel and 16 million tonnes for aluminium. The most dramatic change in materials of prodiuction has been the increase in polymers and resins. Figures for USA and European production of plastics in 1996 are 28 million and 25 million tonnes respectively, thereby significantly exceeding world productions of nonferrous metals.

Clearly, even if proportionately the use of ferrous metals is declining, in absolute terms they still represent the bulk of all engineering and structural materials used in the world today. Hence our concentration on their behaviour is based on the evidence that most engineering components are made of steel, or some other ferrous material.

Design against failure is a complex, multifaceted process, in which we need to examine each facet separately for its influence on the eventual successful avoidance of failure in our design. Figure 2.2 is a schematic view of the design process for a generic engineering component. The *failure focus* identifies the type (or *mode*) of failure the designer will focus on, or specifically design against, given the material characteristics and the operating conditions of the component.

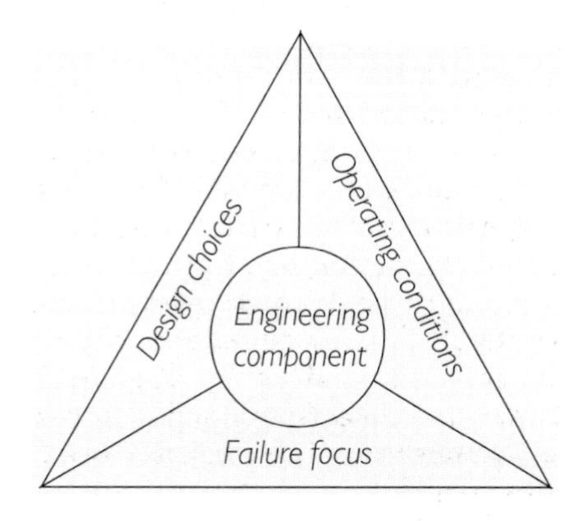

Figure 2.2 The designer's focus in component design

Figure 2.3 is an interaction matrix between various design variables and the primary failure focus of the designer. The marked cells in the matrix indicate where there is a strong interaction between specific design variables and the *mode of failure* the designer is primarily attempting to avoid.

Design variables		Failure focus: Rupture yielding	Excessive deflection	Fatigue	Buckling	Wear	Corrosion	Stres corrosion	Creep
Material properties	S_U, S_y, S'_e E, K_c, H	S_U S_y K_c	E S_y	S'_e S_U	E S_y	H			
Geometry	Load-efficient distribution of materials Stress intensity factors	×	×	×	×				
Load character	Static loading	×	×		×			×	×
	Time-dependent dynamic loading	×		×		×			
Environment	Operating temperature	×	×						×
	Corrosive						×	×	
	Abrasive					×			

Figure 2.3 Interaction matrix between design variables and the primary failure focus of the designer S_u = rupture strength; S_y = yield strength; S'_e = endurance limit; E = elastic modulus; K_c = fracture toughness factor; H = hardness

In what follows we will investigate the several modes of failure and the tools the designer has for selecting component characteristics such that the likelihood of some given mode of failure is minimised.

In general, engineering components can fail by suffering one or more of the following effects:

- the component breaks. This type of failure is generally signified by some active part of our component breaking under load. The technical terms for this type of failure are *rupture*, or *fracture*;
- the component sustains plastic, non-recoverable, deformation. This type of failure is most commonly localised and often leads to eventual rupture of the component. The common technical term for this type of plastic deformation is *yielding*;
- the component experiences a time dependent failure mechanism known as *fatigue*. This type of mechanism is still the subject of intense research, since it is not yet fully understood. Many engineering components are exposed to time dependent, dynamic, loading. Under these conditions the material of construction becomes *tired* and will fail at a significantly lower stress intensity than it could sustain in a static manner;
- the component experiences a form of elastic instability with large lateral deflections under compressive loading. This type of failure is mostly experienced by slender columns under axial loads, and thin walled pressure vessels subject to external pressure (e.g. vacuum vessels). The technical term for this type of failure is *buckling*. A component will buckle under a load significantly lower than the yield compressive load for the same component. Once *buckled*, the component can no longer sustain any significant load;
- at elevated temperatures there is long-term, relatively slow, plastic deformation of the component under steady load. The effect is attributed to the fact that key material properties such as yield strength and ultimate tensile strength are generally obtained from tests of the material at ambient temperatures. Loaded components exposed to higher temperatures, such as gas turbine blades, or heat exchanger piping, suffer this type of slow deformation. In technical jargon this effect is known as *creep*;
- the component experiences elastic or plastic deformation beyond some permitted bound. The technical term for this type of failure is *excessive-deflection*;
- one or more surfaces of the component suffer local failure due to high local rubbing or concentrated loads. This type of failure usually wears away some part of the surface locally and is technically termed *wear*. Also under the same heading is *erosion*, a type of surface damage caused by heavy particles impacting on a surface locally. A typical examples are found in jet exhausts of steam turbines, where the condensing water particles erode the local surface, and in conveyor systems used for transporting abrasive materials. Another effect of wear is a type of localised spalling of the surface resulting from surface fatigue. Most commonly, this type of *wear* failure occurs in rolling element bearings;
- the material suffers some form of degradation or chemical conversion, most commonly through oxidation. The technical term for this type of failure is *corrosion* . It is often the result of unacceptable or unpredictable environmental exposure of materials to corroding agents. It can also be a result of a Galvanic current set up by a combination of electrode potential differences in adjacent materials in the presence of a suitable electrolyte, such as sea water or bodily fluids. The process is often exacerbated by the presence of stresses at the grain boundaries of the metal. This type of corrosion is a significant problem for specific combinations of materials and environments. A typical example is experienced with certain grades of stainless steel in chloride surroundings such as the seaside. This effect is known as *stress corrosion*.

2.1.1 Engineering materials

Most engineering materials are classified by either one of three designations, namely their:

- colloquial or trade name. Examples are: *Teflon* – for the product poly-tetra-fluoro-ethylene (*PTFE*); and *mild steel* – for a range of low carbon steels used in general engineering and designated by a typical 0.1% carbon content (for example: AS 1442-1020, or AISI-SAE 1020);

- standard specification according to national or internationally recognized standards such as Standards Australia (AS), Society of Automotive Engineers (SAE), British Standards Institute (BSI) or American Iron and Steel Institute (AISI);
- trade specification by companies producing special materials, usually prefaced by the company name. Comsteel-444 (19/2), equivalent to AISI-444 ferritic stainless steel, is a good example of this type of designation.

Materials are chosen for a variety of reasons but generally three dominant criteria are used in the selection process:

- first of all, the material *must meet the specified duty* for the product (strength, elasticity or wear resistance for example);
- secondly, the material *must be readily available*. Material suppliers are notorious for listing exotic products which can only be produced if the buyer is prepared to wait many months or is prepared to order a large tonnage;
- thirdly the material *must meet production requirements*. The product must be capable of being produced. It may be exciting to design a product in a special ceramic, but the excitement soon evaporates when it is discovered that the only furnace capable of producing the desired manufacturing conditions is in some remote location, or has inadequate capacity.

Ultimately, material selection is determined in consultation with the producer or distributor of the materials concerned. This will ensure that our material specification for the design is up-to-date.

(a) Iron, steel and cast iron

Probably the most versatile of all engineering materials, iron and steel have many desirable qualities. Iron is abundant in the form of iron ore (haematite or iron-oxide), it is readily reduced from ore to metal, and when alloyed with carbon it provides a vast variety of properties.

When alloyed with carbon, iron is referred to as either *cast iron* ($C>3\%$) or as a *steel* ($C \leq 1.5\%$). The range $1.5\% < C < 3\%$ is not used commonly. Cast iron, as the name implies, is always *cast* into shape (poured in the molten state). Although steel may also be cast into shape, it is generally hot or cold formed (by pressing while still solid), or machined (by cutting, grinding, drilling, or shaped in a host of related machine tooling operations).

Casting irons have a wide variety of excellent properties, some of which may be changed by appropriate heat treatment. Liquid iron is of relatively high *fluidity* and may be cast into quite intricate shapes. Due to the high carbon content in grey cast irons they have excellent wear resistance as well as high vibration damping. These two properties alone make this material most useful for machine tool frames and engine components. Due to uncertainty associated with the quality of the surface finish in cast components, unusually large machining allowances are called for when a component is machined from a normal gravity type of casting.

(b) Plain carbon steel

Most engineering designers will recognize such terms as *mild steel*, *tool-steel* or *spring-steel*. However the variety of steels commercially available for the production of commodities is vast and the range can be encompassed only through consultation with metals handbooks or manufacturers. The range is further complicated by the changes in properties available through heat treatment. This text is intended to provide only an overview of the range of steels. Consequently, the brief survey that follows is broken up into the segments indicated on the *tree of steels* shown on Figure 2.4.

Plain carbon steels range in carbon content from 0.1% to 1.5% with minor additions of manganese phosphorus, silicon and sulphur. To a very large extent the properties of the final product are affected by the way the steel, poured from the furnace, behaves during the cooling process. In the final stage of processing, the molten steel is surrounded by a great deal of free oxygen. This oxygen reacts with the carbon in the steel and forms carbon monoxide (CO). As the steel cools, the CO bubbles rise to the surface, and the resulting metal structure is significantly less homogeneous than it would be without the bubbles. The resulting material is known as *rimming steel*, named after the CO bubbling process (which is referred to as rimming). Rimming steels are suitable for low-strength cheap steel applications. Thin plate used in metal beverage can production is a typical application of rimming steel, where the thinning out process during rolling tends to reduce the effect of the rimming bubbles of CO and generally improve structural integrity.

Where the steel maker wishes to produce steel without the CO bubbles, additives are used at the last stages of production. Quantities of aluminium or silicon combine with free oxygen in preference to the formation of CO. The result is a very still or *killed* cooling process. In contrast to the rimming steel, *fully killed* steels tend to suffer cooling shrinkage. In general the uneven end of the cast steel ingot is cut off and returned to the furnace. The lower volumetric efficiency of this production process results in a more expensive steel, but a steel vastly superior to rimming steel in structural properties.

Current steel-making processes rely on continuous casting and, in general, additives are used to produce a *fully killed* steel. The terms *rimming*, or *semi-killed*, are nowadays only of relevance to the lesser-used, ingot-making processes where the steel is cast in batches (*ingots*).

Practical uses of steels tend to fall into three major categories, namely, as structural members, as components to be heat-treated and as fabricated components. For the structural components the property of *weldability* is often a major consideration in making the proper material choice. Heat treated components on the other hand may require high strength and it is this property which will most influence material choice. Finally, for fabricated components, the property of *machinability* may be of major importance. Each of these properties can be enhanced by proper alloying of the steel in the production process. Many other performance characteristics of steel may also be enhanced by the addition of suitable base elements. One common additive is sulphur, in the range 0.15% to 0.35% to enhance machinability.

The range of steels alloyed with this high sulphur content are referred to as *free-cutting* steels. The heat-treatment of steels is a special subject, but, in general terms, the process is aimed at rearranging the grain structure of the steel to enhance certain properties. Two major properties enhanced are *hardness* and *toughness*, the latter being the ability to absorb energy during shock loading (and hence resist sudden fracture on impact loading).

The presence of carbon in steel is essential for the hardening process. In medium and high carbon steels, sufficient carbon is present for hardening to be achieved through a temperature cycling

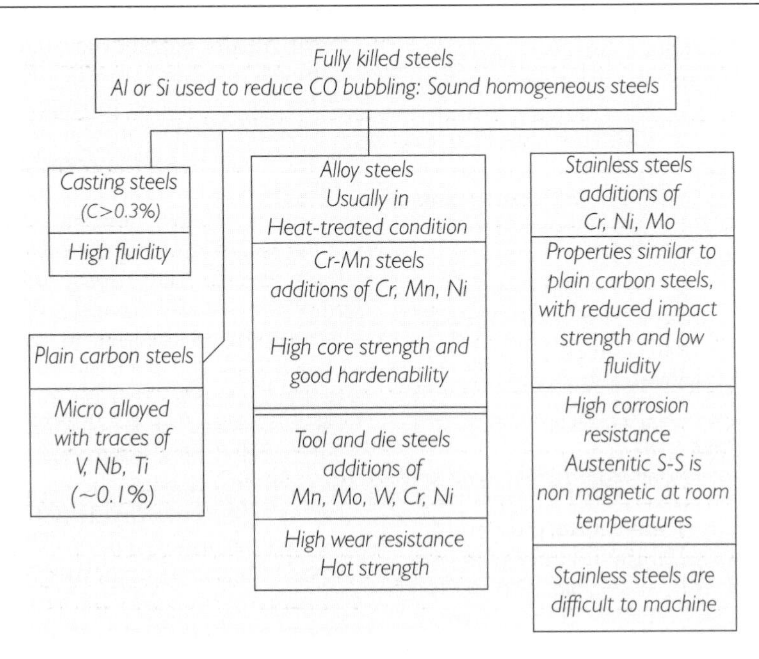

Figure 2.4 The tree of steels

process. This usually results in a very hard crystalline form of the iron-carbon solution (*Banite* or *Martensite*). For steels with a 0.2% or less carbon content the hardening thermal cycle is often preceded by a process of surface carbon absorption or *carburizing*. Surface hardening metals are usually referred to as having been *case-hardened*. Nevertheless, even steels with such low carbon content can be *through hardened* (hardened right through).

Casting steels are similar in properties and composition to wrought steels, but the casting process reduces the impact strength of the steel. Furthermore, the casting steels are less fluid than their cast iron counterparts. The resulting casting is rough with very large shrinkages. For these reasons substantial casting tolerances and machining allowances must be used for cast steels.

(c) High-strength low-alloy (HSLA) steels

"Low-alloy steels have undergone some recent developments that have led to significantly higher strength products. ... structural steels that have yield strengths of greater than 500 MPa ...in contrast to earlier structural steels of only half that figure." (van Vlack,1985)

These HSLA steels are alloyed with very small quantities (0.05%) of vanadium, niobium or titanium to produce fine grain size and finely distributed precipitates within the ferrite. These precipitates reduce the formation of dislocations, and thereby provide the real strength of the alloy. The resulting steel may be used in the un-heat-treated state, making it attractive for use in the design of automobiles, high pressure pipe lines and major structures.

Alloying of metals in general enhances some property of the main constituent. The effects of principal alloying elements are listed in Table 2.1

(d) Stainless Steel

"The term stainless steel denotes a large family of steels containing at least 11.5 percent chromium. They are not resistant to all corroding media... Stainless steel competes with nonferrous alloys of copper and nickel on a corrosion-resistance and cost basis and with light metals such as aluminium and magnesium on the basis of cost and strength-weight ratio. Stainless steel has a number of alloy

Table 2.1 Alloying constituents in steels

Constituent	Influence on parent metal
nickel (Ni)	Toughness and deep hardening;
chromium (Cr)	Corrosion resistance, toughness, hardenability;
manganese (Mn)	Deoxidation (killing) as well as hardenability ;
silicon (Si)	Deoxidation (killing), resistance to high temperature oxidation, raises critical temperature for heat treatment, increases susceptibility to decarburization (loss of surface carbon) and graphitization (formation of free graphite);
molybdenum (Mo)	Promotes hardenability, improves hot tensile and creep strengths;
vanadium (V)	Promotes toughness, shock resistance and hot strength. One of three micro alloying elements;
aluminium (Al)	Deoxidation (killing), promotes fine grain structure;
boron (B)	Increases hardenability;
tungsten (W)	Hardness / hot wear strength in tool steel. Usually as tungsten carbide;
vanadium (V), niobium (Nb) and titanium (Ti)	Used in trace amounts as microalloying elements

compositions and many suppliers. Information on its properties and fabrication can be obtained readily. Sound techniques have been evolved for casting, heat treating, forming, machining, welding, assembling, and finishing stainless steel. It will be found that this material usually work-hardens (which makes machining, forming and piercing more difficult), must be welded under controlled conditions, and under inert gas. It has desirable high strength, corrosion resistance and decorative properties." (Niebel and Draper,1974)

Stainless steels are hardenable by cold working and the properties of the steel may be enhanced through appropriate heat-treatment, cold-working, or a combination of these techniques. The single most important characteristic of stainless steels is their resistance to corrosion. This resistance is due to a thin transparent chromium oxide film which forms on the surface of the metal on exposure to oxygen or air. Although protection is provided against most oxidizing substances, some further surface protection is necessary when stainless steel is used with reducing agents, such as hydrochloric acid. In situations where both hot strength and low surface oxidation rates are needed, a stainless steel with a ceramic coating may be used.

In some applications stainless steels are used for their appearance and aesthetic qualities rather than their corrosion resistance. The surface will accept a high polish and some stainless steels may be readily moulded by mechanical deforming processes.

(e) Special steels

Tool and die steels are a group of materials developed for the forming and machining of metals. Hardness, toughness and high hot strength are the major characteristics of these steels. Carbon content varies from 0.25% to 2.35% and chromium, vanadium and tungsten are the main additives to enhance material properties as described earlier. Typical trade designations for such steels are often descriptive of their intended application:

- *high speed steels*: for general purpose use in drills, cutting tools and highly abrasive cutting conditions;
- *hot work steels*: used for Al and Mg extrusion, press forging dies, shear blades for heavy work, plastic moulding dies, forging dies, high temperature and abrasive conditions, extrusion tooling and copper and brass die casting;
- *cold work tool steels*: general purpose tools, air-hardening and oil-hardening tools, punches, dies, wear resistant tools, such as brick moulds, crusher rolls and blanking tools;
- *shock resistant steels*: shock resistance is a prime property of these tough steels. They are used in pneumatic chisels, shear blades, hand and machine punches, chisels and dies;
- *carbon tool steels*: less expensive, low alloy, high carbon steels, used where service conditions are not severe, for example in scale pivots, gauges, precision engineers tools;
- *mould steels*: medium carbon, chromium alloy steel, for high pressure zinc die casting, plastic moulds and valve spindles. These steels have good corrosion and abrasion resistance.

Table 2.2 – Common nickel alloys

Metal	Ni (%)	C (%)	Cu (%)	Cr (%)	Fe (%)	Rem.	S_u (MPa)	S_y (MPa)
Monel 400 ASTM B 127 B163-5	66.5	0.2	31.5		1.2	Mn,Si	483	193
Incoloy 825 ASTM B163 B407-9	32.5	0.05	0.4	21.0	46.0	Al, Mn, Si, Ti	586	241
Inconel 600 ASTM B163 B166-168	76.0	0.08	0.2	15.5	8.0	Mn, Si	552	242

S_u = Ultimate tensile (rupture) strength; S_y = Yield strength

(f) Nonferrous metals.

The three most common base metals alloyed to form metal products are aluminium, copper and nickel. As well as these, there are many other non ferrous metals used in manufacturing. Typically, titanium alloys are used in aerospace and recreation industries (aircraft components, bicycle frames, golf clubs, sports wheelchairs). Titanium oxide is still the most useful source of white paint. Magnesium alloys are also used in the aerospace industry and in high performance automobiles. Tungsten is used in electric light filaments and in conjunction with carbon, as tungsten carbide tools. Beryllium, a highly toxic metal, finds use in some high energy physics application. However, most of these uses are highly application specific and outside the scope of our discussion.

(g) Aluminium

The alloys of aluminium are particularly useful due to their corrosion resistance, malleability and fluidity in casting applications. High strength to weight ratios available with aluminium make these alloys particularly useful in the aerospace industry. Hardening of the alloy may be achieved by either heat treatment, followed by ageing, or by cold working, resulting in strain hardening. Aluminium may be cast, rolled, extruded, forged, spun or *shot-peen* deformed. *Shot peening* is a process in which small metal balls (*shot*), of a mixed size are blown in a high-pressure air stream against the surface, thereby impact-loading and *flowing* the material by plastic-deformation at the surface of the component.

Apart from steels, aluminium is the single most useful structural metal. However, aluminium is particularly susceptible to galvanic corrosion and stress corrosion. Particular care is needed in applications where aluminium in contact with other metals may be exposed to an electrolyte (e.g. salt water).

(h) Copper

Copper is available as a commercially pure base metal for conductors, or alloyed with zinc (*brasses*), tin, aluminium and silicon (*bronzes*). Brasses are used for moulding and deep drawing. Although some work-hardening takes place during cold-working, the metal is readily annealed at moderate temperatures and successive stages of pressing can result in very large aspect-ratio (*depth-to-diameter ratio*) pressings. Typical examples are cartridge cases and cosmetic containers.

Brass type 360A (American Society for Testing of Materials designation) is the most common type of brass available and it is referred to as *free-machining brass*. This alloy is one of a series of *leaded-brasses* containing about 2% by weight of lead. This alloy is probably the most easily machined metal known, and it is generally used as a benchmark for machinability.

Bronzes are most useful materials where corrosion resistance is important. Main uses are valves, naval pressure valves, ships' propellers and general purpose bearings. Phosphor-bronze is a particularly hard alloy and, in general, it is used in corrosion-resistant spring material. Due to its resistance to sea water corrosion and its high strength, it is also useful as a casting alloy for high power marine propellers. Nickel bronzes are used mainly in food machinery.

(i) Nickel

The most common types of nickel alloys are *Monel* 400 (31.5/66.5 copper-nickel alloy with nickel the main constituent), *Incoloy* 825 (38% to 45% nickel and the remainder iron) and *Inconel* 600 (72% nickel, with 15% Cr and 8 Fe). *Table 2.2* lists some properties for these alloys. Nickel alloys are particularly useful for their corrosion and heat resistant properties. *Monel*, *Incoloy* and *Inconel* are trade names registered by nickel suppliers, but through their wide ranging application the names have passed into common engineering usage. Typical applications are:

- *Monel*: protective plates in chemical, marine and steam plants. Good corrosion resistance to sea water, sulphuric, phosphoric and hydrochloric acids, as well as fatty acids;
- *Incoloy 825*: heating element sheaths, process piping, heat treatment containers;
- *Inconel 600*: furnace parts, heat treatment equipment, electric heater elements.

2.1.2 Corrosion and oxidation of metals

Corrosion (from *corrodere*, Latin for *gnawing away*) is the surface deterioration resulting from a chemical reaction of the component material with its environment. The most common form of corrosion is oxidation, a chemical process that takes place at the surface of materials exposed to oxygen contained in substances such as air or water. Almost all materials will react with oxygen. The process of oxidation involves some exchange of energy, either to or from the oxidation process, and it takes the following form:

$$M + O + Energy = MO \text{ (oxide of material).}$$

If the energy of oxidation is positive, the material remains stable. However, materials with negative oxidation energies will continue to react with oxygen to form oxides in a time and temperature dependent way.

Interestingly, all materials, other than gold, PTFE and some ceramics, have a negative oxidation energy. Consequently, all metals (other than gold) will oxidise at surfaces exposed to the environment. The *downside* of oxidation is that the oxide film formed at the exposed surface is, in most cases, grossly inferior in structural properties to the parent metal. Hence, we need to either protect the surface or make some *allowance* for the time-dependent loss of material properties due to oxidation. Typically, pressure vessel design codes of practice make provision for a *corrosion allowance* in calculating the specified vessel wall thickness.

Since the surface of all components (other than those protected by gold or ceramic plating) will continue to oxidise throughout the life of the component, oxidation protection is a critical design concern. The answer to this concern lies in the experience with materials that remain in stable abundance on the earth's surface, namely, the oxides of metals themselves. This is the *upside* of the oxidation process. We can protect the parent metal by coating the metal surface with metal oxides (in paint) or by *plating* (electro-deposited surface coating) with rapidly oxidising materials such as chromium (Cr), zinc (Zn), tin (Sn), or nickel (Ni). In this way we can *sacrifice* the outer layer to oxidation, while the inner, protected, parent metal remains intact.

Using the above method of protection for the parent metal may still require regular maintenance or re-coating. An attractive alternative is to provide some alloying of the parent metal with low corrosion-rate substances that will ensure long-term protection of the parent metal. Typically alloying steel with chromium will significantly reduce its rate of corrosion even at high temperatures. This notion forms the basis for the production of a range of stainless steels.

In this introductory text we have merely *scratched the surface*, as it were, of a large body of knowledge on oxidation and corrosion. The bibliography identifies some excellent, readable texts on this subject, which should be approached by all engineering designers with cautious respect for its complexity.

It has been estimated by corrosion engineers that the annual losses due to corrosion in some industrialised countries represents between 3% to 4% of gross national product. Chawla and Gupta (1994) offer the following estimates converted to 1982 U.S. dollars:

U.S.A. $US 143 billion;

U.K. $US 30 billion;

Australia $US 4.3 billion .

Chawla and Gupta go on to suggest that between 15% to 20% of these losses could be avoided if proper design precautions were taken. Hence, for these three countries alone, the value of *better corrosion protection design* represents between $US 26 to 33 billion (1982 values). Clearly, design for corrosion protection is a worthwhile consideration in the total design process. However, a more extensive study of this topic is beyond the scope of our interest here.

Some important general issues in corrosion are:

- surface reaction with oxygen is a non-linear function of time and temperature. Hence, oxidation rate decreases with time as the oxide film builds up, but increases in rate with increased temperature;
- while it is desirable to use materials that will oxidise slowly, not all oxide films form a structurally suitable film of protection. For example, aluminium oxide (Al_2O_3) forms approximately 2.5 times faster than iron oxide. Yet iron oxide is significantly inferior structurally to aluminium oxide. Moreover, iron oxide is much more permeable to environmental oxygen than Al_2O_3. Hence, aluminium is a more desirable material to use where oxidation survival is a critical design consideration;
- although oxidation is often referred to as *corrosion,* the term also includes undesired chemical combination of the parent metal with other chemicals, such as sulphur and chlorine. In copper chlorides, the hydrated

form $CuCl_2$-$2H_2O$, well known as the familiar *verdigris,* or green-of-Greece, which forms a decorative patina on bronze statues;

- electrochemical, or *galvanic* corrosion takes place when metals of dissimilar electrode potential are connected in the presence of an electrolyte, that permits the passage of current. The system essentially behaves like a voltaic cell (battery). The rate of galvanic corrosion is partly determined by the difference in electrode potential of the metals in contact (electro-motive-force, or *EMF-series* of metals), conventionally measured relative to the electrode potential of hydrogen. Typically, copper and zinc have EMF potentials of 0.522 and -0.76 volts respectively. When connected in the presence of an appropriate electrolyte (dilute sulphuric acid for example), a current will pass between the zinc *anode* and the copper *cathode.* As a result, the zinc will *sacrificially* corrode and zinc *plating* will be deposited on the copper cathode, thereby protecting it from corrosion. Clearly, where a metal component needs to be protected in the presence of a hostile electrolyte, a *sacrificial* cathode or anode (depending of the relative EMF potentials) provides a useful solution to the corrosion problem. This type of galvanic protection is commonly used in aggressively hostile environments such as chemical plants (Chawla and Gupta, 1994).

2.1.3 Plastics and other composites

As noted in the introduction to this section, plastics represent a rapidly growing manufacturing industry sector. Although their evaluation for structural integrity is outside the scope of this book, we note that designers have access to a great variety of single polymer and composite plastic materials. Various fibres, including glass and carbon filaments, are used as reinforcing in a resin matrix, to make up the broad selection of composites used in many structural applications. These composites are particularly useful where engineered products need to have both complex shapes and high strength to weight ratios. Typical applications are automobile bumper-bars and aircraft wing components. Plastics are durable, often corrosion resistant and in general electrically insulating. In special cases they can have superior structural properties to metals (fatigue life of polypropylene for example). However, generally compared to

Table 2.3
Comparisons of some mechanical properties of various reinforcing fibres and composites
(Data from Crane and Charles, 1984)

Fibre	Tensile strength (MPa)	Tensile Modulus (GPa)	Density (te/m³)
Glass	1500 -4300	70-86	2.5
Carbon	2000 -3500	200 -400	1.7 - 2.02
Boron	35000	420	2.65
Kevlar	2700	60 - 130	1.45
Polyester matrix	20 - 40	1 - 3	1.4 - 2.2
Epoxy matrix	40 - 90	1 - 4	1.6 - 1.9
High carbon steel	2800	210	7.8
GFRP(cloth,50%)	240	14	1.7
GFRP (uni-directional-50% Glass)	1200	50	2.0
Carbon-fiber R.P.– (unidirectional-HT fibre)	1600	129	1.5

metals, plastics are weak and compliant. The most difficult feature of designing with polymers is the time-dependent nature of their structural properties, such as yield strength.

The early use of plastics was due specifically to their easy low temperature mouldability and their relatively low cost for mass-produced components. It is, however, in the composite fibre reinforced form that plastics have come of age as a significant structural material. The earliest reinforcing used was glass, in polyester or epoxy resin for *fibreglass*. As the technology of fibres advanced, metals and carbon reinforced composites became the most advanced structural materials. This development reflects the constant striving by designers for lighter and stronger structures in the automobile, aircraft, space craft and recreational sporting industries. The most desirable property of these composites is the ability to engineer with them almost endless combinations of directional strength and stiffness behaviour. Table 2.3 gives some comparative structural properties of reinforcing fibres used in composites. (Woebcken, 1995 is very useful general reference for a wide range of plastics data.)

2.1.4 Wood and concrete

Wood was probably the first complex structural material used by human beings. Due to its growth pattern, timber is anisotropic and quite variable in structural properties even within one species. The most common mode of failure in timber under compression is the buckling of wood fibre (held in a resin matrix). For this reason, timber is much stronger in tension than in compression. Standardised structural grading allows for predictable performance in timber structures. Laminated, either as beams or plywood, timber properties can be significantly enhanced over the bulk properties. In general, structural timber today is used mainly in the housing industry for roof structures, most commonly as *gang nailed* trusses, or as laminated beams.

Concrete is a mixture of four components, namely, sand, cement, stone and water. The main attraction for its use lies in its ability to *flow* or be moulded into shape. Concrete develops its strength over a period of time after pouring (up to 60 MPa in compression, after 28 days). Due to its brittle nature, concrete is weak in tension and, as a result, in structural use the tensile loads are taken by combining the concrete with reinforcing materials, usually structural grade steel.

2.2 Material selection

On current estimates the number of available material choices for engineering components range between 40,000 to 80,000. Selecting the right material for a specific design is probably the single most important decision facing designers. Unless other constraining factors point the way to a particular choice the following recipe will generally result in a short list of hopefuls from the ocean of materials:

Will it work?: The answer to this question rests with the matching of dominant or primary criteria, such as *strength, hardness, elastic behaviour* (tensile or shear moduli), *hardenability, toughness, magnetic properties, thermal* and *electrical conductivity;*

Will it last?: This question is answered by the important criteria of *corrosion* and *heat resistance*, as well as *resistance to wear, dynamic loading, shock, creep* and *stress corrosion*;

Can it be made?: This question refers to *castability, machinability* and *surface finish performance* or surface enhancement, such as coating, plating and anodizing;

Can it be done within specified limits?: The main constraints of most engineering components or structures are *cost* and *weight*. For these constraints to be met the *designer needs a feel for relative costs and relative weights*.

There are many formal sources available for material selection in hard copy (texts and standards), software (material selector *expert systems*) and manufacturers' catalogues. Ashby (1992) lists 23 database and material selector software sources. In selecting the correct material for any given application, it is expected that designers develop some *wisdom* about the parametric nature of material choices in design.

Typical design parameters are strength to weight ratio (S_y / ρ), in mass limited designs, and stiffness to weight ratio (E / ρ) in designs where structural deflection and mass are both limiting design constraints. It is this type of parametric design that forms the basis of several of the *expert systems* used in material selection software. Ashby (1992) describes the Cambridge Material Selector (*CMS*) software and its selection charts, all based on design parameters chosen by the user. There are several mechanical design examples provided, all of which make use of the CMS charts of design parameters.

A typical, albeit simple, design example is given in Ex 2.1, for a mass limited design.

Ex 2.1

Select the minimum mass cross-section area of a uniform bar of length l carrying a load W as shown in Figure 2.5.

(a) Stress limited design

mass of rod, $m = \rho \times l \times A$,

stress in rod, $\sigma = W / A$.

Failure Predictor: $\sigma_{fail} = S_y$, where σ_{fail} is the value of σ corresponding to yielding failure.

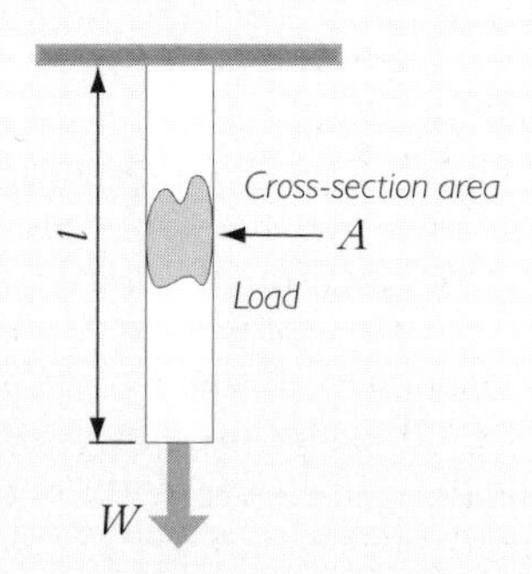

Figure 2.5 Uniform rod in tension; Material properties are: density = ρ; yield strength = S_y; elastic (Young's) modulus = E

Design rule : limit stress to well below the failure value (yielding), by at least an acceptable factor of safety (for stress), F_σ:

$$\Rightarrow \sigma \le \frac{\sigma_{fail}}{F_\sigma}.$$

Therefore

$$\Rightarrow \frac{W}{A} \le \frac{S_y}{F_\sigma};$$

$$\Rightarrow A \ge \frac{WF_\sigma}{S_y} \text{ and } m \ge \frac{\rho l W F_\sigma}{S_y}.$$

By appropriate manipulation of the mass and stress inequalities above, we find that the lowest mass design will require us to choose a material with the largest S_y/ρ ratio.

(b) Deflection limited design

Extension of rod,

$$\delta_l = (load\ /\ stiffness) = W \Big/ \frac{AE}{l}.$$

Failure Predictor: $\delta_{fail} = \delta_{all}$, some limiting allowable value of extension of the bar.

Design rule : limit the extension well below the limiting value, by at least an acceptable factor of safety (for extension), F_δ:

$$\Rightarrow \delta_l \le \frac{\delta_{all}}{F_\delta}.$$

Therefore

$$\Rightarrow A \ge \frac{WlF_\delta}{E\delta_{all}}, \text{ and } m \ge \frac{\rho W l^2 F_\delta}{E\delta_{all}}.$$

Now, again we can manipulate the deflection and mass equations to determine the governing parameter for the *best design*. Here, our material choice is revealed to be the one with lowest ρ/E ratio.

This simple example highlights several issues involved when we need to exercise design judgement in the choice of materials of construction.

Both yielding (plastic deformation of the component) and deflection (elastic deformation of the component) are clearly identified as failure modes to be avoided. Hence, the designer may need to exercise balanced judgement about the best material that will satisfy both requirements.

The design requirement in each case was revealed by the governing *design rule*. Moreover, the design rule, in combination with the constraint on mass limitation, resulted in a materials-based *design parameter* that needed to be maximised or minimised.

In general, our synthesis of other engineering components throughout this text will involve these steps.

In all design cases the *best* choice of material will be identified by one or more design parameters which represent the essence of the constraints on the design. While each design case needs its own specific treatment, as a general rule, where the design parameters lead to some unacceptable design, we must look for some *neglected* or overlooked parameter of constraint.

A simple example is the design of a *SCUBA* (*self-contained underwater breathing apparatus*) tank, subject to internal pressure p, to be carried by a human. This design is mass and strength limited. We can carry out a simple *thought experiment* by considering the tank to be a cylinder of inner diameter D and wall thickness t, we can express the ratio of outer to inner diameter as $\Phi = (D+2t)/D$.

For thin-walled pressure vessels the allowable stress is given by (refer to Section 3.4.3)

$$\tau \approx pD/4t.$$

The mass is proportional to material volume and may be approximated by

$$m = \frac{\rho L \pi D^2}{4}\left(\Phi^2 - 1\right) \approx \rho L \pi D t.$$

Clearly the stress equation suggests that the diameter should be minimised, and the mass equation suggests that Φ should be as close to unity as feasible. However, given the volumetric limitations, we must allow the length of this vessel to increase indefinitely to allow minimising the diameter. The resulting minimum mass thin walled cylindrical tank of a fixed volume is found to be one of infinite length with zero wall thickness and zero diameter. This makes absolutely no sense at all until one realises that the parameter of human proportions (a human must carry the device) have been neglected in the evaluation. It is instructive to manipulate the design relationships of the *SCUBA* tank to determine the appropriate design parameter to be minimised for *best* material choice (ρ / S_Y), and this is left as an exercise for the reader.

2.3 Design for static loading, introducing failure predictors and factors of safety

A study of designing for *static* or *slowly-applied* or *time-independent* loading of engineering components must begin with some basic notions about material behaviour. Design for structural integrity of a component involves the matching of applied loads and the component's capacity to resist these loads. Most design texts commence with some basic concepts in stress analysis, often leaving the crucial balance between stresses, loads and material selection criteria to the designer. In what follows, we explore these crucial relationships in a generic way that applies to the synthesis of a whole range of engineering components.

The essence of good design is judgement. While judgement is developed through experience, it is usually the result of some deep understanding of the underlying principles that govern the behaviour of the system about which we need to exercise judgement. The issues involved in the design synthesis of engineering components for static loading are the choice of:

- material properties;
- load distribution;
- component geometry; and,
- in the elementary cases considered in this text, a simple, tractable analytical model of the component we are to synthesise.

Material properties having structural relevance include yield strength (S_Y), rupture strength (S_U), elastic moduli (E– tensile, and G– torsional), Poisson's ratio (μ), and fracture toughness factor (K_C). Each of these is determined from controlled laboratory tests on test specimens carefully prepared from the material in question. Work on the behaviour of materials under load has been progressing since the early part of the 20th century. It seemed perfectly natural for material scientists to begin the investigation of the macroscopic behaviour of metals by observing and describing the behaviour of single crystals.

The most fascinating aspect of material behaviour was seen to be the commencement of plastic deformation. In a macroscopic sense the material under tensile load was found to develop *sliding* of crystallographic planes. Figure 2.6 shows the type of behaviour proposed schematically.

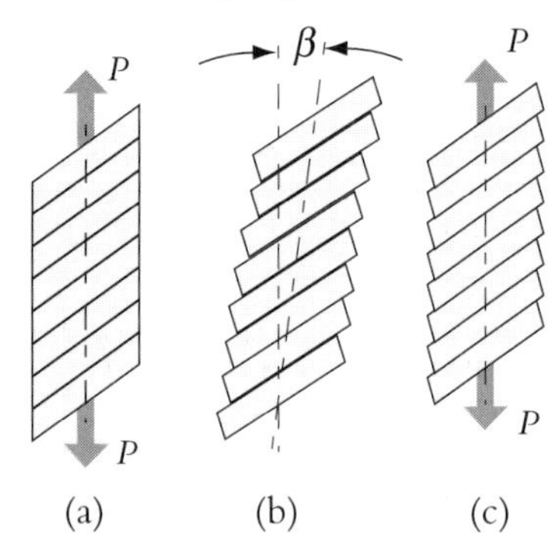

Figure 2.6 Sliding along crystallographic planes in tensile specimen; (a) tensile test specimen under load; (b) planes of material have slid along crystallographic planes; (c) the specimen rotates by angle β, back to original alignment axis. (After Timoshenko, 1956)

This type of failure model was investigated by, among others, such luminaries as L. Prandtl, G.I. Taylor and W. Boas[2.2]. In components made of

2.2 References to these works may be found in Timoshenko, *Strength of Materials*(1956),Princeton:Van Nostrand

polycrystalline materials such as steel, these models describe the very localised plastic deformation that immediately precedes failure. Figure 2.7 shows a typical stress-strain curve for ductile materials. In the failure region ductile materials exhibit sliding of material along crystalline planes at approximately 45° to the tension direction. The slip lines resulting from this process are called *Lueders lines*, after W. Lueders, who first studied this effect in 1854. In ductile test specimens Lueders lines first occur during the strain hardening phase of the stress-strain behaviour of the material. This region is signified by the small dip in the stress strain curve shortly after the yield point (where $\sigma = S_y$; refer to Figure 2.7). Following strain hardening, caused by local deformation in the failure region, the test specimen will progress to fracture at the ultimate tensile stress S_U.

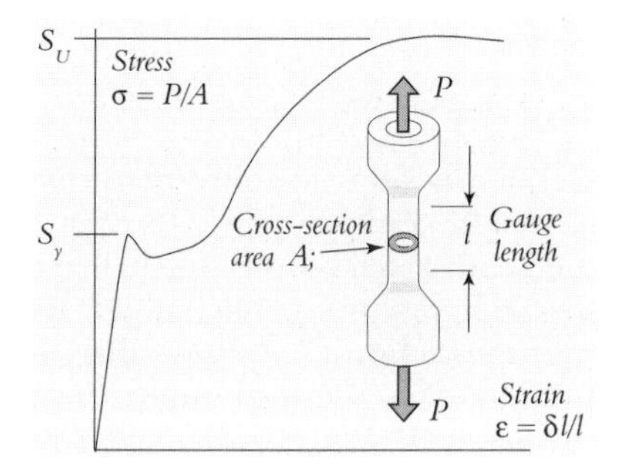

Figure 2.7 Uniaxial tensile test specimen and stress-strain result (schematic)

In essence, it is now recognised that ductile materials, exhibiting such behaviour, actually fail in shear along the shear planes formed by the sliding of material along the crystalline boundaries. Timoshenko (1956, Part II: Chapter X) presents a scholarly, and eminently readable, account of the development of our understanding of failure mechanisms in elastic materials. All serious students of engineering design should take time to read this work.

Whatever the mechanism of plastic deformation, the issue of interest in all testing is the discovery of essential mechanical properties for each material — yield strength, rupture strength and elastic modulus. This elastic modulus is the slope of the stress-strain curve preceding the stress at which the material reaches the limit of its linear elastic behaviour, otherwise known as the *proportional limit*. Since this slope is extremely difficult to measure accurately, the International Congress for Testing of Materials (1906) set the proportional limit at a stress where the material has plastically deformed (sustained *permanent set) by* 0.001%. The currently accepted proportional limit for ductile materials is a permanent set of 0.01%.

Figure 2.7 shows schematically one of the tests commonly used for establishing key metal properties. The uniaxial tensile test is performed on a carefully prepared test specimen that has a highly polished surface and an absence of sudden changes in cross-sectional geometry. A central length l of the specimen is identified as the *gauge-length* over which extensions will be measured as the applied load P is increased.

The resulting uniaxial *stress-strain* graph is used to establish tensile yield strength (S_y) and ultimate tensile strength (S_U) for this particular material. The graph shown in Figure 2.7 is an idealised schematic representation of the behaviour of ductile, work-hardening, materials such as exhibited by many common structural steels. In general, there is an initial, linear, or closely-linear, part of the graph which is referred to as the *elastic* region. The slope of the elastic region is a direct measure of the tensile elastic modulus, or *Young's modulus* of the material. It is generally assumed, not entirely correctly for most materials, that unloading the test specimen during the elastic region will allow it to return to its original geometry. As already noted earlier, yielding, or permanent deformation, commences at the *yield point*.

Figure 2.8 Two tensile test specimens after fracture. The one on the left is ductile and the one on the right is brittle. The differences in material behaviour near the fracture region are clearly visible.

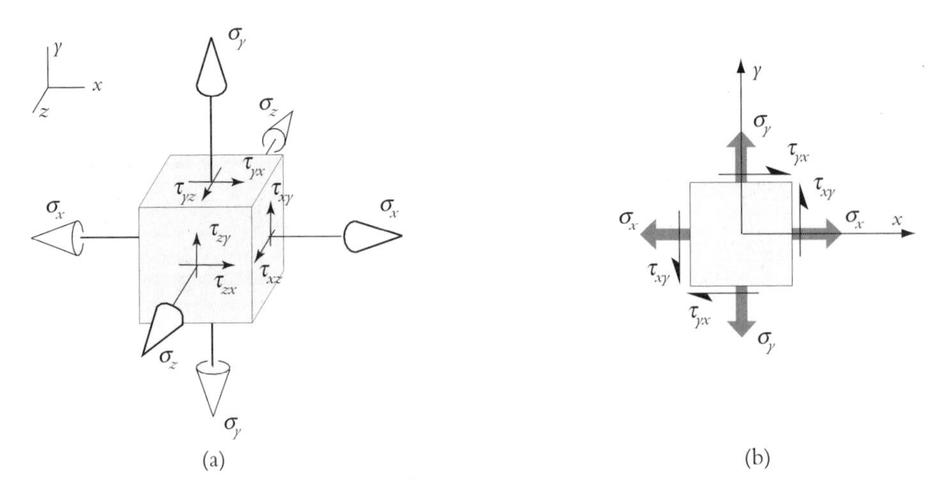

Figure 2.9 Multi-axial stress conditions. (a) a small element of material under the action of a general tri-axial stress system; Shear stresses (shown only on the visible faces of the component) act in pairs to provide shearing moments on the element; (b) bi-axial stress system acting on a plane element.

Not all materials exhibit a clearly identifiable yield point. However, low and medium carbon steels commonly exhibit work hardening, with a corresponding region of strain at constant or even lowered stress, as indicated on Figure 2.7. This characteristic behaviour serves to clearly identify the yield-point for such materials.

In Figure 2.8 we see two fractured tensile test specimens. The ductile specimen shows significant deformation (*necking down*) prior to fracture, while the brittle specimen has no visible deformation in the fracture region. The ductile behaviour is clearly in agreement with the schematic stress-strain behaviour indicated in Figure 2.7. The brittle behaviour serves to warn designers about the sudden nature of fracture experienced with this class of material.

Where the structural behaviour of real engineering artefacts is concerned, life is not nearly as friendly as a well controlled test in the laboratory. Real artefacts are not only subject to multiaxial stresses but, as already noted, they may be also subject to various environmental conditions which serve to degrade the material properties listed in the reference data base. The engineering designer's task then becomes one of making do with the available data and applying design tools to match the data to real-life conditions.

In most real-life design problems, the stress conditions are multiaxial, that is the most highly stressed element of the component to be designed has stresses acting on it in more than one vector direction. Typical examples are shafts with imposed torsion and bending loads, resulting in biaxial stress conditions, and pressure vessels, subject to triaxial stress conditions. Figure 2.9 shows the schematic representation of a small element of material exposed to a general stress field. The terminology used for identifying the various stresses acting is indicated and we need to be absolutely clear about the conventions used for identifying stresses.

For general direct stress we use the symbols σ_x, σ_y, σ_z, respectively and for shear stress τ_{xy}, τ_{yz}, τ_{zx}, as indicated in Figure 2.9.

In Figure 2.10 (a) we consider a small element of material acted on by a general biaxial stress field. We now consider a plane *AB* in this element at some general angle ϕ, shown separately in Figure 2.10 (b). Carrying out the force balance on this partial element we can express the equilibrating direct stress σ_ϕ and shear stress τ_ϕ as given by the two equations in Figure 2.10. We can now differentiate the stress equation to find the maximum and minimum direct stress acting. We find that these values occur for

$$tan\ 2\phi = \frac{2\tau_{xy}}{\sigma_x - \sigma_y}.$$

Substituting into the direct stress equation in Figure 2.10 we find the maximum and minimum direct stresses acting on the small element are

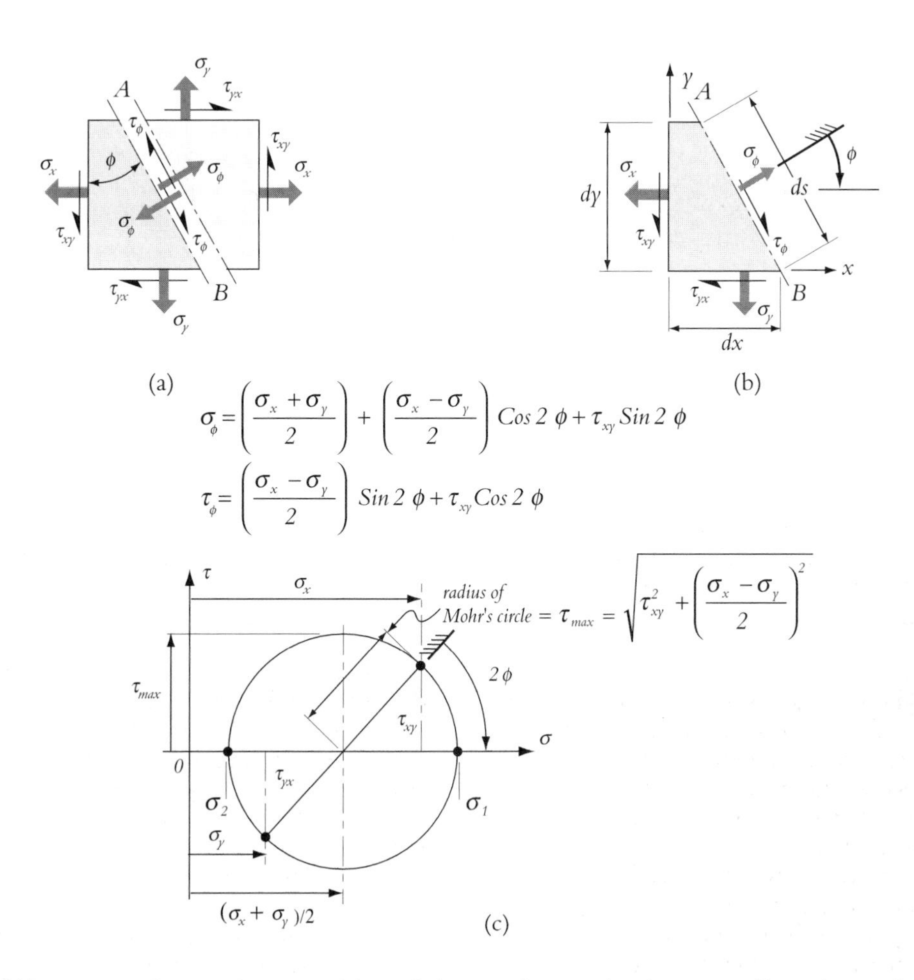

Figure 2.10 The concept of principal stress — (a) small element of material under a general biaxial stress field; (b) free body diagram of a portion of the small element of material, indicating equilibrating direct and shear stresses; (c) Mohr's stress circle.

$$\sigma_1 = \frac{\sigma_x+\sigma_y}{2} + \sqrt{\tau_{xy}^2 + \left(\frac{\sigma_x-\sigma_y}{2}\right)^2},$$

$$\sigma_2 = \frac{\sigma_x+\sigma_y}{2} - \sqrt{\tau_{xy}^2 + \left(\frac{\sigma_x-\sigma_y}{2}\right)^2}.$$

From the shear stress equation of Figure 2.10 we find that on the planes corresponding to these maximum and minimum direct stresses the shear stress $\tau = 0$. We refer to σ_1 and σ_2 as the maximum and minimum *principal stresses*. It is easy to demonstrate that the equations for σ and τ given in Figure 2.10 represent a circle as indicated in Figure 2.10 (c), with the origin at

$$\sigma_O = \frac{\sigma_x+\sigma_y}{2},$$

and radius

$$\tau_{max} = \sqrt{\tau_{xy}^2 + \left(\frac{\sigma_x+\sigma_y}{2}\right)^2}.$$

The original derivation of this graphical representation of stresses is due to Otto Mohr[2.3], after whom the procedure is named.

2.3 Otto Mohr, *Civilingenieur*, p. 113, 1882, quoted by Timoshenko (1955) *Strength of Materials*, Part I, New York: Van Nostrand

Once we move from the simple matter of uniaxially loaded bars (used for determining structural properties) to *real-world* components subjected to more complex load conditions, failure prediction becomes problematic. It is not possible to draw a direct comparison between the behaviour of the *real-world* component and experimentally measured structural performace of uniaxially loaded bars. Since experimental failure mechanisms are almost universally uniaxial, failure prediction in more complex, multiaxial, stress situations becomes a design dilemma. Having been provided with structural data resulting from uniaxial tests, we need to find a way to account for all the stresses acting on our component.

Clearly, one needs a means of combining the multiaxial stresses in some appropriate way to predict how their combined magnitude will influence the failure conditions prescribed for the component to be designed. There are three relatively simple rules for combining multiaxial stresses to predict the yield conditions (or rupture conditions) in an engineering component. These rules are alternately referred to as *failure predictors*, *failure criteria* or *theories of failure*. In this text we will use the term *failure predictors* consistently, to indicate the real nature of these rules.

2.3.1. Maximum principal stress failure predictor (MPFP – sometimes maximum normal stress):

In brittle materials subject to *tensile loading*, failure will occur when the maximum principal stress (positive tensile) reaches the failure criterion (usually rupture, since there is almost no yielding with these types of materials). Typical examples are grey cast iron products and some components of high carbon steels at very low temperatures (subject to low temperature embrittlement).

A key feature of brittle materials is that they tend to fail in tension rather than in shear. This preference is also characterised by the very little deformation in the failure region before rupture. There is almost no sliding of the shear planes in the failure region (no Lueders' lines).

MPFP Design rule :

When a component of brittle material is exposed to a multiaxial stress system, failure due to fracture will occur when the maximum principal stress anywhere in the component exceeds S_U, the ultimate tensile stress observed in the uniaxial laboratory test of the material from which the component is made. Consequently, in a component constructed from brittle material, we must always limit the maximum principal stress anywhere in the component to below S_U.

$$\Rightarrow \sigma_1 \le \frac{S_U}{F_d}.$$

Ex 2.2

When a cylinder of ductile material is loaded in pure torsion T, its failure surface is, in general, a plane section, normal to the axis of the applied torque T. However, a cylinder of brittle material subjected to pure torsion fails along a 45^0 helix. What is the explanation for this spectacularly different behaviour.

Ans:

Referring to Figure 2.11, we observe a glass rod being loaded by a torque T in pure torsion. The material of the rod will oppose the torque by a resisting moment composed of the distribution of stresses across its cross-section. The distribution of stresses is a linear function of the rod's radius, ranging from zero at the neutral axis (the rod's central axis) to maximum at the surface of the rod. Failure will commence at the outer surface of the rod, where the stresses due to the applied torque are the greatest.

We now examine a small element of the rod's outer surface and plot the stresses acting on this element on a Mohr's stress circle, shown on Figure 2.11(b). On the face of the element normal to the x-axis there is a shear stress τ_{xy} acting only. There is zero normal stress acting on this face. This stress condition is seen as point $+\tau$ on the τ-axis of the Mohr's stress circle.

Similarly for the face normal to the y-axis we find its stress representation as $-\tau$ on the τ-axis of the stress circle. We can now draw the stress circle through these points and we find that this circle crosses the σ–axis at $+\sigma$ and $-\sigma$ as seen in Figure 2.11(b). The implication is that, should we rotate the coordinate axes chosen to examine our original surface element through 45^0 (i.e. one half of the $2\phi = 90°$ rotation on the Mohr's circle), we would observe only direct stresses acting on these faces of the element. This state of affairs is indicated in Figure 2.11(c). Clearly, the tensile load, $+\sigma$, resulting from the applied pure torsion is the

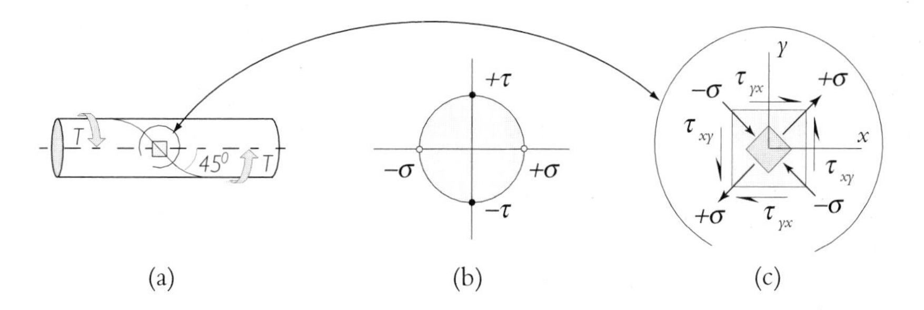

Figure 2.11 Cylinder of brittle material subjected to pure torsion. (Ex. 2.2)

greatest principal stress acting on the small element of the rod's surface and according to the maximum principal stress failure predictor (*MPFP*), failure will commence due to this stress. Consequently, the failure surface on this brittle rod will be the 45° helical surface indicated on Figure 2.11(a).

As a simple follow-up exercise it is instructive to show how a ductile rod exposed to pure torsion might fail.

2.3.2 Maximum shear stress failure predictor (MSFP):

In ductile materials failure occurs by shearing along the shear planes, at approximately 45° to the tension direction. When the component is exposed to multiaxial stress conditions, the maximum shear stress in the material is calculated from the combination of the largest and smallest principal stresses acting. Referring to Figures 2.12(a) and 2.12(b), for the triaxial stress case we find:

$$\tau_{max} = \frac{\sigma_1 - \sigma_3}{2}.$$

For the biaxial stress case, we may use precisely the same expression $\tau_{max} = (\sigma_1 - \sigma_3)/2$, where σ_3 is taken to be the lesser of σ_2 and zero. We must not neglect the third stress direction even in the biaxial stress case.

We note that conventionally principal stresses are designated from largest to smallest as

$$\sigma_1 > \sigma_2 > \sigma_3 \cdot$$

In both cases we need to find the principal stresses acting. This can be done by applying a Mohr's circle analysis to the known direct and shear stresses acting on our small element of material under consideration. This procedure is indicated schematically in Figure 2.12.

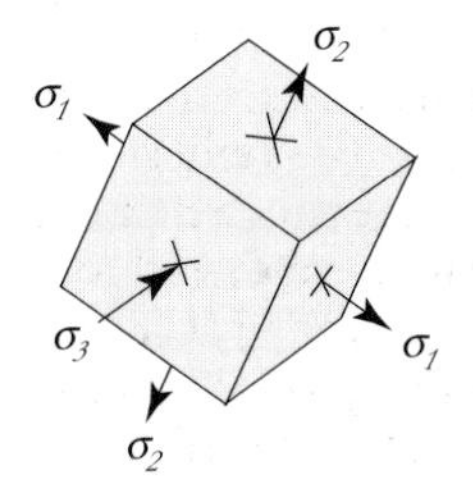

(a) *Small element rotated to principal stress axes*

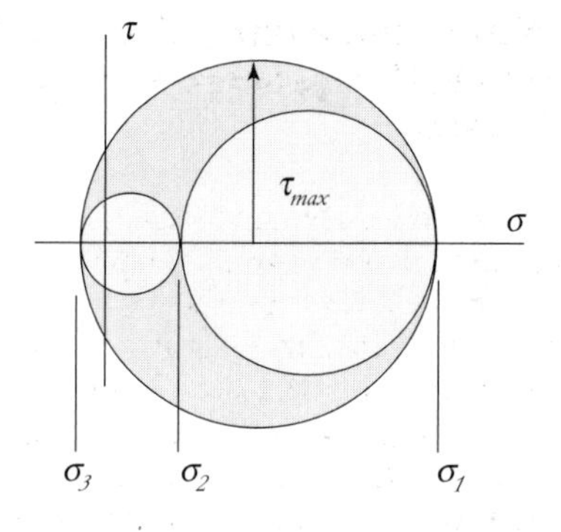

(b) *Mohr's stress circles for triaxial stress field*

Figure 2.12 Principal stresses for a triaxial stress field

To avoid any confusion about which principal stresses to combine in determining the maximum shear stress acting, we will invariably assume that we are dealing with a general triaxial stress field.

In the tensile test specimen at yielding ($\sigma_1 = S_y; \sigma_2 = \sigma_3 = 0$) we find:

$$\tau_{yield} = \frac{S_Y}{2}.$$

Combining the above equations suggests that, according to *MSFP*, failure in yielding will occur when:

$$\tau_{yield} = \frac{\sigma_1 - \sigma_3}{2}.$$

MSFP Design rule:

In a component, constructed from a ductile material, exposed to a multiaxial stress system, failure due to yielding will occur when the maximum shear stress, τ_{max}, anywhere in the component, exceeds τ_{yield}, the shear stress at yielding observed in the uniaxial laboratory test of the material from which the component is made. Consequently, in a component constructed of ductile material, we must always limit the shear stress anywhere in the component to below τ_{yield}.

$$\Rightarrow \tau_{max}\left(= \frac{\sigma_1 - \sigma_3}{2}\right) \le \frac{\tau_{yield}}{F_d}\left(= \frac{S_Y}{2F_d}\right).$$

Ex 2.3

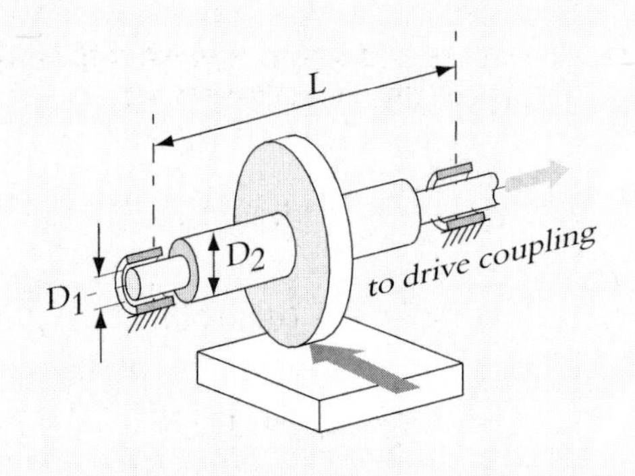

Figure 2.13 Loads generated by grinding process; D_1 = 10 mm, D_2 = 20 mm

It is interesting to examine the nature of *MSFP* applied to practical design situations. In general, we are interested in components so loaded that they experience multiaxial stress conditions. Some typical examples are beams loaded in combined bending and torsion and thick walled pressure vessels subject to internal pressure.

Figure 2.13 shows a grinding process schematically. A block of steel is being ground flat by a 300 mm diameter grinding wheel rotating at 6,000 rpm. The vertical force of contact between the steel plate and the grinding wheel is negligible. However, the force generated in the cutting process, tangential to the grinding wheel, is 100 N. Draw bending-moment and shear-force diagrams for the drive shaft. Identify the expected modes of failure and specify the minimum shear stress property of the shaft material for this design. The design factor of safety F_d = 2.3.

The uniaxial stress system is likely to crop up in many of our experiences in designing for structural integrity. Transversely loaded beams and axially loaded columns represent common examples of uniaxial stress systems. However, we need to avoid complacency when dealing with problems of structural integrity. There are many cases of simple engineering components subjected to general stress fields. Figures 2.14 and 2.15 show typical examples.

Ex 2.4

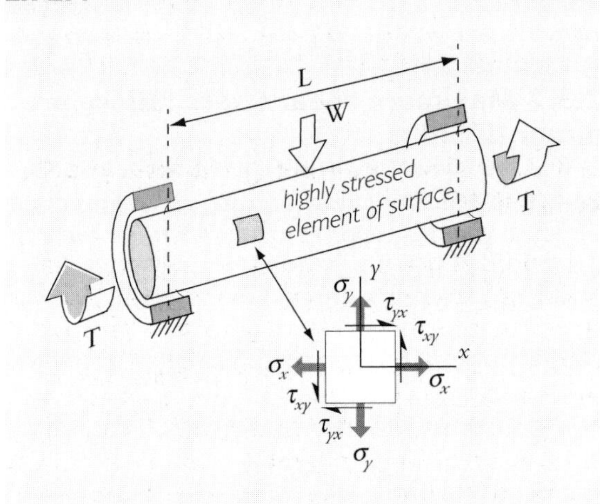

Figure 2.14 A shaft (engineering component capable of transmitting rotational power) experiencing biaxial stress

(a) For the shaft of diameter D mm and span L, loaded as indicated in Figure 2.14, derive the stress system acting on the surface element shown, in terms of the applied load W.

(b) For the thin-walled pressure vessel, internal diameter D_i and wall thickness t, exposed to internal pressure p, derive the stress system acting on the surface element shown, in terms of p. Assume that the loads imposed by the supports are negligible compared to those due to the internal pressure p. [*Hint*: Draw a simple free-body diagram of a unit length of one half of the cylindrical shell and consider its equilibrium under the action of the applied pressure and the equilibrating forces provided by the stresses in the walls of the vessel.]

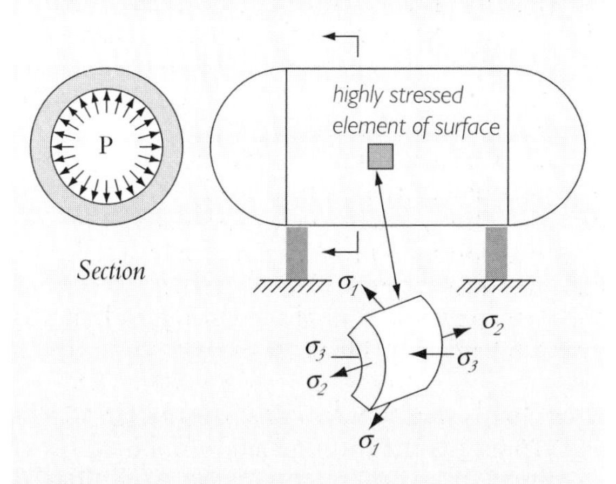

Figure 2.15 A cylindrical pressure vessel (engineering component capable of storing fluid under pressure) experiencing triaxial stress

2.3.3. Distortion energy failure predictor (DEFP):

DEFP is sometimes referred to as the *von Mises-Hencky theory* of failure. The need for developing a theory different from the simple *MSFP* resulted from observations that components subjected to uniform hydrostatic pressure (equal tension or compression), in all three axial directions, exhibited significantly higher yield stresses than that predicted by *MSFP*. Originally it was proposed that yielding took place when the *shear strain energy* stored in the component was the same as the shear strain energy at yielding in the uniaxial tensile specimen. It is an interesting historical fact that J. C. Maxwell, in a letter dated 1856, to W. Thompson (later Lord Kelvin of Largs), expressed the view that *when distortion energy in a material reaches some limit, the material will begin to give way*[2.4].

The strain energy content of a component subjected to a triaxial stress system may be considered as consisting of two independent parts, energy due to volume change (*dilation*) and energy due to *distortion*.

2.4 Timoshenko (1956, ibid)

Figure 2.16 shows a small cube of material being dilated by loads acting on its faces in the three coordinate directions. We designate the loads as P_x, P_y, P_z, and the area of each face over which the loads act as A and the edge length as L. Consider the small cube of material being extended by P_x over the small distance l_x from its undistorted shape.

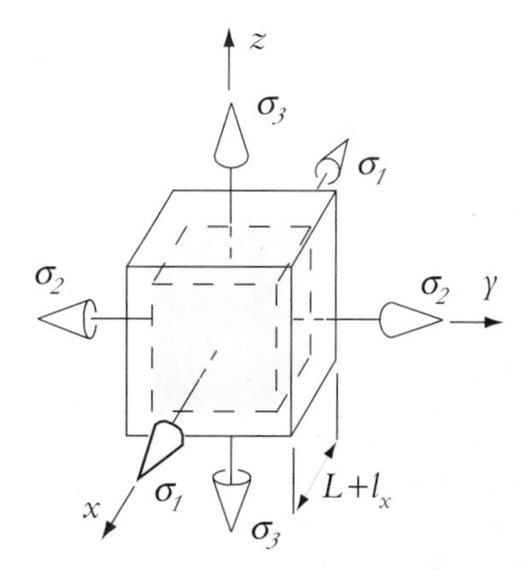

Figure 2.16 Schematic view of the dilation process

The three stresses acting in the three coordinate directions are:

$$\sigma_1 = (P_x / A) > \sigma_2 = (P_y / A) > \sigma_3 = (P_z / A).$$

Designating Poisson's ratio as μ, we can identify the strains due to the three stresses as:

$$\varepsilon_1 = \frac{\sigma_1}{E} - \frac{\mu}{E}(\sigma_2 + \sigma_3);$$

$$\varepsilon_2 = \frac{\sigma_2}{E} - \frac{\mu}{E}(\sigma_3 + \sigma_1);$$

$$\varepsilon_3 = \frac{\sigma_3}{E} - \frac{\mu}{E}(\sigma_1 + \sigma_2).$$

The strain energy stored in the small element due to the extension in the x-direction is $U_x = P_x l_x/2$. It is instructive to show that this expression is a result of the familiar elastic strain energy stored in a spring of stiffness k, extended over a distance x from rest; i.e.

$$U = kx^2/2.$$

Then the strain energy stored per unit volume is found to be

$$u_{total} = \frac{P_x l_x}{2AL} + \frac{P_y l_y}{2AL} + \frac{P_z l_z}{2AL},$$

$$= \frac{\sigma_1 \varepsilon_1 + \sigma_2 \varepsilon_2 + \sigma_3 \varepsilon_3}{2}.$$

Substituting for the respective strains from above results in

$$u_{total} = \frac{\sigma_1^2 + \sigma_2^2 + \sigma_3^2}{2E} - \frac{\mu}{E}(\sigma_1\sigma_2 + \sigma_2\sigma_3 + \sigma_3\sigma_1).$$

However, we are only interested in the strain energy of *distortion*, which results from the three stresses acting on the faces of the cube being unequal. Hence we need to subtract the strain energy of *dilation* from the total.

We find the dilation component of the strain energy by considering the average stress acting on all three faces of the small cube

$$\sigma_{ave} = \frac{\sigma_1 + \sigma_2 + \sigma_3}{3}.$$

The strain energy due to this average stress acting on the faces of the cube is

$$u_{dilation} = \frac{3\sigma_{ave}^2}{2E} - \frac{\mu}{E}(3\sigma_{ave}^2) = \frac{3\sigma_{ave}^2}{2E}(1 - 2\mu).$$

The resulting strain energy of distortion is then

$$u_{distortion} = u_{total} - u_{dilation},$$

$$= \frac{1+\mu}{3E}\left[\frac{(\sigma_1 - \sigma_2)^2 + (\sigma_2 - \sigma_3)^2 + (\sigma_3 - \sigma_1)^2}{2}\right].$$

Next we need to consider the distortion strain energy stored in the test specimen subjected to simple tension (Figure 2.7). Using the above expression we find at yielding

$$u_{distortion,y} = \frac{1+\mu}{3E}(S_Y^2).$$

DEFP Design rule :

For components constructed of ductile material and subject to multiaxial stress conditions, according to the distortion energy failure predictor (DEFP), the component will fail when the distortion energy in the component reaches the distortion energy in the uniaxial (simple tensile) test specimen of the same material at failure. Consequently, in a component constructed of ductile material, we must always limit the distortion energy anywhere in the component to be less than this failure value, i.e.

$$u_{distortion} \leq u_{distortion,failure}$$

For yielding we get for $DEFP_{yield}$:

$$\frac{1+\mu}{3E}\left[\frac{(\sigma_1 - \sigma_2)^2 + (\sigma_2 - \sigma_3)^2 + (\sigma_3 - \sigma_1)^2}{2}\right]$$

$$\leq \frac{1+\mu}{3E}(S_Y^2).$$

We define

$$\sigma_{vM} = \sqrt{\frac{(\sigma_1 - \sigma_2)^2 + (\sigma_2 - \sigma_3)^2 + (\sigma_3 - \sigma_1)^2}{2}},$$

where σ_{vM} is sometimes called the *von Mises' stress* and represents the most general combined stress condition in a component subject to multiaxial stresses. Conventionally our safety factor is applied to this stress. From the *DEFP design rule* we see that *to prevent yielding in a component constructed of ductile material, we must limit the von Mises stress such that*

$$\sigma_{vM} \leq (S_Y / F_d).$$

Ex 2.5

For the shaft and the pressure vessel shown in Figures 2.14 and 2.15 respectively, compare the *MSFP* and *DEFP* design rules. In each case derive the von Mises stress and maximum shear stress in terms of the applied loads and compare the design rules in terms of these applied loads.

2.3.4 Comparison of the three failure predictors

As we have already noted, the mechanical properties of materials are usually determined by relatively simple uniaxial tensile or compressive tests under ideal laboratory conditions. Material failure under more complicated stress conditions have been studied experimentally in very few exceptional cases (see Timoshenko, 1956). The failure predictors described in this chapter have been specifically developed in order to predict material behaviour under complicated general stress conditions. The history of failure predictors dates to the late 19th and early 20th century.

The first practical failure predictor was an early version of the *MPFP* associated with the name of the Scottish engineer and physicist William J. M. Rankine (1820-1874) and it is sometimes referred to as Rankine's theory. As a practical engineer and investigator of steam engines, Rankine probably needed such a failure predictor for his engine designs. As originally proposed it was applicable to both ductile and brittle materials. Eventually, once more sophisticated failure predictors were developed, it was recognised that Rankine's theory had some serious shortcomings for ductile materials. The second oldest failure predictor, now also regarded as unacceptable, was the maximum strain theory, offered by Saint-Venant (c. 1860) Both, *MSFP* and *DEFP* had been proposed as early as 1904, together with comparative studies of all the available failure theories.

In comparing the three failure predictors, described in this chapter, we will need to consider the behaviour of a small element of material under general triaxial stress conditions as shown in Figure 2.9(a). We can plot the failure conditions on σ_1 / S_y and σ_2 / S_y axes as shown in Figure 2.17.

According to *MPFP* the element will fail in yielding whenever any of the three acting principal stresses reaches S_y. This condition is represented by the dashed square *ABCD*. In this comparison we have assumed that yielding in tension and compression both occur at the same absolute value of S_y.

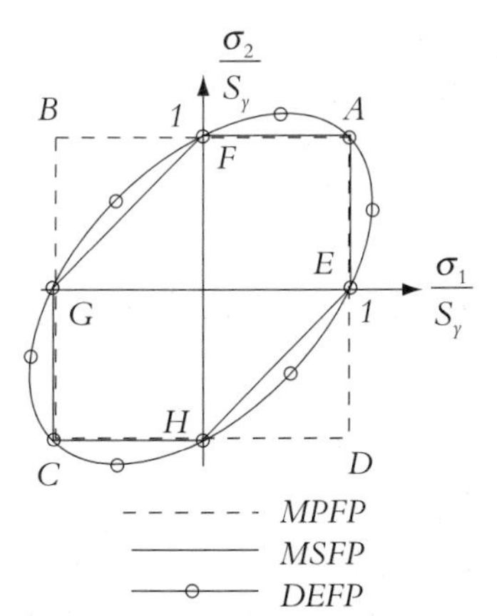

Figure 2.17 Comparison of the three failure predictors

According to *MSFP*, when σ_1 and σ_2 both have the same sign, failure is reached along the boundaries *EAF* and *GCH*. Along these boundaries the two failure predictors, *MPFP* and *MSFP* are in agreement. However, when any two of the three stresses acting have different signs (i.e. tension in one axis and compression in another axis), *MSEP* predicts failure along the boundaries *FG* and *HE*. In the second and fourth quadrants of Figure 2.17, *MPFP*, relative to *MSFP*, seriously overestimates the stresses at which failure will occur.

DEFP is represented by the elliptical boundary in Figure 2.17. In general, this is the boundary that most closely agrees with experiment. However, since the irregular hexagon representing *MSFP* lies inside the ellipse representing *DEFP*, the results from *MSFP* will be conservative.

This representation only deals with two dimensional stress systems. The 3D stress case may be approached similarly, where the *MSFP* and *DEFP* yield surfaces correspond to a regular hexagonal prism and a circular cylinder respectively, as indicated in Figure 2.18.

The axes of both cylinders are aligned along the line of hydrostatic stress ($\sigma_1 = \sigma_2 = \sigma_3$). The cylindrical surfaces shown in Figure 2.18 enclose the elastic regions of behaviour for the material. Anywhere outside the cylinders the material will experience plastic deformation. The shapes of Figure 2.17 are identified as the intersection of these yield surfaces with the $\sigma_1/S_y\ ;\sigma_2/S_y$ plane.

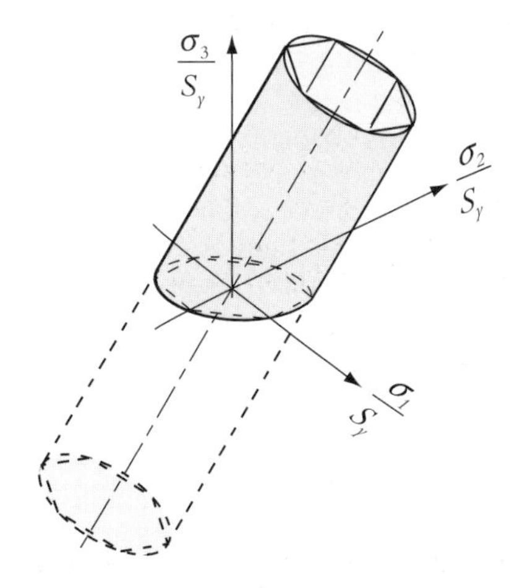

Figure 2.18 3D yield surfaces

Ex 2.6

Show that at yielding in pure torsion:

MSFP predicts $\tau_{max} = 0.5\ S_y$, while

DEFP predicts $\tau_{max} = 0.577\ S_y$,

whereas in pure tension both predict $\tau_{max} = 0.5\ S_y$. [Hint: refer again to Figure 2.11]

2.3.5 Factors of safety

The *design factor of safety*, F_d , is used for handling systematically the many and varied uncertainties associated with specific design situations. It is most commonly applied to designs in one of two ways. F_d may be used to reduce the known strength property of a chosen material (for example S_y / F_d), to result in an *allowable-stress* or *working-stress*. When applied in this way, F_d is referred to as a *stress factor of safety*. Alternatively F_d may be used to increase the predicted or estimated applied load (for example $F_d \times W$), to result in a *safe*-load or *working*-load. In this type of application F_d is referred to as the *load factor of safety*. In this book we will consistently use *stress factors of safety*, since they are less prone to dangerous misinterpretation.

Although we refer to F_d as a factor of *safety*, this by no means signifies that a particular design is *safe* once we have applied F_d to it. In fact, many practising engineers refer to F_d, disparagingly, as a *factor of ignorance* about the precise nature of various influences that will impact on the ultimate structural integrity of a specific component. A general classification of these various influences is provided in Table 2.4.

Earlier in this chapter we made reference to the need for engineering judgement, and how judgement is often the result of wisdom developed about some specific field of design through years of practice. The choice of a proper value for F_d is entirely the result of good engineering judgement.

We hasten to admit that there is a degree of arbitrariness about the choice of the various influence factors listed in Table 2.4. Inexperienced novice designers find this aspect of design most unsettling, having spent considerable intellectual skill in developing a substantial constitutive model[2.5] of the component to be designed. Many experienced designers develop *rules of thumb* (refer to Chapter 1 on the subject of engineering estimation) for estimating overall values of F_d. As a general rule we note that for stationary structures and components (for example pressure vessels and their support systems) an overall value of F_d between 2.0 and 4.0 is commonly accepted practice. For components where mass and inertia are criteria of operational performance (for example, internal combustion engines and air-frames) the constitutive models of structural performance are generally more precisely determined. Moreover, substantial care is taken to select and test material properties in these applications. Consequently, a value of F_d between 1.5 and 2.0 is more likely to be appropriate. Steel ropes, made up of many strands of high tensile steel wire, are used in cranes or in structural support applications, where failure could be life threatening. Any minor damage to the steel wires can result in substantial strength reduction in the rope. Hence, for steel ropes F_d can be as high as 12.

These *rules of thumb* are the result of considerable experience in the design of engineering structures and components. Where no such experience exists, Table 2.4 suggests some data, the application of which still offers considerable intellectual and creative challenge. In fact, once the embodiment of a component has been determined, apart from the development of an appropriate constitutive model of the design problem, the choice of F_d presents the only genuinely creative challenge to the designer.

In many applications the value of F_d is determined by codes of practice. For example, the *Australian Steel Structures Code, AS 1250*, recommends implicitly that $F_d = 3$ for fillet welds. The codes of practice for pressure vessel design and for gear design recommend the use of a *safe working stress*. In some manufacturing organisations internal codes of practice will recommend, or in some cases mandate, appropriate working stresses or *safe* loads. The factors of safety *implied* by these recommended or mandatory codes of practice are referred to as *implicit* factors of safety.

Occasionally in this book we will make reference to the *actual factor of safety*, F_a, as distinct from the *design factor of safety*, F_d. This actual factor of safety, F_a, represents the hard edge of reality, and is usually unknown to the designer, unless a failure occurs or a component is tested to destruction. F_a is the ratio between applied loads and the value of these loads that will cause failure of the component. On the other hand, F_d is the designer's estimate of what F_a should be.

2.5 A *constitutive model* is derived using the available *constitutive equations* of a specific doctrine. In our context the most commonly applied constitutive equations are those of mechanics. However, we must not neglect the likelihood of applying constitutive equations of fluid flow or electro-magnetism in our models.

Finally we sound a note of warning. The risk-averse designer may be tempted to calculate the operating stresses in a component from the carefully constructed constitutive model and then apply an *arbitrarily huge* factor of safety. After all, metal is cheap, when compared to financial or injury loss, or the perceived loss of reputation, incurred when components fail. However, as we noted at the beginning of this chapter, studying failure will concentrate our minds on improved, and eventually successful designs. We regularly advise engineering students that *"if we never design a component that fails in service, we have not succeeded in design"*. In mass limited leading edge design, such as yacht or motor racing, it is a well accepted practice to design components such that they *just avoid failing*. This can only be achieved through careful field-testing under actual racing conditions. This is risky design at its best. Perhaps the realistic designer's task lies between these two extremes.

Ex 2.7

(a) A large chandelier is suspended from the ceiling of a major opera house. The suspension is in the form of a steel bar. Assuming that the connections of the suspension bar to the ceiling and the chandelier, have been substantially over-designed, estimate an appropriate factor of safety for the suspension bar.

(b) In a chemical plant an incinerator is used to burn industrial waste, occasionally including some small amounts of highly toxic by-products. The chimney of the incinerator is a mild-steel tube 500 mm in diameter, 20 m in height and it is expected to be subject to substantial wind loading. Consequently, stay-cables are used to share the wind loads on the chimney. You may consider the chimney mounted by a flange bolted to the top of the

Table 2.4 Estimates of explicit factors of safety

Factor	*Relevance to the design*	*Range of values*
	Consequence of failure	
F_0	*Seriousness of failure*	1 – 1.4
	Uncertainties associated with load estimates	
ℓ_1	*Magnitude of the load*	1.0 – 1.6
ℓ_2	*Rate of load application (shock loading?)*	1.2 – 3.0
ℓ_3	*Load sharing between elements of the component*	1.0 – 1.6
	Material and modelling related uncertainties	
s_1	*Variations in material properties*	1.0 – 1.6
s_2	*Manufacturing uncertainties*	1.0 – 1.6
s_3	*Environmental and operational uncertainties (temperature, corrosion)*	1.0 – 1.6
s_4	*Effects of stress concentrations (analytical values)*	*can be high*
s_5	*Reliability of mathematical model*	1.0 – 1.6

Design factor of safety $F_d = F_0 \times \ell_1 \times \ell_2 \times \ell_3 \times s_1 \times s_2 \times s_3 \times s_4 \times s_5$

incinerator structure. Identify some probable modes of failure for the chimney and estimate a suitable design factor of safety for the chimney.

(c) At some monuments and scenic locations safety-rails are used to protect visitors. The safety-rail is usually constructed from heavy steel pipe, with the horizontal sections welded to vertical supports secured to the ground. The ground connection is either in the form of heavy flanges, welded to the bases of the vertical supports and bolted to the ground, or the base of the vertical supports are securely embedded in the ground. Estimate the *worst credible accident* that might befall this type of safety-rail. Identify the modes of failure for a typical rail section and estimate an appropriate design factor of safety.

2.4 Assignment on failure prediction and factors of safety

The aims of this brief assignment are:

(a) to clarify the ideas of *mode of failure*, *failure predictor*, and *factor of safety*; and

(b) to give practice in the design of simple engineering components to resist failure.

2.4.1 Background

A *service table* is commonly located in front of each passenger seat in commercial aircraft to provide a foldaway surface for supporting meal trays, writing or reading materials, drinks and various other articles. Figure 2.19 illustrates the approximate form of one of the two support arms for such a service table. The aluminium from which these arms are made has the following materials properties:
$S_y = 145$ MPa ; $S_u = 250$ MPa ; $E = 70$ GPa.

2.4.2 The design problem

(a) Use any necessary assumptions to estimate the maximum bending moment experienced by the arm when the table is subjected to your chosen value of the design force, W.

(b) Is a *failure predictor (theory of failure)* required in order to predict the failure of these arms by yielding? Why or why not?

(c) List the possible modes of failure for the major elements of the service table.

(d) On the basis of reasoned argument, decide upon a suitable factor of safety to be used in the design of the arm to resist yielding in bending. Tabulate your calculation. Does the resulting factor of safety seem reasonable?

(e) Design the arm to resist yielding. Ignore details of the end connections.

(f) To ensure the suitability of the table for certain uses (e.g. meals and writing), it is desirable not only that the table and its support arms avoid any gross structural damage, but also that the vertical displacement of the table's near edge under the design loads be kept to within, say, 10 mm. Indicate the method by which you would incorporate this second mode of failure into your design process. There is no need to perform the calculations for this part of the design.

(g) Suggest reasons why designers might take more care with such devices than with (say) the design of seats in railway carriages or street-cars (trams) for structural integrity.

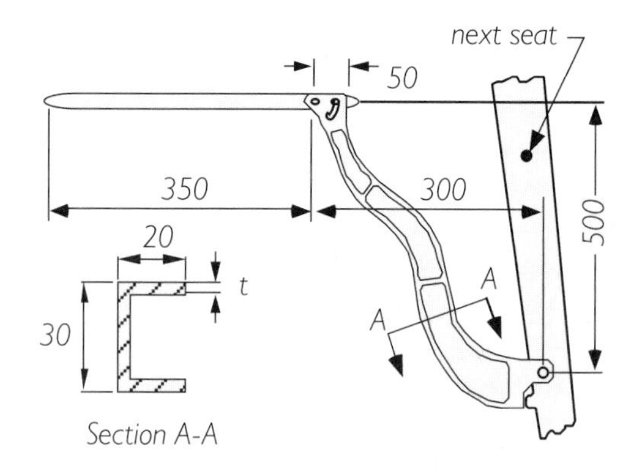

Figure 2.19 Aeroplane service table (schematic, not to scale), dimensions in mm

Engineering Design Associates

Project Title : **AIRLINE SERVICE TRAY**

Date 02:14:98

Page 1 of 7

Ckd. tim

(a) Estimation of design force, W, & max. bending moment

Wide range of service loads, involving a fair amount of educated guesswork. Begin by tabulating some likely forces to get a feel for the magnitude of W :–

SOURCE OF LOAD		FORCE
– meal tray	< 5 kg	50 N
– books	< 1 kg	10 N
– writing (activity)		20 N
– elbow (proportion of torso weight)		~100 N
– briefcase / travel bag	~20 kg ?	~200 N
– sitting	~100 kg ?	~1000 N — (excessive!)

[It might be interesting to have a class discussion in the process of tabulating the above figures. Encourage the students to estimate "reasonable" worst-case values, but not to be too conservative. Uncertainties in the magnitude (& location) of W can be accounted for in the factor of safety, F_d.]

On the basis of the above list, choose : $W = \left(\frac{200\text{ N}}{2}\right) = 100\text{ N}$

Note that we are designing one of the table arms, so the "design force" W needs to be the force on one arm, hence the division by 2 (since 2 arms). This is very easily forgotten !!

[It doesn't matter if students choose a value different from the above — the main aim is that they consider the matter intelligently.]

Engineering Design Associates

Project Title : **AIRLINE SERVICE TRAY**

Date 02:14:98

Page 2 of 7

Ckd. Tim

Arbitrarily/conservatively choose the line of action of vertical force W to be 50 mm from the outer tip of the unfolded table,

i.e. 600 mm from the lower hinge

W, 'P', (50), 300, 300, 'Q'

$$M_P = 100\ \text{N} \times 300\ \text{mm} = 30 \times 10^3\ \text{N.mm}$$

$$M_Q = 100\ \text{N} \times 600\ \text{mm} = 60 \times 10^3\ \text{N.mm}$$

Note that the moment M varies along the arm PQ, and is a maximum ($= M_Q$) at 'Q'.

(b) Comments on the need for a theory of failure

No, a theory of failure (i.e. a means for combining the various stresses in a multi-axial stress situation into a single quantity, to be used for predicting failure) is not necessary here. This is because the table arms are subject only* to BENDING stresses, which are UNIAXIAL. Therefore the results of (uniaxial) laboratory tests (to determine S_y) can be applied directly: the component will begin to yield when the bending stress equals the yield stress, S_y.

(* Note: the above statement assumes that shear stresses are negligible compared to bending stresses. An experienced designer will usually make this assumption for "long-ish" beams like the table arm. Its validity can be checked later. If shear stresses are not ignored, then the stress situation is multi-axial, and a theory of failure would be needed to predict the onset of yielding.)

"Theories of failure" — such as the Maximum Shear Stress Theory of Failure (Tresca), and the Maximum Shear Strain Energy Theory of Failure (Von Mises) — are only necessary in cases where the results of uniaxial laboratory tests are to be applied to situations involving MULTI-AXIAL stress.

(They may be used for uniaxially stressed components but in such cases they merely reduce to the trivial result that yielding begins when $\sigma = S_y$.)

Engineering Design Associates		Date
Project Title :	**AIRLINE SERVICE TRAY**	02:14:98
		Page 3 of 7
		Ckd. Jim

(c) Possible modes of failure for service table

1. Yielding of table top in bending
2. Fracture " " " " "
3. Yielding of support arm " "
4. Fracture " " " " "
5. Shearing failure of hinge pin at 'Q'
6. " " " pins at 'P' (2 pins)
7. Tearing (shear failure) of slot at 'P'
8. Compressive failure of reaction point at 'Q'
9. Excessive deflection of table (vertical)
10. " " " " (lateral sway)
[11. Buckling of lower (compressive) flange of arm] —— (students may not know of this mode.)
12. Excessive friction (& squeaking ?!) in hinges

[Note that students may only think of a few of the above, or they may even come up with others. Three or four should suffice, just so they get the general idea.]

(d) Estimation of safety factor, F_d

In the following determination of F_d, the main issue is that the size of the sub-factors $l_1, l_2, l_3, S1, \ldots S_5$ depends in each case upon the level of uncertainty concerning the relevant quantity, not upon the magnitude of the quantity itself.

[The values given in the table below are my subjective estimates, lying within ranges commonly used. The high value ($F_d = 6$) results from the great uncertainty concerning the loads on the arm. If a more conservative value of W were employed in the first place, then values of F_d as low as 3 or 4 might be OK!
Really, students just need to wrestle with the vagueness and uncertainty of this — there's no other way out!]

Engineering Design Associates

Project Title : **AIRLINE SERVICE TRAY**

Date 02:14:98

Page 4 of 7

Ckd. tim

Factor	Description	Value	Usual Range
F_o	consequences of failure (not serious)	1·0	[1·0 → 1·4]
l_1	uncertainty in load magnitude	1·5	[1·1 → 1·6]
l_2	" " rate of load applic'n (dynamic loads)	1·3	[1·2 → 3·0]
l_3	" " sharing of load between members	1·5	[1·1 → 1·6]
s_1	uncertainty due to material props (value of S_y)	1·1	[1·1 → 1·6]
s_2	" " " manuf. variations (casting accuracy)	1·1	"
s_3	" " " environmental effects (erosion, dents)	1·1	"
s_4	" about stress concentrations (near ends)	1·5	[(analytical can be high)]
s_5	" in mathematical model (eg. 2dim'l forces)	1·1	[1·1 → 1·6]

Overall $F_d = F_o \, l_1 l_2 l_3 \, s_1 s_2 s_3 s_4 s_5 = 6{\cdot}42$

$\Rightarrow$ $F_d \approx 6$

(e) Design the arms to resist yielding

Design inequality :– $\sigma_{B,MAX} \leqslant S_y / F_d$ ——— (1)

... where $\sigma_{B,MAX}$ = maximum bending stress in the arm.
($\sigma_{B,MAX}$ will occur at the outer fibres of the arm)

In general, $\sigma_B = \dfrac{My}{I_{xx}}$; $\sigma_{B,MAX} = \dfrac{M\, y_{MAX}}{I_{xx}}$ ——— (2)

but $\sigma_{B,MAX}$ will vary along the length of the arm due to varying ratios of M, y_{MAX} and I_{xx}.
Ultimately the design would need to be checked at various locations along the arm — at least at the lower and upper extremities (near 'Q' and 'P').
For this initial design exercise, we will consider the arm's cross-sectional design only near the lower end, where the dimensions given in detail 'section A-A' are assumed to apply.

Engineering Design Associates	Date
Project Title: ***AIRLINE SERVICE TRAY***	02:14:98
	Page 5 of 7
	Ckd. tim

Hence $M = M_Q = 60 \times 10^3$ N.mm — M

$d = 30$, $b = 20$, t

$y_{MAX} = \left(\frac{d}{2}\right) = 15$ mm.

[Note: "design the arm" in the context of this problem essentially boils down to determining a value for t, since all other design variables are given.]

The problem now is to express I_{xx} in terms of t. Students may need some help with this, although they should be familiar with it from last year. Two approaches are presented below — the first is more elegant, the second is more general ...

(i) I_{xx} by subtraction :

Since the channel section is symmetrical about the x-x axis, the easiest way of finding I_{xx} of the overal section is to subtract the 2nd moment of area of the inner "negative area" rectangle from that of the outer rectangle.

i.e. $I_{xx} = \dfrac{bd^3}{12} - \dfrac{b_o d_o^3}{12}$

where $b_o = (b - t)$

$d_o = (d - 2t)$ ← (note: two thicknesses!)

$$\Rightarrow \quad I_{xx} = \frac{bd^3}{12} - \frac{(b-t)(d-2t)^3}{12} \quad \text{—— (3)}$$

I_{xx}

It is possible, of course, to expand and rearrange the above equation (a quartic in t), but that is not necessarily very useful ...

$$I_{xx} = \frac{t}{12}\left[t^3(-8) + t^2(4).(2b+3d) + t(-6d).(2b-d) + (d^2).(6b+d)\right]$$

Rearranging our earlier design inequality, and inserting the expression for $\sigma_{B,MAX}$, we can write:

$$\frac{M\,y_{MAX}}{I_{xx}} \leqslant \frac{S_y}{F_d} \quad \Rightarrow \quad I_{xx} \geqslant \frac{F_d\,M\,y_{MAX}}{S_y} \quad \text{—— (6)}$$

Now, $\dfrac{F_d\,M\,y_{MAX}}{S_y} = \dfrac{6.0 \times 60 \times 10^3 \text{ N.mm} \times 15 \text{ mm}}{145 \text{ N.mm}^{-2}}$ ← (note: $1\text{ MPa} = 1\frac{\text{N}}{\text{mm}^2}$)

$= 37,242 \text{ mm}^4$.

Engineering Design Associates **Project Title :** **AIRLINE SERVICE TRAY**	Date 02:14:98
	Page 6 of 7
	Ckd. tim

So from (6), we require $\boxed{I_{xx} \geqslant 37,242 \text{ mm}^4}$

where I_{xx} is evaluated by means of equation (3) [or (5)]

[The computational challenge can be approached in various ways at this stage. My preference would be for a few trial calculations with guessed values for t (since this would help us develop a feel for the effect of t), but students might need some cajoling to do it this way. They hate guessing!!]

required I_{xx}

From equation (3), trial calculations with $b = 20$ mm & $d = 30$ mm

$\Rightarrow$ for $t = 5$ mm ; $I_{xx} = 35\,000 \text{ mm}^4$ (insufficient)

$t = 10$ mm ; $I_{xx} = 44\,167 \text{ mm}^4$ (OK)

$t = 6$ mm ; $I_{xx} = 38\,196 \text{ mm}^4$ (just OK)

Hence choose $\boxed{t = 6 \text{ mm}}$.

t

It could be argued that for a slender arm of outer dimensions = 20 × 30 mm, a section thickness of 6 mm seems somewhat heavy. An alternative would be to relax some (artificial) constraints so that some other design parameter can be varied. Say we took $d = 40$ mm, since it makes more sense to increase d instead of b. Then we would find that the arm could comfortably be given a thickness of $t = 3$ mm and still be well within strength requirements.

(f) Discussion of design for acceptable vertical deflection

Students should mention the need to model displacement mathematically (using beam deflection theory) then to apply the deflection constraint to the design which resulted from the stress constraint (i.e. (e) above). If deflection "governs" the design, then a new arm thickness will need to be assigned on the basis of this second mode of failure.

The discussion of deflection analysis, outlined above, can be purely qualitative (for the sake of this project anyway). Students shouldn't do any deflection calculations.

(g) More care is taken on an aeroplane component than on a similar tram component because weight is a more serious design consideration on an aeroplane.

Engineering Design Associates

Project Title : **AIRLINE SERVICE TRAY**

Date 02:14:98

Page 7 of 7

Ckd. tim

Addendum:

(ii) I_{xx} by subsidiary areas & IP axis th'm :

Students may need to be reminded about the "parallel axis theorem" ...

centroid of area, 'G'

G G

area, A

$\bar{y}$

x x

(distance from centroid 'G' to XX axis)

(a statement of the parallel axis theorem)

Let I_{GG} = (2nd moment of area of A about its own centroidal axis GG), then

$$I_{xx} = I_{GG} + A\bar{y}^2 \quad \text{—(4)a.}$$

Suggested breakup of C-section into 3 sub-areas:—

b_2 b_1 d_1 A_1 A_2 $\bar{y}_1$ d_2 x x ($\therefore \bar{y}_2 = 0$) (same as A_1)

Overall,

$$I_{xx} = \frac{b_2 d_2^3}{12} + 2\left(\frac{b_1 d_1^3}{12} + A_1 \bar{y}_1^2\right) \quad \text{—(4)b.}$$

where $b_2 = t$; $d_2 = d$

$b_1 = (b-t)$; $d_1 = t$

$y_1 = \left(\frac{d-t}{2}\right)$

Hence, $I_{xx} = \frac{td^3}{12} + 2\left[\frac{(b-t)t^3}{12} + (b-t).t.\left(\frac{d-t}{2}\right)^2\right]$

$$\Rightarrow \quad I_{xx} = \frac{td^3}{12} + \frac{(b-t)t^3}{6} + \frac{t(b-t)(d-t)^2}{2} \quad \text{—(5)}$$

... which is a nasty little quartic in t, but that's OK!

Equations (3) & (5) can be shown to be identical (I hope...)

2.5 Design for dynamic loading (fatigue)

"All machine and structural designs are problems in fatigue because the forces of Nature are always at work and every object must respond in some fashion." Carl C. Osgood (1979)

It has long been recognised by engineers that component failure due to time dependent loading occurs at significantly lower stresses than that observed under static loading. In the preface to a bibliography on material and component fatigue, John Mann notes (Mann, 1970):

"Investigations relating to the failure of materials and structures under repeated or cyclic loads have been in progress for over one hundred and twenty-five years."

While Mann goes on to record publications of experimental work and other observations on fatigue by W. A. J. Albert (1838, on ropes), and W. J. M. Rankine (1843, on the failure of railway axles), it is now generally agreed that the term *fatigue* owes its introduction to failure mechanisms to Jean-Victor Poncelet in 1829[2.5]. The first systematic studies of failure mechanisms under cyclic loading were conducted by Anton Wöhler (1819-1914). Wöhler's work was initially aimed at investigating fatigue failures in railway axles (Timoshenko, 1953). Consequently, as a significant first step in this failure investigation, Wöhler needed to establish the load patterns to which the axles were exposed in operation. This approach, described in Wöhler's historic 1858 article[2.6], is common practice in fatigue investigations today. Wöhler went on to investigate the fatigue failure mechanism in a rotating beam machine as well as in reversed bending (Wöhler 1858). Figure 2.20 shows Wöhler's rotating beam test schematically.

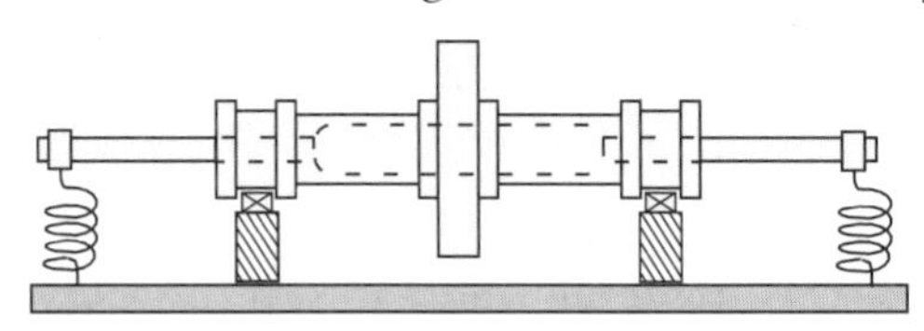

Figure 2.20 Wöhler's rotating beam test for the fatigue limit of railway axles (After Timoshenko, 1953)

In the rotating beam test the central hollow shaft and flywheel were rotated at constant speed and a load was applied to the two *beam* specimens at each end, by an adjustable spring, connected to the rotating specimen through a bearing. However, fatigue tests take a long time to run, since the component needs to be subjected to many cycles of stress before fracture. Wöhler recognised the difficulty of running such tests with full size axles and at normal operating speeds of around 15 RPM. Hence, he used specimens of reduced scale and ran the tests at double speed, thereby being able to generate substantial fatigue data in a reasonable time. Wöhler also recognised that in order to establish practical design working stresses for a wide range of materials and geometries, the limiting fatigue failure stress (*endurance limit*) needed to be related to static load test data. In our brief introduction to design to resist fatigue failure, we will make use of such relationships. However, as a general rule, designers must seek fatigue data from tests that expose components to cyclic loads comparable to those *'seen'* by the component in practice. Typically, vehicle components exposed to cyclic loading, are tested by applying loads observed on a vehicle test-track. Air-frames are tested for endurance by exposure to wind-gust data observed in test flights. Experience with fatigue failure has shown that life estimates or safe load estimates are much less precise than strength calculations (Collins, 1993). Moreover, results from fatigue testing of materials exhibit appreciable statistical spread, even when closely similar test conditions are used. Consequently, all design predictions for fatigue resistance must be treated as cases in statistical reliability.

Studies of fatigue failure have identified three stages of the process, namely *crack initiation*, followed by *crack growth* and finally *fracture*. It is now generally agreed that all failure mechanisms commence with some local material discontinuity or crack. The generation of such a discontinuity is termed the crack-initiation phase of fatigue failure. In general, it is also an accepted notion in fatigue design that all materials have some form of discontinuity or *micro-crack* which acts to initiate the fatigue failure process (see for example Collins, 1993, or Mitchell and Landgraf, 1991). Given the appropriate condition necessary for crack initiation, once initiated the crack will continue to grow with each further cycle of loading applied to the component. Once the crack has grown to a level where the material section can no longer resist the applied loads, the component will fracture.

Where the applied load is sufficient to cause plastic or permanent local deformation of the component, either due to load level or due to some geometric stress concentration, cyclic application

2.5 Timoshenko (1953), *loc. cit.*

2.6 Mann (1970), *loc. cit.*

of such loads will propagate the crack to failure after relatively few cycles of application (of the order of one to 10^5 cycles of load application, Collins, 1993). In this type of *low-cycle* fatigue, the material suffers cumulative damage at the tip of the crack as the component progresses to failure. Cumulative damage and crack-growth rates are the main objectives of the study of fatigue life prediction based on fracture mechanics.

For component load levels that limit deformation to the elastic region, fatigue failure may still occur at relatively large numbers of cycles of load application (in the order of greater than 10^5 cycles). Crack-growth and cumulative damage in this type of *high-cycle* fatigue is still the subject of intense research, and it is also the domain of fatigue life prediction based on fracture mechanics. In this introductory text we focus our interest on simple design procedures that take appropriate account of time-dependent load applications, rather than on fatigue failure mechanisms. Nevertheless, some component design situations call for the investigation of fatigue failure due to design error. In such cases it is important to recognise some clear *telltale* signs of fatigue failure. Figure 2.21 shows a typical rotating beam type of fatigue failure. This is exactly the style of failure investigated by Anton Wöhler. The rotating beam was exposed to a load which was fixed in direction Hence the extreme fibres of the beam suffered reversed tension and compression as the beam rotated through each full revolution.

Figure 2.21 Typical fatigue failure surface of a rotating beam

The crack, having been initiated at some surface irregularity or geometric stress concentration, progressed in stages through the section. This crack-growth progression is indicated by the lighter striated area of the cross section seen in the lower part of the figure. The striae are colloquially referred to as *beach marks*, a commonly recognisable *telltale* sign associated with fatigue failure. The *beach-mark* section of the failure surface is generally smoother than the final fracture section, also clearly seen at the top of Figure 2.21. This relative smoothness is partly due to the opening and closing of the fracture surface during repeated load reversals.

2.5.1 Factors influencing the onset of fatigue failure

Failure in fatigue is the result of the interaction of many complex influences, including the magnitude and cyclic character of the applied load. As already noted, crack initiation can occur as a result of some material defect or due to a geometric stress concentration. Hence surface finish and component geometry are key influences in determining structural integrity under cyclic loading.

The cyclic nature of the applied load

Figure 2.22 shows the generic terminology used in describing cyclic loading. Although the loading indicated here is sinusoidal, more general load application is still commonly described by the parameters of the stress-time curve indicated in Figure 2.22. In this context we recognise three main types of stress-time curves, namely *fluctuating*, *fully-reversed* and *repeated* stress-time cycles.

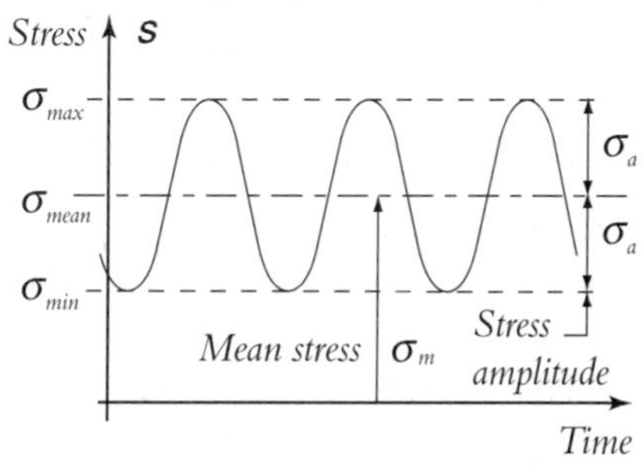

Figure 2.22 Fluctuating load cycle indicating generic stress-time terminology

The nature of stress-time curve is described by the relative magnitude of the mean stress. *Fluctuating load cycles* have all three (mean, minimum and maximum) stresses non-zero. *Fully-reversed stress*, as implied by the name, results from loading where the signs of minimum and maximum stress change during the load cycle, and the stress alternates symmetrically about a mean value of zero. *Repeated loading* implies a zero minimum stress. Figure 2.23 indicates some typical load cycles encountered in practice. Figure 2.24 shows some typical engineering components with their associated load-time curves.

Ex 2.8

Examine your surroundings, either at home, work, or at some sporting venue, or even possibly at the gymnasium where you work out to maintain your physical fitness. Identify ten components in these surroundings that could, in your opinion, suffer fatigue failure if not properly designed. For each of these components suggest the most likely load cycles they are exposed to in service. Wherever possible, estimate reasonable service load cycles and plot these on load-time axes. *Note*: you are not expected to do any detailed calculations for this exercise.

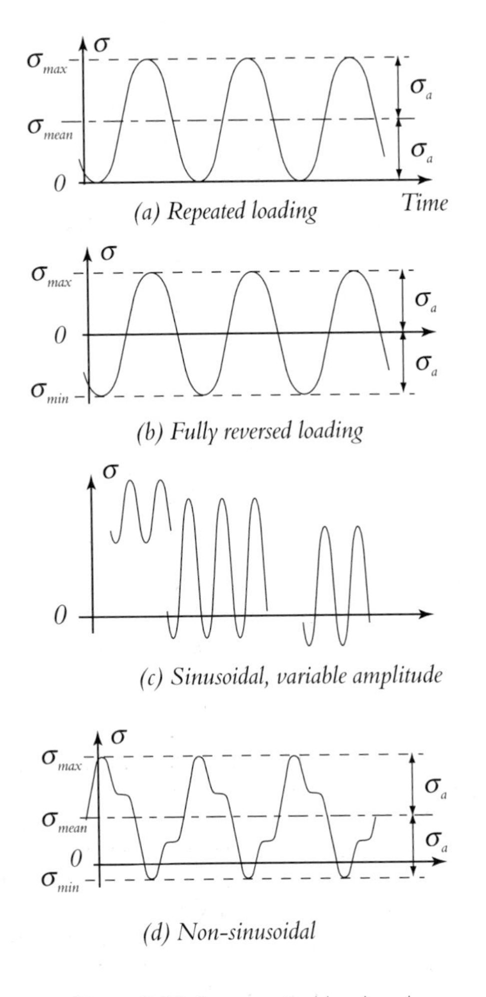

Figure 2.23 Some typical load cycles

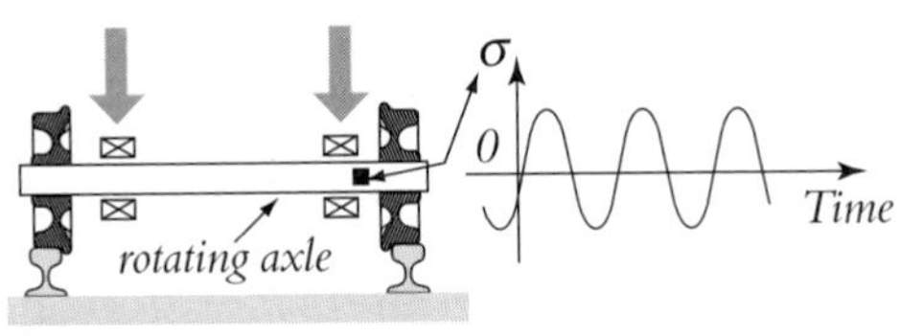

(a) Railway axle; sinusoidal load cycle

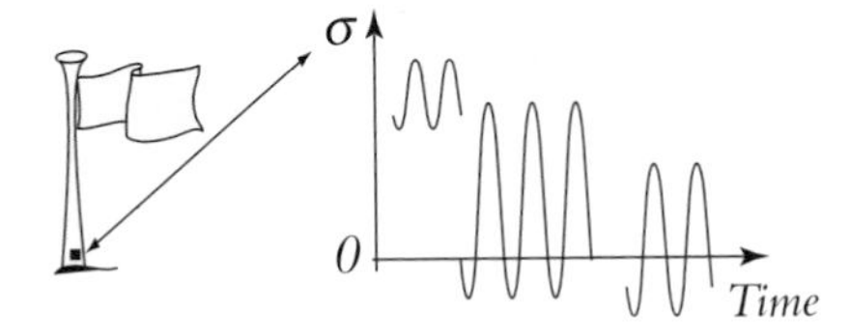

(b) Fluttering flag; sinusoidal, variable amplitude load

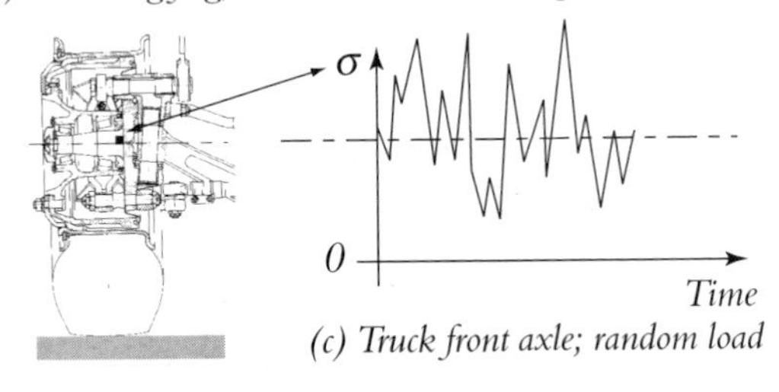

(c) Truck front axle; random load

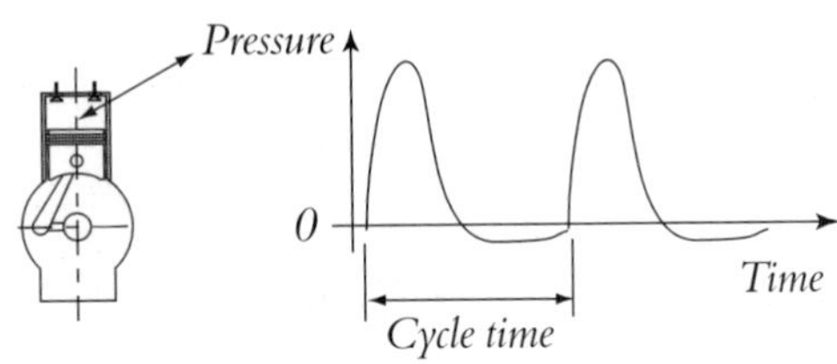

(d) Internal combustion engine; repeated load cycle

Figure 2.24 Examples of practical load cycles

Material behaviour under cyclic loading

A question that has long puzzled fatigue researchers is the effect of accumulated damage on the component subject to cyclic loading. As noted earlier, estimation of cyclic load behaviour from *static* test data is, at best, a gross simplification of the fatigue failure process. During the first phase of fatigue failure, components accumulate damage at the microscopic or grain boundary level. This aspect of fatigue behaviour is dealt with by dislocation theory (see for example Collins 1993, for an introduction to dislocation theory and fracture mechanics). Once a crack has formed, damage is accumulated at the crack tip. The stress field at and around the crack tip is invariably triaxial, even in uniaxially loaded elements, so deducing crack growth phenomena from uni-axial static stress strain data is unacceptable. Figure 2.25 shows the behaviour of elastic materials when subjected to cyclic loading and unloading.

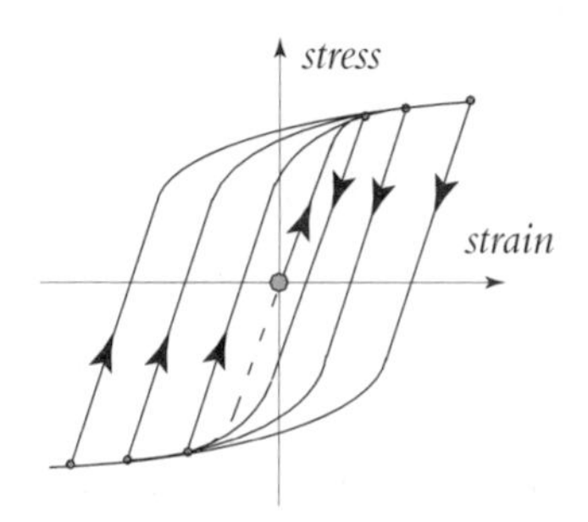

Figure 2.25 Cyclic stress strain curves (schematic)

The exaggerated hysteresis loops indicate that during straining the material suffers accumulated damage. During the load cycle, the material suffers both elastic and plastic strain. The sum of these two strains is referred to as total strain and experiments with a wide range of elastic materials has shown that the most reliable predictor of fatigue life is the total strain (see for example Manson, 1965 and Fuchs and Stephens, 1980). Manson and Fuchs and Stephens have proposed relatively simple formulas for relating total strain to fatigue life (Shigley, 1986), however, as Shigley notes:

[these equations are] *"of little use to the designer. The question of how to determine the total strain at the bottom of a notch or discontinuity has not been answered. There are no tables or charts of strain concentration factors in the literature."*

In part to overcome these difficulties, Shigley goes on to propose an approximate empirical approach based on converting stress amplitude (known from the applied cyclic loading) to plastic strain (suffered in the plastic region). However, these empirical rules are hardly different from estimating cyclic load behaviour (S_e') from bulk strength data (static strength behaviour S_U).

Experimental results from cyclic load tests are usually presented in two common forms indicated in Figures 2.26 and 2.27. Stress level plotted against numbers of cycles to failure are called *S-N diagrams* (*S* for stress, *N* for the number of full load-reversal cycles to failure).

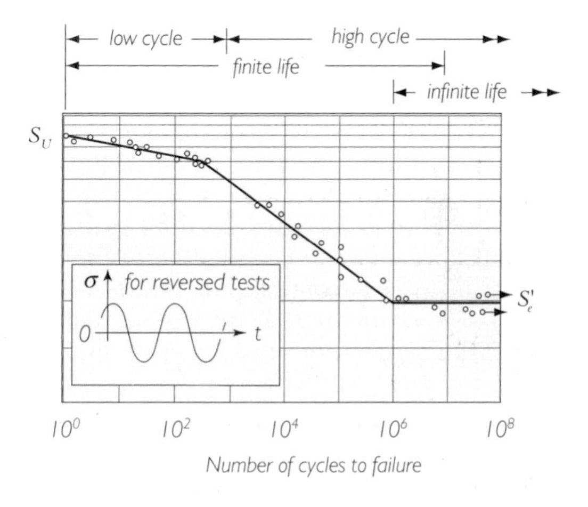

Figure 2.26 S-N Data for fully reversed axial fatigue tests; Data is for normalised chromium-molybdenum steel, S_U range 800 to 860 MPa, S_e'=338 MPa (After NACA Tech. Note 3866, Dec. 1966)

Figure 2.26, plotted for a normalised, chromium-molybdenum steel, indicates the range of fatigue failures suffered by the components tested. The points on the graph also indicate the large numbers of tests needed to obtain a single *S-N* curve. Unfortunately, the logarithmic scale of the ordinate doesn't do justice to the real scatter of the points relative to the *S-N* regression line.

Due to the consistently uncertain nature of the fatigue failure of any test specimen, even when exposed to ideally similar test conditions, any inference from fatigue test data needs to be modified by measures of reliability statistics. Figure 2.27 illustrates the inherent uncertain nature of fatigue failures on a *P* (probability of failure)-*S-N* diagram, plotted for extruded 2024 aluminium alloy.

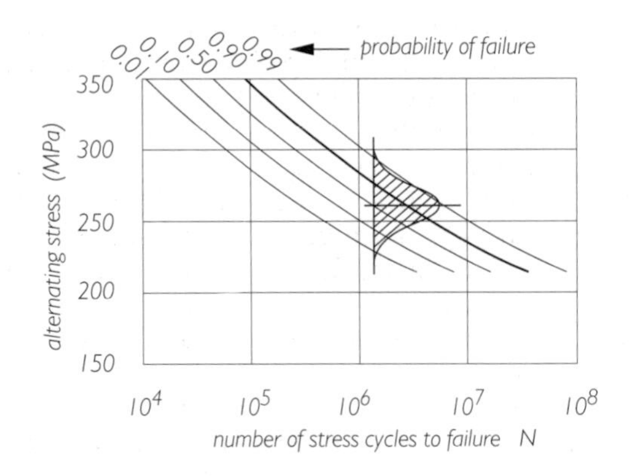

Figure 2.27 P-S-N Data for fully reversed axial fatigue tests; Data is for extruded 2024 aluminium alloy, $S_U = 470$ MPa; un-notched specimens on rotating beam test.

In practice, it is rare to find components exposed to constant amplitude load cycles. Hence, to predict a safe fatigue life, *S-N* data similar to that shown in Figure 2.26 are used in estimating accumulated damage to a component subjected to a spectrum of load cycles. The most common method for estimating accumulated fractional damage during real life load cycles is due to A. Palmgren, later modified by M. Miner, the *Palmgren-Miner linear damage rule*, or *Miner's rule* (see for example Shigley, 1983; Collins, 1993; Palmgren, 1924; Miner, 1945). The procedure used is illustrated schematically in Figure 2.28. Consider a material exposed to some varying sequence of stress levels, $\pm S_1$ for n_1 cycles, $\pm S_2$ for n_2 cycles, $\pm S_3$ for n_3 cycles, and so on as indicated in Figure 2.28(a). Figure 2.28(b) shows the number of cycles N_1, N_2, and N_3 required to fail the material at cyclic stress levels $\pm S_1$, $\pm S_2$, and $\pm S_3$ respectively. *Miner's rule* states that failure of the material will occur whenever the *cumulative damage*

$$\frac{n_1}{N_1} + \frac{n_2}{N_2} + \frac{n_3}{N_3} + \cdots \geq 1 .$$

Another significant feature of fatigue performance, seen in Figure 2.26, is that at some level of stress, the test specimens appear to survive cycles of load applications $N > 10^6$.

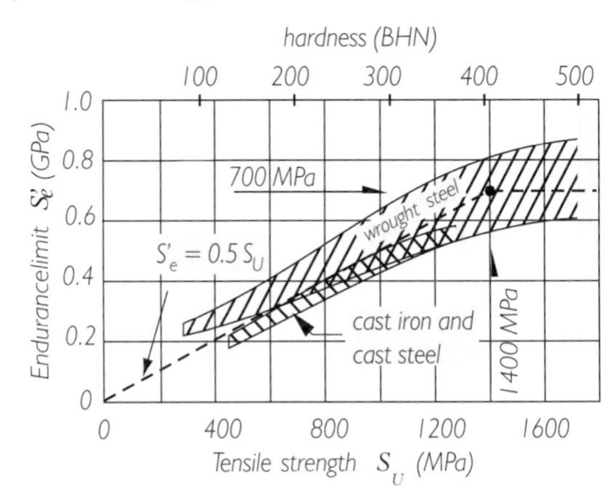

Figure 2.29 Scatter band of observations of the relationship between endurance limit and ultimate tensile strength for a range of steels and cast iron. (After Lipson and Juvinall, 1961)

The stress level for which test specimens experience a high probability of survival beyond $N = 10^6$ is called the *ideal endurance limit*, denoted

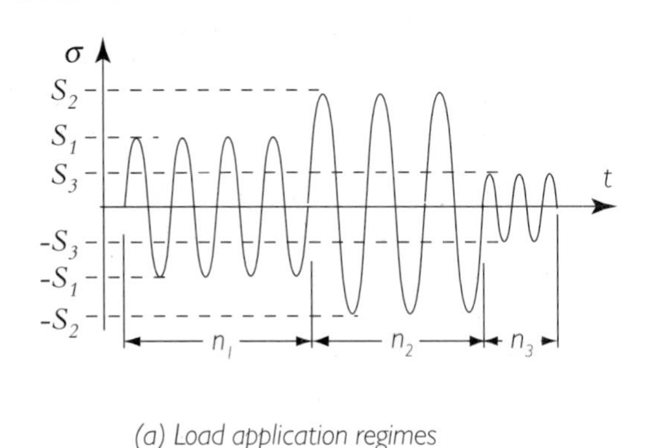

(a) Load application regimes

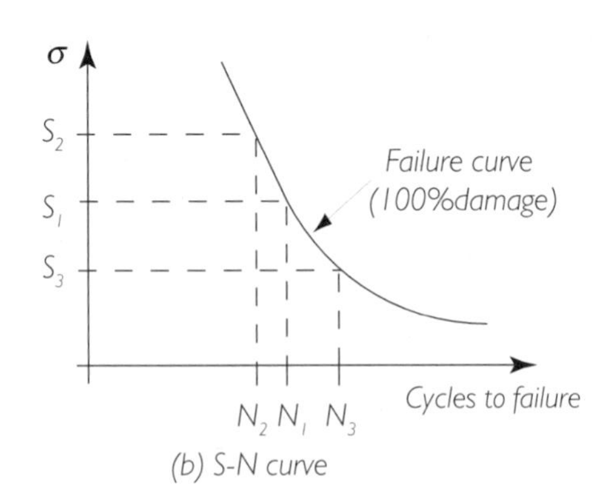

(b) S-N curve

Figure 2.28 Cumulative damage due to cyclic loading

S'_e. The *ideal* signifies that the result is for *smooth, un-notched* test specimens tested in ideal laboratory conditions. Figure 2.29 is a plot of endurance limit data against S_U for a range of steels and cast iron.

For most steels, up to approximately S_U = 1400 MPa, a fairly linear relationship can be seen to exist between S_U and S'_e, whereby $S'_e \approx 0.5 S_U$. In the absence of experimental values for S'_e, this provides a reasonable approximation.

Effect of geometry (stress concentrations)

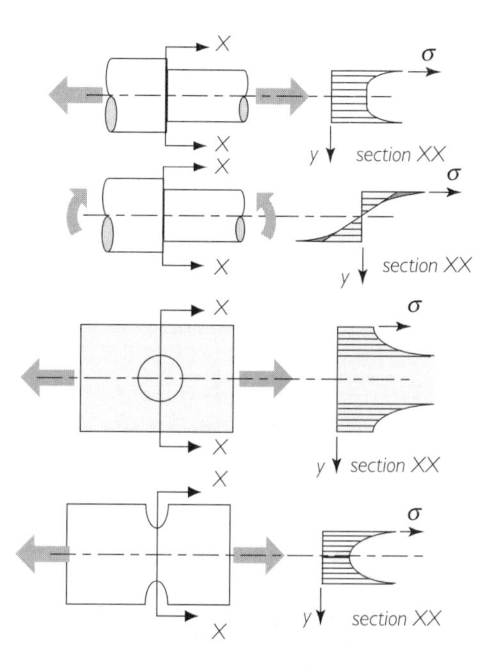

Figure 2.30 Some typical stress concentrators and their associated stress distributions. The scales of stress distributions are exaggerated to indicate approximate shapes only.

In his studies of fatigue failures, Wöhler already recognised the significance of component geometry, and its effect on initiating failure by fatigue. As noted earlier, fatigue behaviour is more closely related to total strain than to component stress. Local straining is commonly experienced at geometric stress concentrations (notches and holes) in components essentially designed for *static loading*. This type of local straining is said to *redistribute* stresses in components.

High strain sites in a component are high probability sites for crack initiation. Consequently, in designing against fatigue failure, we need to make appropriate allowance for stress concentrators by modifying S'_e with an appropriate geometric stress concentration factor.

The scale of stress concentrations shown in Figure 2.30 have been exaggerated to indicate the considerable increase in stress intensity at changes in geometry. In *real* components, the effect of stress concentration is highly localised. Nevertheless, the effect of the highly localised increase in stress intensity is to generate sites for crack initiation in fatigue failure.

We define two factors for modifying endurance limit data from ideal, smooth un-notched, specimens.

$$K_t = \frac{actual\ \ maximum\ \ stress}{nominal\ \ stress}, \text{ and}$$

$$K_f = \frac{endurance\ \ limit\ \ of\ \ notched\ \ specimen}{endurance\ \ limit\ \ of\ \ un-notched\ \ specimen}.$$

K_t is generally obtained from elastic theory, or photo-elastic studies of suitable models of the component under investigation (Timoshenko and Goodier, 1982). Alternatively, localised stress increase due to changes in geometry may be found from finite element analysis (see for example Samuel and Horrigan, 1995). Figure 2.31 shows one of a number of graphs obtained for test specimens with geometric stress concentrators. The curves are a function of both the component geometry and the nature of the cyclic loading applied. In general, K_t must be established from data obtained under representative loading conditions.

K_f (the subscript f in this case denoting the relation of this factor to a fatigue, or endurance limit) is estimated from the combinations of K_t and another factor called *notch sensitivity index*, q. Notch sensitivity is a material property, and is a function of material bulk strength S_U and another geometric variable called *notch radius*. The *notch radius* is a measure of the notch size on the test specimens used for determination of q, defined as

$$q = \frac{K_f - 1}{K_t - 1}.$$

This is generally rewritten as

$$K_f = 1 + q(K_t - 1).$$

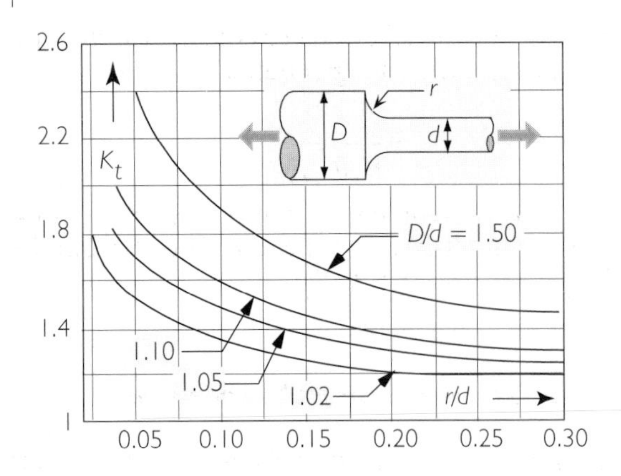

Figure 2.31 Round shaft with fillet in cyclic tension/ compression. Geometric stress concentration factor. (After Peterson, 1974)

Notch sensitivity index q is shown on Figure 2.32 for a range of steels and one aluminium alloy. Clearly, the effect of a notch increases with the bulk strength of steel. It is interesting to compute the effect of a 0.5 mm radius (commonly used for fillets in manufacturing) on a simple component exposed to reversed bending, as would be the case with the axles of railway wagons (these were the first components investigated by Wöhler in his 1858 studies[2.7]).

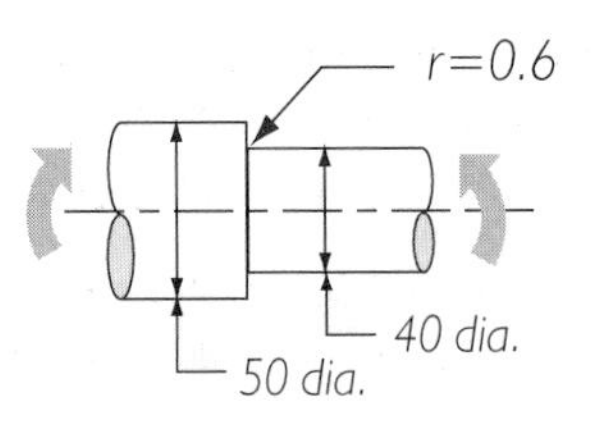

Figure 2.33 A simple engineering component subjected to reversed bending (dimensions shown in mm)

Consider the simple component shown in Figure 2.33 undergoing cyclic reversed bending. With the geometry shown, we find from reversed bending data (similar to the data shown on Figure 2.31) that for $r = 0.5$ mm, $K_t = 2.6$.

From Figure 2.32 we find that the notch sensitivity index q, for two steels, one at S_U = 1400 MPa (a very high strength steel, such as *spring steel* or a nickel alloy), and another at S_U = 700 MPa (a relatively cheap structural steel).

$$q_{1400} = 0.9, \text{ and } q_{700} = 0.7;$$

$$K_{f700} = 1+0.7(2.6-1) = 2.12;$$

$$K_{f1400} = 1+0.9(2.6-1) = 2.44.$$

If we now approximate S_e' to 0.5 S_U, we find that once this value is modified for stress

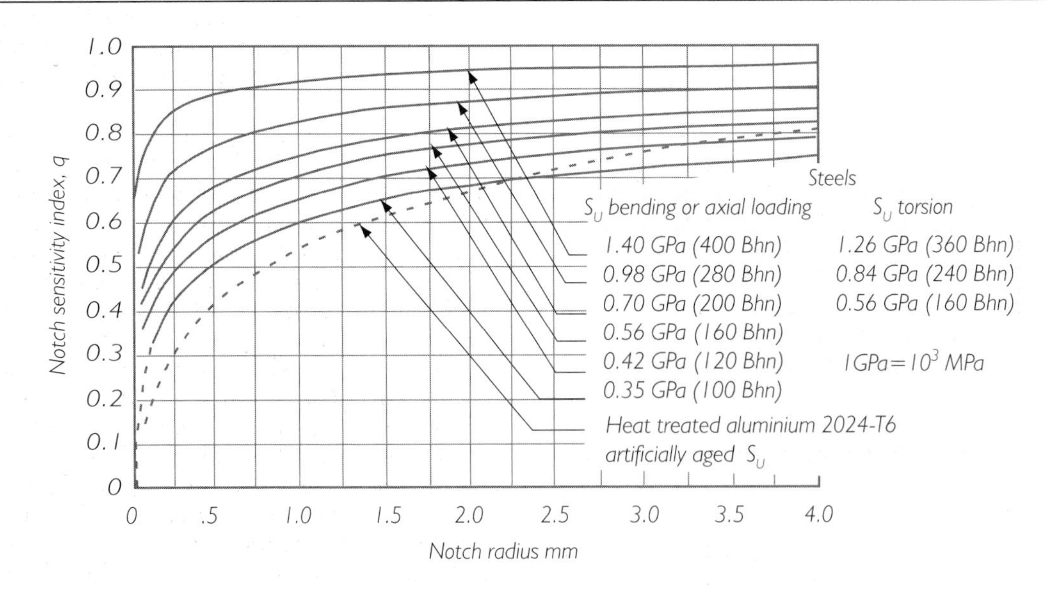

Figure 2.32 Notch sensitivity index for steels and AL 2024 alloy. (AfterJuvinall, 1967)

2.7 Wöhler, A. 1858, *loc.cit.*

concentration, the resulting S_e (not allowing for surface finish or factor of safety)

$$S_{e1400} = 0.7/2.44 = 287 \text{ MPa};$$

$$S_{e700} = 0.35/2.12 = 165 \text{ MPa}.$$

So, even though we may double the bulk strength of this component, we only gain a 74% increase in endurance limit. Moreover, once we take surface finish factors into account, the difference is further reduced.

Effect of surface finish

We also need to make appropriate allowance for surface finish on the component. Whilst geometric stress concentrators represent macro-scale changes in geometry, surface defects, acquired in manufacturing, may be similarly regarded as stress concentrators, but at the microscopic level.

The effect of surface finish on endurance limit is expressed in terms of a surface finish factor K_s, as shown on Figure 2.34.

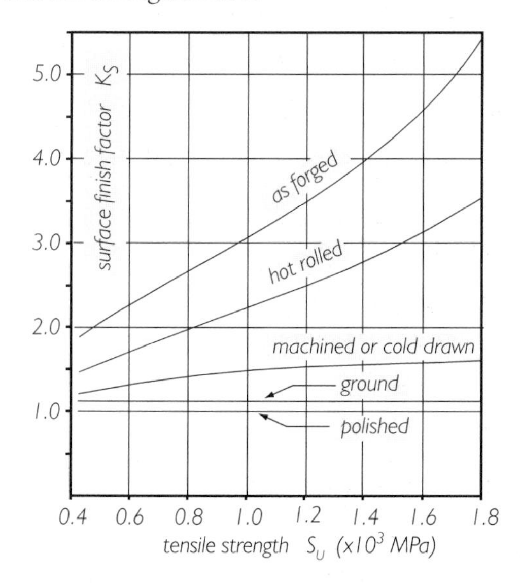

Figure 2.34 The effect of surface finish on endurance limit

This figure, the result of compiling data from various sources, shows the importance of both surface finish and its influence on endurance limit with increased bulk strength S_U. Both factors K_f and K_s serve to reduce S'_e, the endurance limit for smooth un-notched specimen, resulting in a practically useful design endurance limit

$$S_e = \frac{S'_e}{K_s K_f F_d}.$$

S_e may be regarded as the realistic endurance limit of an actual component, as distinct from the ideal endurance limit, S'_e.

Other influences on endurance limit

So far we have dealt with only the most significant factors influencing endurance limit. There are a number of other influences which need to be taken into account in critical designs. We have already noted the influence of corrosive environments on bulk strength properties of materials. In studies of reversed bending of sharply notched mild-steel specimens immersed in air and oil, a significant improvement (approximately 50%) in endurance limit was found for oil immersed specimens over that found in air. Moreover, when specimens were immersed in brine, the endurance limit disappeared altogether, with specimens continuing to fail as the applied stress amplitude was reduced to zero (Frost et al. 1974).

The effect of increased operating temperature, although not as significant as corrosion, serves generally to reduce or completely eliminate the endurance limit. At high temperatures, for example in gas turbines, the effect of creep-failure becomes more significant (Collins, 1993).

The size of a component has a slight but predictable effect of reducing the endurance limit. This effect is based on the theory that imperfections (crack initiation sites) per unit volume remains constant in material resulting from a given manufacturing process. Hence the probability of finding imperfections increases with the gross dimensions of the component (Collins, 1993; Frost et al., 1974). Surface coating (plating) and fretting (microscopic movement between components in contact) are other significant factors serving to reduce the endurance limit.

Our objective in presenting this section on fatigue behaviour of materials is to illustrate the highly complex nature of fatigue failure. In what follows we make use of some of these ideas in a simplified design procedure for uniaxial, high-cycle fatigue.

2.5.2 A simplified design procedure for uniaxial fatigue

In practice, fully reversed load cycles with zero mean stress are rare. Consequently, the designer needs to take account of the effect of non-zero mean stress on the fatigue life of a component. Although Wöhler was the first to study material behaviour systematically under cyclic loading, the first to offer design rules for components experiencing time dependent loads with non-zero mean stress were J. Goodman and W. Gerber in the late 19th century[2.8]. Figure 2.35 shows typical cyclic loading experimental data plotted as *mean stress* versus *peak-stress-at-failure*. Goodman assumed the endurance limit to vary linearly from 0.5 S_U at zero mean-stress to zero at a mean stress of S_U. This assumption is indicated by the linear relationship on Figure 2.35. Gerber fitted a parabola to the available data, indicated as the *Gerber assumption*, also passing through ± 0.5 S_U at zero mean-stress.

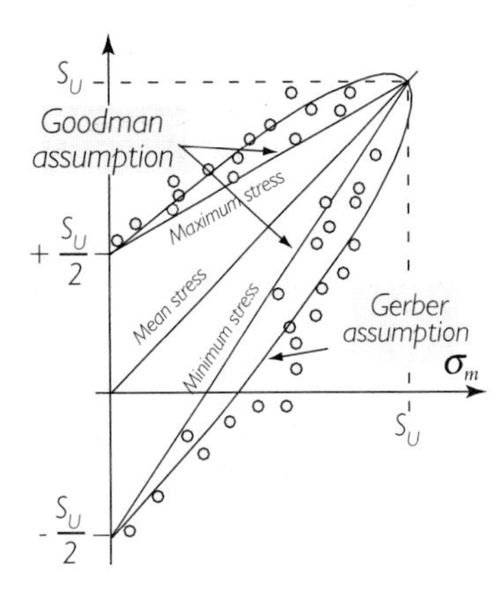

Figure 2.35 Goodman diagram, showing both Goodman's and Gerber's assumed curve fits

Once reliable zero-mean stress endurance limit data, S'_e, became available, Goodman's original assumption of ± 0.5 S_U at zero mean stress was modified to make the original Goodman line pass through ± S'_e at zero mean stress. This line, commonly used for infinite-life design against fatigue is referred to as the *modified Goodman* line.

As we have seen in Figure 2.29, Goodman's original assumption of ± 0.5 S_U at zero mean stress was a very good practical assumption. In design practice, where fatigue data for a specific application is unavailable, and where we are prepared to use a substantial factor of safety, 0.5 S_U is commonly employed as an approximation to the endurance limit S'_e.

Results of laboratory tests on smooth, un-notched, test specimens are also plotted on an *A-M* diagram (*A* for stress amplitude, and *M* for mean stress). Figure 2.36 shows an *A-M* diagram for steel specimens, where the axes are ratios of stress σ_m/S_U (mean stress divided by ultimate tensile stress) and σ_a/S_U (stress amplitude divided by ultimate tensile stress) respectively.

The curves drawn on the *A-M* diagram are for *constant life* data at various numbers of load cycles to failure, ranging from $N = 1$ (static load behaviour) to $N >> 10^6$ (*infinite life*). Clearly, a single load excursion to S_U will result in component failure on either axis of the *A-M* diagram. Hence, any test-specimen that has some combinations of applied mean stress (σ_m)and alternating stress amplitude (σ_a) lying inside the *A-M* diagram, is estimated to fail at some life given by the nearest life curve above the point *representing this combination of stresses*. However, for design purposes we are only interested in values of combined stress that lie below the *infinite life* curve on the *A-M* diagram.

The *infinite life* curve is a regression drawn through the available data and its similarity to the *Gerber assumption* line indicated on Figure 2.35 is clearly recognisable. Another important feature of cyclic load behaviour, also notable on Figure 2.36, is that compressive mean stress has significantly less effect than tensile mean stress on the endurance limit. This effect is attributed to the observation that fatigue failure in compressive loading is the result of a different fracture mechanism than that observed in tensile fatigue failures. We postulate here that in most conservative design situations we will be concerned with crack propagation due to tensile loading damage during cyclic loading. Hence, for design purposes we can redraw the first quadrant of the *A-M* diagram as indicated on Figure 2.37.

From the perspective of wishing to avoid failure by fatigue, the shaded portion of Figure 2.37(b) would seem the appropriate *design space*, not yet modified for K_s, K_f and F_d. Noting that one bound of our *design space* is S_U, on the mean stress, σ_m axis, will alert designers to the need to consider failure by mean stress yielding, S_y. This was first considered by C.R. Soderberg (Soderberg, 1935), in a conservative way, replacing the modified

2.8 Quoted in Frost, Marsh and Pook (1974)

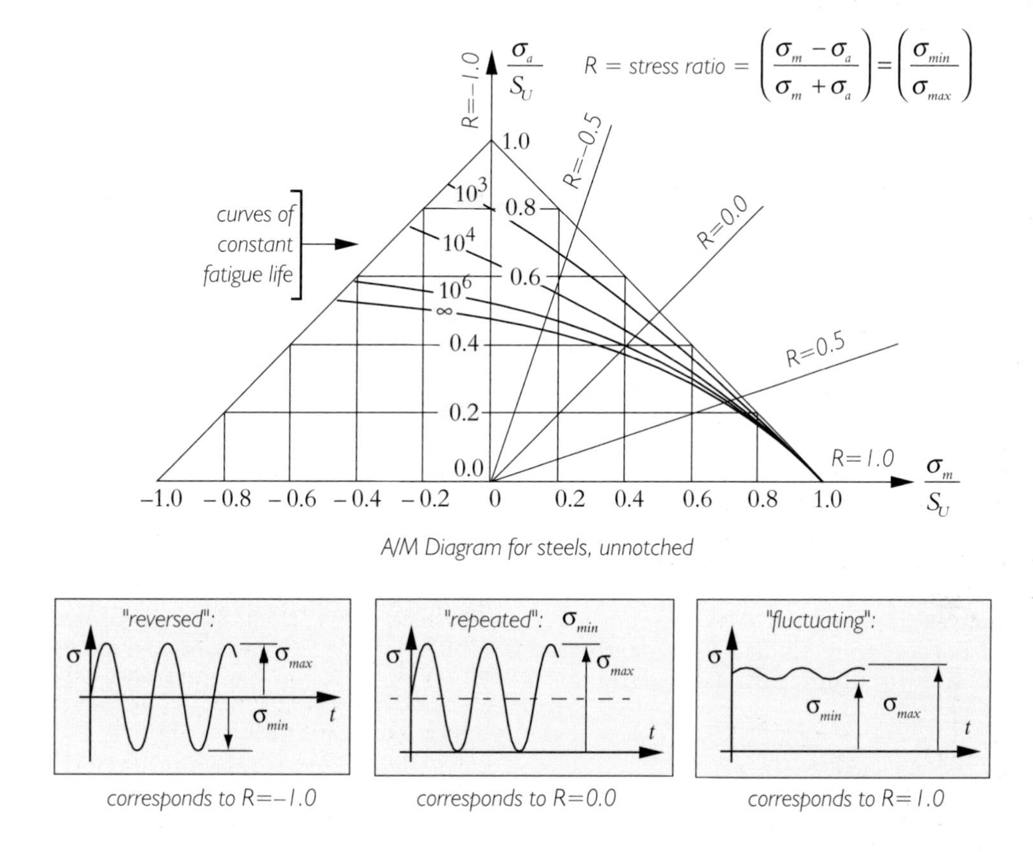

Figure 2.36 A-M diagram for smooth un-notched steels, showing load lines with varying stress ratios

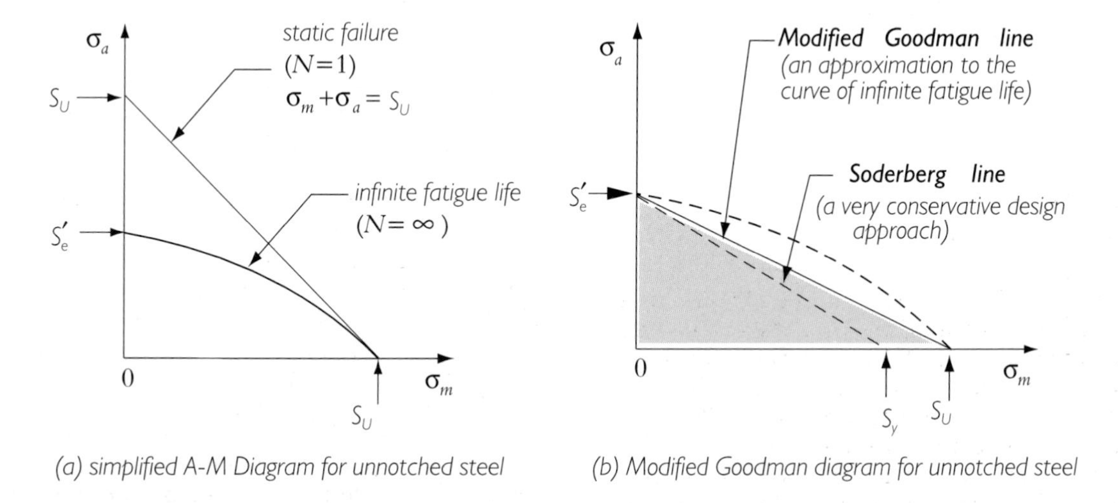

Figure 2.37 Approximations to the A-M diagram, used in fatigue design for infinite life

Goodman line in Figure 2.37(b) with one passing through S_y on the mean stress axis (the *Soderberg line*).

Practical design requires a somewhat less conservative approach to that suggested by Soderberg. This is shown on the *A-M* diagram of Figure 2.38, constructed for a steel component. Figure 2.38(a) shows the *modified Goodman* line constructed from infinite life data for a specific steel material. This line is further modified by stress concentration and surface finish factors K_f and K_s, resulting in a *fatigue failure line* for the steel component. The *yield failure line* is also shown on this diagram. Figure 2.38(b) shows the *A-M* diagram redrawn with both the fatigue and yield failure lines modified by a design factor of safety F_d, resulting in the shaded *safe design* space for this component. All designs with combined mean and alternating stress amplitudes falling inside this *safe design* space will survive infinite cycles of applications of the specified design loads.

An example of applying the *A-M* diagram to design will demonstrate its use for selecting component size in simple, uniaxial, fatigue applications.

Ex. 2.9

Figure 2.39 shows one end of a small rectangular marine diesel connecting rod. The loading on this steel component may be regarded as cyclic uniaxial loading. The design task is to determine the actual factor of safety, F_a, that has been applied to the 20 mm square section connecting rod. The material chosen for this application is structural grade mild-steel, with S_U = 600 MPa and S_y = 400 MPa.

The load P is estimated to fluctuate between 10 kN and 30 kN.

From a chart similar to Figure 2.31, but in this application for flat bars in cyclic axial loading, we estimate K_t as follows:

$r/D = 3/20 = 0.15$; and $D/d = 30/20 = 1.5$.

Hence, for a 600 MPa steel in reversed axial loading we find

$$K_t = 1.87.$$

From Figure 2.32 for this grade of steel we find the notch sensitivity index,

$$q = 0.82.$$

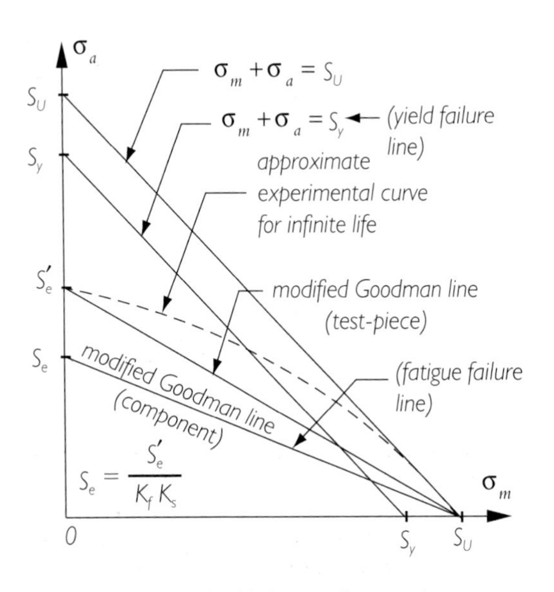

(a) Construction of A-M diagram for a steel component

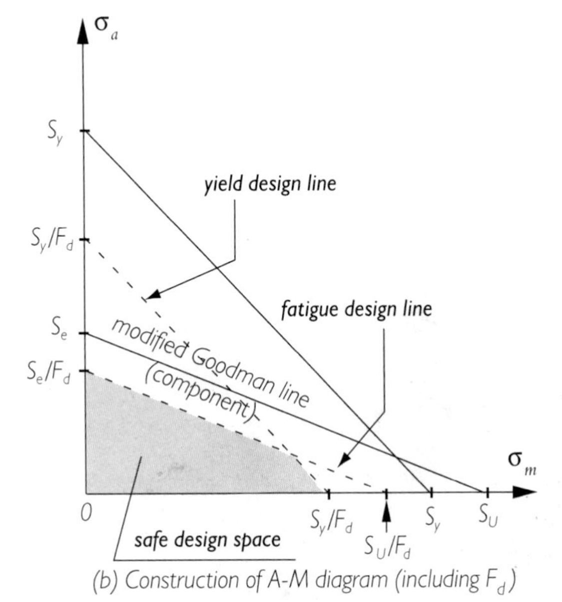

(b) Construction of A-M diagram (including F_d)

Figure 2.38 Constructing the A-M diagram for the design specification of a steel component

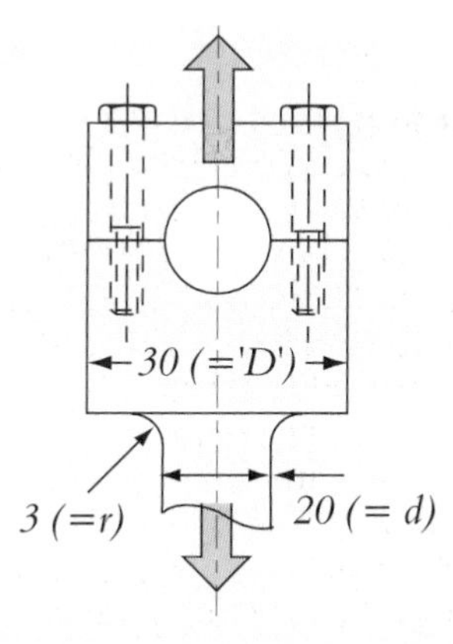

Figure 2.39 Mild-steel connecting rod (dimensions in mm, not to scale)

This leads to

$$K_f = 1 + q(K_t - 1),$$
$$= 1 + 0.82(0.87),$$
$$= 1.71.$$

We expect that this component will have a machined finish and from Figure 2.34 we estimate $K_s = 1.33$.

We estimate the endurance limit as

$$S_e' = 0.5 S_U = 300 MPa$$

$$\Rightarrow S_e = \frac{S_e'}{K_s K_f} = \frac{300}{1.33 \times 1.71} = 132 MPa \cdot$$

The remainder of this exercise is the graphic construction of the *A-M* diagram. From the applied loading we get

Mean load = (30 + 10)/2 = 20 kN,

Load amplitude = (30 – 10)/2 = 10 kN.

Hence, the slope of the load-line on the *A-M* diagram will be 1:2. It is absolutely crucial to recognise that the load-line has been constructed from *load data*, rather than *stress data*. Clearly, to establish the slope of the load line we require only the ratios of the two load or stress components and this ratio ($\sigma_m : \sigma_a = W_m : W_a$) is independent of the geometry (yet to be chosen) of our component. Figure 2.40 shows the *A-M* diagram constructed for this component.

Reading from Figure 2.40 the values of allowable stress in the 20 mm square rod are found to be

$$S_{a(all)} = 91.5 \text{ MPa and } S_{m(all)} = 183 \text{ MPa.}$$

[*Note*: these results can be found either by using an accurate plot of Figure 2.40, or by solving for the intersection of the two straight lines, the *modified Goodman line* and the *load line*. In most general practical designs, the graphical approach is recommended]

We can now compare the allowable stress amplitude and mean stress values to the actual stresses experienced by the component in service. These actual stresses are:

$$\sigma_a = 10\text{kN}/(20 \times 20 \text{ mm}^2) = 25 \text{ MPa, and}$$
$$\sigma_m = 20\text{kN}/(20 \times 20 \text{ mm}^2) = 50 \text{ MPa.}$$

Consequently we interpret these figures as corresponding to an actual factor of safety F_a of 91.5/25 = 183/50 =3.66 for both mean load and alternating load amplitude.

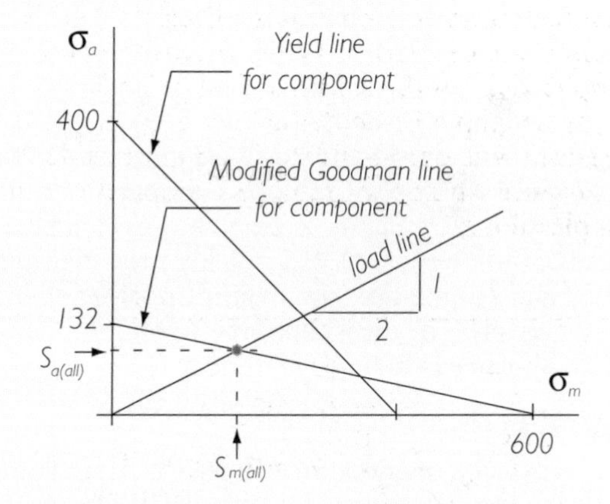

Figure 2.40 A-M diagram for connecting rod, Ex. 2.9

Ex. 2.10

Riveting is a common forms of joining process used where welding might cause thermal damage to the parent materials to be joined. A typical section through a riveted component is shown in Figure 2.41. A small riveting press is used in the aircraft industry for riveting aluminium components. The cycle of operation is to clamp the pre-drilled parts to be riveted, insert the rivet from a magazine and then apply a hammering process to *upset* the rivet head. The press and its cycle of operation is indicated schematically in Figure 2.42. Establish the stresses in the most highly stressed part of the press and select a suitable dimension for this component.

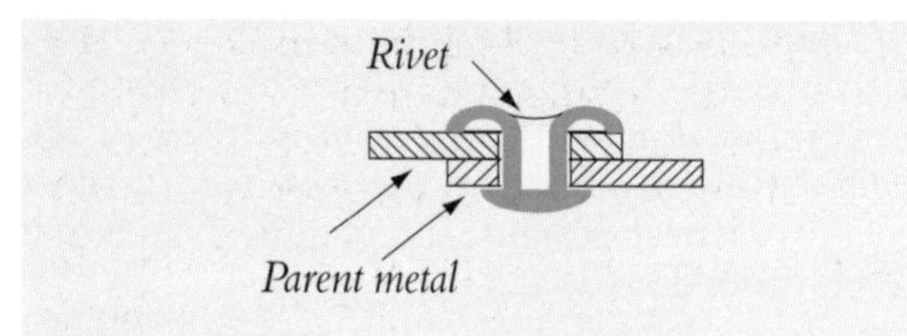

Figure 2.41 Riveted plates (schematic)

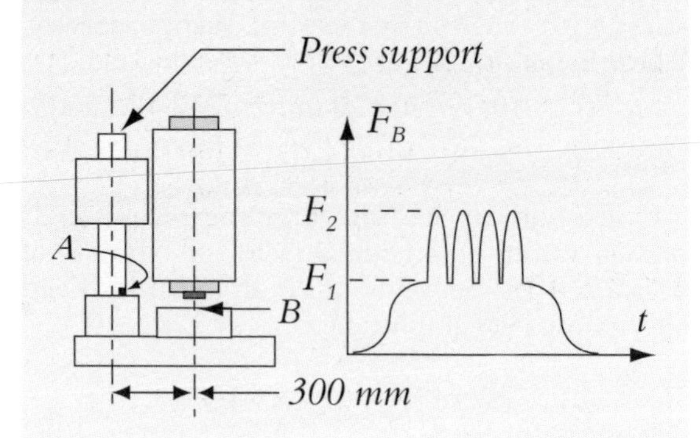

Figure 2.42 Riveting press and typical load cycle (schematic)

One of the most highly stressed locations on this press is at the base of the press support indicated at location A on Figure 2.42. The press load cycle is also indicated, with F_1 (= 1 kN) the clamping load and F_2 (=2 kN) the maximum riveting load respectively. The press support is a solid rod whose diameter needs to be determined.

Ex. 2.11

A stationery manufacturer needs a design for a stapler return spring. The material chosen is a spring steel with S_U = 1,100 MPa in the heat treated condition. Figure 2. 43 shows the proposed design of the spring, with the dimensions B and t to be chosen.

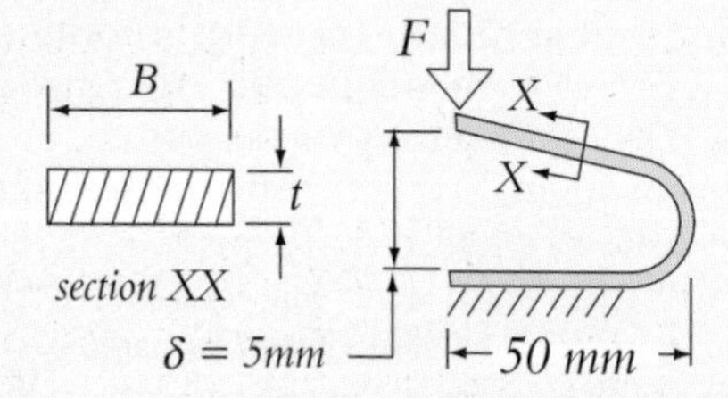

Figure 2.43 Stapler spring (schematic)

The return spring force, F, is required to be approximately 20 N, at the maximum deflection, of the spring $\delta = 5mm$. Select appropriate values for the dimensions of the spring. You may assume that $K_t = K_f = K_s = 1$ and $F_d = 1.2$.

2.6 Assignment on design to resist fatigue failure

The aims of this brief assignment are:

(a) to give practice in designing engineering components to resist failure when the applied load varies with time;

(b) to clarify notions of stress amplitude and illustrate various features of the A-M diagram; and

(c) to provide some experience of exercising judgement in the use of data.

2.6.1 Background

Modern bicycle wheels are complex structures consisting of a hub, a light (and therefore flexible) rim, and numerous wire spokes — the standard number of spokes being 36. Each of these is pre-tensioned by means of a small threaded nut inside the rim so that the wheel remains rigid, and so that net compression of any spoke is avoided under normal operating conditions. Spokes occasionally break after an extended period of service.

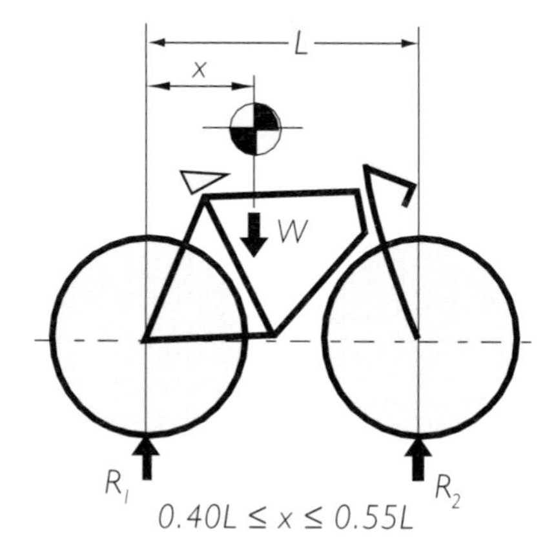

Figure 2.44 Bicycle frame and approximate load distribution

Your task will be to design the spokes to resist fatigue failure. Figure 2.44 shows a bicycle frame schematically, indicating the assumed location of its centre of gravity, for the purposes of this assignment. Figure 2.45 shows a bicycle wheel, indicating the loads imposed on it.

It is surprisingly difficult to analyse the forces inside the wheel structure. Nonetheless if you were to investigate the effect on the spokes of an applied radial load P between the hub and the ground, you would discover that changes in spoke forces are as shown in Figure 2.45. This shows that spokes directly above the hub experience an increase in tensile force of $\Delta T = \lambda_1 P$, whereas spokes directly below it experience a decrease in tensile force of $\Delta T = \lambda_2 P$. Spokes in between these positions experience intermediate changes in tensile force (with $\Delta T = 0$ for horizontal spokes). The values of λ depend on the number of spokes, the geometry of the spoke pattern used, and the dimensions of the hub, but for standard bicycle wheels

$$\lambda_1 = +0.04 \text{ and } \lambda_2 = -0.08.$$

Not all bicycle riders are the same shape or size. Figure 2.44 describes the range of positions over which a rider's centre of gravity can be located, and Table 2.5 below lists some data describing the range of body masses for a human population.

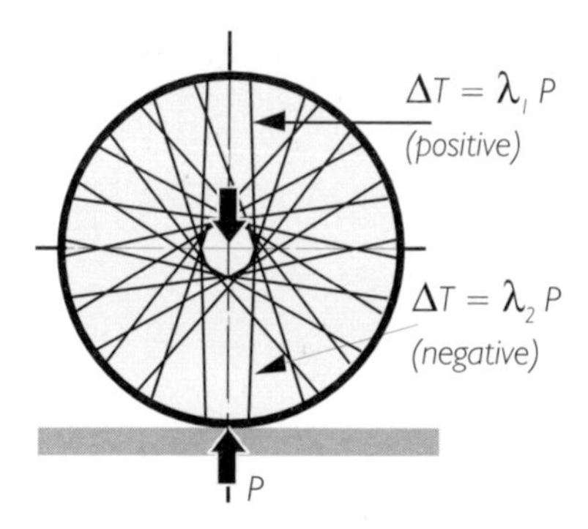

Figure 2.45 Wheel loads on bicycle

2.6.2 The design problem

(a) Determine the value of the initial (pretension) force T_o which would be needed to ensure that the tension T in the spokes is always greater than $0.3 \times T_o$.

(b) Sketch an accurately scaled graph of the stress cycle against time as the wheel is ridden through several revolutions. Label key quantities.

(c) Explain why an *A-M* diagram is needed to carry out a fatigue design for the spokes. Using data provided below, sketch the appropriate *A-M* diagram accurately to scale;

(d) Evaluate the slope of the load line corresponding to stresses on these spokes. What is the significance of the load line? How is it used?;

(e) Design the spokes (i.e. ascertain the required diameter). Use a factor of safety of $F_d = 2.0$.

(f) Discuss the effect of using a higher pretensioning force on the spokes.

2.6.3 Data

Table 2.5
Adult bicycle riding population

	Percentile		
	2.5%	50%	97.5%
Male	58	74	95
Female	43	61	89

(Average weights of U.S. adult population kg)

The wire material used to manufacture the spokes has the following properties:

$S_U = 880$ MPa $\qquad S_y = 760$ MPa

$S'_e = 450$ MPa $\qquad E = 200$ GPa

(c) Taking notch size to be equal to 0.5 mm, the notch sensitivity factor, q for this material may be taken from Figure 2.32.

(d) The theoretical stress concentration factor, K_t depends on the details of the rim and hub connections, since these are the locations of greatest geometrical discontinuity. Its value may be taken to be 2.6 in this case.

Engineering Design Associates		Date
Project Title :	**BICYCLE SPOKE DESIGN**	09:14:98
		Page 1 of 5
		Ckd. fdr

Design for "worst case" of 97.5 percentile male riding with centre of mass nearest to rear wheel ($x = 0{\cdot}40 \times L$).

$\Rightarrow \quad P = R_1 = W\left(\frac{L-x}{L}\right)$ (moments about front axle)

$= mg\left(1 - \frac{x}{L}\right) = 95\,\text{kg} \times \frac{9{\cdot}81\,\text{N}}{\text{kg}} \times (1 - 0{\cdot}4)$

$= \boxed{559\ \text{N}}$ — **P**

From given values of λ_1 [upper spokes] & λ_2 [lower spokes] we have

$\Delta T_{upper} = \lambda_1 P = +0{\cdot}04 \times 559 = \boxed{+22{\cdot}4\ \text{N}}$ — **ΔT_{upper}**

$\Delta T_{lower} = \lambda_2 P = -0{\cdot}08 \times 559 = \boxed{-44{\cdot}7\ \text{N}}$ — **ΔT_{lower}**

Hence

$T_{max} = T_0 + \Delta T_{upper} = (T_0 + 22{\cdot}4)\,\text{N}$

$T_{min} = T_0 + \Delta T_{lower} = (T_0 - 44{\cdot}7)\,\text{N}$

and $(T_{max} - T_{min}) = 22{\cdot}4 + 44{\cdot}7 = \boxed{67{\cdot}1\ \text{N}}$ — **$(T_{max} - T_{min})$**

[Note: $(T_{max} - T_{min}) = P(\lambda_1 - \lambda_2) = \frac{1}{2} T_a$, where T_a is the "load amplitude" on the spokes]

$\Rightarrow \quad T_a = \boxed{33{\cdot}5\ \text{N}}$ — **T_a**

(a) We require $T \geq 0{\cdot}3 T_0$ always (in order to provide some kind of safety margin against spokes going into compression & hence buckling) This is equivalent to requiring $0{\cdot}3 T_0 = T_{min}$

$= T_0 - 44{\cdot}7$ [from before]

$\Rightarrow \quad T_0(1 - 0{\cdot}3) = 44{\cdot}7$

$\Rightarrow \quad T_0 = 44{\cdot}7/0{\cdot}7 = \boxed{63{\cdot}9\ \text{N}}$ — **T_0**

(b) We can now sketch the graph of force T versus time, which is proportional to the graph of stress. [i.e. axial stress in the spoke $\sigma = T/A$, where A is the spoke cross-sectional area.]

Engineering Design Associates

Project Title: **BICYCLE SPOKE DESIGN**

Date 09:14:98

Page 2 of 5

Ckd. fdr

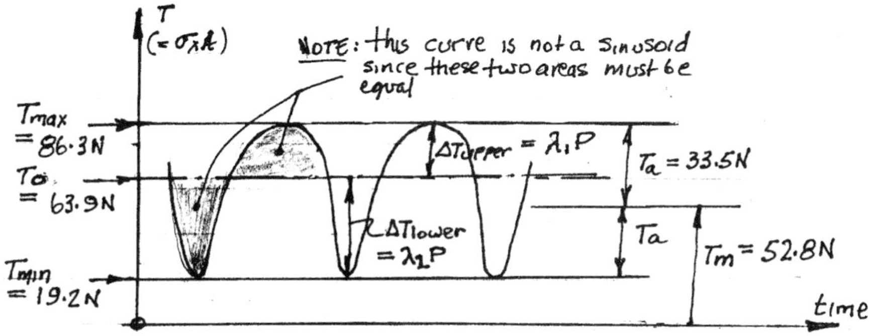

T $(=\sigma_T A)$

NOTE: this curve is not a sinusoid since these two areas must be equal

$T_{max} = 86.3N$

$T_o = 63.9N$

$T_{min} = 19.2N$

$\Delta T_{upper} = \lambda_1 P$

$\Delta T_{lower} = \lambda_2 P$

$T_a = 33.5N$

T_a

$T_m = 52.8N$

time

On the above graph we see that

$$T_m = \tfrac{1}{2}(T_{max} + T_{min}) = \tfrac{1}{2}(86.3 + 19.2) = \underline{52.8}$$

[NOTE: $T_m \neq T_o$ - this may be confusing for students]

(c) An A-M diagram ("Amplitude-Mean") is required in order to investigate fatigue resistance simply because the mean stress is non zero, and we need to take account of both the σ_a & σ_m values.

The following quantities require evaluation in order to sketch the A-M diagram:—

K_s : from graph, for $S_u = 880$ MPa ($= 0.88$ GPa), $k_a = 0.69$ (for cold-drawn components)

$\Rightarrow K_s = 1/k_a = 1/0.69 = \underline{1.45}$

q : from graph, for $S_u = 0.88$ GPa, $r = 0.5$mm

$q \approx 0.75$

K_f : $K_f = 1 + q(K_t - 1) = 1 + 0.75(2.6 - 1) = 2.2$

$$S_e = \frac{S_e'}{k_s k_f} = \frac{450 \text{ MPa}}{1.45 \times 2.2} = \underline{141 \text{ MPa}}$$

$S_e/F_d = (141/2.0) = \underline{71 \text{ MPa}}$

$S_u/F_d = (880/2.0) = \underline{440 \text{ MPa}}$

$S_y/F_d = (760/2.0) = \underline{380 \text{ MPa}}$

T_m

K_s

q

K_f

S_e

S_e/F_d

S_y/F_d

S_u/F_d

Engineering Design Associates

Project Title : **BICYCLE SPOKE DESIGN**

Date 09:14:98

Page 3 of 5

Ckd. fdr

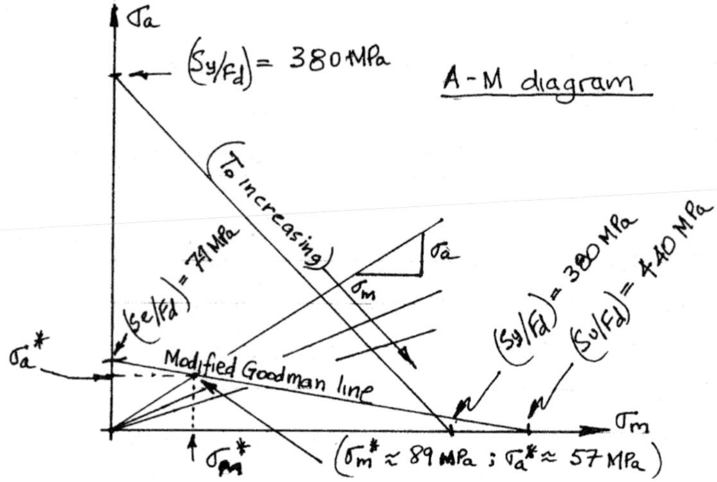

Note: Some students prefer to check the intersection of the load line $[\sigma_a = (T_a/T_m)\sigma_m]$ and the modified Goodman line $[\sigma_a + (S_e/S_u)\sigma_m = (S_e/F_d)]$. The intersection occurs at

$$\sigma_m^* = \frac{S_e/F_d}{(T_a/T_m) + (S_e/S_u)} \quad (\simeq 88.7 \text{ MPa}) \text{ and}$$

$$\sigma_a^* = (T_a/T_m)\,\sigma_m^* \quad (\simeq 56.2 \text{ MPa})$$

σ_m^*

The intersection point represents the largest possible values of σ_m & σ_a which will satisfy the design requirement of being smaller than that combination of σ_m & σ_a which would cause failure within a given safety factor.

(d) Load line : By definition the slope of the load line is (σ_a/σ_m)

$$\frac{\sigma_a}{\sigma_m} = \frac{T_a}{A} \times \frac{A}{T_m} = \frac{T_a}{T_m} = \frac{33.5}{52.8} = 0.634$$

$\left(\frac{\sigma_a}{\sigma_m}\right)$

The "load line" includes all points on the A-M diagram whose (σ_a, σ_m) coordinates form a ratio (σ_a/σ_m), which is the same as the ratio of "load amplitude" (T_a) to "mean load" (T_m).

Engineering Design Associates **Project Title :** **BICYCLE SPOKE DESIGN**	Date 09:14:98
	Page 4 of 5
	Ckd. fdr

[It is worth noting that for a component subjected to reversed bending, the load line slope is determined by the ratio of "moment amplitude" to "mean moment"]

By varying the cross-sectional geometry of our component (in this case A), we simply cause the (σ_m, σ_a) coordinates to move up or down along the load line. Hence, we can use the load line to determine the component geometry (in this case A), such that it will withstand the applied loads during its service life.

(e) From A-M diagram we have $\sigma_m^* = 89$ MPa

We require $\sigma_m \leq \sigma_m^*$ (Solid circular cross section $A = \pi d^2/4$)

$$\Rightarrow T_m \leq \sigma_m^* A$$

$$\Rightarrow A = \frac{\pi d^2}{4} \geq T_m / \sigma_m^*$$

$$\Rightarrow d \geq \sqrt{\frac{4\,T_m}{\pi \sigma_m^*}} = \sqrt{\frac{4 \times 52.8\,N}{\pi \times 89\,MPa}} \quad \left(1\,MPa = 1\,\frac{N}{mm^2}\right)$$

$$= \underline{0.87\,mm}$$

d

(f) In discussing the effect of increasing T_o, it is helpful to refer back to the load v. time graph. Notice that T_a is determined from ΔT_{upper} & ΔT_{lower}, which are controlled by the applied load P, and so T_a is independent of T_o. Hence, increasing T_o has the effect of increasing T_m by the same amount. The result of this change is to reduce the slope of the load line.

Engineering Design Associates

Project Title : **BICYCLE SPOKE DESIGN**

Date 09:14:98

Page 5 of 5

Ckd. fdr

At first sight we might feel that we have increased σ_m^*, and consequently we might expect a SMALLER cross-sectional area. However, increasing σ_o has increased σ_m (by the same amount), and when this increase is accounted for, we find that the area required to carry the load is in fact larger. [In our example doubling σ_o results in a cross-section diameter of approximately 1mm, or an increase of about 15% and an increase of about 32% in area] The increase in diameter is acceptable, since we would almost certainly use a wire diameter greater than 0.87mm found in (e).

This effect of increasing mean stress to reduce the influence of load fluctuation is quite subtle and since it is commonly used in engineering where the uncertainties of fatigue failures can not be tolerated [eg aircraft engine bolts] it is worth spending time in dealing with it.

2.7 Design for contact loading

Perhaps the most commonly recognised triaxial stress situation experienced in engineering is due to contact loading. Contact stresses arise in all engineering applications where two elastic bodies make contact for the purpose of transmitting loads. Typical examples are gear teeth and rolling element bearings. Heinrich Hertz, a German physicist (1857-1894), was the first to solve the problem of the stress field developed between two spheres in contact in 1881[2.9]. Eventually Hertz was responsible for early investigations of fatigue failures due to contact loading between bridge support plates and rollers and train wheels and rails. Additionally, he is also credited with the notion of hardness measurement determined by the indentation of a sphere into a plate. This was to become the hardness test eventually associated with the name of Johan August Brinell (1849-1925), a Swedish metallurgist, who first displayed his hardness testing machine at the Paris exhibition of 1900.

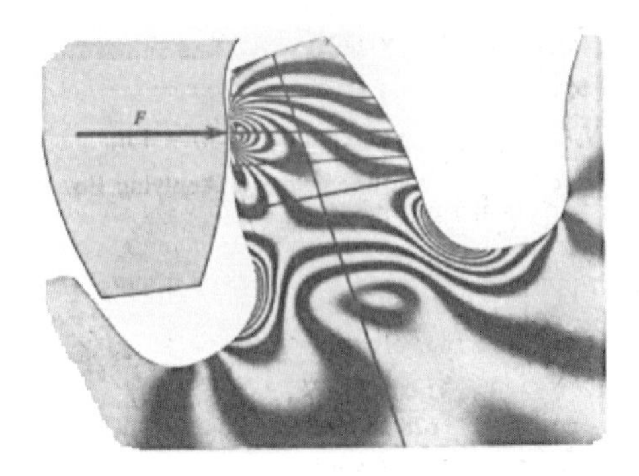

Figure 2.46 Photo-elastic model of gear teeth in contact. The density of fringes indicate stress level

Figure 2.46 is an indication of the nature of contact stresses experienced in gear teeth during normal operation. The photo-elastic model shows fringes increasing in density at the contact point, corresponding to a sharp local increase in stress there. Figure 2.47 shows two ball bearing races that have suffered surface damage due to contact loading. The upper race in this figure exhibits heavy surface damage resulting from uneven loading and due to shock loading from vibration while stationary. The lower race exhibits typical surface fatigue damage from rolling contact over an acceptably long service life. Figure 2.48, showing the wheels of a gantry crane rolling on a crane bridge, is an example of wheel-to rail contact.

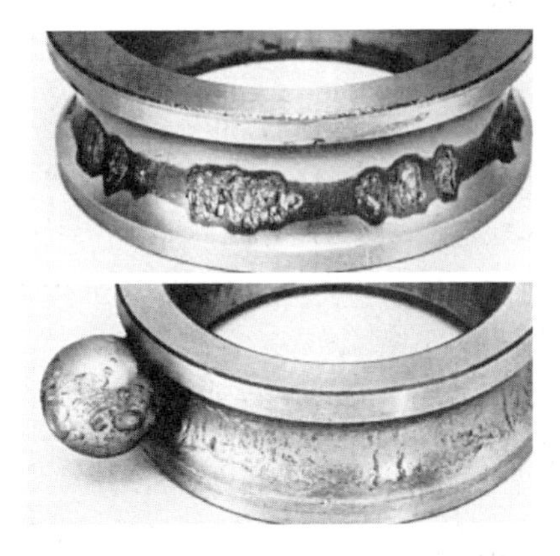

Figure 2.47 Ball bearing races showing surface damage failure resulting from contact loading. (After Braun, 1987)

Figure 2.48 Wheel and rail contact on crane bridge

These examples illustrate some of the practical situations where contact stresses can play a significant part in determining the structural integrity and ultimate safe service life of engineering components.

In order to understand contact phenomena, we classify the various practical engineering situations where these phenomena occur. As a precursor to contact stress analysis we need to classify the majority of contact load situations into generic

2.9 H. Hertz, *J. Math.* (*Crelle's J.*), vol. 92, 1881, quoted by Timoshenko, 1953 *loc.cit*

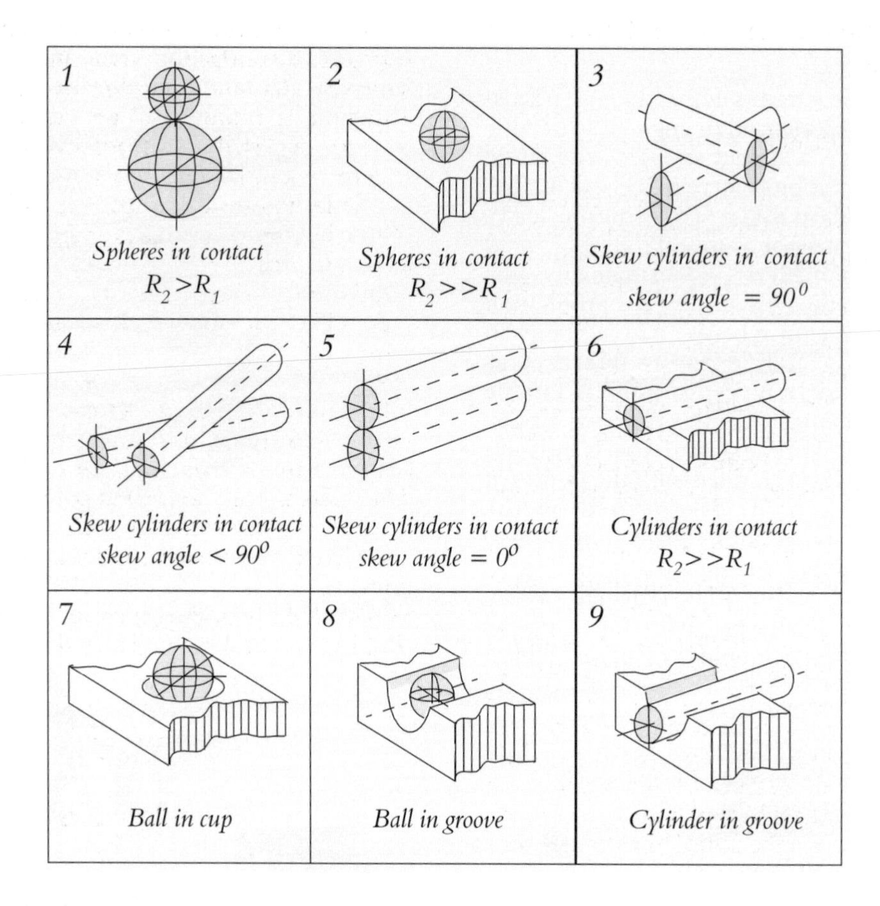

Figure 2.49 A generic classification scheme for contact phenomena

contact types. Figure 2.49 shows nine generic contacts categorised by both the analytical procedure used to derive contact stresses and by the special geometry of contact. Cases 1 to 3 represent *spheres* in contact, since in each case the contact area between the two bodies is a circle. Cases 4 to 6 are classified as a more general type of contact, with the two bodies represented by ellipsoids. As the larger semi-axes of the contacting ellipsoids tend to infinity, the contact surface degenerates to a rectangle.

Finally, cases 7 to 9 represent a special class of contact where the two contacting surfaces partially or wholly *conform* to one another. The result of this conformity of surfaces is manifested in significantly larger contact areas than that observed in the first six cases and commensurately reduced stresses over the contact area.

Clearly, cases 5, 6, 8 and 9 represent contact load cases commonly encountered in practice. Case 6 contact is equivalent to a wheel rolling on a rail (as seen in Figure 2.48), while case 8 is representative of contact loading experienced in ball bearings.

Ex 2.12

(a) Which case in Figure 2.49 represents gear teeth in contact?

(b) List ten engineering applications familiar to you, where contact loading may be represented by the generic cases shown in Figure 2.49. Match your contact load case to the ones shown in Figure 2.49. Can you identify a contact load case characterised by case 7 (ball in cup)?

2.7.1 The nature of contact between elastic bodies

(a) Spheres in contact

We examine the nature of contact between elastic bodies in terms of the stresses that develop and eventually lead to failure. We also examine the modes of failure and the design parameters that influence how failures occur. Failure in contact can result from quasi-static loading resulting in local plastic deformation. Appropriately, this mode of failure is called *failure in bearing* , and it includes a class of failures colloquially referred to as *brinelling*, in which one of the two bodies in contact partially embeds itself into the other, plastically deforming it.

Another common mode of failure due to contact loading occurs when one body impacts the other. Both of these modes of failure have been investigated by several workers (see for example Timoshenko and Goodier, 1982). Investigations in the theory of elasticity deal with pressures and displacements over the contact surface, as well as elastic impact between two spheres along their centre lines. Here we present a brief summary of the elastic solution to spherical bodies in contact, using the steps described in Timoshenko and Goodier.

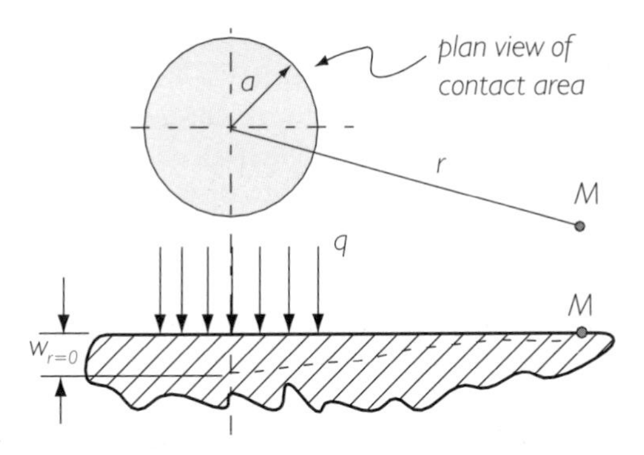

Figure 2.50 Elastic behaviour of a semi-infinite solid under the action of a uniform pressure 'q' over a circular boundary radius 'a'. (After Timoshenko and Goodier, 1982)

The more general case of two elastic ellipsoidal bodies in contact was also originally solved by Hertz. However the overall approach to all of these contact problems is the same. It entails an understanding of the distribution of deflection and stress that results from a concentrated force acting normal to the surface of a semi-infinite elastic body. These results for a point load (or, at least, a load acting on an infinitesimal area) are then extended to cases of distributed loads by the principle of superposition. In this process the effects of stress and deflection are integrated over the bounded area of the distributed load. The original problem of a constant pressure distribution was solved by a number of workers including Joseph Boussinesq (1842-1929), a student of Saint-Venant, Horace Lamb (1849-1934) and A. E. H. Love (1863-1940)[2.10].

Consider the elastic deformation of a semi-infinite solid under uniform contact pressure over a circular area of radius a, as shown in Figure 2.50. The deflection at the general point M at distance r from the centre of the bounding circle, over which q acts, is expressed in terms of the complete elliptic integrals (Jahnke and Emde, 1945). However, for a point on the bounding circle, or at its centre, these simplify to the following expressions for deflection w:

$$w_{r=a} = \frac{4(1-\mu^2)qa}{\pi E},$$

$$w_{r=0} = \frac{2(1-\mu^2)qa}{E},$$

where E and μ have their usual meanings of elastic modulus and Poisson's ratio respectively.

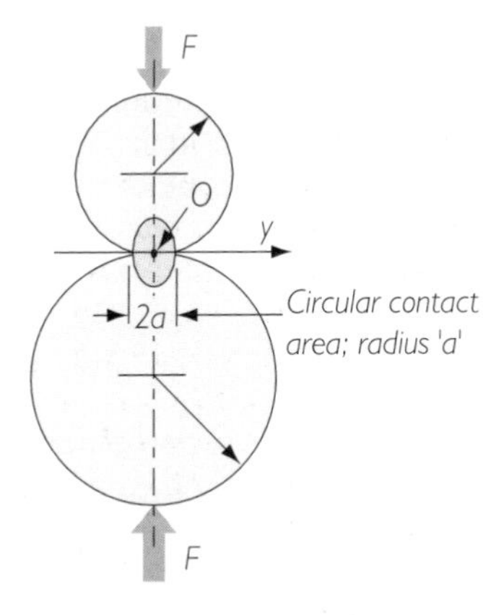

Figure 2.51 Two spheres in contact under the action of contact force F

Several related problems of engineering interest are solved in elastic theory, among them the case of a rigid *circular die* acting on a semi-infinite

2.10 Cited in Timoshenko and Goodier, *loc.cit.*

boundary (rather than the constant pressure distribution q dealt with above). This was the problem originally solved by Boussinesq. In this case the deflection is constant over the area of the rigid die and the resulting pressure distribution is of interest. Similarly, when two spheres come into contact under load, their contact surfaces will deflect so that every point of the contact surface on one sphere will come into contact with its corresponding point on the surface of the other sphere and the contacting surface will be common to both spheres. This situation is depicted schematically in Figure 2.51.

Geometry: Notation used in the analysis of contact behaviour of two spherical elastic bodies under the action of contacting force F is shown in Figure 2.52, which corresponds to the situation before F is applied.

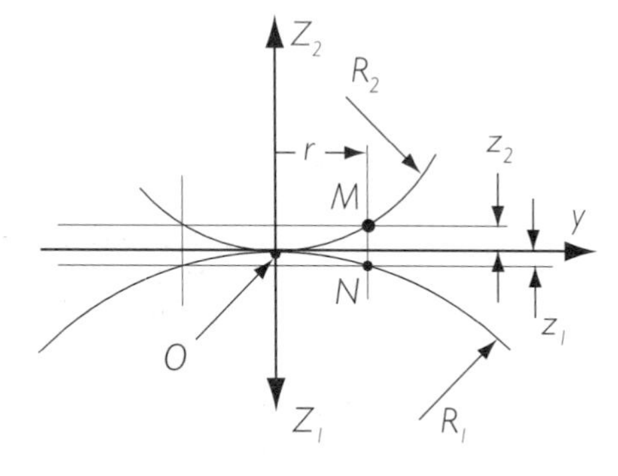

Figure 2.52 Enlarged view near contact point O

Here the area near the common contact tangent passing through O at the contact point has been enlarged for clarity. In the contact region where r, *and* consequently z_1 and z_2, are small relative to the sphere radii R_1 and R_2, we can approximate to show that

$$z_1 \approx \frac{r^2}{2R_1} \text{ and } z_2 \approx \frac{r^2}{2R_2}.$$

The total distance between the latitudinal planes passing through M and N on the two undistorted spheres is

$$z_1 + z_2 = \frac{r^2(R_1 + R_2)}{2R_1R_2} = \Gamma r^2,$$

where Γ is a factor depending on the geometry of the contact surfaces alone. Traditionally, Γ is defined such that R_1 and R_2 are positive when the surfaces are convex (as drawn in Figure 2.52):

$$\Gamma = \frac{R_1 + R_2}{2R_1R_2}. \qquad (2.1)$$

For flat surfaces, set $R_2 = \infty$, so that

$$\Gamma_{\text{flat}} = \frac{1}{2R_1}.$$

If one surface — say R_2 — is concave (forming a *spherical seat*), then although the definition of Γ in *(2.1)* is unchanged, it is now calculated as

$$\Gamma_{\text{concave}} = \frac{R_1 - |R_2|}{-2R_1|R_2|}.$$

Note that the resulting ratio will always be positive, since both numerator and denominator are negative, and $|R_2|$ must necessarily be greater than R_1 for a physically realisable contact.

Deformations: When the force F pushing the two spheres together is increased from zero, the contact between the two bodies will change from being a mere *point*, to being a *surface of contact*, having finite area and a tiny circular boundary. We designate the radius of this *surface of contact* as a, noting that a will be very small in comparison with R_1 and R_2. Clearly, a will increase in some fashion as F increases, and a pressure (having some yet-to-be-determined distribution) will be exerted across the *surface of contact*. We expect this distributed pressure will give rise to some local deformations and stresses in the spheres, and that these will attenuate at distances remote from the contact region.

Since a is so small compared to R_1 and R_2, it is reasonable to model the stress and deflection behaviour of the spheres in the vicinity of the contact region in terms of a distributed pressure on the flat surface of a *semi-infinite solid*. The deflection of the spheres' surfaces near the contact region will look just like that shown in Figure 2.50, except that the pressure distribution, q, will be non-uniform. Moreover, mathematical relationships relating a point force on the surface to the deflections everywhere in the *semi-infinite solid* can be employed here to solve the pressure and deflection distributions. These mathematical relationships present results for both stress ($\sigma_r, \sigma_z, \sigma_\theta, \tau_{rz}$) and deflections throughout a *semi-infinite body*. For example, the depression of the surface due to a concentrated load, dF, acting on the surface at some radial distance r from a point of interest is given by:

$$dw = \frac{dF(1-\mu^2)}{\pi E r}. \qquad (2.2)$$

Equation *(2.2)* is the basis of determining, by integration, surface deformations in *semi-infinite solids* (including our contacting spheres) due to distributed loads.

The following *thought-experiment* may be helpful in visualising the deflections of the spheres due to the localised contact pressure, q. Imagine that the spheres are in initial light contact as shown in Figure 2.52, and that they are being supported firmly at points remote from the contact region, so that the parts of these bodies that are far away from contact region can be regarded as being fixed in space. Then imagine that the distributed load, q, is imparted, equal and opposite, to the *surface of contact* (radius a) of each sphere, not by contact with its partner sphere, but by some equivalent agent interposed between the spheres. In this *thought-experiment*, the contact regions of the two spheres will separate, and the surfaces of the spheres will be locally deflected as shown in Figure 2.53.

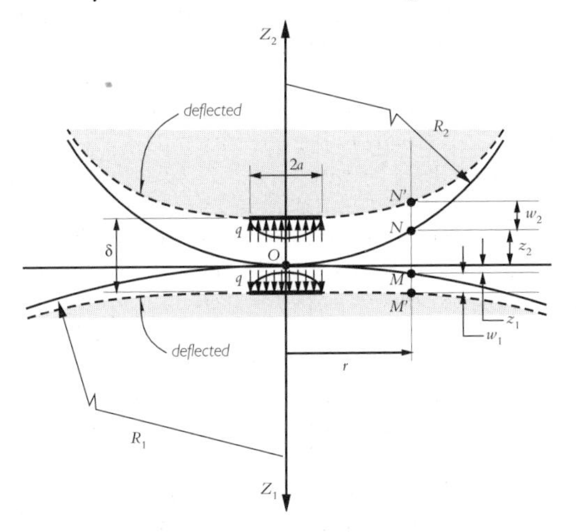

Figure 2.53 Sphere deflections due to contact

As the two spheres deform under the action of the contacting force, F, points like M will suffer local deformation w_1 in the direction Z_1. Similarly, points like N will suffer local deformation w_2 in the direction Z_2. Whereas the initial separation of points N and M was ($z_1 + z_2$), it has now increased to ($z_1 + z_2$) + ($w_1 + w_2$); this applies to any pair of points at radius r from the Z-axis. It is clear from Figure 2.53 that when we consider points lying on the surface of contact (i.e. for $r<a$), this distance ($z_1 + z_2$) + ($w_1 + w_2$) is equal to the net separation of the contact surfaces, δ. But this net *separation* is only a figment of our *thought-experiment* — the contact surface does not separate at all. In fact, δ is the net *approach* of the two spheres, as experienced by points far from the contact region. We can therefore state:

$$\delta - (w_1 + w_2) = (z_1 + z_2) = \Gamma r^2. \qquad (2.3)$$

We now examine the deflection of points inside the circular region of radius a. By integrating equation *(2.2)* over this surface of contact, it is readily shown that the net deflection between two points M and N is given by

$$(w_1 + w_2) = \frac{\kappa_1 + \kappa_2}{\pi} \iint q \, ds \, d\psi, \qquad (2.4)$$

where

$$\kappa_1 = \frac{1-\mu^2}{E_1} \text{ and } \kappa_2 = \frac{1-\mu^2}{E_2} \qquad (2.5)$$

are frequently-occurring constants that depend on the material of each sphere, and s and ψ are geometrical variables defined in Figure 2.54.

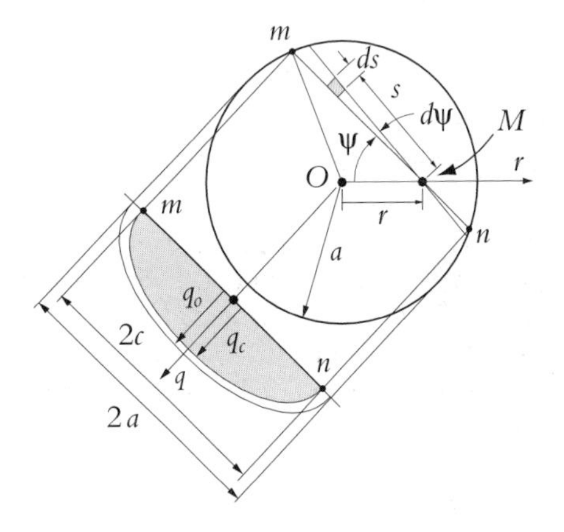

Figure 2.54 Integration of distributed forces

Hertz demonstrated that the pressure, q, must have a *semi-spheroidal* distribution over this surface of contact. The variation of q across each chord like mn is semi-elliptical, with the maximum pressure, q_o, acting at the origin.

After introducing the semi-spheroidal pressure distribution into *(2.4)*, integrating, and combining with *(2.3)*, we obtain the following results in terms of q_o :

$$w = \frac{1-\mu^2}{E}\frac{q_0\pi}{4a}\left(2a^2 - r^2\right) \qquad (2.6)$$

so, from *(2.3)*

$$\frac{\left(\kappa_1+\kappa_2\right)q_0\pi}{4a}\left(2a^2 - r^2\right) = \delta - \Gamma r^2 .$$

Since this equation is satisfied for any *r*, the assumed *q* distribution will be correct, as long as the terms agree. Therefore:

$$\delta = \frac{\left(\kappa_1+\kappa_2\right)q_0\pi a}{2} \qquad (2.7)$$

$$a = \frac{\left(\kappa_1+\kappa_2\right)q_0\pi}{4\Gamma} \qquad (2.8)$$

For the two spherical bodies, pressed together by a force *F*, we can calculate the value of q_0 , the maximum pressure, by integrating equation *(2.4)* over the contact surface and equating it to *F*. The resulting relation is

$$q_o = \frac{3F}{2\pi a^2}. \qquad (2.9)$$

The following simple example will illustrate the application of these contact equations.

Ex 2.13

A steel ball of diameter *D* metres is resting on a flat steel surface, as shown on Figure 2.55. Both the ball and the surface area of the same material, mild steel, with E = 210 GPa and μ = 0.3. Find the value of *D* such that the maximum pressure under the ball just exceeds 300 MPa (the local yield strength of the material).

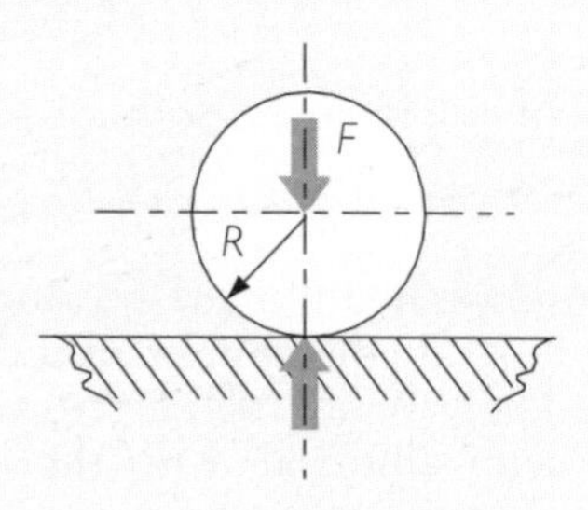

Figure 2.55 Steel ball on flat steel plate

The density of steel is 7.8 tonne/m³, so we get for the mass of the ball *m* and the resulting force *F*

$$m = 7.8\times10^3\frac{\pi D^3}{6}\text{kg}; F = 9.81m \text{ N},$$

and hence, from *(2.9)*

$$q_o = \frac{3F}{2\pi a^2} = 1.914\times10^4\frac{D^3}{a^2}. \qquad (2.10)$$

In this case both the ball and the surface are of the same material and we get for the elastic constants in *(2.5)*

$$\kappa_1 = \kappa_2 = \kappa = \frac{1-0.3^2}{210\times10^9} = 4.33\times10^{-12}\left[\frac{\text{m}^2}{\text{N}}\right].$$

The geometric constant Γ depends only on the values and signs of the radii R_1 and R_2 of the two spheres in contact, as expressed by *(2.1)*. In this case both radii are of the same sign and since for the flat surface $R_2 \to \infty$, we get

$$\Gamma = \frac{1}{D}.$$

From (*2.8*)

$$a = \frac{\kappa q_o \pi D}{2} = 6.802\times10^{-12} q_o D. \; (2.11)$$

Eliminating *a* from (*2.10*) and (*2.11*) above, we get

$$q_o = 7.45\times10^8 D^{4/3},$$

and for q_o =300 MPa, D ≈ 0.5 m.

(b) general contact between ellipsoidal bodies

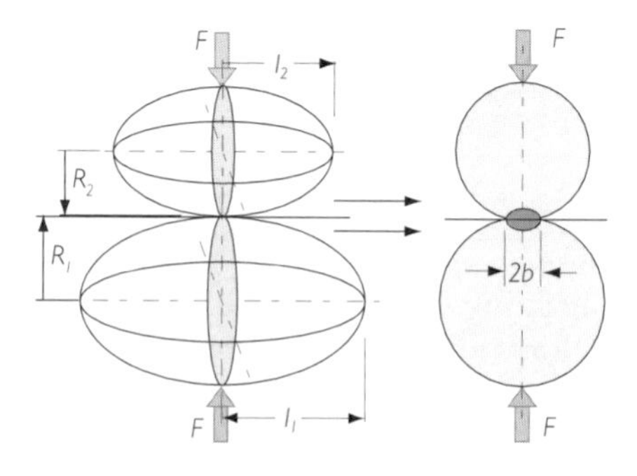

Figure 2.56 General ellipsoidal bodies in contact

This problem was solved by Hertz in 1881 and the solution to this more general result has practical application to skew cylinders in contact, as exemplified by case 4 in figure 2.49. A typical example of such a case is offered by *crowned* helical gear teeth. However, most commonly the solution of this problem is applied to cylinders rolling on other cylinders, with axes aligned, or cylinders rolling on a flat surface. Figure 2.56 shows the two ellipsoids in contact.

As the two semi-axes, l_1 and $l_2 \Rightarrow \infty$, the two ellipsoids degenerate to two contacting cylinders of radii R_1 and R_2 respectively, as shown in Figure 2.56. In this process, the area of contact degenerates from an ellipse to a rectangle of half-width b. Then by arguments similar to those used for finding the contact area and compression for two spheres, the following equations are derived for the two cylinders:

$$b = \sqrt{\frac{4F(\kappa_1 + \kappa_2)R_1R_2}{\pi\ell\,(R_1 + R_2)}}, \qquad (2.12)$$

$$\delta_c = \frac{F(\kappa_1 + \kappa_2)}{\pi\ell} \times \left\{\frac{2}{3} + ln\frac{4R_1}{b} + ln\frac{4R_2}{b}\right\}^{2.11}. \qquad (2.13)$$

Here F/ℓ is the force per unit length applied to the two cylinders in contact, κ_1 and κ_2 are the elastic constants defined earlier for two spheres in contact, and δ_c is defined for cylinders, similarly to spheres, as the approach distance of the two cylinders during the elastic deformation. The distribution of pressure over the contact rectangle of width $2b$ is found to be elliptical, and integrating this pressure distribution over the contact surface results in

$$q_o = \frac{2F}{\pi\ell b}, \qquad (2.14)$$

where q_o is the maximum pressure developed on the contact surface.

(c) Stress fields generated by contacts

Having so far discussed only the nature of the contact between the two bodies, we now need to examine the resulting stress fields and their effects in design.

Figure 2.57 shows the stress fields developed in the contact region for both spheres and cylinders in contact. For spheres in contact, the shearing stress τ_{max}, on which design is normally based, occurs slightly below the surface. Its value for μ = 0.3, is $0.31q_o$ at a depth of approximately $a/2$. For cylinders (μ=0.3) in contact, the maximum shearing stress $\tau_{zy} = 0.304q_o$ occurs at $z = 0.78b$.

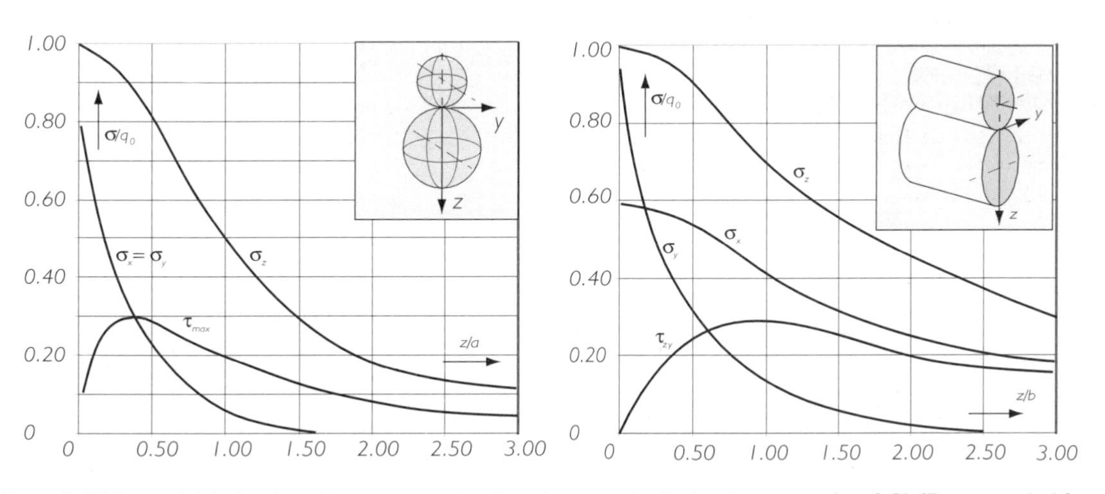

Figure 2.57 Stress fields developed in contact region for spheres and cylinders in contact (μ=0.3) (Figures scaled from Timoshenko and Goodier, 1985)

2.11 See, Warren Young (1989) *Roark's Formulas for Stress and Strain*, New York:McGraw Hill

Stress field calculations are based on the theory of elasticity (see for example Timoshenko and Goodier, 1985, articles 138 and 139), and the first to calculate these stress fields for spherical bodies in contact were M. T. Huber and A. N. Dinnik in 1904 and 1909 respectively[2.12].

In both the spherical and cylindrical contacts, significant tensile stresses develop at the periphery of the contact surface. In design situations, where brittle materials are used, these may be the governing stresses, rather than the compressive or shear stresses developed on the central axis of the contact region.

2.7.2 Some applications

Contact phenomena are of greatest interest in rolling element bearings, railed vehicle wheel design and in gear tooth contact. As seen from equation (*2.1*), contact geometry plays a significant part in the pressures and stresses developed in the contact region. The value of the geometry variable, Γ, will be determined by the class of contact defined in Figure 2.49. Referring to that figure, we note that for contact cases 1 to 6 the contacts are *non-conforming*. In these cases the contact radii of curvature are in opposite directions, and hence, according to the derivation used for Γ, of the same sign. Cases 7 to 9 are *conforming* contacts, and in these cases, due to the curious derivation used, the radii of curvature have opposite signs. The overall effect is that *conforming* contacts result in larger areas of contact and consequently, reduced local pressures and stresses. Typically, for two cylinders in *non-conforming* contact where $R_1 = 2R_2 = R$, we find

$$\Gamma = \frac{(R_1 + R_2)}{2R_1R_2} = \frac{3}{4R},$$

and from (*2.12*)

$$b = \sqrt{\frac{2F(\kappa_1 + \kappa_2)4R}{3\pi\ell}}.$$

Similarly, for two cylinders in *conforming* contact ($R_1 = 2R_2 = R$),

$$\Gamma = \frac{(R - 2R)}{2R(-2R)} = \frac{1}{4R}, \text{ and}$$

$$b = \sqrt{\frac{2F(\kappa_1 + \kappa_2)4R}{\pi\ell}}.$$

The two alternative arrangements are shown in Figure 2.57.

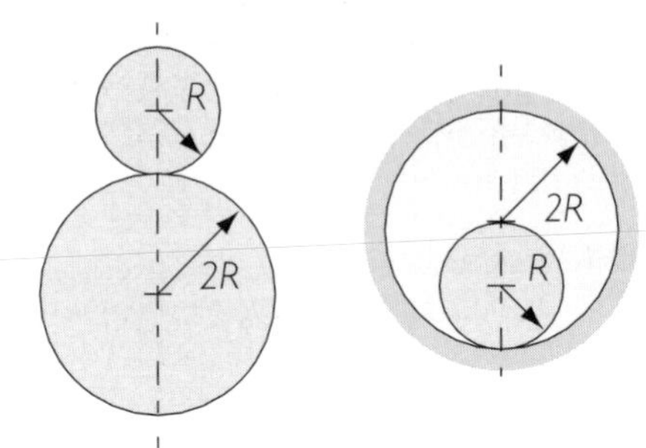

Figure 2.58 Two cylinders in 'non-conforming' and 'conforming' contact

Clearly, the area of contact is larger for the *conforming* contact, and the resulting stresses are lower for this case. This is the type of contact improvement applied by Ernest Wildhaber in 1926, when he conceived and patented the notion of convex-concave gear profiles. Later, in 1956, M. L. Novikov published work on similar conforming profile gears, and although the two gear profiles have some geometric differences, they have become known as Wildhaber-Novikov gears (Ishibashi and Yoshino, 1985).

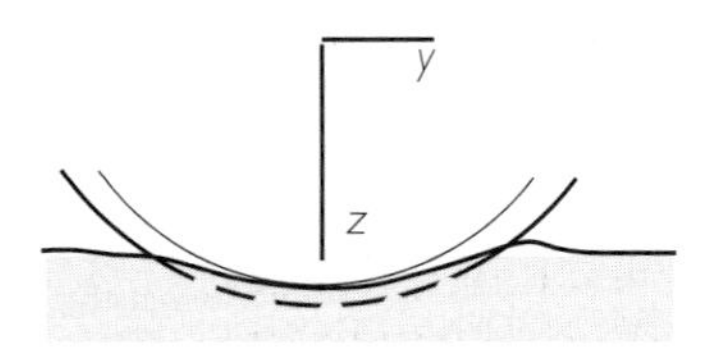

Figure 2.59 Schematic representation of a ball under load rolling on a flat surface (After Palmgren, 1959)

In rolling element bearing applications the pioneering work on contact phenomena is due to Arvid Palmgren (Palmgren, 1959). Figure 2.59 is a schematic interpretation of a ball, under load, rolling on a flat surface. This condition is typical of all rolling element bearing behaviour. Once rolling commences, a small *bow-wave* of material is generated ahead of the rolling element, as seen

2.12 Cited in Timoshenko and Goodier, *loc. cit.*

in Figure 2.59. During motion, the rolling element is continually *climbing the bow-wave* (in other words, the resultant contact force has a slight rearward component resisting the motion), resulting in rolling friction and heat generation. When the rolling element is a ball, such as in ball bearings, uneven friction on the surface of the ball makes it rotate about an axis not perpendicular to the plane of bearing rotation, causing further rolling friction and efficiency loss.

Ex. 2.14

(a) The front wheel of a bicycle used for mountain racing has a steel hub of 26 mm internal diameter. The designer of the bearing in this case decides to use 3 mm diameter steel balls to carry the load (there are two bearings in each hub). Figure 2.60 shows one half of the hub schematically.

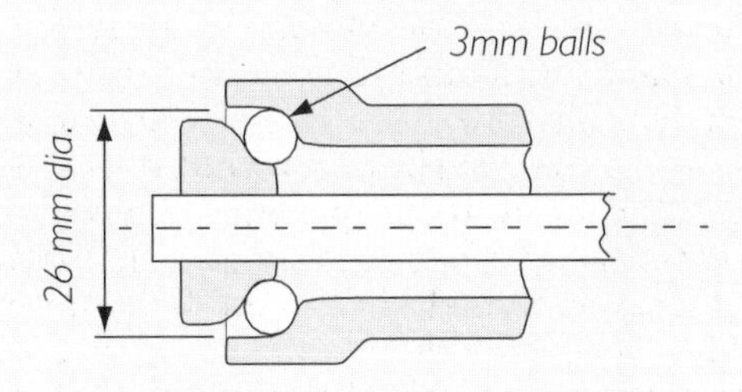

Figure 2.60 One half of a mountain bike hub (schematic only)

You may assume that the bearing arrangement is a 45° angular contact. Estimate the failure load for this bearing and compare your result with the data for human riders given in the assignment of Section 2.6. What would be an appropriate factor of safety for this application.

(b) A light rail vehicle has 8 steel wheels, 0.5 m diameter each. You are expected to select an appropriate wheel and rail material for this application. The load to be carried is estimated at 5 tonne. Select an appropriate material and estimate a suitable factor of safety.

(c) In an instrument, three identical cylindrical aluminium rods, with hemispherical ends, are used to support the 200 kg instrument on a flat steel surface. Calculate the appropriate diameter for the hemispherical ends of the aluminium rods. You may assume that each of the three legs of this instrument support carry equal weight.

Ex. 2.15

Figure 2.61 is a schematic view of a type of sliding door track. The track material is aluminium and the balls are similar to those used in steel ball-bearings (usually manufactured from very high tensile strength materials, hardened and ground. Each track has four such balls supporting the door. Your task is to determine the maximum weight of door these tracks are capable of supporting without failure.

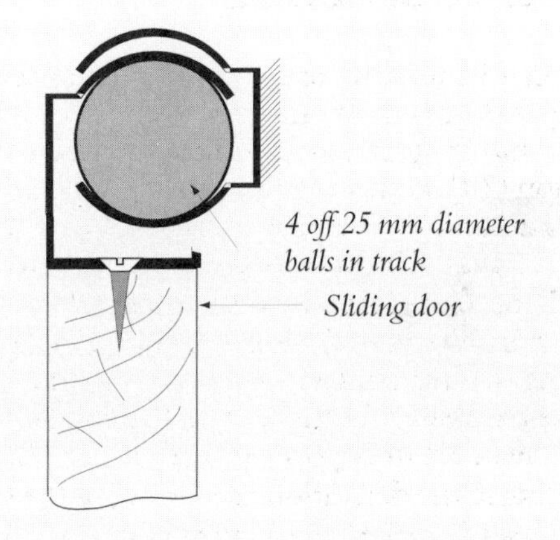

Figure 2.61 Schematic section through sliding door track.

Ex. 2.16

A steel ball rests on a flat surface. Determine the diameter of the ball needed to just yield the surface on which it rests. Carry out the same determination for steel (S_y=600 MPa), titanium (S_y=400 MPa), aluminium (S_y=100 MPa).

Ex. 2.17

Some delivery vans use sliding doors to allow a rapid entry into the vehicle for passengers and goods. Figure 2.62 is a schematic view of a proposed sliding door support system for such a vehicle (a postal van). The door weighs 90 kg and it will be supported on two sets of glide rollers rolling on the top of the door track. Two further sets of glide rollers, on the bottom of the door track are used to stabilise the motion of the door during operation, but provide no vertical reaction.

You have been commissioned by an auto parts supplier to design the rail and rollers for this application. You should use readily available cheap materials, typically *mild steel* (CS 1040 or similar), perhaps case hardened rollers if necessary.

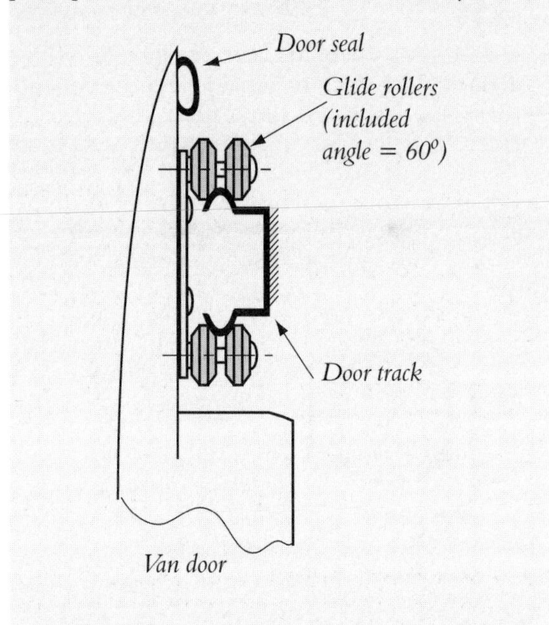

Figure 2.62 Proposed delivery van sliding door system

2.8 Chapter Summary

In this chapter we have examined material properties and failure mechanisms as well as the complex interactions between materials, loads and environmental variables that need to be considered in design. We have briefly described common engineering materials and their uses and we presented three important failure predictors to be used in design for static loading, namely

- the maximum principal stress failure predictor (*MPFP*) for brittle materials;
- the maximum shear-stress failure predictor (*MSFP*) for ductile materials; and
- the distortion energy failure predictor (*DEFP*), where a more precise selection of material size is required for a ductile component.

In design applications where time-dependent loads are imposed on components, we have presented an introduction to design against *fatigue*. Finally, we have briefly introduced *Hertz theory* and described a range of important contact phenomena.

It is important to recognise that we have focused our attention on introducing simple design procedures for a range of commonly encountered situations in the design specification of engineering components.

3

DESIGN SYNTHESIS OF GENERIC ENGINEERING COMPONENTS

Actioni contrarium semper et aequalem esse reactionem: sive corporum duorum actiones in se mutuo semper esse aequales et in partes contrarias dirigi. Newton, Principia Mathematica (1687) [To every action there is always opposed an equal reaction: or, the mutual actions of two bodies upon each other are always equal, and directed to contrary parts]

Having prepared ourselves with a survey of the relevant notions of failure predictors, modes of failure, and the mechanics of rigid bodies, we are now ready to embark upon the designer's journey of discovery. This journey will take us to the heart of what engineering designers do: the design synthesis of real objects.

The activity is called *synthesis* because it entails the bringing together of ideas and specific choices about the solution to a problem so that something new (the plans describing the solution, or in common parlance, the *design*) is created. Design synthesis begins by being broadly conceptual, concentrating on the overall type or form of the solution. The designer continues the process of synthesis by thinking and making decisions about the solution in greater and greater detail, considering its configuration and the arrangement of its parts. Ultimately someone (the designer or his/her delegate) will take the process of design synthesis to its very conclusion by specifying all aspects of the solution (including its shape, material and method of manufacture) in a highly detailed, integrated fashion. At that stage, someone else can take on the task of physical synthesis (that is, making/constructing/manufacturing the designed object).

Being involved in the process of design synthesis can be very satisfying — even exciting! There are few professional experiences more rewarding for the design engineer than seeing some real artefact, which began as a twinkle in their mind's eye or a sketch on their note pad, entering the world.

But we must begin at a humbler, more prosaic level. The emphasis in this chapter and the next will be on the design synthesis of just the basic kinds of structural component introduced in Section 1.3 (Structural distillation): *columns*, *vessels*, *bolted-joints* and the like.

The procedure for design synthesis of engineering components involves the following stages:

— establish the loads to be resisted;

— identify the geometric constraints imposed on the component;

— select candidate materials and geometries that seem appropriate to the component to be synthesised;

— develop a mathematical model that will realistically represent the interaction between the imposed loads and the resisting forces and moments offered by the component;

— identify credible modes of failure and apply the relevant failure predictors and design rules in choosing the final component geometry;

— wherever possible, combine design variables into design parameters to allow some optimisation of the final component.

All of these steps require the exercise of design judgement. This is a kind of design wisdom developed through experience. Perhaps the most significant influence on the development of design judgement is practice in design. In the words of golfing great Gary Player, remarking on a 'lucky' golf shot, *"the more I practise, the luckier I get"*.

We seek to develop some simple rules for the design of a range of generic engineering components. These rules will be derived from simple physical and mathematical models of the behaviour of various basic components in an *approximately first-principles* manner. In each case we will adopt the general design method for structural design, as outlined in earlier chapters. In some cases (see for example the discussion of design against axial tension in Section 3.4.2) we will make that general design method explicit. At every step our goal is to build the confidence of the novice designer, and to *de-mystify* the process of design for structural integrity.

3.1 Shafts: bending and torsion

The first of our generic engineering components is the shaft. Shafts transmit torque between rotating components of machinery. They are ubiquitous in engineering practice. In general, the shaft is to mechanical transmission of mechanical energy what the electric wire is to the transmission of electrical energy.

3.1.1 Historical notes

The invention of the wheel did not automatically imply the invention of the shaft. The first wheels were used to grind grains, and had no shafts, but instead were rotated directly by hand. Even the application of the wheel to transport did not bring the shaft into human history — the first wheeled vehicles were pushed or pulled by beasts (probably people!), and possessed *axles* rather than *shafts*. Eventually, wooden shafting appeared as societies developed technologies in which mechanical power needed to be transmitted over distance from energy sources such as animals, wind or flowing water.

Shaft design became an important matter during the Industrial Revolution[3.1]. The advent of steam power and reciprocating engines to drive massive arrays of machinery, rail vehicles and ships, meant that shafts were required to withstand larger power densities, of a more dynamic nature, and over a larger numbers of cycles, than ever before.

Significantly earlier than the Industrial Revolution, George Bauer, the German metallurgist who took the name of Agricola, published sketches of water driven lifting equipment used for de-watering mines.[3.2] A typical example is shown in Figure 3.1. The balls of rag, marked G, would *mop-up* the water and when lifted to the surface would be squeezed dry. In the figure, the shaft used is a square section piece of timber, marked A.

Steam engines were the main forms of prime movers used in the new industrialised communities of England in the early 19th century. Major contributions to this development came from:

- James Watt (1736-1819), inventor of the modern steam engine;
- Matthew Boulton (1728-1809), who financed Watt's engine development;
- Richard Trevithick (1771-1833), the first to apply high pressure steam power to locomotion (1801); and
- George Stephenson (1781-1848), the inventor of the railway locomotive (1825) whose famed *Rocket* is shown in Figure 3.2.

Figure 3.1 Dewatering system for mines (after Agricola, 1530)

3.1 The term generally refers to the change from handicraft and cottage industry, to manufacturing in factories (see for example McPherson, 1944, Milner, 1977, Gale, 1952)

3.2 Bauer, G. (Agricola) 'De re Metallica', dealing with mining and smelting, was published in 1530.

While locomotives and most other transport vehicles made use of axles (a shaft with no torque being transmitted), it was mainly in the energy transmission industry that shaft design had its genesis.

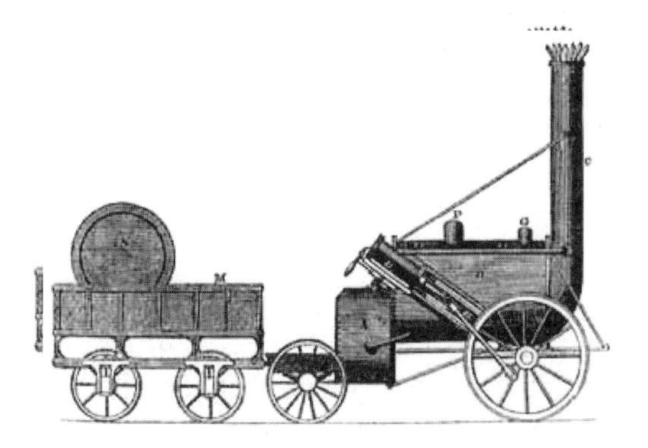

Figure 3.2 George Stephenson's Rocket, the winner of the first railway locomotive competition held on the Liverpool-Manchester line in 1829. The Rocket reached a speed of 58 km/hr. (Figure reprinted by permission from the archives of the Science Museum London)

As indicated in Agricola's drawing (Figure 3.1), the transmission of energy from a primary source, water or steam, uses the shaft to deliver the energy in a form useful to do work. It is entirely due to the widespread utility of shafts for the transmission of mechanical energy that shaft design has become an essential part of the designer's toolkit. As a contrast to these early examples, Figure 3.3 shows a modern application of shafts in a compact two-stage gearbox.

Figure 3.3 Two stage gearbox (David Brown, Series 'H'-heavy duty speed reducer)

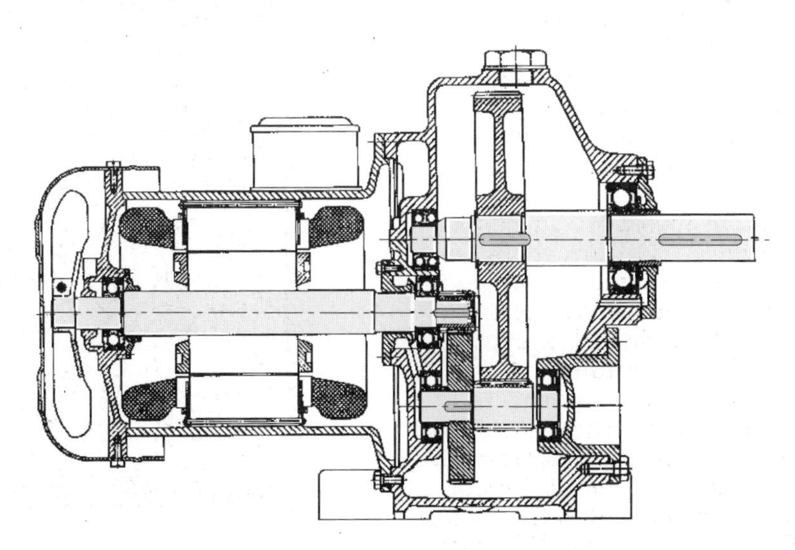

Figure 3.4 Close coupled electric motor

3.1.2 Factors affecting shaft design

Let us immediately consider some examples of engineering hardware to get a feel for the structural issues confronting us in shaft design.

Example 1 — gearbox

Figure 3.4 shows a cross-sectional elevation through an electric motor close-coupled to a two-stage reduction gearbox. There are three shafts in this system. We consider them one by one.

The first shaft, on the left, serves several purposes: it supports the electric motor's armature winding between bearings; it supports the motor's air cooling fan at its left-hand end; and it serves as the input shaft to the gearbox at its right-hand end. As such, it carries a small input gear, called a *pinion*, to the right of its right-hand bearing, and is fastened to it by means of a *key* mounted in a *keyway*. The *pinion* meshes with a larger diameter gear fixed to the second of the three shafts — the intermediate shaft of the gearbox.

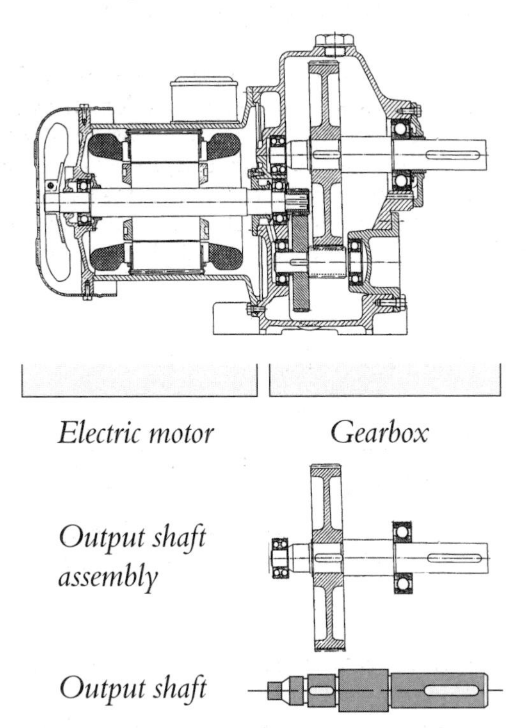

Figure 3.5 The geared motor 'dissected'

This shaft has gear teeth machined around its midsection to form the second-stage pinion. This pinion meshes in turn with the larger second-stage gear that is attached to, and rotates with, the larger output shaft protruding to the right of the gearbox. It is instructive to identify each of the three shafts, the six bearings, the two pinions and the two gears in Figure 3.4.

Focusing now on the *output shaft*, we ask a typical, pessimistic, structural question. Where will the output shaft break? There are several factors bearing on this question, and the would-be designer needs to develop a feel for them almost intuitively. Firstly, consider the loads on the shaft. They are of two types: torsional and transverse. When gears mesh together to transmit power, they develop large radial forces that tend to separate the gears, so that the output shaft in Figure 3.5 experiences a large upwards force component from the gear as it is pushed upwards by the meshing pinion beneath it. Moreover, there will be a tangential force on the gear tooth as it transmits the torque from gear to pinion. This tangential force results in an equal and opposite force acting horizontally on the central plane of the shaft. The two horizontal forces form a *couple* corresponding to the torque being transmitted. The shaft responds to these forces in much the same way as a beam, by developing internal bending moments to resist the forces. The two sets of transverse forces (the applied loads and their reactions) are indicated schematically in Figure 3.6.

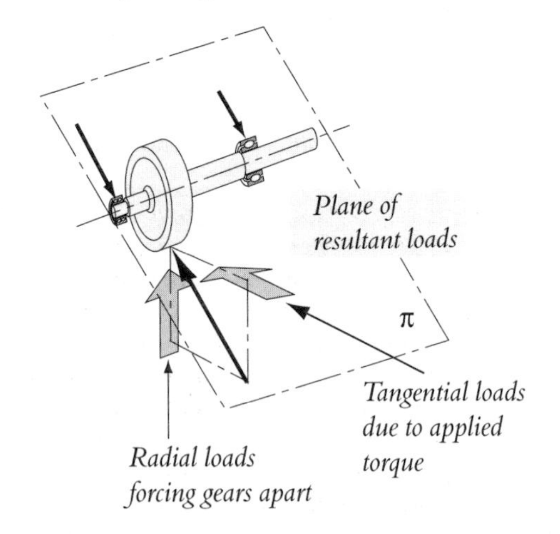

Figure 3.6 Resolved loads on output shaft

The transverse loads combine to generate one set of reactions on the bearings, as indicated in Figure 3.6. For this rather simple case of gear tooth loading, all the *action* of shaft forces and bending moments takes place in the resultant plane π. Hence, so far as the shaft is concerned, it can only *see* the resultant bending moments in this plane. Since the bearings can not take any substantial part in resisting the applied moments, as a conservative approximation we will assume that for transverse loads the shaft behaves essentially as a simply supported beam.

The bending moments acting in the resultant plane are shown in the bending moment diagram sketched below the shaft cross-section in Figure 3.7(a), and can be seen to be a maximum near the central plane of the gear. At first sight it may be conjectured that the shaft will break somewhere near here.

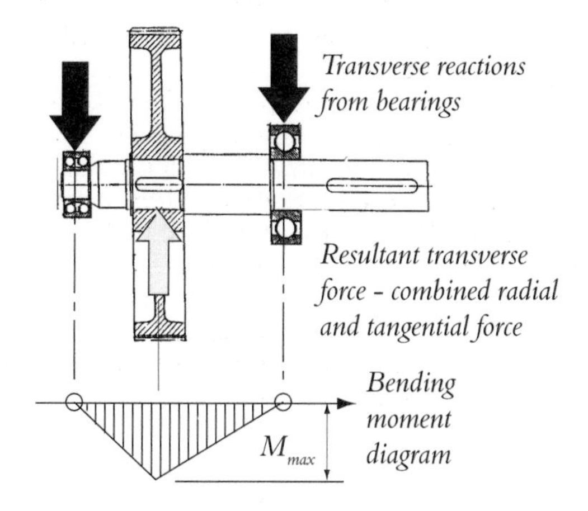

Figure 3.7(a) Bending moments on output shaft

But the shaft is not only being *bent*; it is also being *twisted*. That is the shaft's primary purpose — to transmit torque from the gear to the outside world. When in use, the gearbox will be connected to some other piece of machinery by means of a coupling, attached to the output shaft by yet another keyway. *Essentially the shaft transmits torque between keyways that are located where both the gear and the coupling are attached to it: see Figure 3.7(b).*

So the designer needs now to ask, *"What is the worst combination of bending and torque experienced by the shaft?"* But there are yet further complications. The shaft is not of uniform diameter. In fact, this output shaft has five distinct diameters along various sections of its length, in a stepped fashion typical of shaft design. Large diameter regions will generally be stronger than small diameter regions, and this must be taken into account when considering the most severely stressed region of the shaft.

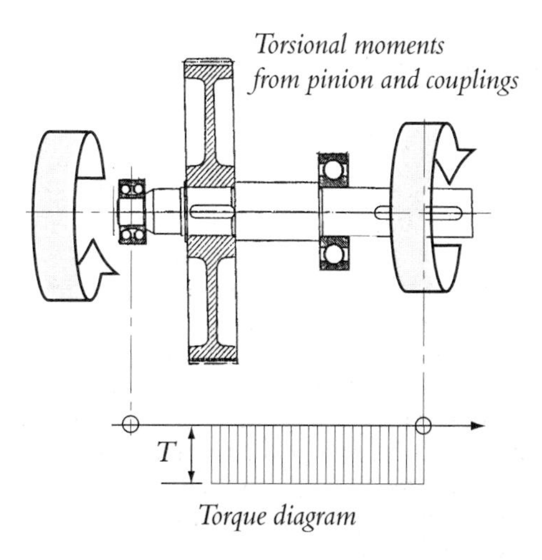

Figure 3.7(b) Torsional moments on output shaft

Finally, the designer must take account of the effect of *stress concentrations*. These sudden changes, or discontinuities, in the geometry of an object have the effect of increasing the peak stress in their vicinity. In this shaft, they are most clearly represented by the steps in shaft diameter, and by the keyways themselves, as highlighted in Figure 3.7(c).

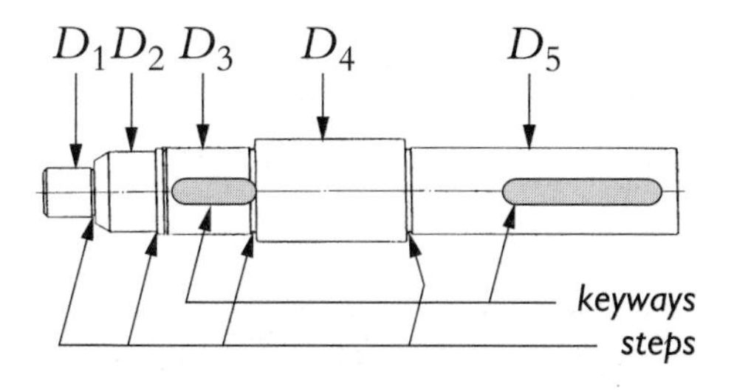

Figure 3.7(c) Stress concentrations and shaft geometry

So, all things considered, where is the shaft most likely to break? Our guess is just to the right of its connection to the gear, but the thorough designer will consider the strength of the shaft at various locations along its length.

Example 2 — solids pump

Mixed-flow pumps have an axial inlet and a radial outlet, as indicated on a typical example in Figure 3.8. Where these pumps are expected to pass heavy sludge and even solids, the required driving power can be substantial. In the example shown, the drive to the pump unit is transmitted through a vee-belt drive, with the pulley mounted on the right hand end of the shaft.

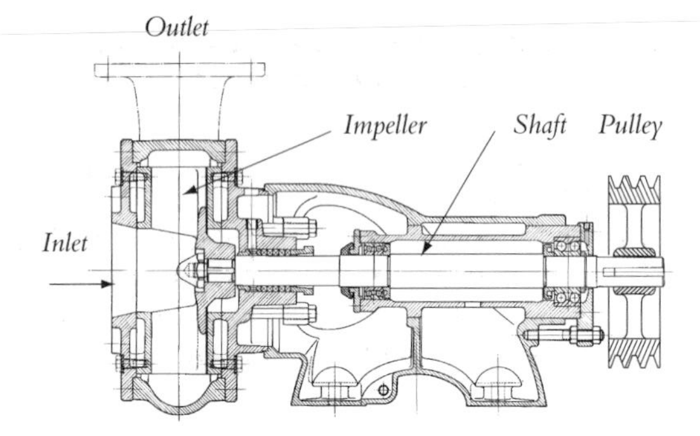

Figure 3.8 Typical arrangement of a mixed-flow pump used for pumping heavy sludge including solids

Figure 3.9 is an engineer's sketch showing a simplified representation of the forces exerted on the shaft of the sludge pump. These loads arise from several different forces, and we will consider each one of them in turn.

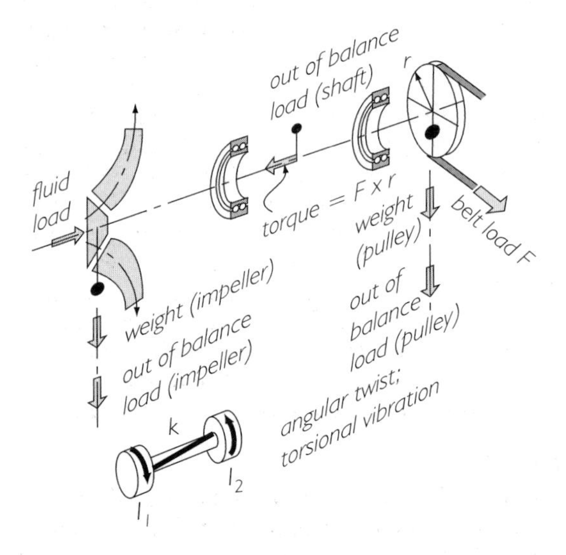

Figure 3.9 Schematic of forces on sludge pump impeller and shaft

- The weight of the heavy impeller acts gravitationally downwards at all times. Since the shaft is *overhung* from its bearings by some distance, this weight will give rise to bending moments within the shaft which are a maximum at the bearings. These moments cause bending stresses which are compressive at the bottom of the shaft, and tensile at the top of the shaft. Material on the surface of the shaft will move alternately from regions of compression to regions of tension as the shaft rotates. The configuration is very similar to that of a rotating beam fatigue test (see Chapter 2.4);
- The weight of the pulley acts in the same manner as the weight of the impeller;
- *Out-of-balance forces*: Each of the rotating components (impeller, pulley and shaft) will potentially impose so-called *out-of-balance* forces upon the shaft. These are dynamic forces caused by the centre of mass of the component located away from the central axis of rotation, usually due to manufacturing inaccuracies. These forces can be very large and in some high speed machinery they are corrected by *dynamic balancing*. They depend on the mass of the component, the radial location error in the centre of mass, and the square of the rotational speed. It is noteworthy that out-of-balance forces, unlike weight force and other unidirectional forces, *rotate with the shaft* and therefore do not give rise to alternating stresses;
- The shaft is driven around by belt tension acting at the periphery of the pulley;
- All of the above forces act laterally to the long axis of the shaft, and are resisted by lateral reaction forces at the bearings;
- The energy required to accelerate the fluid outwards from the pump centreline (and hence pressurise the fluid) is provided by the drive belt, which twists the shaft. The torsional moment may be calculated from the shaft power and the rotational speed:

$$P = T\omega \quad = 2\pi NT/60,$$

where T is the torque, ω is the angular velocity (radians per second), N is the rotational speed (revolutions per minute), and P is the power.

- The impeller of a centrifugal pump experiences axial forces from the stream of fluid (even for fans which move air) as the fluid changes direction and moves radially outwards from the impeller;
- The impeller and pulley systems each have rotational inertia, and the shaft has rotational elasticity (it is, in effect, a torsional spring). The shaft will experience angular twist during its operation, and may even suffer from torsional vibration. These are additional issues for the shaft designer to consider.

The first six of the above effects are loads that may be represented in bending moment and torque diagrams in much the same manner as for the previous example. They give rise to stresses, and these will have a peak value at some location along the shaft. If we assume that the fan's shaft is solid with a diameter, d, at the location of the peak stress, then the peak stresses there at the surface of the shaft will be as stated in Figure 3.10.

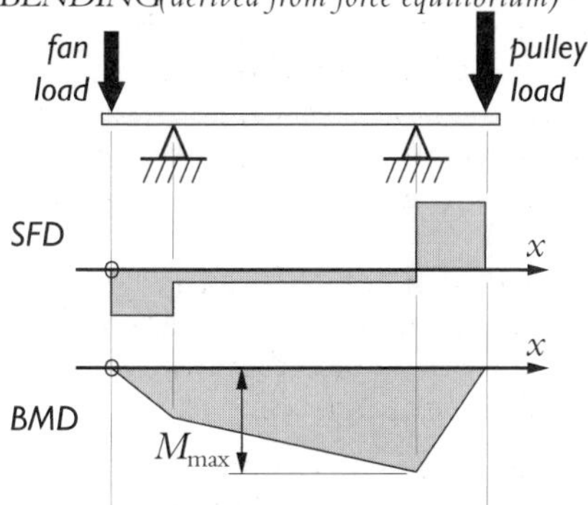

stresses arising from BENDING:

$$\sigma_{B,\max} = \frac{M_{\max} y_{\max}}{I_{zz}} = \frac{32 M_{\max}}{\pi d^3}$$

stresses arising from TORSION:

$$\tau_{T,\max} = \frac{T r_{\max}}{I_P} = \frac{16 T}{\pi d^3}$$

Figure 3.10 Loads and peak stresses on fan shaft

Note that the stress relationships shown above make use of some common section properties for solid circular shafts in which both $y_{max} = r_{max} = d/2$. Throughout this topic we will be making frequent use of these section properties, recalled here for convenience in Equations *(3.1)* and *(3.2)*.

Second moment of area of a solid circular section about a transverse (bending) axis:

$$I_{zz} = \frac{\pi d^4}{64}. \qquad (3.1)$$

Polar second moment of area of a solid circular section about a central (torsional) axis:

$$I_P = \frac{\pi d^4}{32} \qquad (3.2)$$

Most shafts in practice are both solid and circular, although exceptions are not uncommon. For hollow circular shafts, it is often convenient to define the hollowness ratio of the shaft as:

$$\beta = \left(d_i / d\right),$$

in which case the expressions for the second moments of area become, respectively,

$$I_{zz} = \left(1-\beta^4\right)\frac{\pi d^4}{64} \text{ and } I_P = \left(1-\beta^4\right)\frac{\pi d^4}{32}.$$

Ex 3.1

The *tail-shaft* of a rear-wheel drive automobile is usually a hollow tube for transmitting engine power to the rear wheels. Consider a high performance sedan, weighing 1.66 tonne fully loaded, with a load limit on the rear axle of 975 kg. The engine power is rated at 85 kW and the maximum engine torque available is 168 Nm at 3900 rpm. Acceleration from 0 to 100 km/h is 11.7 seconds. Wheel radius is 300 mm and the gear ratios in the four-speed gearbox are 4.23:2.45:1.45:1.2. A further 3.2:1 gear reduction takes place in the differential. Estimate the *worst credible loads* to which this tail shaft will be subjected during normal operation. Clearly identify any assumptions made in arriving at your estimate.

Ex 3.2

Figure 3.11 shows an *auger*, a type of earthmoving equipment used for digging post holes. This device is usually driven by a *power take-off* from a standard farm tractor, through a bevel gearbox as indicated schematically in the figure. The tractor power available is 200 kW, at 1800 rpm, with a reduction of 4:1 in the bevel gearbox. You have been asked by a manufacturer to estimate the worst credible loads to which the central shaft of the auger will be subjected in normal operation. Clearly state any assumptions and major uncertainties in deriving this estimate.

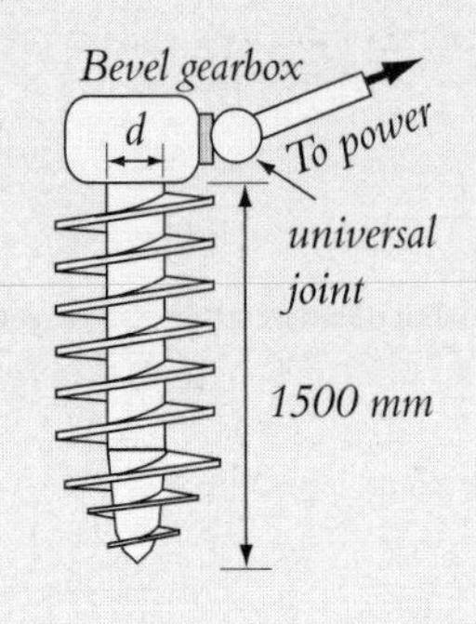

Figure 3.11 Posthole auger schematic (not to scale)

3.1.3 Modes of failure for shafts

Typically shafts are rotating elements that may be subjected to large, combined (bending and torsional) loads which usually vary with time. As a consequence, they are prone to the following modes of failure:

- yielding due to excessive static stresses;
- fracture due to excessive static stresses;
- fatigue fracture due to excessive dynamic stresses;
- excessive lateral deflection;
- excessive torsional deflection.

Each of these is considered in the following sections.

3.1.4 Design of shafts to resist yielding and fracture

The stress state in shafts is generally biaxial, arising from the action of combined bending and torsion. Necessarily, therefore, the designer must use a failure predictor (see Chapter 2.3) in determining the likelihood that a shaft will fail due to excessive stress. Recall that a failure predictor is a function of the principal stresses $(\sigma_1, \sigma_2, \sigma_3)$ of a multi-axially stressed component, and component failure is predicted to occur when that function of stresses reaches a particular value. That particular value is usually derived from uniaxial stress tests in the materials laboratory.

The three major failure predictors available to the shaft designer are the Maximum Principal Stress Failure Predictor (*MPFP*), the Maximum Shear Stress Failure Predictor (*MSFP*), and the Distortion Energy Failure Predictor (*DEFP*). The last two of these (*MSFP* and *DEFP*) are commonly used for shafts. Simply stated, shaft design formulae are generally based on the notion that failure in the shaft occurs when the following equalities are true:

Failure by yielding

- *MSFP*

$$\frac{\sigma_1 - \sigma_3}{2} = \frac{S_y}{2} \tag{3.3}$$

- *DEFP*

$$\sqrt{\frac{(\sigma_1 - \sigma_2)^2 + (\sigma_2 - \sigma_3)^2 + (\sigma_3 - \sigma_1)^2}{2}} = S_y \tag{3.4}$$

Failure by fracture (static stress)

- *MSFP*

$$\frac{\sigma_1 - \sigma_3}{2} = \frac{S_U}{2}$$

- *DEFP*

$$\sqrt{\frac{(\sigma_1 - \sigma_2)^2 + (\sigma_2 - \sigma_3)^2 + (\sigma_3 - \sigma_1)^2}{2}} = S_U$$

Failure by fatigue fracture (dynamic loading)

- *MSFP*

$$\frac{\sigma_1 - \sigma_3}{2} = \frac{S_e}{2} \tag{3.5}$$

- *DEFP*

$$\sqrt{\frac{(\sigma_1 - \sigma_2)^2 + (\sigma_2 - \sigma_3)^2 + (\sigma_3 - \sigma_1)^2}{2}} = S_e \tag{3.6}$$

Clearly, the left-hand side of each of these equations represents either the maximum shear stress, $\tau_{\max}$, in the case of *MSFP*, or the *von Mises stress*, σ_{VM}, in the case of *DEFP*. These concepts were discussed in Section 2.3. They are quoted above in their full three-dimensional form to avoid ambiguity, although they are often stated in a two-dimensional form in relation to shafts since the stress state in shafts is usually biaxial ($\sigma_2 = 0$).

Design rules for shafts are generally based on the third of the above sets of equalities — Equations *(3.5)* and *(3.6)*. This is because of the predominance of *dynamic loading* in shaft applications (especially the case of *reversed bending*), and because *fatigue loading* represents the most restrictive form of the above equations (that is, *failure is predicted at the lowest values of the stress functions*).

3.1.5 Principal stresses in shafts

Shafts in engineering practice mostly have a circular cross-section, and are mostly solid. Determination of the biaxial stress state of an element on the surface of a shaft in combined bending and torsion is therefore straightforward, as shown in Figure 3.12. Of course, the designer needs to work with principal stresses in applying the failure predictors. However, neither the *x*-direction nor the *y*-direction in Figure 3.12 are principal stress directions since, by definition, a principal stress is one which acts normal to a plane upon which the shear stress is zero.

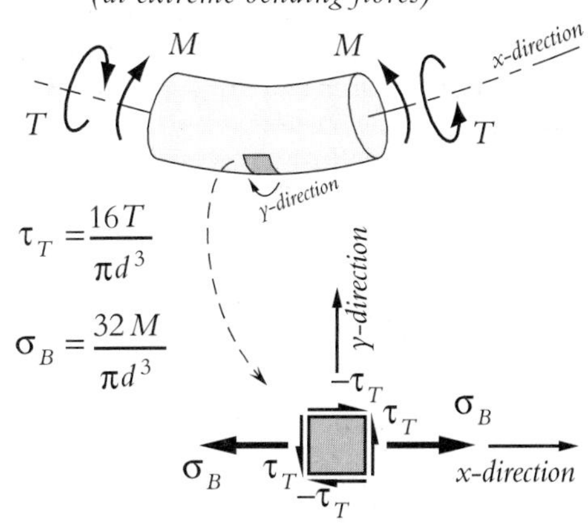

Figure 3.12 Biaxial stress state of a shaft element

Fortunately there is a simple graphical construction, which must be at the fingertips of every structural engineer, that allows working stresses such as those in Figure 3.12 to be quickly translated into their corresponding principal stresses.

A summary of the Mohr's circle construction

Refer back to Figure 2.10 where we examined a small element of material with both direct and shear stresses acting on it. We found the normal and shear stresses, σ_ϕ and τ_ϕ respectively, acting on a plane *AB* through this element at some angle ϕ to the reference direction. It is easily shown that

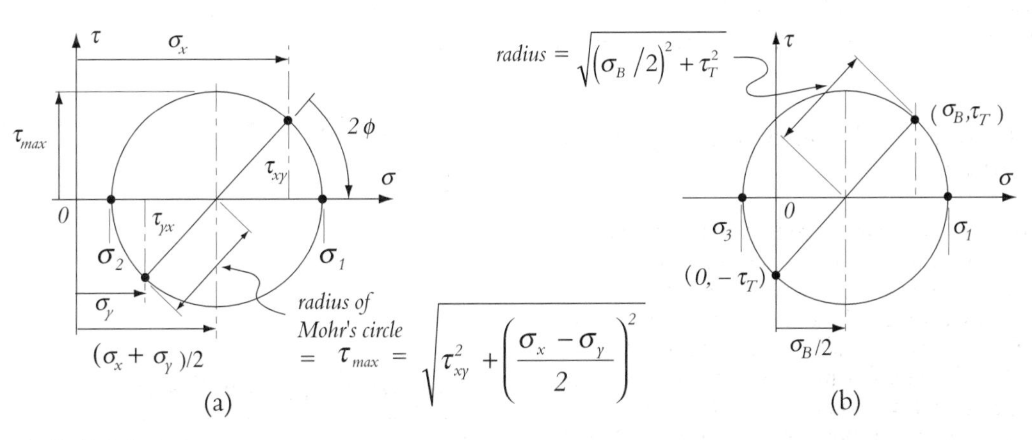

Figure 3.13 Mohr's circle construction

the variation of normal and shear forces acting on planes oriented at various angles ϕ to the reference direction, follows the equation of a circle. This is shown as the general Mohr's stress circle in Figure 2.10(c) and redrawn here as Figure 3.13(a)

The Mohr's circle construction is helpful in determining the principal stresses in a shaft. The (σ, τ) coordinates corresponding to the x- and y-direction faces of the stress element, being (σ_B, τ_T) and $(0, -\tau_T)$ respectively, define two points on the circumference of a circle symmetrical about the (*horizontal*) normal stress axis. The radius of the Mohr's circle is then found to be:

$$\text{radius} = \sqrt{(\sigma_B/2)^2 + \tau_T^2}$$

and its centre is at the point$(0, \sigma_B/2)$ along the horizontal axis, as shown in Figure 3.13(b).

Recall that for a solid shaft, the worst-case values of σ_B and τ_T are

$$\sigma_B = \frac{32M}{\pi d^3} \qquad (3.7)$$

and

$$\tau_T = \frac{16T}{\pi d^3}. \qquad (3.8)$$

By simple construction from the Mohr's circle, we can use these values to obtain the following expressions for the maximum and minimum principal stresses.

$$\sigma_1 = \frac{\sigma_B}{2} + \sqrt{(\sigma_B/2)^2 + \tau_T^2}$$

$$= \frac{16}{\pi d^3}\left(M + \sqrt{M^2 + T^2}\right) \qquad (3.9)$$

$$\sigma_2 = 0 \qquad (3.10)$$

$$\sigma_3 = \frac{\sigma_B}{2} - \sqrt{(\sigma_B/2)^2 + \tau_T^2}$$

$$= \frac{16}{\pi d^3}\left(M - \sqrt{M^2 + T^2}\right) \qquad (3.11)$$

Also it is immediately apparent that:

$$\tau_{max} = \sqrt{(\sigma_B/2)^2 + \tau_T^2}$$

$$= \frac{16}{\pi d^3}\sqrt{M^2 + T^2}. \qquad (3.12)$$

3.1.6 Failure prediction for dynamic loading

We are now able to substitute the above values of σ_1, σ_2 and σ_3 from Equations *(3.9)*, *(3.10)* and *(3.11)* into the failure predictors for dynamic loading embodied by Equations *(3.5)* and *(3.6)*. This is a tedious but algebraically easy task, and it would be instructive for the reader to try it. Eventually the following two predictions of failure are obtained. They are strikingly similar — in fact they only differ by the factor "3/4" in front of the torque term — and they suggest, as usual, that *DEFP* is a slightly less conservative failure predictor than *MSFP*.

- *MSFP* predicts failure when

$$S_e \le \frac{32}{\pi d^3}\sqrt{M^2 + T^2}.$$

- *DEFP* predicts failure when

$$S_e \le \frac{32}{\pi d^3}\sqrt{M^2 + \frac{3T^2}{4}}.$$

The designer's approach now follows our well-established procedure of design for structural integrity. Based on these failure predictions, we ensure that the *actual* stress value on the right-hand side of one or other equation is always less than the *limiting* stress value on the left-hand side, by at least some acceptable factor of safety, F_d. Therefore we can write the following two *design inequalities*:

- *MSFP* design rule:

$$\frac{S_e}{F_d} \ge \frac{32}{\pi d^3}\sqrt{M^2 + T^2} \qquad (3.13)$$

- *DEFP* design rule:

$$\frac{S_e}{F_d} \ge \frac{32}{\pi d^3}\sqrt{M^2 + \frac{3T^2}{4}} \qquad (3.14)$$

Usually it is the shaft diameter that serves as the primary design variable, so Equation *(3.14)*, for example, would normally be used in the form:

$$d^3 \ge \frac{32F_d}{\pi S_e}\sqrt{M^2 + \frac{3T^2}{4}}.$$

Design rules incorporated in national or industry-based design codes are essentially variations on the theme of the above two design inequalities. It will be instructive at this point to consider the shaft design equations from a representative national code, *AS1403 — the Australian Standard for shafts*.

Shaft Design Equations from AS1403

Case 1 — (general) fully reversed torque:

$$d^3 \geq QK_{sf}\sqrt{\left(M^*\right)^2 + \frac{3T^2}{4}}; \qquad (3.15)$$

Case 2 — pulsating torque:

$$d^3 \geq Q\sqrt{\left[K_{sf}\left(M^*\right)\right]^2 + \frac{3T^2}{4}\left(\frac{1+K_{sf}}{2}\right)^2}; \qquad (3.16)$$

Case 3 — steady torque:

$$d^3 \geq Q\sqrt{\left[K_{sf}\left(M^*\right)\right]^2 + \frac{3T^2}{4}}; \qquad (3.17)$$

where

$$Q = \frac{10F_d}{S'_e};\ K_{sf} = K_s K_f;\ M^* = M + \frac{Fd}{8}.$$

Notes:

- These design equations have been adapted slightly to suit the symbols used in this book, and to make the common elements between the three versions more obvious.
- F is a tensile axial load applied to the shaft. This load results in stresses superimposed on the stresses due to bending. Hence:

$$\sigma_B = \frac{32}{\pi d^3}\cdot M$$

$$\sigma_A = \frac{F}{A} = \frac{4F}{\pi d^2} = \frac{32}{\pi d^3}\cdot\frac{Fd}{8}$$

$$\sigma_B + \sigma_A = \frac{32}{\pi d^3}\cdot\left(M + \frac{Fd}{8}\right).$$

- The number '10' appears just because

$$\frac{32}{\pi} \approx 10;$$

- The three cases are identical for $K_s = K_f = 1$.
- The factors K_s and K_f are precisely those defined in Section 2.4.1, where they are used to modify the *material endurance limit*, S'_e, to become the *component endurance limit*, S_e. Recall:

$$S_e = \frac{S'_e}{K_s K_f}.$$

It is interesting to notice that the form of the three equations differs only on the basis of how the two *fatigue factors*, K_s and K_f, are applied. The intention of the equations is to apply these factors only to dynamic loads. The assumption in all three equations is that the bending moment, M, is always a dynamic, reversed load, and is always effectively multiplied by K_s and K_f. In *Case 1 – fully reversed torque*, the torque term, T, is also so multiplied. In *Case 3 – steady torque*, the torque term is not affected by the dynamic loading factors. *Case 2 – pulsating torque*, represents a kind of *halfway* situation between the other two cases, in which the torque term is multiplied by a reduced form of the K_s and K_f factors.

3.1.7 Failure by excessive lateral deflection

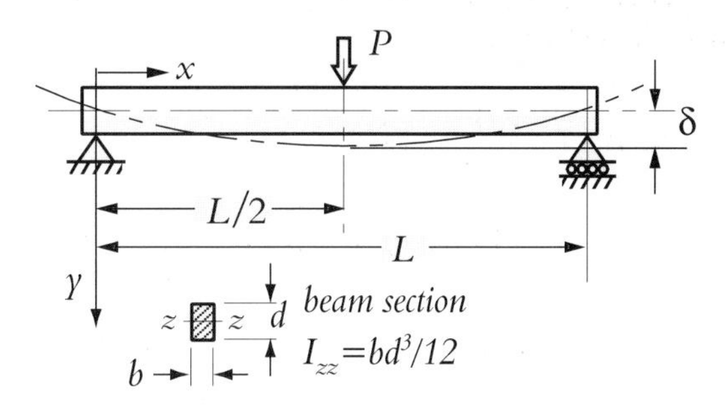

Figure 3.14 Deflection due to bending

In its deflectional response to a transverse load, a shaft acts just like a beam. Shaft deflection formulae for a wide variety of transverse loading situations are tabulated extensively in the engineering literature, and may be combined by the principal of superposition. Some are provided in Appendix D of this book. The simplest case is that of simple bearing support and a centrally loaded beam as shown in Figure 3.14. For this we obtain:

$$\delta = \frac{PL^3}{48EI_{zz}}.$$

In practice, the need to maintain adequate clearances within rotating machinery, or to maintain the accuracy and alignment of meshing components, or to avoid excessive vibration forces, will often place limits on the acceptable value of

$\delta < \delta_{allowable}$.

3.1.8 Failure by excessive torsional deflection (excessive twist)

Excessive twist is not a common design consideration. It often arises indirectly, however, in connection with torsional vibration problems — a shaft that is unduly flexible in torsion will generally have a low natural torsional frequency.

A good measure of the torsional flexibility of a shaft is the angular twist per unit length:

$$\phi = \frac{\theta}{L} \qquad (3.18)$$

for a given applied torque. The quantity ϕ is a function of the torque, T, the polar second moment of area, I_p, and the material's shear modulus, G.

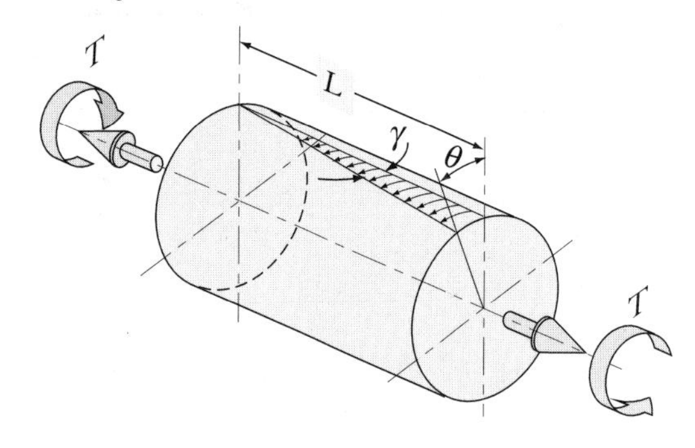

Figure 3.15 Rotational deflection due to torsion

From Figure 3.15 we can write $(\theta d/2) = (\gamma L)$, so ϕ can be expressed as:

$$\phi = \frac{2\gamma}{d}.$$

But from Hooke's Law in torsion,

$$\gamma = \frac{\tau_T}{G},$$

and from the familiar expression for torsional shear stress at the surface of a shaft we have

$$\tau_T = \frac{T(d/2)}{I_P}.$$

Substituting these into the definition of ϕ in Equation *(3.18)* yields:

$$\phi = \frac{T}{GI_P}, \qquad (3.19)$$

where

$$G = \frac{E}{2(1+\mu)}.$$

Note that ϕ is not an angle, but a *twist per unit length* — mathematicians will recognise this as the quantity properly called *torsion*. Normally we are interested in the total deflection, θ, particularly in control applications.

For a steel shaft of diameter d and length L, the total deflection is:

$$\theta = L\phi = \frac{LT}{GI_P} \approx \frac{26.5LT}{Ed^4}.$$

3.1.9 Summary of design rules for shafts

(a) Design against fracture under combined dynamic bending and torsion

- Equations *(3.9)*, *(3.10)* and *(3.11)* predict the principal stresses under combined loading. How they are applied by the designer depends on the choice of failure predictor. *DEFP* is commonly used for shafts. *MSFP* is slightly more conservative. Consult and use established national or industry-based shaft design codes, which present design formulae such as Equations *(3.15)*, *(3.16)* and *(3.17)* for specific applications. In general these codes adopt the following approaches.
- Using *Distortion Energy Failure Predictor (DEFP)*:

Model

$$\sigma_{VM} = \frac{32}{\pi d^3}\sqrt{M^2 + \frac{3T^2}{4}}.$$

Require $\sigma_{VM} < S_e$.

Rule:

$$\frac{S_e}{F_d} \geq \frac{32}{\pi d^3}\sqrt{M^2 + \frac{3T^2}{4}}. \qquad (3.14)$$

- Using *Maximum Shear Stress Failure Predictor* (*MSFP*):

Model

$$\tau_{\max} = \frac{16}{\pi d^3}\sqrt{M^2 + T^2}. \qquad (3.12)$$

Require $\tau_{max} < S_e/2$.

Rule:

$$\frac{S_e}{F_d} \geq \frac{32}{\pi d^3}\sqrt{M^2 + T^2}. \qquad (3.13)$$

(b) Design against excessive deflection

- Lateral deflection:

Model — Deflection model must be derived from beam theory to suit a particular application. It is not possible to generalise about this, as the shaft bending layout and resulting deflection model may be quite complicated. The simplest case is that of a simply supported, centrally loaded shaft, for which:

$$\delta = \frac{PL^3}{48EI_{zz}}.$$

Require $\delta < \delta_{allowable}$.

- Torsional deflection:

Model

$$\theta = L\phi; \quad \phi = \frac{T}{GI_P}. \qquad (3.19)$$

Require $\theta < \theta_{allowable}$.

Ex 3.3

Select component dimensions (outside diameter and wall thickness) for the tailshaft and the auger described in Ex. 3.1 and 3.2. A structural grade steel, $S_y = 340$ MPa and $S_U = 560$ MP a may be used. In each case use an explicit procedure, as indicated in Section 2.3.5, to derive a suitable design factor of safety F_d.

Ex 3.4

Figure 3.16 shows a rock cutting machine schematically. The cutting tool is fitted to the machine spindle and it cuts the rock when the arc of the tool tip intersects the specimen sliding underneath the spindle. The material chosen for this shaft application is a grade *R4* steel with $S_U = 1000$ MPa and $S_y = 740$ MPa.

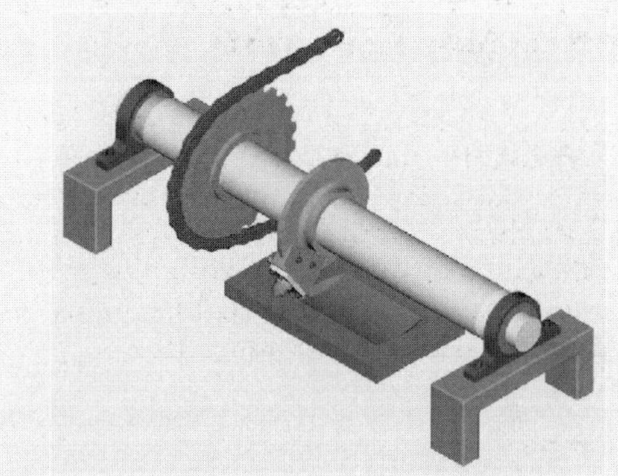

(a) *General view of rock cutting machine*

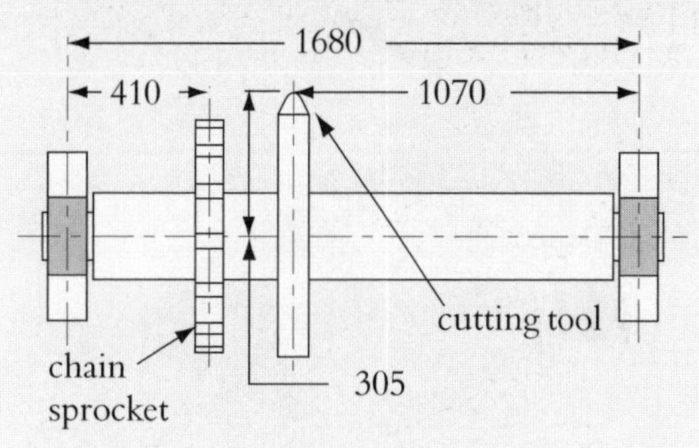

(b) *Plan view (schematic only) dimensions are mm*

Figure 3.16 Rock cutting machine

The maximum tangential load on the cutting tool is estimated at 52.5 kN. The chain drive sprocket has a pitch diameter of 280 mm, and the motor driving the system has an output power of 60 kW.

(a) Estimate the diameter of the shaft needed for this application. You may use a design factor of safety $F_d = 4$ (is this too conservative?).

(b) Suggest a relatively simple design change that might significantly reduce the shaft size required yet maintains the existing bearing span and cutter position.

Notes:

- Chain sprocket drives generate resultant tangential forces *normal* to the direction of the drive. You will need to integrate the tangential forces acting on the sprocket periphery to determine the resultant tangential forces generated.
- Your choice of drive direction (not identified) should result in the most conservative design.

3.2 Assignment on shaft design

3.2.1 Aims

The aims of this assignment are:

(a) to give practice in shaft design calculations;

(b) to provide an example of design against multiple modes of failure; and

(c) to present novice designers with the challenge of a design task specified in a less structured fashion than in previous assignments.

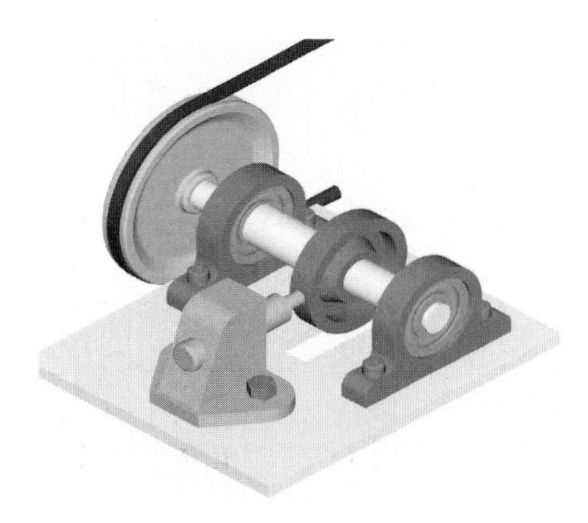

Figure 3.17 Wear testing machine

3.2.2 Background

A metallurgical wear-testing machine is being designed to facilitate the experimental determination of the rates of wear for metal surfaces in rubbing contact. The standardized wear components consist of a *pin* in radial contact with the cylindrical surface of a 200 mm diameter disc. The disc is rotated at a known speed, the torque required to rotate it is accurately measured, and the *wear rate* is defined in terms of the rate at which the pin's length reduces.

The general arrangement of the main components is shown in Figure 3.17. Note that for reasons of compactness it has been decided to locate the disc between the shaft bearings. However, the *optimum* (best) position of the drive motor, and hence the direction of the pulley force on the shaft, has not yet been decided. The two alternatives being considered are (see Figure 3.18):

Case 'A' - with the drive motor on the opposite side of the shaft to the pin;

Case 'B' - with the drive motor on the same side of the shaft as the pin.

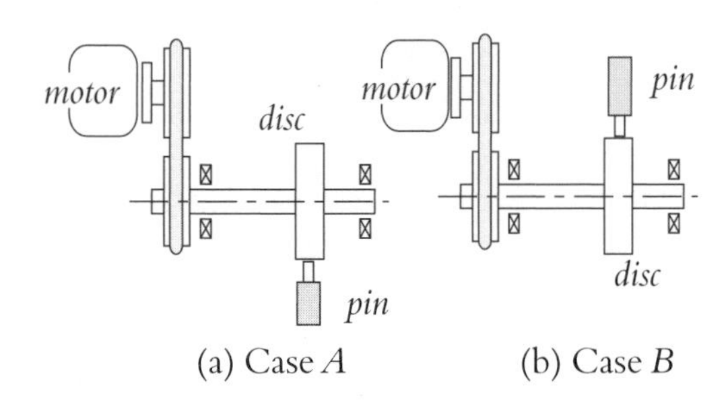

Figure 3.18 Plan views of the two wear testing machine arrangements (not to scale)

3.2.3 The design problem

Your present task is to determine what influence, if any, the position of the drive motor has on the shaft design, and then design the main shaft for acceptable deflection and infinite fatigue life when subjected to bending and torsional effects under the following design conditions :

- maximum applied pin force (radial on disc), $F_r = 50$ N;
- maximum tangential rubbing speed (at pin), $V = 60$ m/s;
- worst-case coefficient of sliding friction, $\mu = 1.20$ (*yes, greater than unity*!);
- average pre-tension in pulley drive-belt, $P_{ave} = 40$ N;
- maximum allowable radial deflection at pin, $\delta = 0.10$ mm.

Various decisions have already been taken, and these are indicated in Figure 3.19 schematically.

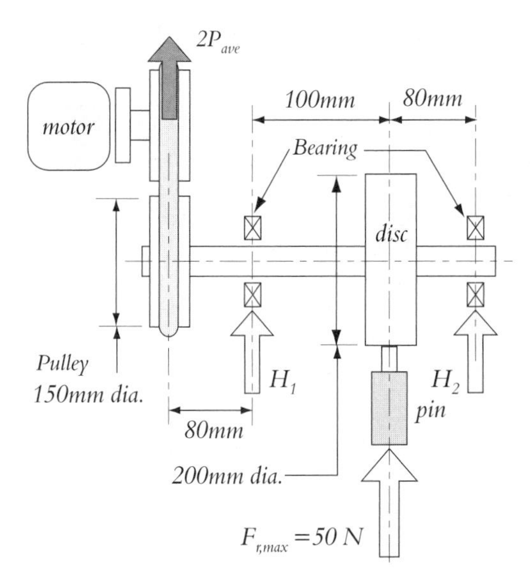

Figure 3.19 Wear testing machine (schematic plan view, not to scale)

The shaft diameter, d = ? (to be determined)
F_d = 3.0; K_s = 1.15; K_f = (unknown)
E = 210 GPa; S_y = 320 MPa; S_U = 480 MPa
(Use a reasonable assumption to estimate S_e').

Notes (and supplementary assignment questions)

(a) Base the shaft design calculations on appropriate standard formulae.

(b) Note that it would be reasonable to base these preliminary calculations on a shaft of constant diameter. The detailed geometry of the shaft is of course as yet unspecified. Explain your method of handling the geometrical factor K_f in view of this.

(c) The shaft is manufactured from a commercial grade of mild steel. The whole rotating assembly is statically and dynamically balanced so that out-of-balance forces and moments are small and their effect on shaft deflections and stresses can be neglected.

(d) It can be shown (*after some pain and suffering*) that the radial deflection of the shaft in line with the pin when both F_r and P_{ave} are acting in the directions shown in Figure 3.20(a), is given by:

$$\delta = \frac{F_r}{EI}\frac{a^2(a+b)}{3} - \frac{2P_{ave}}{EI}\frac{abc}{6}.$$

For the alternative direction of P_{ave}, merely change the sign of the P_{ave} *term*.

(e) Comment on any important assumptions or simplifications made in your analysis.

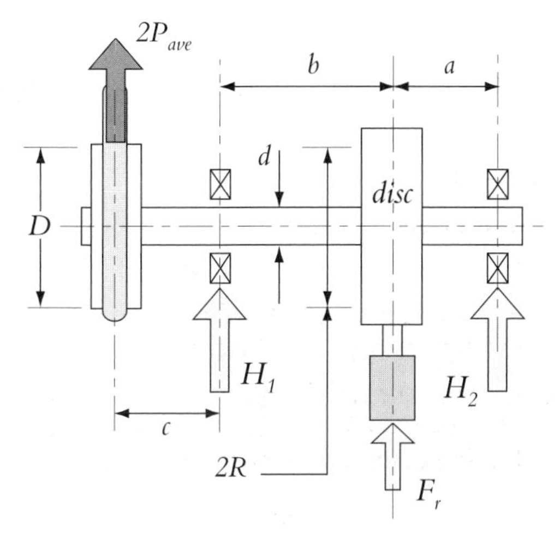

(a) Plan view – radial loading on disc

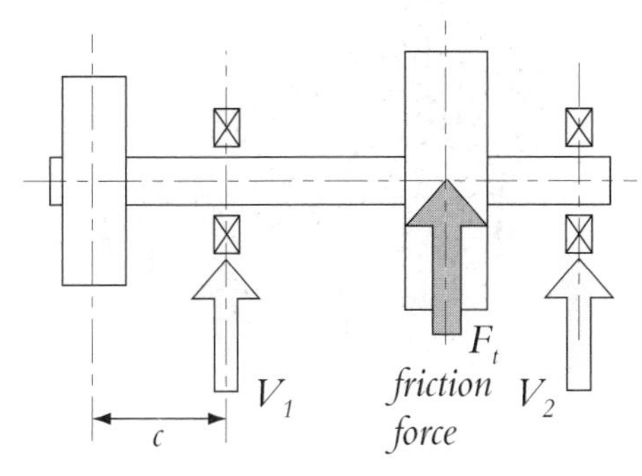

(b) Elevation – tangential loading on disc

Figure 3.20 Forces acting on wear tester

Engineering Design Associates

Project Title : **Wear testing machine**

Date 04:3:99

Page 1 of 5

Ckd. evm

SHAFT DESIGN — WEAR-TESTING MACHINE

Subdivide this open-ended design problem into sub-tasks:

(a) shaft stress analysis — case 'A', — case 'B'

(b) determine required d to satisfy stress (fatigue)

(c) shaft deflection analysis — case 'A', — case 'B'

(d) determine required d to satisfy deflection

(e) choose between case 'A' and case 'B'

(f) determine d

Key issues to be considered:

1) Bi-axial bending — will need to consider bending of the shaft in two separate planes, and then combine the bending moments

2) Multiple failure modes — the notion of a "governing mode of failure" will assist in coordinating the requirements of two modes of failure (fatigue & excessive deflection).

(a) Shaft Stress Analysis

(1) Bending in Horiz. Plane

case 'A'

$\sum M_1 = F_r \cdot b + H_2 (a+b) - 2P_{ave} \cdot c = 0$

$\Rightarrow H_2 = \dfrac{2P_{ave} \cdot c - F_r \cdot b}{(a+b)}$

$= \dfrac{80 \times 80 - 50 \times 100}{180} = 7.8\ N$

$\Rightarrow H_1 = -(80 + 50 + 7.8) = -137.8\ N$

(check: $\sum M_3 = \ldots = 0$, as reqd)

$M_{1H} = 80\,N \cdot 80\,mm = 6400\ N.mm$

$M_{2H} = 7.8\,N \cdot 80\,mm = 622\ N.mm$

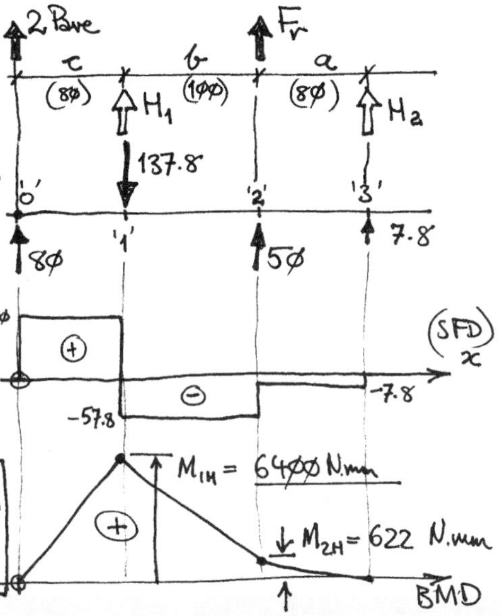

2Pave, F_r, c (80), b (100), a (80), H_1, H_2, 137.8, '0', '1', '2', '3', 80, 50, 7.8

+80, (SFD), x, −57.8, −7.8

$M_{1H} = 6400$ N.mm, $M_{2H} = 622$ N.mm, BMD

Engineering Design Associates

Project Title : **Wear testing machine**

Date 04:3:99

Page 2 of 5

Ckd. evm

[sht. 2 of 5]

case 'B'

$$\Sigma M_1 = F_r.b + H_2'(a+b) + 2P_{ave}.c = 0$$

$$\Rightarrow H_2' = \frac{-2P_{ave}.c - F_r.b}{(a+b)}$$

$$= \left(\frac{-80\times80 - 50\times100}{180}\right) = -63.3\text{N}$$

$$\Rightarrow H_1' = (80 - 50 + 63.3) = 93.3\text{ N}$$

(check: $\Sigma M_3 = \ldots = 0$, as req'd)

$M_{1H}' = 80\text{N} \times 80\text{mm} = 6400\text{ N.mm}$

$M_{2H}' = 63.3\text{ N} \times 80\text{mm} = 5067\text{ N.mm}$

80, 63.3, '0', '1', '2', '3', 93.3, 50

+63.3, +13.3, (SFD), x, -80

(BMD), x, $M_1 = -6400$N.mm, $M_{2H}' = -5067$ N.mm

(ii) Bending in Vert. Plane

(cases 'A' & 'B' the same)

F_t; c (80), b (100), a (80); V_1, V_2

$$\Sigma M_1 = F_t.b + V_2(a+b) = 0$$

$$\Rightarrow V_2 = -\left(\frac{b}{a+b}\right).F_t$$

$$= -\left(\frac{100}{180}\times60\right) = -33.3\text{ N}$$

$$\Rightarrow V_1 = -(60 - 33.3) = -27.7\text{ N}$$

$M_{1V} = 0$

$M_{2V} = 33.3\text{ N} \times 80\text{ mm}$

$= 2667\text{ N.mm}$

26.7, (miss Ms. Court), 60

+33.3, (SFD), x, -26.7

(BMD), x, $M_{1V} = 0$ N.mm, $M_{2V} = 2667$ N.mm

Engineering Design Associates

Project Title : **Wear testing machine**

Date 04:3:99

Page 3 of 5

Ckd. evm

(iii) Vector Sum of BM's

case 'A' @'1' : $M_{1A} = \sqrt{M_{1H}^2 + M_{1V}^2} = \sqrt{6400^2 + 0^2} = 6{,}400$ N.mm ⊛

@'2' : $M_{2A} = \sqrt{M_{2H}^2 + M_{2V}^2} = \sqrt{622^2 + 2667^2} = 2{,}738$ N.mm

case 'B' @'1' : $M_{1B} = \sqrt{M_{1H'}^2 + M_{1V}^2} = \sqrt{6400^2 + 0^2} = 6400$ N.mm ⊛

@'2' : $M_{2B} = \sqrt{M_{2H'}^2 + M_{2V}^2} = \sqrt{5067^2 + 2667^2} = 5{,}726$ N.mm

Note : ⊛ — in both case 'A' & case 'B', the worst (largest) BM is 6,400 N. mm, occuring at section '1' along the shaft.

∴ $M_{MAX} = 6400$ N.mm (regardless of case 'A'/'B')

(iv) Shaft Torque

$T = F_t \times R = 60\,N \times 100\,mm = 6000$ N.mm

(same for cases 'A' & 'B')

(b) Stress-based Calculation of d

Use "steady-torque" equation from course text, (same formula as that in AS1403):

$$d^3 \geqslant \frac{10 F_s}{S_e} \sqrt{\left[K_s K \left(M_q + \frac{F_q d}{8}\right)\right]^2 + \frac{3 T_q^2}{4}}$$

where $M_q = 6400$ N.mm (case 'A' & case 'B')

$T_q = 6000$ N.mm

$F_q = 0$ N

$K_s = 1.15$ (surface finish factor)

$K = 1.5$ ← [assume a reasonable value, corresponding to steps in the shaft and their associated stress concentration factors. Note that this term is equivalent to "$K_f = 1 + q(K_t - 1)$" in fatigue terminology]

Engineering Design Associates

Project Title : **Wear testing machine**

Date 04:3:99

Page 4 of 5

Ckd. evm

Use $F_s = 3.0$

$$"S_e" = S_e' \simeq \frac{S_u}{2} \simeq \frac{480}{2} = 240 \text{ MPa}$$

$$\therefore \quad d \geq \left\{ \frac{10 \times 30}{240} \sqrt{(1.15 \times 1.5 \times 6400)^2 + \frac{3 \times 60000^2}{4}} \right\}^{\frac{1}{3}}$$

$= \boxed{11.51 \text{ mm}}$ (for cases 'A' and 'B')

(C) Shaft Deflection Analysis

case 'A':

$$\delta = \frac{1}{3EI}\left[F_r\left(\frac{a^2b^2}{a+b}\right) - P_{ave}\cdot(abc)\right]$$

$$= \frac{1}{3EI}\left[50 \times \left(\frac{80^2 \times 100^2}{180}\right) - 40 \times (80 \times 100 \times 80)\right] \text{ N.mm}^3$$

$$= \frac{1}{3EI}\left[17.8 \times 10^6 - 25.6 \times 10^6\right]$$

$$= -\frac{7.82 \times 10^6}{3EI}.$$

case 'B':

$$\delta = \frac{1}{3EI}\left[F_r\left(\frac{a^2b^2}{a+b}\right) + P_{ave}(abc)\right]$$

(change sign from case 'A' & case 'B')

$$= \frac{1}{3EI}\left[17.8 \times 10^6 + 25.6 \times 10^6\right] \text{ N.mm}^3$$

$$= \frac{+43.4 \times 10^6}{3EI}$$

In either case, $|\delta| = \frac{K_{def}}{3EI}$ where (A) $K_{def} = 7.82 \times 10^6$ N.mm³

(B) $K_{def} = 43.4 \times 10^6$ N.mm³

Write $\boxed{\delta \leq \frac{\delta_{MAX}}{F_D}}$ as our design inequality, where

(actual def'n) (max. allowable def'n)

F_D = factor of safety for deflection (note: F_D not specified in the problem statement.)

$$\Rightarrow \boxed{\delta_{MAX} \geq \frac{K_{def}}{3EI}\cdot F_D}$$

Now, $I = \frac{\pi d^4}{64}$ $\Rightarrow$ $\delta_{MAX} \geq \frac{K_{def}F_D}{3E\left(\frac{\pi d^4}{64}\right)}$

$$\Rightarrow d^4 \geq \frac{64 K_{def} F_D}{3E\pi\,\delta_{MAX}}$$

So we require:

$$\boxed{d \geq \left\{\frac{64 K_{def} F_D}{3E\pi\,\delta_{MAX}}\right\}^{\frac{1}{4}}}$$

Engineering Design Associates **Project Title :** **Wear testing machine**	**Date** 04:3:99
	Page 5 of 5
	Ckd. evm

(d) Deflection-based Calculation of 'd'

Substituting the values of K_{def} for cases 'A' & 'B', we obtain:

case 'A' : $d \geqslant \left\{ \dfrac{64 \times 7.82 \times 10^6 \text{ N. mm}^3}{3 \times 210 \times 10^3 \text{. N.mm}^{-2} \times \pi \times 0.1 \text{ mm}} \right\}^{\frac{1}{4}} \cdot (F_D)^{\frac{1}{4}}$

$\geqslant 7.1 \text{ mm} \cdot (F_D)^{\frac{1}{4}}$ (for $F_D = 3.0$, $d \geqslant 9.33$ mm)

case 'B' : $d \geqslant \left\{ \dfrac{64 \times 43.4 \times 10^6 \text{ mm}^4}{3 \times 210 \times 10^3 \times \pi \times 0.1} \right\}^{\frac{1}{4}} \cdot (F_D)^{\frac{1}{4}}$

$\geqslant 10.9 \text{ mm} \times (F_D)^{\frac{1}{4}}$ (for $F_D = 3.0$, $d \geqslant 14.3$ mm)

(e) Selection of Configuration

Case 'A' is preferable with respect to deflection.
Stress favours neither one case nor the other.
⇒ choose case 'A' as the best configuration.

(f) Determination of Shaft Diameter, d

Stress is the governing design criterion,
[$d \geqslant 9.33$ mm for deflection, with $F_D = 3.0$, whereas
$d \geqslant 11.51$ mm for stress, with $F_S = 3.0$.]

Hence, $d \geqslant 11.51$ mm ⇒ $d = 12$ mm
(ie. next rounded dimension up)

Note: at this stage, designer would need to go back and check this assumed value of K (K = stress concentration factor), to ensure that it was conservative (ie. higher than actual).

3.3 Mechanical springs: bending and torsion

The essential purpose of a spring is to provide a predictable deflection, or rate of deflection, when a load is applied to it. A good spring is resilient — it returns to its original position when the load is removed. In abstract terms, a spring can be thought of as an energy storage device. It converts mechanical work (derived perhaps from the kinetic energy of some moving mass, or from the thermal energy of some expanding gas, or from the chemical energy in someone's muscles) into stored elastic potential energy. This is really the higher energy state of the spring material that arises when its atoms are distorted slightly from their natural positions relative to each other. But let us not be distracted by the solid-state physics. Springs just like to be pushed around, and get back into shape afterwards!

Although springs come in a wide array of shapes and sizes, the most common category — and the one examined here — is that of helical coils.

3.3.1 Historical note

Springs have been used throughout human history and prehistory. Figures 3.21 and 3.22 show two early examples of so-called *beam springs*. In Figure 3.22, a sapling is bent in order to store energy (for an animal trap, say), and acts as a simple cantilever beam. The recurve bow of Figure 3.22 aided tribesmen under various Mongolian khans to conquer the greater part of Asia and Europe during the 12th and 13th centuries, and is still used by sports people today. Its double-curved shape enables the bow to store and impart a great deal of energy for its size, so that it is ideally suited to archery from horseback.

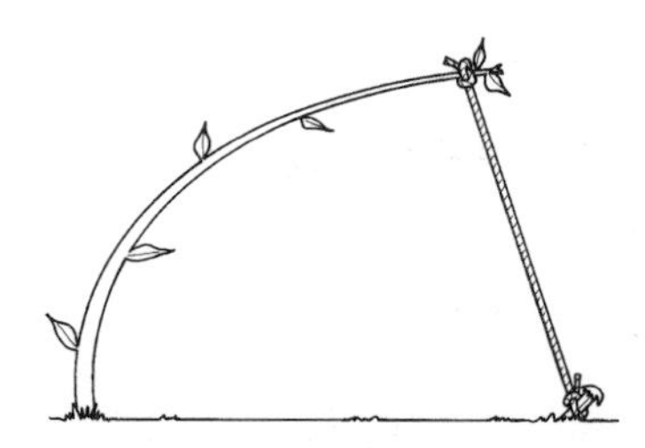

Figure 3.21 Tree catapult

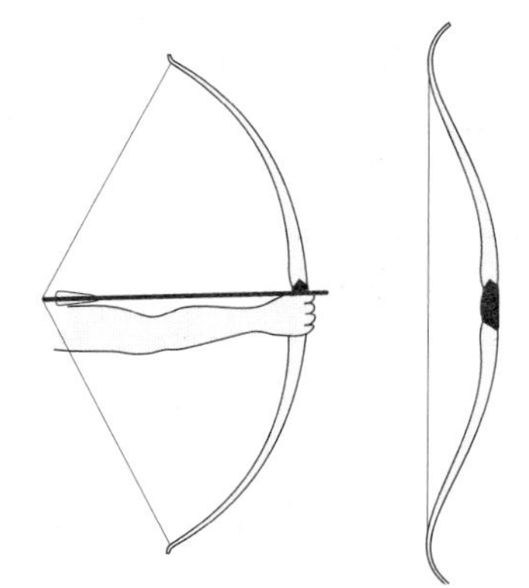

Figure 3.22 Recurve bow

The *ballista* was an ancient engine of war used by both the Greeks and Romans to hurl stone missiles. A typical example of a *ballista* is shown in Figure 3.23. It employed at its core a torsion spring formed from twists of heavy rope. Helical (coil) springs did not enter into engineering usage until much later, when the late middle ages permitted the fabrication of more sophisticated steel components.

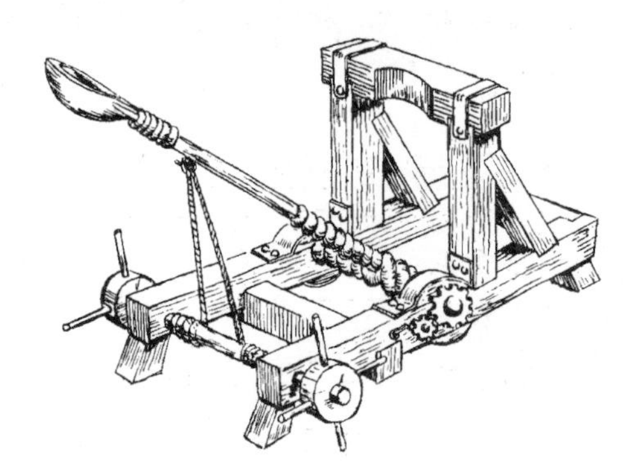

Figure 3.23 Roman Ballista

Probably the earliest form of steel springs were the *leaf springs* used in the suspension of vehicles, particularly those used for human transport. These springs came into common usage in horse-drawn carriages and made the easy transition to the *horseless* carriage, the automobile, in the late 19th and early 20th century. The photo in Figure 3.24 shows a type of suspension used in many early automobiles. Although leaf springs continue to be used in some passenger vehicles, their use in most automobiles has now been replaced by various forms of coil spring suspension. Figure 3.25 shows a coil spring suspension used on a modern 4-wheel drive vehicle. The combined spring and damper arrangement shown, with the damper inside the coil spring, is commonly referred to as a *MacPherson Strut*, after its inventor.

Figure 3.24 Front suspension on 1910 Isota-Fraschini designed by Ettore Bugatti

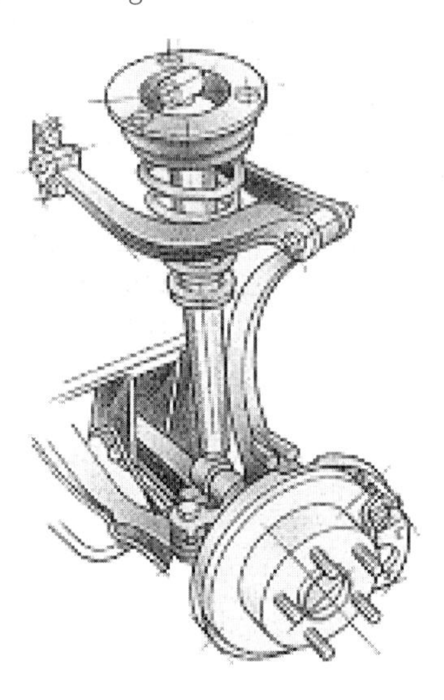

Figure 3.25 Artist's impression of a modern coil spring suspension on a 4-wheel drive automobile

3.3.2 The nature of mechanical springs

1. Linear elastic phenomena

All solid materials exhibit a linear elastic region in which, at least for relatively small deflections, the deflection is linearly proportional to the applied load. This is true whether the load and its corresponding deflection are axial (tension or compression), transverse (bending), shear, or torsional. It took Robert Hooke (1660) to formulate this notion into a *law*:

$$F = kx$$

in which the symbol k is the constant of proportionality, called the *stiffness*, that relates the force, F, to the displacement, x, from the spring's neutral position. For torsional systems, the expression is

$$T = k_T\theta$$

where T and θ are the torque and angular displacement, respectively, and k_T is the *torsional stiffness*. Stiffness is the cardinal characteristic of a spring. The SI units for k are newtons per metre (N/m), and newton-metres per radian (N.m) for k_T.

The elastic potential energy stored in a spring is $U = \int F dx$ (or alternatively $U = \int T d\theta$). For linear springs these integrals evaluate to $U = \frac{1}{2}kx^2$ (which is easily predicted, since the average spring load during deflection from rest to x is $\frac{1}{2}kx$), or $U = \frac{1}{2}k_T\theta^2$ for torsional springs.

2. Common types of mechanical spring

Figure 3.26 shows some of the more commonly encountered forms of mechanical spring. Helical springs are used in both compressive and tensile applications. A helical compression spring such as the one in Figure 3.26(a) is immediately recognisable from its almost cylindrical outer shape. The flattened ends provide a stable platform for transmitting compressive force. As we shall discover shortly, the primary loading experienced by the wire in such a coil is *torsional* about the helical centreline of the coiled wire. Despite this, the spring itself is *not* referred to as a torsion spring, but as a *compression* (or *tension*) spring, since this term reflects the nature of the spring's external loading and displacement.

Leaf springs, seen in Figure 3.26(b), are often used to provide suspension in vehicles such as railway and motor cars. The stack of compliant plates, or *leaves*, is strapped together around the centre, and acts as a set of beam springs in parallel. The unit is thicker towards the middle of its span where the bending moment is greatest.

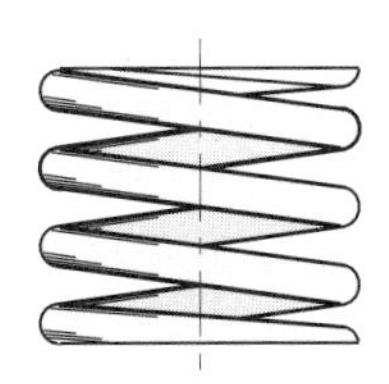

(a) *Simple helical compression spring*

(b) *Leaf spring*

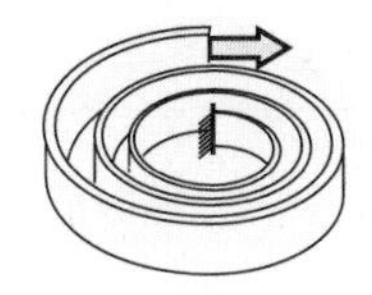

(c) *Clock type torsion spring*

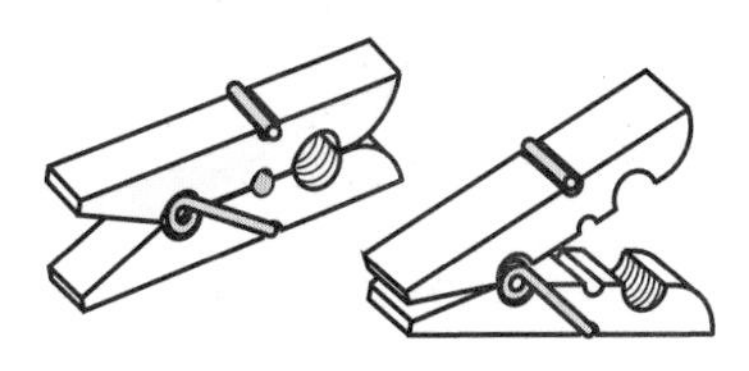

(d) *Clothes peg*

Figure 3.26 Examples of various types of springs

In Figure 3.26(c), a classic *clockwork-type* torsion spring is depicted. It is interesting to note that this type of spring is really just an extension of a single leaf spring, with the long *leaf* coiled into a spiral. Indeed, the leaf is a spirally curved beam, and experiences pure bending at every point along its length as the central shaft is wound about the spring's torsional axis.

Helical coils can also be employed as torsion springs, as exemplified by the clothes-peg spring of Figure 3.26(d). The central axis of the helical coil is always the torsional axis of the spring. We expect that most readers will possess sufficient domesticity to be familiar with this device!

Again it is noteworthy that the helical torsion spring in the clothes-peg example of Figure 3.26(d), similar to the spiral leaf in Figure 3.26(c), is subjected only to bending along its whole helical length. Contrast that to the torsional loading on the wire of the helical compression spring in Figure 3.26(a). When considering the stress state of the wire in helical coil springs, the following slightly counter-intuitive statement is true:

"The wire of a helical torsion spring is in bending, whereas the wire of a helical compression (or tension) spring is in torsion."

3. Some special types of springs and their applications

Additional examples, some commonplace and some unusual, are shown in Figures 3.27 to 3.35. They are presented not so much to seek after *design rules* for each case, but to demonstrate the diversity of *things that are springs*, and to pose the kinds of challenges that face designers in modelling the behaviour of springs in the face of such diversity. In fact, *design rules* do not exist in the literature for many of the weird and wonderful spring configurations confronting the designer. However an insightful engineer, with a modicum of structural mechanics theory, can develop quite good mathematical models to predict spring stiffness and other spring characteristics *from first principles* as required.

Clothes pegs are essentially compliant fasteners: they must be sufficiently flexible to accommodate a range of cloth thicknesses without generating too great a variation in clamping force. The old-style clothes pegs of Figure 3.27 achieve this by means of flexible *legs*, which are in effect beam springs.

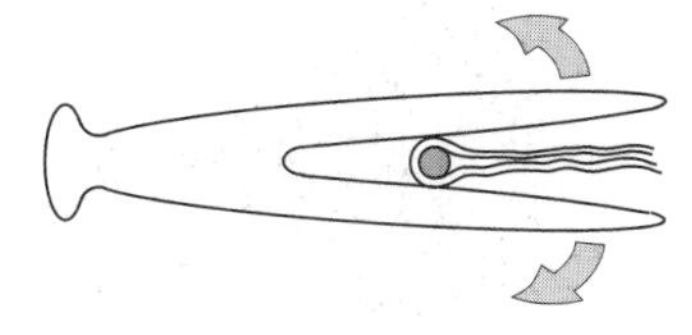

Figure 3.27 Wooden clothes peg

So how would the designer predict the performance of the humble, solid, wooden peg? Probably no peg manufacturer has ever done this by analysis — in all likelihood a skilled model-maker would have carved a few dozen pegs and sorted out a good shape by trial and error. But we suspect that is because the manufacturers had no enthusiastic engineer handy! The mathematical modelling is easy and the design variables are few.

Ex. 3.5 Peg stiffness

Assuming it is turned from a known species of wood (so the outer shape will have circular cross-sections), and has a symmetrical slot, develop a model for the structural behaviour of the wooden peg shown in Figure 3.27. List the key design variables. What are the key performance measures? Suggest and justify any reasonable simplifications to the mathematical model.

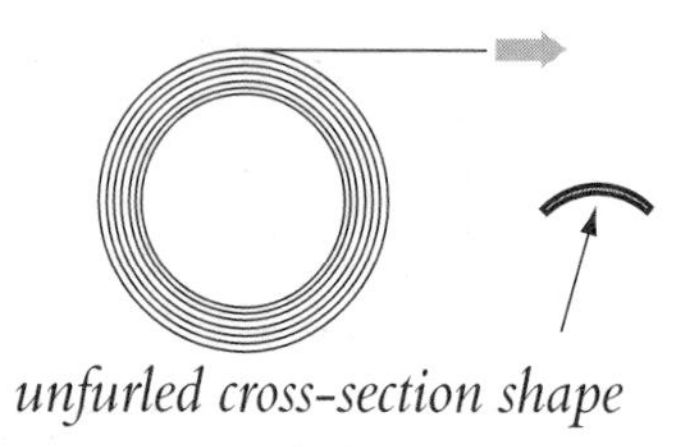

Figure 3.28 Negator spring ($k_T \approx 0$)

Most metal tape measures are of the type illustrated in Figure 3.28. Any metal tape that is wound onto the hub must maintain a flat cross-section, but as it unfurls it can resume its normally concave cross-section. Any tendency of the spiral spring to remain furled (due either to it being formed with a coiled repose shape, or being provided with a separate torsional return spring in some applications) is therefore negated by the forces arising from the metal tape as it becomes concave. Ideally, the resultant effort required to furl or unfurl the tape is minimal, and the effective torsional stiffness is zero.

Belleville springs are shallow conical frusta popular in specialised machine applications due to their compactness, their adaptability and their non-linear force/deflection characteristics (Figure 3.29). They are shown in various configurations in Figure 3.30. Belleville springs can easily be arranged in parallel to increase overall stiffness (with all cones facing the same way, as in the top-left figure), or in series to provide greater flexibility (alternate cones facing in opposite directions, as in the bottom-right figure), or in combinations of both.

This raises the question of how spring stiffnesses might be combined when springs are arranged in series or parallel configurations.

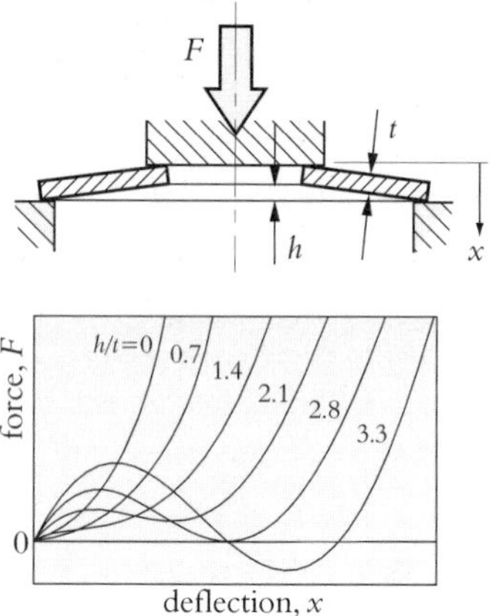

Figure 3.29 Disc or Belleville spring and force/deflection characteristics

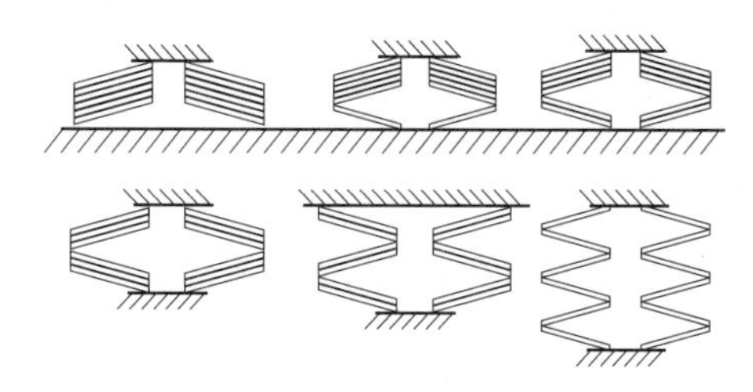

Figure 3.30 Belleville springs may be applied in series or parallel

Figure 3.31(a) shows two springs in series. In this configuration, each spring is subjected to the full applied axial load, and their deflections sum together so that the overall stiffness is reduced. Their combined stiffness can be easily calculated to be:

$$k_{TOTAL} = \left[\frac{1}{k_1} + \frac{1}{k_2} \right]^{-1} . \qquad (3.20a)$$

Figure 3.31(b) shows the same two springs in parallel. Here, each spring is constrained to undergo the same axial deflection as the other, but shares only a proportion of the total applied load, so their axial forces sum together and the overall stiffness is increased. Their combined stiffness is simply:

$$k_{TOTAL} = k_1 + k_2 . \qquad (3.20b)$$

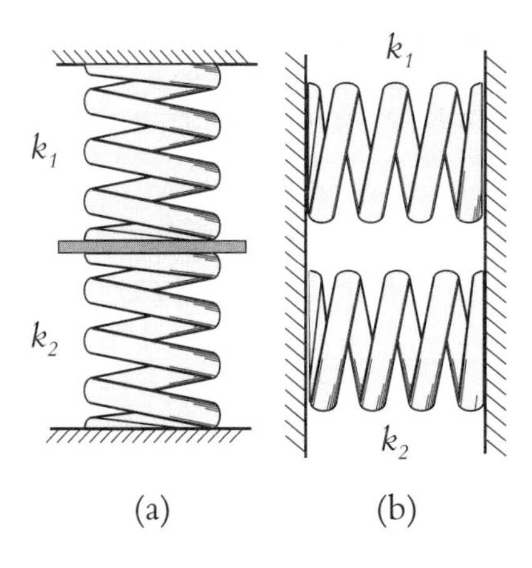

Figure 3.31 Springs in series and parallel

These results can be extended in a similar way for any number of combined springs. As a less obvious example of combined spring behaviour, consider the *diaphragm spring* shown in Figure 3.32, consisting of a series of radial beam springs cantilevered out from a relatively stiff outer conical ring. Each cantilever looks and behaves like a little diving board.

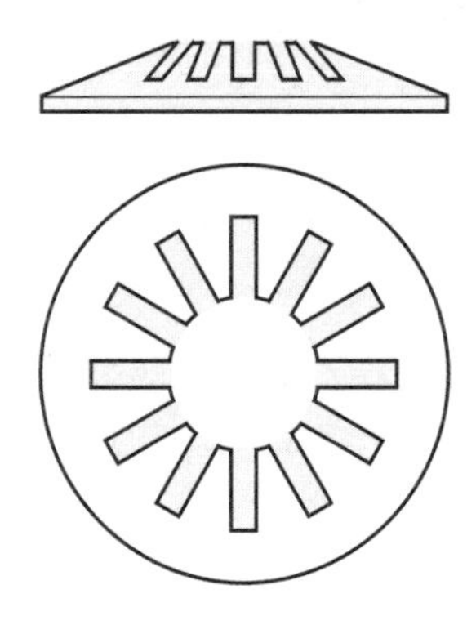

Figure 3.32 Diaphragm spring (k=constant)

Ex. 3.6 Diaphragm spring stiffness

Define the major design variables for the diaphragm spring shown in Figure 3.32, and arbitrarily choose some appropriate dimensional or material property values for each. Develop a model for the structural behaviour of the spring, simplifying your mathematical model where necessary. What is the best way to handle the small flexure of the outer conical ring? Calculate the overall stiffness.

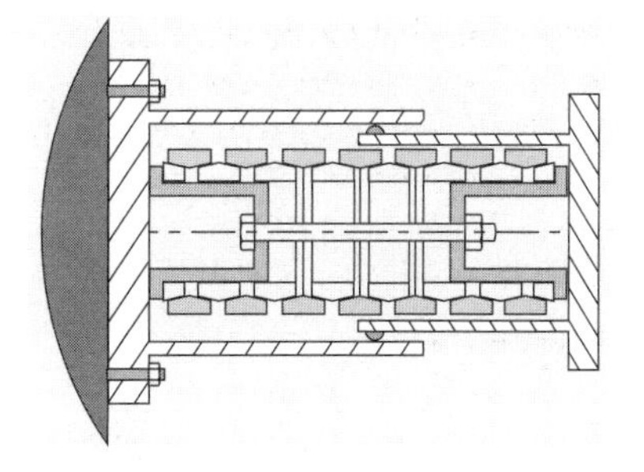

Figure 3.33 Energy absorbing application — railway buffer

Didn't you always want to know what lay inside those strange looking domed protruberances on the ends of railway carriages where they impact with their neighbours? Figure 3.33 reveals the secret, and it is indeed elegant. Railway buffers contain a series of rings in an alternating sequence: smaller diameter *compression rings* that have an outward-facing vee profile, and larger diameter *tension rings* that have an inward-facing vee profile. The rings are clamped loosely together by means of a central holding bolt between end fittings, with the conical faces of alternating rings meshing together as shown. Now imagine what happens when the buffer is subjected to a compressive impact. Axial compression squeezes the stack of rings together, and the oblique faces of the vee profiles tend to slide on one another. In effect, the compression rings get pushed into the tension rings. Because the opposing conical faces on each vee profile are symmetrical, the resultant force distribution on each ring is radial — radially inwards on compression rings, and radially outwards on tension rings. The effect is that the stack deflects axially, and dissipates a lot of energy in frictional heating as it does so.

Ex. 3.7 Railway buffer stiffness

This is a tricky one! Develop a model for the structural behaviour of the railway buffer shown in Figure 3.33, simplifying your mathematical model where necessary. You will need to assume a value for the coefficient of friction for greased steel-on-steel, but it will be rather small — say, $\mu \approx 0.1$. You will also need to consider how to relate the radial load on a ring to its radial strain. Calculate the overall axial stiffness of the buffer. Given the outer size constraints, what are the most effective design variables to employ in controlling the buffer's stiffness?

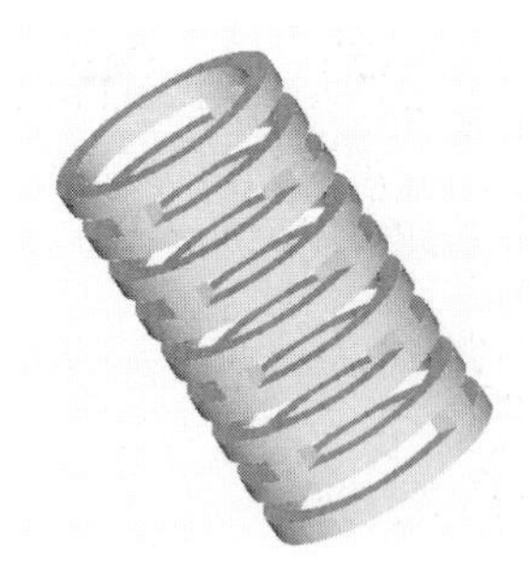

Figure 3.34 Special type of spring — plastic moulding application

The injection moulding process is capable of producing complex shapes quite cheaply. The example shown in Figure 3.34 is a special type of compression spring that has been fabricated in plastic using this process. It is actually a form of stacked beam spring.

Ex. 3.8 Plastic moulded spring behaviour

Using sketches to illustrate your answer, explain how the spring in Figure 3.34 works and describe how you would predict its stiffness. Be as specific as you can, but do not perform any detailed analysis.

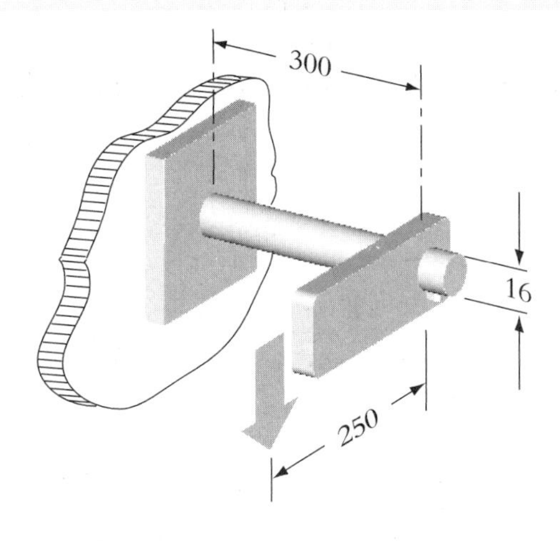

Figure 3.35 Simple torsion bar

Simple torsion bars are widespread in engineering equipment. A shaft driving a crank is really just a torsion bar; a wheel brace tightening a nut on a vehicle's wheel is really just a torsion bar; and Section 1.4.1 explored the possibility of considering that part of the common paper-clip is really just a torsion bar! Usually, however, the term is reserved for a fixed component that provides torsional support to other parts of a device, whilst at the same time providing some elastic compliance. Torsion bar suspension systems on motor cars are very widespread.

Ex. 3.9 Modelling a torsion bar

The overall cantilevered length of the torsion bar in Figure 3.35 is 300 mm, and its diameter is 16 mm. The crank fitted to it is made from solid rectangular bar. Both the crank and torsion bar are steel with E = 210 GPa, and both are horizontal initially. A vertical force is applied to the crank 250 mm from the centreline of the torsion bar. Calculate the stiffness of the unit. What proportion of the deflection at the load is due to torsion, and what proportion is due to bending? (Ignore the bending of the crank.)

3.3.3 Linear compression springs, static loading

1. Nomenclature

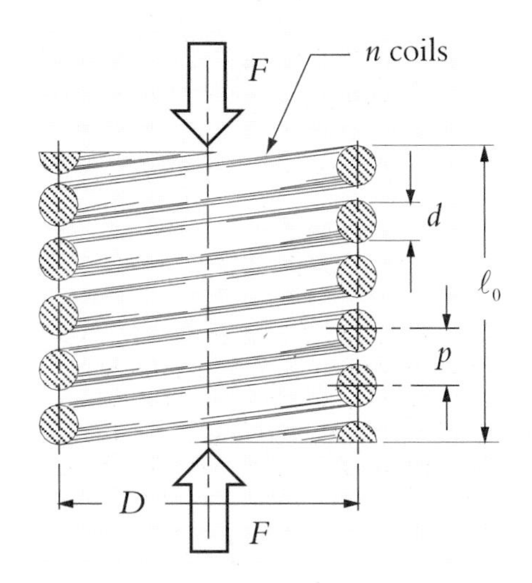

Figure 3.36 Simple compression spring

The coil geometry of any helical spring is defined uniquely by four variables:

d = wire diameter;

D = coil pitch diameter (the diameter of the

helix formed by the wire centreline, or equivalently the average of the outer and inner diameters of the coil);

p = linear pitch (the axial distance separating two adjacent coils); and

n = number of active coils.

Designers will principally manipulate variables d, D and n to achieve the desired operation and design requirements. Other dependent variables are sometimes used in describing the spring, for example:

ℓ_0 = free length of the spring.

This nomenclature is illustrated in Figure 3.36. We also define the following coordinate variables:

x = displacement of the end of the spring from its neutral position;

s = helical distance along the coil of wire, measured from the end of the spring.

2 Modelling stress in compression springs (static loading)

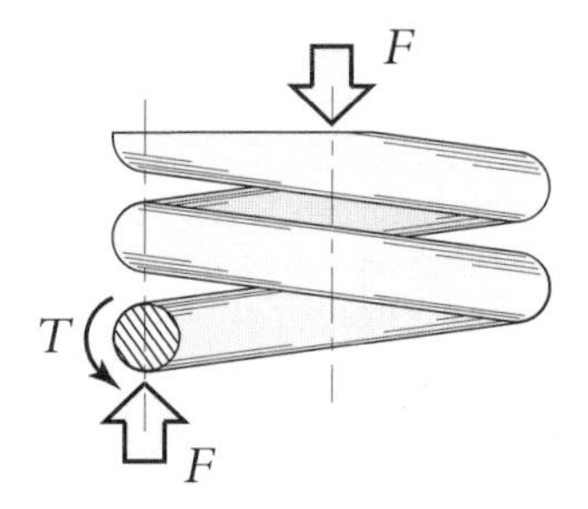

Figure 3.37 Free body diagram of a few "working" coils

No matter where the helical coil is notionally *cut* to reveal inner forces and torques in the wire, a free body diagram like the one in Figure 3.37 is obtained. Working from this we obtain the torque, T, acting about the wire axis due to the axial compressive force, F, as:

$$T = \frac{FD}{2}$$

The cut face of the wire is therefore exposed to shear stresses from two sources: that due to the shear force F acting parallel to the cross-section section of area $A = \pi d^2/4$, and the torsional shear stress arising from T, which varies with wire radius r as $\tau_T = Tr/I_P$.

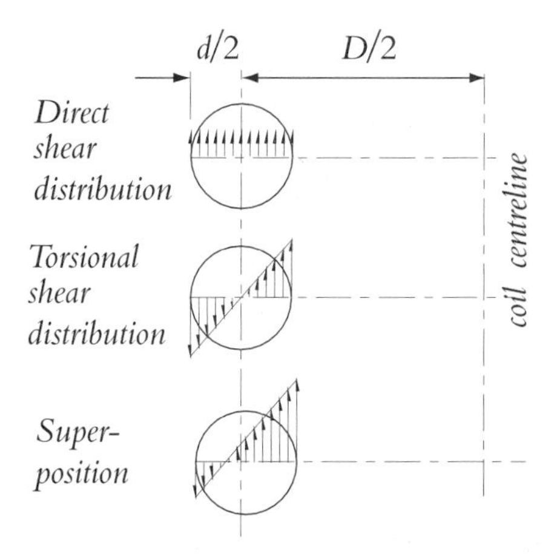

Figure 3.38 Superposition of the two types of shear stress in the spring section

These two shear stresses combine additively all over the wire cross-section, but the extreme cases occur along the horizontal diameter of the wire, since this is where the shear stresses both act in a vertical direction. Vertical shear stresses are plotted along this diameter in Figure 3.38.

The shear stress (positive upwards) acting on the inner and outer surfaces of the section is:

$$\tau = \pm \frac{Td}{2I_P} + \frac{4F}{\pi d^2}.$$

Recall that the polar second moment of area of the wire is

$$I_P = \frac{\pi d^4}{32}.$$

Therefore

$$\tau_{max} = \frac{8FD}{\pi d^3} + \frac{4F}{\pi d^2} = \frac{8FD}{\pi d^3}\left(1 + \frac{0.5}{C}\right),$$

where the coiling ratio is defined as

$$C = \frac{D}{d}. \qquad (3.21)$$

The parameter C is sometimes called the *spring index*, and is used widely within helical spring literature as an indication of the curvature of the coil.

It can be seen from Figure 3.38 that the maximum shear stress acts on the inner surface of the wire, since this is where the two superimposed shear stresses augment each other. The same result applies to tension springs also.

The previous equation for τ_{max} is sometimes rewritten as:

$$\tau_{max} = K_{shear}\frac{8FD}{\pi d^3}, \qquad (3.22)$$

where

$$K_{shear} = \left(1+\frac{0.5}{C}\right) = \left(1+\frac{d}{2D}\right). \qquad (3.23)$$

The parameter K_{shear} is the shear-stress multiplication factor for helical springs. It is a useful means of predicting the maximum *nominal* shear stress (unmodified by the effects of stress concentrations). As we shall see in Section 3.3.4 on cyclically loaded springs, stress concentration effects further elevate the actual peak stress in the vicinity of the inner surface of the wire. However these effects are highly localised and tend to disappear when the material of the spring yields locally (*sets*), so they are not critical for statically loaded springs.

3. Modelling deflection in compression springs (static loading)

We use the symbol s to denote coordinate distances along the helical centreline traced out by the wire of a coil spring. The relation between torsional shear strain and rotation for a short length, ds , of this wire is just like that for any shaft in torsion:

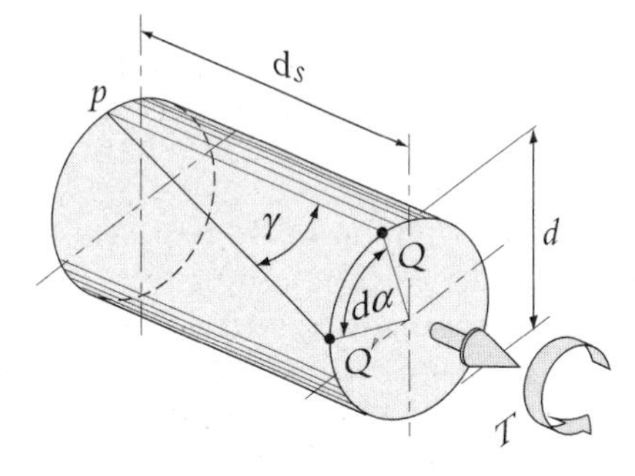

Figure 3.39 Shear strain and rotation in an elemental length of a helical spring

In the exaggerated figure above, torque T produces a small rotation dα in the element of length ds, by means of the shear strain γ in which the longitudinal line PQ rotates to PQ'. Since the small arc length QQ' (the circumferential deflection over the small axial distance ds) is subtended by both γ and dα, it must be that $\gamma \cdot \mathrm{d}s = \mathrm{d}\alpha \cdot d/2$. Now by applying Hooke's Law for circular bars to the shear stresses and shear strains at the surface of spring wire, we obtain:

$$\gamma = \frac{\tau}{G} = \frac{Td}{2I_p}\cdot\frac{1}{G}$$

or

$$\gamma = \frac{8FD}{\pi G d^3}, \qquad (3.24)$$

where

$$G = \frac{E}{2(1+\mu)}. \qquad (3.25)$$

The material constant G is the shear modulus of elasticity (approximately 90 GPa for most steels). Integrating:

$$\alpha = \int_0^{n\pi D}\frac{2\gamma}{d}ds = \int_0^{n\pi D}\frac{16FD}{\pi d^4 G}ds$$

$$= \frac{16FD^2 n}{d^4 G}. \qquad (3.26)$$

In Equation *(3.26)*, α is the total torsional rotation of the spring wire due to the applied force F. However, any rotation, dα, of wire at a radius $D/2$ from the central axis of the coil corresponds to an axial compression of dx = dα x $D/2$ in the coil. Therefore the overall compression of the coil spring is given by:

$$x = \alpha \cdot D/2 = \frac{8FD^3 n}{d^4 G}. \qquad (3.27)$$

The spring *rate* k (or its *stiffness*) is the force, F, required to cause a deflection x of the force at the centre of the coil. It is trivial now to re-express Equation *(3.27)* in terms of the spring *stiffness*, k:

$$k = \left(\frac{F}{x}\right) = \frac{d^4 G}{8D^3 n}. \qquad (3.28)$$

A similar result may be obtained by energy methods (*Castigliano's theorem*).

3.3.4 Stresses in linear compression springs, dynamic loading

Springs are very often subject to dynamic loading. Frequently the spring deflection will vary cyclically with time over a large number of cycles. Our approach to the prediction of stresses in helical compression springs in the previous section remains valid for dynamic loading, except that there is a further effect to be considered. Additional account must be taken of the effects of stress concentration due to wire curvature.

We found in Section 3.3.3 that the torsional and direct shear stresses may be superimposed to get the maximum shear stress in the coil wire, and that this occurs at the inner surface of the coil spring. This is precisely the location of a localised stress concentration phenomenon first reported and analysed by A. M. Wahl[3.3], who found that the shear stress on the inside surface of a curved bar is increased due to its curvature.

The Wahl correction factor for stress concentration due to curvature is a function of the geometric spring constant $C=D/d$, and is of the form:

$$K_{Wahl} = \frac{4C-1}{4C+1} + \frac{0.615}{C}, \qquad (3.29)$$

which is inclusive of the influences of direct shear (F/A) and torsional shear $(Td/2I_p)$ as well as wire curvature. For large C, this is only marginally different from the value of K_{shear}, given in Equation *(3.23)*. In practical spring manufacture, C usually takes values ranging from 6 to 12.

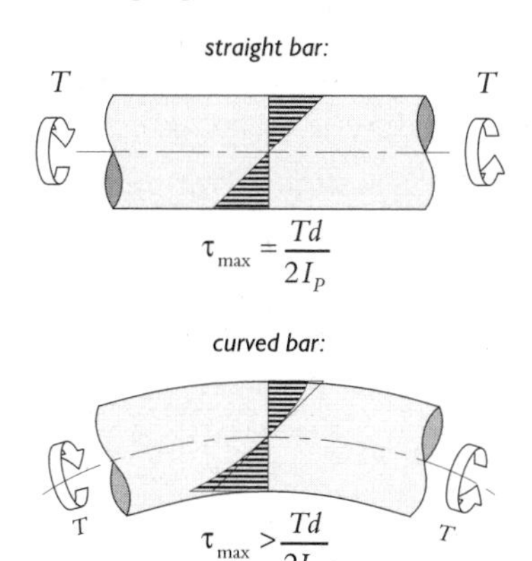

Figure 3.40 Stress concentration due to wire curvature

The Wahl correction factor was originally developed to account for stress concentration due to curvature in round bars under torsional loading. It is a shear-stress multiplication factor for helical springs, and is used in the equation

$$\tau_{peak} = K_{Wahl}\frac{8FD}{\pi d^3}, \qquad (3.30)$$

to predict the highly localised peak shear stress near the inner surface of the spring wire. This peak stress disappears due to local yielding once the spring is *set*, so it is not such an important issue under static loading. But consider the situation under dynamic loading (cyclic variation with time). If the initial spring load and deflection is such that τ_{peak} exceeds S_{sy} (the shear stress value corresponding to the onset of yielding) by a small amount, then, naturally enough, the spring wire will yield locally in the vicinity of this peak stress. The metal on the innermost surface of the wire will experience microstructural sliding of its grain boundaries — it will *flow* (invisibly) — until the strains associated with the local peak stress are *relieved*. In this condition the spring is said to be *set*. When the load applied to the spring diminishes again (say to zero), the deflection of the spring away from its *set* configuration will cause new peak stresses to develop at the inner wire surface, this time in the opposite direction to the first set of stresses. If the change in spring deflection is great enough, the new peak stresses may be sufficient to cause yielding again, reversing the first *set*. In the extreme, the spring will be repeatedly *set* and *reset*, suffering localised reversed yielding, and it will very soon fail.

For this reason it is essential for dynamically loaded springs that the value of τ_{peak} remain below S_{sy}. It is also highly desirable that such springs be designed to have stresses below the *shear stress endurance limit* (S'_{se}) so they will have a satisfactory dynamic life.

Unfortunately there is very little data on the value of S'_{se}. The best is that of F.P. Zimmerli[3.4], whose results suggest, interestingly, that the shear stress endurance limit is independent of material, size and tensile strength for all spring steels up to 10 mm in diameter. The only relevant issue is whether or not the spring wire has been *peened*. Peening is a process of bombarding the surface with fine steel particles. This process tends to remove surface defects as well as work hardening the surface, but is not always feasible, especially for small springs.

Zimmerli's results may be summarised as follows:

- for unpeened springs, $S'_{se} = 310$ MPa;
- for peened springs, $S'_{se} = 465$ MPa.

Springs subject to fluctuating loads should be designed against fatigue by means of the approach familiar from Chapter 2.4, but using shear

3.3 See Wahl, A.M. Helical Compression and Tension Springs, *ASME Paper A–38, J. Appl. Mech.* 2(1):1935

3.4 F.P. Zimmerli , 'Human failures in spring applications', *The Mainspring*, publication of Associated Spring Corporation, Bristol, Connecticut, No. 17, August-September 1957

properties (S'_{se}, S_{sy} and S_{sU}) to construct the *A-M diagram*. Shear stresses associated with dynamic loading, and hence the mean and amplitude of such fluctuating stresses (τ_a and τ_m), should be calculated on the basis of Equation *(3.30)*, using K_{Wahl}.

3.3.5 Other practical issues for compression springs

1. Buckling

Compression springs can buckle (like a column), but for $(\ell_0/D) < 4$ this is not a serious problem. Guiding the spring can overcome such buckling problems generally.

2. Bottoming

Compression springs can *bottom out* — they become solid when the coils are fully in contact with each other. In this way the stress in the section may be limited.

3. Binding

The average coil diameter, D, increases when a spring is compressed (and likewise decreases when it is stretched). This may lead to problems if the spring is required to operate in a cavity, so the designer must ensure sufficient clearance is provided. The worst case occurs when the spring is fully *bottomed*, in which case the outer diameter is:

$$D_{O,\max} = \frac{\sqrt{\pi^2 D^2 + p^2 - d^2}}{\pi} + d \cdot$$

3.3.6 Tension springs

The stress and deflection relationships presented above are generally applicable to helical tension springs also. However, in tension springs the stresses in the end supports may be more significant than in the coils due to stress concentrations. Cunning design of the end loops of wire can alleviate this problem to some degree. Some typical end connections in extension springs are shown in Figure 3.41.

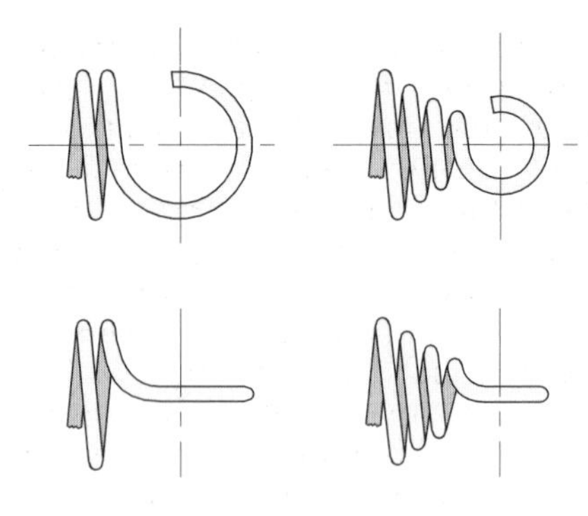

Figure 3.41 End designs for tension springs

3.3.7 Torsion springs

As discussed in Section 3.3.3, most torsion springs are, from a structural analysis point-of-view, curved beams in bending. Stresses in torsion springs are found from the theory of bending of curved beams. In general

$$\sigma_{B,\max} = K_B \frac{M}{Z_c}, \qquad (3.31)$$

where K_B is a correction factor to allow for stress concentrations in the bending stress distribution induced by curvature, and Z_c is the section modulus of the curved *beam*. For round wire,

$$Z_c = \frac{I_{ZZ}}{c} = \frac{\pi d^4/64}{d/2} = \frac{\pi d^3}{32}.$$

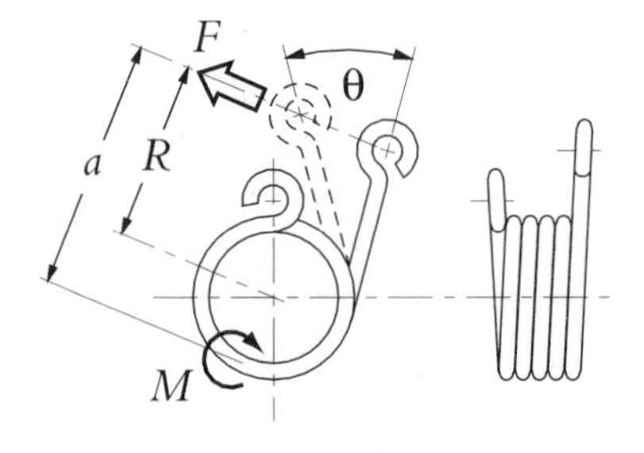

Figure 3.42 Bending moment in a torsion spring

3.4 F. P. Zimmerli, 'Human failures in spring applications', *The Mainspring*, publication of Associated Spring Corporation, Bristol, Connecticut, No. 17, August-September 1957

Torsion springs arise in a wide variety of shapes, and employ many techniques for exerting the torque, $T=FR$, about the torsional axis of the spring. The arrangement in Figure 3.42 is a typical one. As in this figure, it is often convenient to express the maximum bending moment, $M=Fa$, experienced by the wire as a product of some contact force, F, and a maximum perpendicular distance, a, to the wire centreline (a is the *bending moment-arm*, not to be confused with the *torsional moment-arm*, R). The maximum bending stress for a round-wire torsion spring can then be written:

$$\sigma_{B,\max} = K_B \frac{32Fa}{\pi d^3}. \quad (3.32)$$

Bending stress concentrations are similar in concept to those described for shear stress in Section 3.3.4. They are again most severe at the inner fibre of the wire cross-section. A conservative approximation to Wahl's formulation of the stress concentration there is given by:

$$K_B = \frac{4C-1}{4C-4}. \quad (3.33)$$

The total rotation of a torsion spring (having elastic modulus E, mean coil diameter D, wire diameter d, and n working (active) coils, with a force F acting on it about a torsional moment arm R) can be calculated from strain energy methods to be:

$$\theta = \frac{64FRDn}{d^4E}.$$

Some care needs to be taken in applying this equation, since:

(a) it uses the torsional moment, FR, not the maximum bending moment, Fa;

(b) the mean coil diameter decreases by the ratio $n/(n+\theta)$ when the spring is wound;

(c) the number of active coils should be adjusted to account for the flexural contribution of the spring's straight ends (if any);

(d) large inaccuracies can arise from inter-coil friction; and

(e) practical tests show that, by ignoring the effect of wire curvature, the equation underestimates the true rotation by about 6%.

The torsional stiffness, k_T, is the torque per radian of rotation. Introducing a correction factor, C_{curv}, to account for wire curvature, k_T may be found from the previous equation.

$$k_T = \frac{T}{\theta} = C_{curv} \frac{d^4E}{64Dn}. \quad (3.34)$$

Often the stiffness of torsion springs is expressed as a *spring rate*, k'_T, which is the torque per turn (2π radians) of the spring. Combining $C_{curv}\, 2\pi/64$ into a single constant, we obtain:

$$k'_T = \frac{d^4E}{10.8Dn}. \quad (3.35)$$

Ex. 3.10 Automatic garage door

An automatic garage door consists of several plate assemblies that are stored horizontally 3.2 m above the garage floor in a roller-track near the ceiling, and can be wound down along the curved track into their vertical *door* position by means of a small, geared electric motor. The motor is connected to the door plates through drive chains (exactly like a bicycle chain), which it drives with a long shaft and sprockets having a pitch diameter of 50 mm, as shown in Figure 3.43.

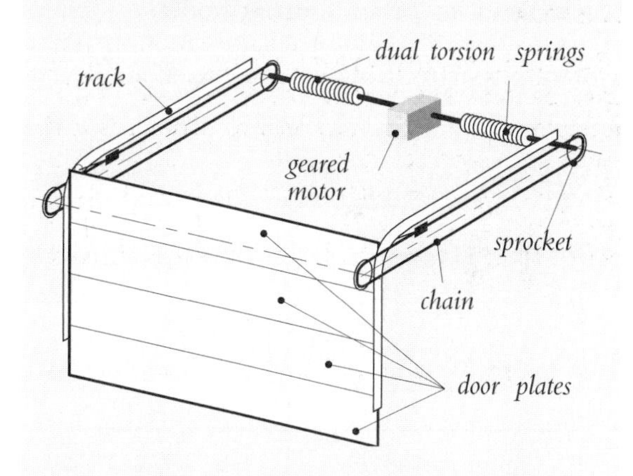

Figure 3.43 Garage door schematic

The plates have a total mass of 160 kg. The geared motor unit has plenty of torque capability to overcome friction in the track system, but not to support the weight of the door plates when they are nearly closed. For this purpose, two large helical torsion springs are fitted around the shaft between the geared motor and the chain's drive sprockets. Design the torsion springs to take the full weight of the door in the *almost-closed* position.

3.3.8 Properties of materials used for springs

Table 3.1 gives the properties for some commonly used spring materials. The following needs to be noted:

(a) Tensile strength varies with heat treatment and wire diameter. It is higher at small diameters.

(b) Temperature and shock loading are critical factors in material choice.

(c) Since failure is by shear, we need a value for yielding in shear: S_{sy}

(d) But S_{sy} is not always well reported. If tabulated values are not available, designers may use the following approximate values with a suitable safety margin: $S_{sy} \approx 0.577\, S_y \approx 0.433\, S_U$.

3.3.9 Vibrations and natural frequency

Springs are usually the means by which vibration is imparted to other inertial bodies — their restoring force drives the oscillatory motion in free vibration systems. But a spring is itself an inertial body, and the designer needs to be aware of the dangers of exciting vibratory motions within the spring at resonant frequencies.

For a spring supported between two parallel plates, its natural frequency (in Hz) is:

$$f = \frac{1}{2\pi}\sqrt{\frac{k}{m}}.$$

The mass m of the spring may be found from:

$$m = \frac{\pi^2 d^2 D n \rho}{4}.$$

Where a spring is compressed very rapidly (as in the valve gear of an engine), parts of the spring can *bump* against adjacent coils before the whole spring has time to adjust itself to the average compression. The resulting wave of motion is called *surging* and it can occur at frequencies as high as the 13th harmonic of the spring.

Typically an engine at 5000 rpm (83.3 Hz) has its valve springs operating at 2500 cycles/minute (41.7 Hz). The 13th harmonic of this motion is at 13x41.7=542 Hz. The spring must be designed to have a natural frequency well above this value — at least 15 or 20 times the forcing frequency is recommended.

Table 3.1 Spring material properties

Name	Designation	Description	S_U / MPa (2 mm wire)
Music wire (0.8-0.95%C)	ASTM A228–51	The best and toughest material used for springs; 0 °C < temp < 120 °C	2000
Oil tempered wire (0.6-0.7%C)	ASTM A229–41	General purpose spring steel; 0 °C < temp < 180 °C	1700
Chrome vanadium	ASTM A231–41	Most popular alloy spring steel —good for shockloading; 0 °C < temp < 220 °C	1800
Inconel	ASTM X–750	Clock springs	1300
Phosphor bronze	ASTM B–159	Commutator brush springs	900

Ex. 3.11 Simple compression spring example

The following information is available for a typical helical compression spring.

- material is music wire: $S_U = 2000$ MPa; $S'_{se} = 310$ MPa - unpeened;
- free length $\ell_0 = 105$ mm;
- outside diameter $D_0 = 14.3$ mm
- wire diameter $d = 2.24$
- working (active) coils $n = 21$
- preload = 45 N; maximum load = 225 N

Determine factor of safety against fatigue, natural frequency and surge frequency.

Ex. 3.12

(a) The recurve bow shown in Figure 3.22 is capable of imparting substantially greater energy per unit extension to a projectile than the simple *longbow*. Explain this using energy principles.

(b) The ballista (or catapult) shown in Figure 3.23 imparts energy to a projectile by winding up a torsion spring made up from heavy elastic ropes. A movie director has hired you to advise on a realistic ballista to be used in a period drama. Develop a simple mathematical model for the ballista, that will allow you to select a torsion spring necessary to hurl a 10 kg projectile over a distance of 100 m. Clearly explain all your design assumptions.

3.3.10 Summary of design rules for springs

(a) Compression and tension springs

For compression and tension springs, Equations *(3.28)* and either *(3.22)* or *(3.30)* are usually manipulated to achieve the desired spring behaviour.

Static loading:

Model

$$\tau_{\max} = K_{shear} \frac{8FD}{\pi d^3}, \qquad (3.22)$$

where

$$K_{shear} = \left(1 + \frac{0.5}{C}\right); \; C = \left(\frac{D}{d}\right). \qquad (3.23)$$

Require $\tau_{\max} < S_{sy}$.

Dynamic loading:

Model

$$\tau_{\text{peak}} = K_{Wahl} \frac{8FD}{\pi d^3}, \qquad (3.30)$$

where

$$K_{Wahl} = \frac{4C-1}{4C+1} + \frac{0.615}{C}. \qquad (3.29)$$

Require $\tau_{\text{peak}} < S'_{se}$.

Also check *A-M* diagram for τ_a and τ_m.

Also check natural vibration frequency for resonance.

Linear stiffness:

Model

$$k = \left(\frac{F}{x}\right) = \frac{d^4 G}{8D^3 n}. \qquad (3.28)$$

Also check compression springs for buckling, bottoming, binding.

(b) Torsion springs

Bending stresses:

Model

$$\sigma_{B,\max} = K_B \frac{32Fa}{\pi d^3}, \qquad (3.32)$$

where

$$K_B = \frac{4C-1}{4C-4}. \qquad (3.33)$$

Require $\sigma_{B,\max} < S_y$ or S'_e — as applicable.

Rotational stiffness ('spring rate'):

Model

$$k'_T = \frac{d^4 E}{10.8Dn} \text{ (per turn).} \qquad (3.35)$$

3.4 Columns: design for axial loading

In this section, we consider the design of *columns* — structural elements whose purpose is to resist axial compression. For many engineering components we can treat structural integrity for axial compression and tension in similar ways, with some special exceptions. One exception already noted was the difference experienced by structures subjected to dynamic loading (refer to the *A-M* diagram Figure 2.36). *Columns* represent another exception, in that they are designed to resist failure by elastic instability , *buckling*, rather than yielding or fracture.

Buckling is a form of failure due to elastic instability, in which a component might fail at loads significantly below that expected to be carried by the cross section from elementary failure theory.

The first to investigate the elastic deflection curves of beams was Leonhard Euler (1701-1783). While Euler's name remains directly associated with column buckling, it is worth noting that the problem was suggested to him by Daniel Bernoulli (1700-1782), who also contributed to the theory of elastic curves[3.5]. Bernoulli recognised Euler's mathematical capabilities, specifically his knowledge of the calculus of variations (*isoperimetric method*), which Euler used in deriving the curve shape for a beam under axial loading (*column*). J. L. Lagrange (1736-1813) later extended Euler's work to show that Euler's equation has a sinusoidal solution curve, with half wave length proportional to the length of the beam. Although his direct contribution now appears shrouded in history, it was Lagrange who was actually responsible for the classical equation we now refer to as the *Euler buckling curve*.

As more and more use is made of high strength alloys, the cross sectional areas of components are being reduced, and elastic instability has become an important consideration in the design of engineering components.

The first to study buckling of thin rectangular bars was Ludwig Prandtl, who wrote his doctoral thesis on this subject. (*Kipperscheinungen*, Dissertation Munich,1899).

The study of this subject was considerably extended in the early part of this century by several workers including Timoshenko (buckling of I-beams, axially loaded curved sheet panels and torsional buckling of members with thin cross section) and R. von Mises (cylindrical shells under combined axial and lateral pressure; *VDI*, vol 52, p750, 1914).

The buckling of thin tubes under pure bending was studied by W. Flügge, Ing. Arch. Vol. 3, p463, 1932.

3.4.1 Identifying columns

Together with beams, columns constitute the earliest and most enduring structural components employed by humans to fabricate their built environment.

Figure 3.44 Columns and beams in the Neolithic circle at Stonehenge, England

Figure 3.45 Columns of the Temple of Isis at Aswan, Egypt.

3.5 L. Euler, 1759, Sur la force des colonnes, *Mem. Acad. Berlin*, V. 13, quoted in Timoshenko 1953

Beams support loads transversely, and columns support loads axially, and the two are used generally in combination. It is difficult to say very much concisely about the role of the column in ancient architecture — it is so generic, and suffused throughout human experience. The *post and lintel* construction seen in many ancient monuments, such as those depicted in Figures 3.44 and 3.45, typify the relationship between the beam and the column. The columns of stone or brick that survive from antiquity are massive, squat affairs, and always vertical.

Columns are not always like that — the wooden *pushing pole* used to propel a punt on the River Cam in Figure 3.46 is a good example of a column, and it must be long, slender and light to promote portability.

Figure 3.46 Pole as column

This highlights the structural designer's chief challenge in relation to columns: how to make them light and slender, whilst still strong enough to fulfil their pushing function safely.

Columns are rarely found in isolation. In complex structural systems, like the timber frame of a house, or the truss structure of an electricity transmission tower, or a suspension bridge, columns are used in conjunction with many other components such as beams, plates and tension elements.

Ex. 3.13 Find the column(s)

Identify the columns (structural members subject to axial compression) in the timber house frame, electricity transmission tower and suspension bridge shown in Figure 3.47.

Our concern in this chapter will be mainly with axial compression. We begin however by widening the topic to include both tensile and compressive forms of axial loading. This is not because we suspect novice designers to be unfamiliar with the very simple case of tensile stress — rather the reverse! Because design to resist axial tension is already a familiar problem, we take this opportunity to exemplify and develop a General Design Method applicable to all structural integrity design problems.

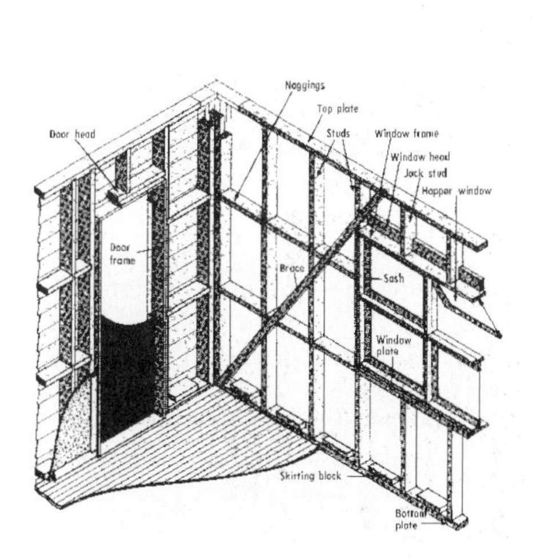

(a) *Stud wall in building*

(b) *Transmission tower*

(c) *Suspension bridge*

Figure 3.47 Columns in complex structures

3.4.2 Design against failure in tension — an example of a General Method of Design against structural failure

The structural integrity of tensile elements is the simplest possible case that an engineering designer can be faced with. After all, the stresses experienced by simple tensile links are very similar to those in uniaxial tension tests conducted in the materials laboratory to establish basic structural properties such as yield stress and ultimate tensile strength.

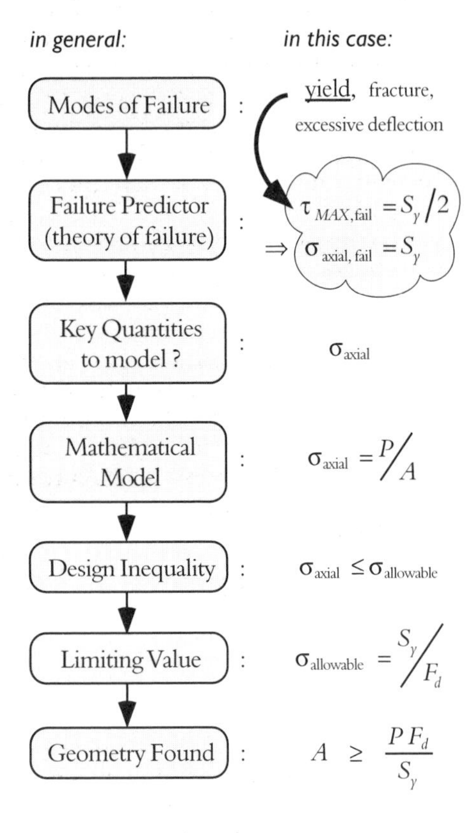

Figure 3.48 Flow Chart of the General Method of Design against structural failure

In designing against the structural failure of a particular element, the designer first considers its various possible modes of failure. In this case of an element whose task is to sustain axial tension, there are three principal modes of failure: failure by yielding in tension, failure by tensile fracture, and functional failure by excessive deflection (extension). We will consider only one of these (yielding), although in practice all likely modes of failure would need to be reviewed separately.

The next step in the General Method of Design is to choose a failure predictor. Failure predictors enable the designer to take the results of uniaxial tension tests in the laboratory and apply them to the more complex multiaxial stress states of real world components. So failure predictors, sometimes called *theories of failure*, may be needed for any of the modes of failure which involve prediction of the component's strength. In this trivial case of design against yielding in tension, the need for a failure predictor is somewhat redundant because there is no translation required between the loads imposed in laboratory tests and the kind of loading to which the component is subjected: they are both uniaxial. Nevertheless for illustration let us choose a failure predictor and apply it. An appropriate failure predictor for ductile materials is the *MSFP* (maximum shear stress failure predictor). This predicts that the component will fail when the maximum shear stress anywhere within it reaches the value of maximum shear stress existing in a uniaxial test specimen at the instant of failure, namely $S_y/2$.

$$\tau_{MAX,\text{fail}} = S_y/2$$

This is equivalent to saying that the axial stress at the point of failure is equal to S_y:

$$\sigma_{\text{axial,fail}} = S_y$$

— a rather obvious result. It does however highlight for us that the axial stress is the key quantity the designer must model. Fortunately for the designer, in this case it is a simple challenge to model this key quantity! The axial stress is expressed in terms of the applied axial tension, P, and the cross-sectional area, A.

$$\sigma_{\text{axial}} = P/A$$

Continuing with the sequence of steps in our General Method of Design, the next step is crucial — and common to all constraint-based design processes. The designer writes down a design inequality, which establishes an upper or lower limit on the modelled quantity. In this case we can simply state that the axial stress must always be less than or equal to some allowable value. That allowable value is chosen to be the failure value moderated by some factor of safety, F_d.

$$\sigma_{\text{axial}} \le \sigma_{\text{allowable}}$$

$$\sigma_{\text{allowable}} = S_y / F_d$$

Having established all of the above quantities and relationships, the designer is now in a position to determine the geometry of the component, since the cross-sectional area can be expressed as a function of the material properties and applied load.

$$A \geq \frac{P F_d}{S_y} \qquad (3.36)$$

The several stages of our General Method of Design for structural components are set out in the flow chart of Figure 3.48.

In this and later chapters, be alert for the development of design methods along these general lines as we consider a wide variety of engineering components.

Ex. 3.14 A little quiz on failure predictors

Would the choice of a different failure predictor (for example, *DEFP* instead of *MSFP*) alter the resulting design inequality in the above case of a component loaded in tension? Why or why not?

3.4.3 Design against failure in compression

Columns are structural elements defined by two main characteristics: their *shape* (they generally possess a long axis), and the nature of the *loads* to which they are subjected (they experience axial compressive forces). Like many common engineering components, they are called by a range of names, including column, strut, brace and prop. If we regard their compressive axial loading as a given common to all columns, then the main basis for distinguishing between various instances of column is their length, or more particularly, their slenderness. Slenderness is a defining characteristic of a column which we will later describe quite precisely, but for the moment let us use *slenderness* loosely to express a column's length relative to its width.

3.4.4 Modes of Failure for columns

Columns that are long and slender fail by buckling — an elastic phenomenon characterised by large lateral deflections perpendicular to the direction of the applied load. When a long column buckles, its material does not experience yielding at all (at least in the initial stages of that buckling event), but rather bows out sideways (Figure 3.49).

Figure 3.49 Buckling of a long, slender column.

Buckling failure is a form of elastic instability, and as such is usually accompanied by all the characteristics of that class of failure, including suddenness, lack of forewarning, and severity of structural displacement. The buckled column may suddenly lose a large proportion of its ability to resist axial deflection, and so if the axial force persists the column will collapse very quickly. If however the axial force is removed before the column collapses completely, then the elastically buckled column may spring back to its undeformed shape, sustaining little or no permanent damage.

At the other end of the spectrum, consider what happens to columns that are relatively *short* and *stumpy*. In the extreme, a very short column acts just like a solid block which under excessive compression will simply experience compressive yielding when the stress reaches the material's yield limit (Figure 3.50).

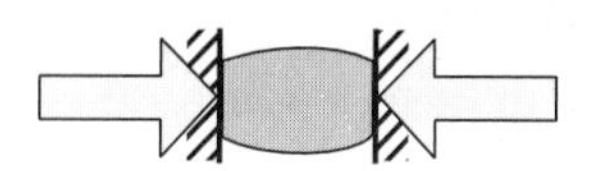

Figure 3.50 Buckling of a short, stumpy column.

Between these extremes of very long and very short columns, we may speak of columns having intermediate slenderness. These exhibit a combined failure mode that involves both yielding and large lateral deflections.

We seek to develop a simple model for the behaviour of columns that will encompass this range of behaviours.

3.4.5 Buckling of Long Columns

As already noted in the introduction to Section 3.4, the mathematical description of buckling behaviour in long columns was first developed by the Swiss mathematician Leonhard Euler. His analysis assumes the buckling phenomenon to be elastic, and so uses elastic beam theory to establish an interaction between axial force and the corresponding bending moment along the span of the column. (This bending moment is developed within the column when the axial force is offset from the column's neutral axis due to lateral deflection.) The familiar mathematical derivation is based on the generic long column shown in Figure 3.51, and is set out below with standard symbols.

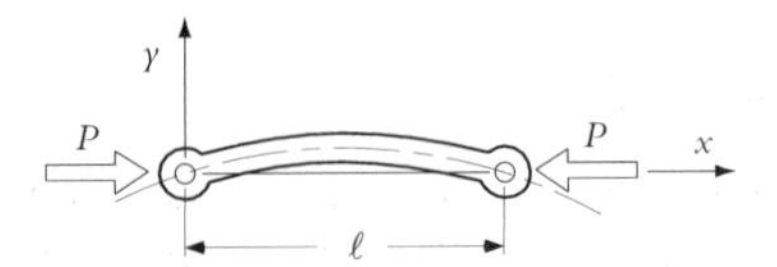

Figure 3.51 The generic long column.

Figure 3.52 shows the left-hand end of the long column in Figure 3.51 as a free body, sectioned to reveal the internal bending moment M set up when the axial force P is offset by lateral deflection y.

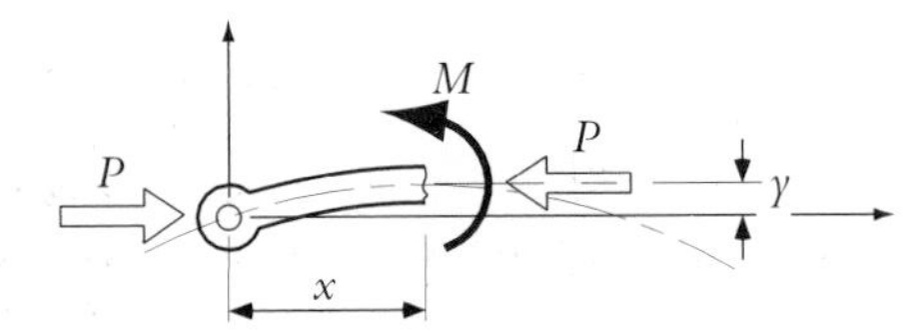

Figure 3.52 Free body for Euler buckling model.

Moment equilibrium dictates that the bending moment is given by $M = -Py$. From elementary elastic beam theory, we know that

$$\frac{M}{EI} = \frac{1}{R} \approx \frac{\mathrm{d}^2 y}{\mathrm{d}x^2},$$

where $1/R$ is the curvature of the column. The above two relationships provide us with the defining differential equation for Euler buckling:

$$\frac{\mathrm{d}^2 y}{\mathrm{d}x^2} + \frac{P}{EI} y = 0, \qquad (3.37)$$

for which a general solution is

$$y = C_0 \cos\sqrt{\frac{P}{EI}}x + C_1 \sin\sqrt{\frac{P}{EI}}x.$$

When the first boundary condition ($y=0$ at $x=0$) is applied to this equation, we obtain

$$0 = C_0 \cos(0) + C_1 \sin(0) = C_0,$$

and when this zero value for C_0 is combined with the second boundary condition ($y=0$ at $x=\ell$), the result is

$$0 = C_1 \sin\sqrt{\frac{P}{EI}}\ell.$$

The above equation is satisfied either by the trivial solution that $C_1=0$ (corresponding to a *column* which *remains straight*), or by the argument of the sine function being some integer multiple of π, that is:

$$\sqrt{\frac{P}{EI}}\ell = n\pi,$$

or

$$P = \frac{n^2\pi^2 EI}{\ell^2}. \qquad (3.38)$$

Values of P given by Equation *(3.38)* lead to indeterminate values of C_1, and therefore to unrestrained lateral deflections (i.e. *buckling*!). The smallest non-zero value of axial force P to cause buckling occurs when $n=1$. This force is termed the *critical buckling load*.

The outcome of this analysis is that for the arrangement shown in Figure 3.51, the critical buckling load of the column is found to be:

$$P_{cr} = \frac{\pi^2 EI}{\ell^2}. \qquad (3.39)$$

A clear interpretation of the critical buckling load is important. Mathematically, P_{cr} is the load at which the deflection, y, first becomes indeterminate. Physically, this corresponds to the onset of large deflections. In other words, P_{cr} is the load that will cause buckling. It is therefore a measure of the column's strength in resisting failure by buckling. The structural safety of a column can be ascertained by comparing the critical buckling load, P_{cr}, to the actual load, P, to which the column is subjected.

3.4.6 Some key terminology

Although P_{cr} is an important direct measure of the axial load-carrying capacity of a column, it is not the most commonly used measure of column strength. This is usually expressed in terms of the quantity (P_{cr}/A), derived as follows.

By definition, the second moment of area of the section, I, can be expressed as:

$$I \equiv Ar^2 \qquad (3.40)$$

where r is the *radius of gyration* (discussed later in Section 3.4.8). Therefore we may rewrite Euler's equation as

$$\left(P_{cr}/A\right) = \frac{\pi^2 E}{\left(\ell/r\right)^2}. \qquad (3.41)$$

where

(P_{cr}/A) = *column stress* (since it possesses the units of *stress*), and

(ℓ/r) = *slenderness ratio* (a key property of the column's *shape*).

Once again it is important to understand the notion of column stress clearly. It is not the actual stress in the column, but rather the value of axial stress that would lead to buckling failure in a long column. It is therefore a useful index of the buckling strength of a given column (analogous to S_y for tension), and is a function of the column's material properties, (E), and, its slender shape (ℓ/r). Indeed the slenderness ratio, (ℓ/r), is precisely what we mean when we speak of the slenderness of a column.

Equation *(3.41)* clearly states an inverse-square relationship between column stress and slenderness ratio, and this is shown below in Figure 3.53.

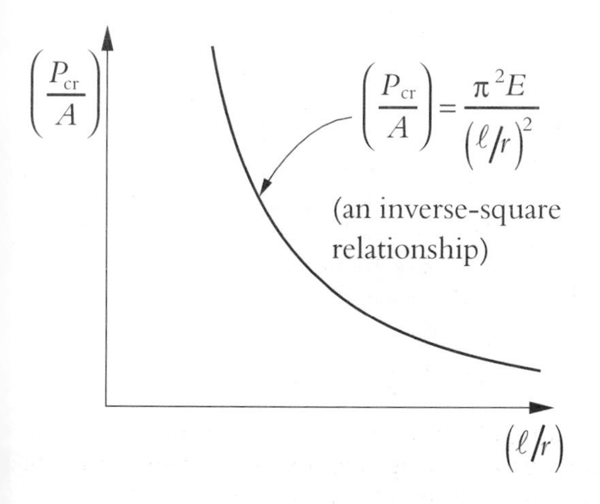

Figure 3.53 Euler buckling of long columns.

3.4.7 Effective length

So far we have said nothing about the manner in which a column is supported at its ends. The generic column of Figure 3.51 was shown with pinned simple supports at either end. This was found to give rise to a buckled shape in the form of a half sinewave of length ℓ. Let us distinguish between the effective length, ℓ, and the actual length of a column, L, and define the ratio between the two:

$$m \equiv \left(L/\ell\right). \qquad (3.42)$$

Furthermore, we take the effective length ℓ to correspond always to the half-sinewave length of the deflected shape. Therefore the ratio m is equal to unity in the case of a simply-supported column (with two pinned ends). The value of the ratio m for a given column depends upon the nature of its end supports, sometimes referred to as the *end fixity* of the column. The theoretical value of this ratio for four common cases is shown below in Table 3.2.

We note two interesting features of this table. Firstly the values of m which are greater than 1 are non-conservative. The shorter effective lengths of cases C and D rely upon the resistance to rotation of the built-in ends to provide a stiffening influence on the column. In practice, fully fixed end supports are difficult to achieve, and judgement must be used in these cases. Shigley (1986) recommends a more conservative value of $m \approx 1.1$ for both fixed-pinned and fixed-fixed column end conditions (cases C and D).

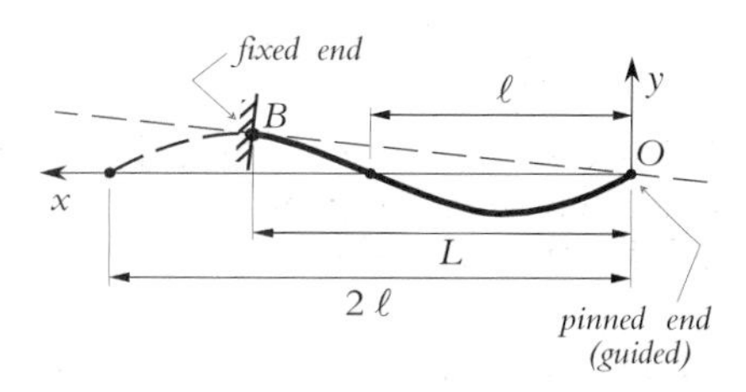

Figure 3.54 Geometry of a fixed-pinned column

Secondly the value of effective length at approximately 0.7 of the actual length in case C arises from an intriguing piece of geometrical analysis presented below. For some reason, use of the ratio $m=\sqrt{2}$ for this case has crept into the literature, particularly in textbooks written by civil engineers! We make a claim for $m = 1.43030$, but perhaps the distinction is academic.

Table 3.2 Effective length and end fixity

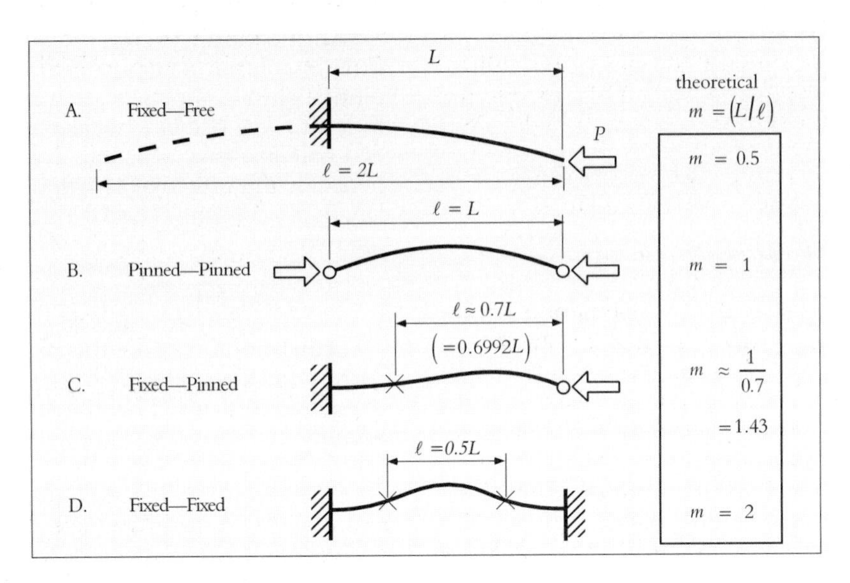

In Figure 3.54, we define the origin O of the deflected column shape (a *sine curve*) to be at the pinned end and the fixed (*built in*) end to be at the point B. Since the pinned end is guided to lie along a line perpendicular to the built–in wall, line OB must be tangent to the sinusoid at B. The equations to the sine curve and the line are easily written down, as stated in Equations *(3.43)* and *(3.44)*, respectively.

$$y = C_1 \sin(\pi x/\ell), \qquad (3.43)$$

$$y = C_2 x\,. \qquad (3.44)$$

The coordinates of point B must satisfy both these equations, so

$$C_1 \sin(\pi L/\ell) = C_2 L. \qquad (3.45)$$

Furthermore, the line and curve are tangent at point B (their slopes must be equal there), so

$$\frac{\pi C_1}{\ell}\cos\left(\frac{\pi L}{\ell}\right) = C_2\,. \qquad (3.46)$$

Now by substituting Equation *(3.46)* into *(3.45)*, cancelling C_1, and using the definition of m from *(3.42)*, we find that Equation *(3.46)* becomes:

$$\tan(\pi m) = \pi m\ .$$

The smallest non-zero value of πm which satisfies this requirement is 4.49340946 radians, and therefore we determine the theoretical value of m for the case of a fixed-pinned column to be

$$m = 4.49340946/\pi = 1.43030.$$

3.4.8 Radius of Gyration

From previous Equation *(3.40)* we can write the definition of the radius of gyration as

$$r \equiv \sqrt{\frac{I}{A}}\,. \qquad (3.47)$$

Just as the principal second moment of area, I, and the section area, A, are properties of the section, so too is the radius of gyration. Note also that the radius of gyration pertains to a particular buckling axis through the section. For instance we should usually distinguish $I_{zz} \equiv A(r_{zz})^2$ from $I_{yy} \equiv A(r_{yy})^2$ — the two are only equal in certain special cases such as round or square cross-sections. Since in general $I_{zz} \neq I_{yy}$, a column will not have the same radius of gyration about its two principal section axes.

Care is necessary in describing these terms. For example, in the expression $I_{zz} \equiv A(r_{zz})^2$, r_{zz} is the radius of gyration *about the z-axis* of the column's cross-section. This of course corresponds to bending (during buckling-induced deformation) about the z-z neutral axis, and therefore buckling deflections in the y-direction (or equivalently, *in the x-y plane*).

Compare these expressions with the basic definitions of second moment of area such as

$$I_{zz} \equiv \int y^2 dA \text{ and } I_{yy} \equiv \int z^2 dA,$$

where dA is, in the case of I_{zz}, an element of area located at a distance y from the z-z neutral axis. This suggests that a physical interpretation of radius of gyration is as shown in Figure 3.55, where the section is replaced by an equivalent section having the same second moment of area, I_{zz}, and area, A, but with all of that area concentrated at a uniform distance r_{zz} from the neutral axis. The radius of gyration, r_{zz}, is therefore the *half-height* of this equivalent section. In the case of a solid rectangular section of height h, it is easily shown that the radius of gyration is $h/\sqrt{12}$, or roughly $0.289h$. For a circular section of diameter d, the radius of gyration is $d/4$.

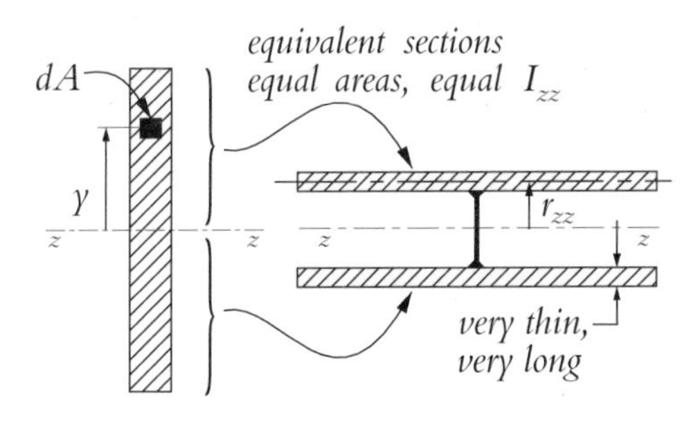

Figure 3.55 Radius of gyration interpreted.

A subtle fact concerning the concepts of radius of gyration, effective length and slenderness ratio is that each of them may have a different value about the two principal axes of a section. Just as a column's two lateral buckling directions may have different corresponding radii of gyration, so too they may have different effective lengths, ℓ, as is illustrated in the following diagram (Figure 3.56).

The column shown is only guided vertically at its mid-span and so may experience vertical deflections there. Therefore the buckling shape in the vertical plane is a half sinewave, whereas the buckling shape in the horizontal plane is a full sinewave, effectively pinned in the middle. In general it is best to treat both ℓ_{zz} and r_{zz} (for vertical buckling), and ℓ_{yy} and r_{yy} (for lateral buckling), as being independent. So a column needs to be defined by two separate slenderness ratios — one for each axis.

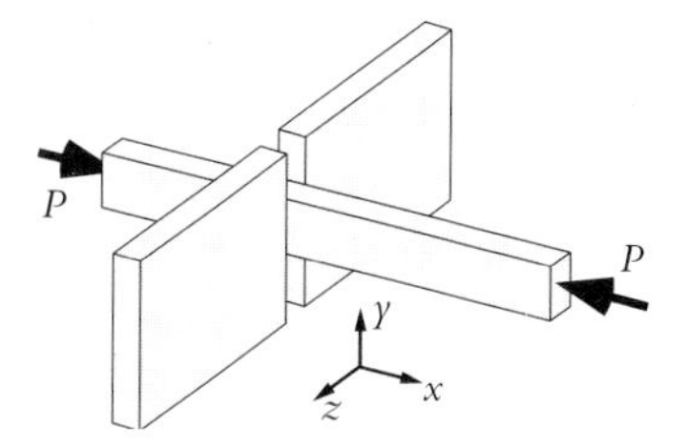

Figure 3.56 Different effective lengths for two buckling directions.

Ex. 3.15 Preferred direction of buckling

Assume that the column in Figure 3.56 has a rectangular cross section whose height is twice its width, and that the vertical guide located midway along the column allows unrestricted vertical displacement but zero horizontal displacement, and that the ends of the column are pinned. As the axial force P increases without limit, in which direction will the column buckle first?

3.4.9 Behaviour of short and intermediate columns

Euler buckling theory clearly fails to describe short columns, since it predicts infinite resistance to compressive loads for $(\ell/r) \to 0$. We know that very short columns fail (by *yielding*) when the axial compressive stress reaches S_y. Many empirical and semi-empirical methods have been proposed for matching the experimental data. The *Johnson Parabola* is one of these curve-fitting methods. It is an inverted parabola in the $[\ell/r - P/A]$ plane, symmetric about the point $(0, S_y)$ and tangent to the Euler curve.

Figure 3.57 represents a model for predicting the failure of columns across the full spectrum of slenderness ratios, from very short to very long columns. The *Johnson-Euler* curves on this diagram represent combinations of column stress and slenderness ratio that correspond to structural failure of the columns.

The coordinates of the transitional point between the Johnson parabola to the Euler curve are solved by satisfying the slope requirements of both curves at that point simultaneously. It is noteworthy that the slenderness ratio associated with this tangency point, $(\ell/r)^*$, is a function of material properties alone.

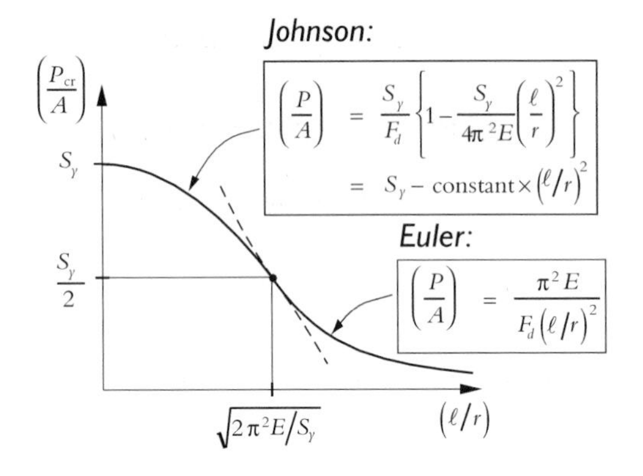

Figure 3.57 Conjunction of Johnson and Euler curves.

$$\left(\ell/r\right)^* = \sqrt{\frac{2\pi^2 E}{S_y}} \tag{3.48}$$

We refer to this value as the transitional slenderness ratio, since it represents the transition from the Johnson parabola to the Euler curve. (The fact that the vertical coordinate of this point equals $S_y/2$ may be regarded as a happy mathematical coincidence.)

3.4.10 Design Method

It only remains now to introduce a *factor of safety* into our calculation. Let us define a *design curve* as being a fraction ($=1/F_d$) of the Johnson-Euler curves, that is:

$$\left(\frac{P}{A}\right)_{\text{design}} = \frac{1}{F_d}\left(\frac{P_{cr}}{A}\right).$$

For structural safety we require the actual stress in the column, (P/A), to be less than or equal to the design value. That is:

$$\left(\frac{P}{A}\right) \le \left(\frac{P}{A}\right)_{\text{design}}.$$

This constraint can be represented graphically as a safe design space on our $[\ell/r - P/A]$ plane, like the shaded area on Figure 3.58. Depending on whether the column slenderness ratio lies either to the left or the right of the transitional value, $(\ell/r)^*$, the safe design space is defined by one of the two following inequalities:

Johnson:

$$\left(\frac{P}{A}\right) \le \frac{S_y}{F_d}\left\{1 - \frac{S_y}{4\pi^2 E}\left(\frac{\ell}{r}\right)^2\right\} \tag{3.49}$$

Euler:

$$\left(\frac{P}{A}\right) \le \frac{\pi^2 E}{F_d\left(\ell/r\right)^2} \tag{3.50}$$

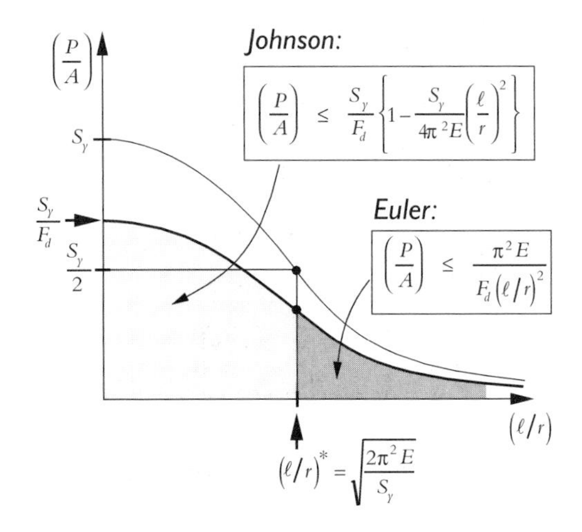

Figure 3.58 Safe design space on graph of column stress versus slenderness ratio.

3.4.11 Design Algorithms for Columns

In designing a column, the sequence of steps pursued by the designer is governed by the type of design problem being solved. Most commonly, the designer begins with a material and some functional requirements such as length, nature of end constraints, load to be supported, and an appropriate factor of safety. From these it is required to determine the minimum geometry of the column cross-section.

Another common situation is where the designer is required to *rate* a column, that is, to determine its safe allowable load based on its known dimensions, material, and required factor of safety. A sequence of calculations fulfilling both these objectives is shown in Figure 3.59.

Other permutations of design parameters and design variables are also possible. Their formulation is left as an exercise for students.

3.4.12 Eccentric Loading

Columns are never perfectly straight. The idealised Euler analysis presented earlier assumed an initially straight column, however a more realistic model would include an initial eccentricity of the load with respect to the geometrical centreline of the column, due to load misalignment or some manufacturing inaccuracy in the column (Figure 3.60).

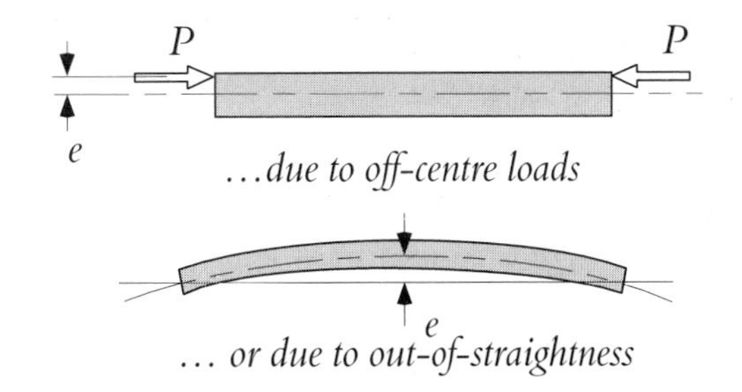

Figure 3.60 Sources of eccentricity.

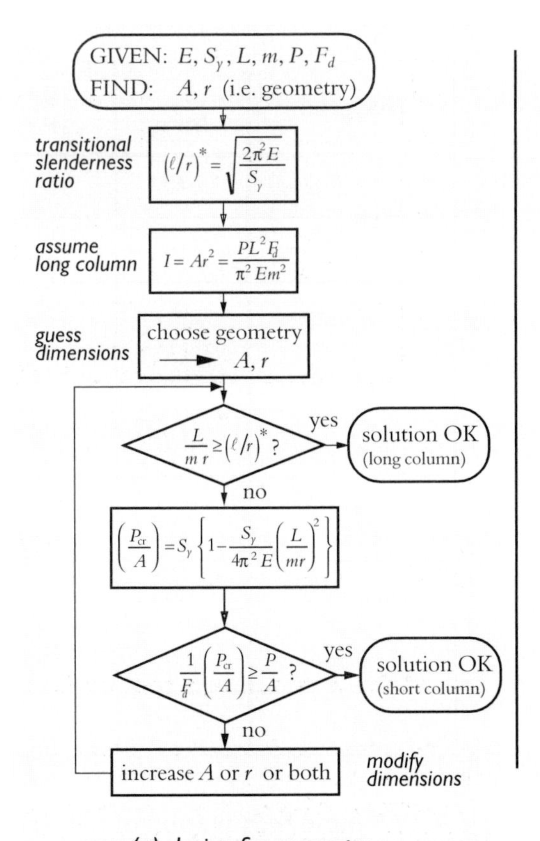

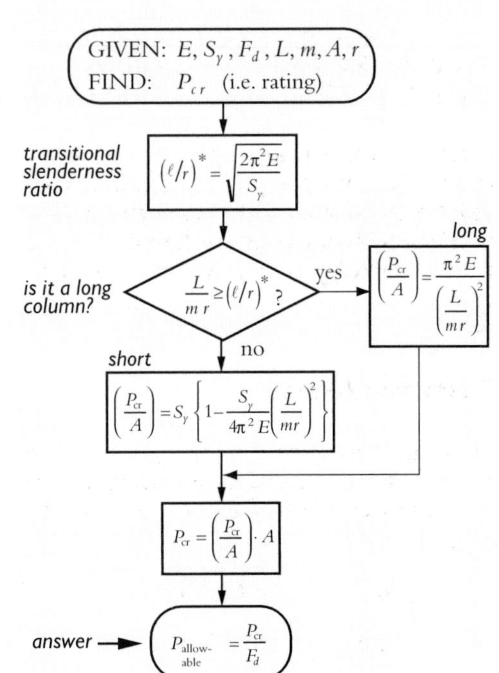

Figure3.59 Design algorithms for columns

The analysis of this kind of eccentric loading goes beyond the needs of our introductory treatment, but it is worth knowing about. It is covered extensively in textbooks, and gives rise to the so-called *secant formula* as a solution. This derives from a differential equation of the following form:

$$\frac{d^2y}{dx^2}+\frac{P}{EI}\left(y+e\right)=0, \qquad (3.51)$$

where the eccentricity, e, is a known function of x. It is instructive to compare this differential equation with the non-eccentric form stated in Equation *(3.37)*.

If e is assumed to be a constant, the design inequality resulting from the solution to Equation *(3.51)* is found to be:

Secant:

$$\left(\frac{P}{A}\right)\leq\frac{S_y}{F_d}\left\{1+\left(\frac{ec}{r^2}\right)sec\left[\left(\frac{\ell}{r}\right)\sqrt{\frac{P}{4AE}}\right]\right\}^{-1} \qquad (3.52)$$

where c is the *extreme fibre distance* of the column cross-section (that is, the largest perpendicular distance from the neutral axis, z-z, of the section to the section's perimeter — for a circular section of diameter d, $c=d/2$; for a rectangular section of height h, $c=h/2$). The quantity (ec/r^2) is termed the *eccentricity ratio*. The relationship between the two quantities (P_{cr}/A) and (ℓ/r) implied by Equation *(3.52)* is shown in Figure 3.61.

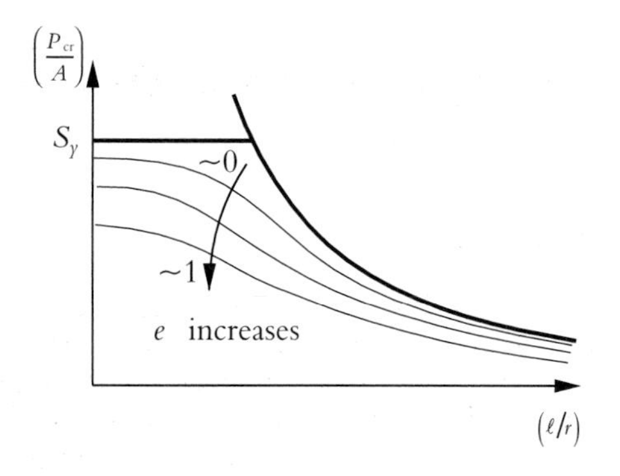

Figure 3.61 Failure characteristics for eccentrically loaded columns.

Note however that it is by no means easy to plot such a graph, since the term (P/A) occurs on both the left and right-hand sides of Equation *(3.52)*, so that the mathematical relationship is only implicit.

Design by this method requires the designer to estimate a reasonable value for the eccentricity, e, which is difficult. Furthermore, the form of Equation *(3.52)* means that application of the secant formula is rather complex and awkward to use for design purposes. Nevertheless the secant formula is the basis of several modern structures codes.

Ex. 3.16 Transitional slenderness ratio for various materials

Evaluate the transitional slenderness ratio for each of the materials tabulated in Table 3.3. (Note that these material property values are representative and approximate only. In practice the yield stress values — and to a lesser extent, the elastic moduli — will depend on details of composition and processing for a given material.)

Table 3.3 Column material properties

Material	S_y MPa	E GPa
brass (cold rolled)	640	120
phosphor bronze	770	120
beryllium copper	1380	120
spring steel	1300	210
stainless steel	1000	200
mild steel (hot rolled)	300	210
aluminium	240	70
magnesium	260	45
titanium	750	115

Ex. 3.17 Simple formulation of a differential equation

Derive the *eccentric column* differential equation given in Equation *(3.51)*. [*Hint*: this is not hard! Look back at the derivation of Equation *(3.37)* if you get stuck.]

3.4.13 Further extensions: beam columns

Designers, like politicians, should know what they don't know even better than what they do know. On this basis it is worthwhile mentioning a class of column design problems which remains beyond the scope of this text. Columns that are subjected simultaneously to both axial compression and lateral bending are referred to as *beam-columns*. Their analysis is difficult owing to the interaction between bending deflections and buckling moments. Their design is treated in standard codes and texts on elastic instability, and should be approached with due humility.

3.4.14 Design Rules for axially loaded components

(a) Tension elements

Static loading:

Model $\sigma_{\text{axial}} = P/A.$

Require $A \geq \dfrac{P F_d}{S_y}.$ (3.36)

Fatigue may be a necessary consideration. Refer to Chapter 2.4.

Extension:

Model: $\Delta L = \dfrac{F}{k}$, where $k = \dfrac{AE}{L}.$

Require: $\Delta L \leq \delta_{\text{allowable}}.$

(b) Compression elements (columns)

Long columns (non-eccentric):

Model: $P_{\text{cr}} = \dfrac{\pi^2 EI}{\ell^2}$ (3.39)

for $(\ell/r) \geq \sqrt{\dfrac{2\pi^2 E}{S_y}}.$ (3.48)

Require: $\left(\dfrac{P}{A}\right) \leq \dfrac{\pi^2 E}{F_d(\ell/r)^2}.$ (3.50)

Short columns (non-eccentric):

Model:

$$\left(\frac{P}{A}\right)_{\text{inelastic buckling}} = S_y - \text{constant} \times \left(\frac{\ell}{r}\right)^2$$

for $(\ell/r) < \sqrt{\dfrac{2\pi^2 E}{S_y}}.$ (3.48)

Require

$$\left(\frac{P}{A}\right) \leq \frac{S_y}{F_d}\left\{1 - \frac{S_y}{4\pi^2 E}\left(\frac{\ell}{r}\right)^2\right\}. \quad (3.49)$$

Eccentric columns:

Model requires:

$$\left(\frac{P}{A}\right) \leq \frac{S_y}{F_d}\left\{1 + \left(\frac{ec}{r^2}\right) sec\left[\left(\frac{\ell}{r}\right)\sqrt{\frac{P}{4AE}}\right]\right\}^{-1} \quad (3.52)$$

(c) Beam columns:

Complicated — consult reference texts for specific cases.

3.5 Assignment on column design

3.5.1 Background

In the petrochemical and process industries, it is common to use *jacketed pipe heat exchangers* to achieve the transfer of heat between two fluids. These are much cheaper than standard multi-tube or plate heat exchangers, although they are not as effective. As their name suggests, jacketed pipe heat exchangers consist of an inner pipe (the *shell*) surrounded by an outer pipe (the *jacket*). They enable the fluid in the inner pipe to be surrounded by an annulus of hotter or colder fluid, without ever allowing the two fluids to mix.

You are a consultant engaged in the design of the heat exchanger shown schematically in Figure

3.62. Several of these exchangers are required for a petroleum refinery extension. They will be installed with their axes oriented vertically, so bending of the pipes under self-weight is not an issue.

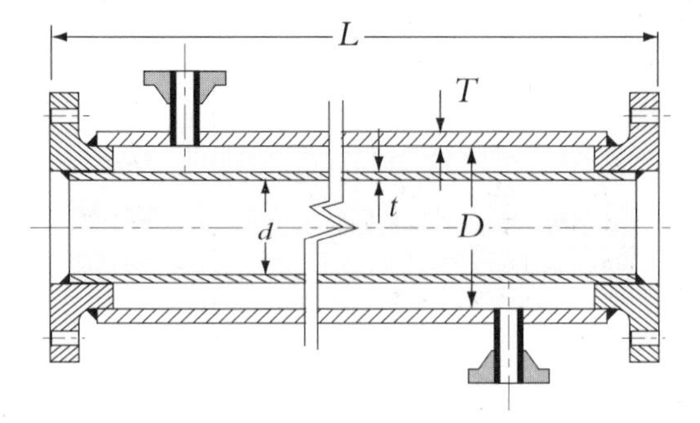

Figure 3.62 Jacketed heat exchanger

You have already *finalized* all of the design variables listed on the sketch, as follows:

- length between flanges: $L = 20{,}000$ mm (it is usual to specify the dimensions of pressure vessels entirely in millimetres);
- inside diameter and thickness of outer jacket: $D = 260.35$ mm; $T = 6.35$ mm;
- inside diameter and thickness of inner shell: $d = 154.06$ mm; $t = 7.11$ mm;

  (the strange dimensions above correspond to a *10-inch Schedule 20* pipe and a *6-inch Schedule 40* pipe respectively, and are available standard sizes — even in this metric age! They have been *hard converted* from imperial dimensions.)
- mean metal temperature of outer jacket: $\theta_j = 80°$ C
- mean metal temperature of inner shell: $\theta_s = 150°$ C

Furthermore, due to the corrosive nature of the inner fluid you have decided to make the entire exchanger from *stainless steel type 316*, for which the following property data is known:

- yield strength: $S_y = 205$ MPa
- Young's Modulus: $E = 196{,}000$ MPa
- coefficient of thermal expansion:

$\alpha = 10.3 \times 10^{-6}$ (° C)$^{-1}$

You are aware that differential thermal expansion is a frequent cause of stress problems in pipes and heat exchangers, and you have decided to check this vessel for its resistance to buckling.

3.5.2. The design problem

(a) Calculate the axial forces arising from differential thermal expansion within the exchanger. Which component of the vessel is subject to the possibility of buckling failure? Are additional measures required in order to prevent this mode of failure?

(b) Design a system of lateral restraints to resist axial buckling. You will need to specify the number and location of such restraints. Note that it is desirable to minimize any obstruction to the flow of fluid within the exchanger, although some such obstruction may be inevitable.

(c) How would your vessel be fabricated? Is it buildable? Briefly itemize the welding sequence, perhaps using a sketch or two for clarity.

(d) Supposing all of the previously listed design variables (i.e. the ones described as *finalized* above) were now regarded as *negotiable*. What alternative or additional measures could the designer take to avoid a buckling failure? (Don't do any calculations for this, just discuss the options qualitatively — there are quite a few of them.)

3.5.3. Additional information

(a) Use $F_d = 2.0$ as your factor of safety against failure due to compressive buckling.

(b) It is reasonable in this case to regard the compressed components of these exchangers as having *fully built in* ends — the welded joint between pipe and flange supports the end of the pipe in such a way as to prevent any rotation, particularly when bolted to adjacent exchangers or other equipment. One needs to be very careful about applying this assumption, as truly irrotational supports are rare in practice. When in doubt, assume a pinned joint, since this is always conservative.

Engineering Design Associates

Project Title : **Jacketed heat exchanger**

Date 07:4:99

Page 1 of 7

Ckd. evm

JACKETED PIPE — Design for Axial Loading (i.e. "COLUMNS")

Differential Thermal Expansion This may bamboozle students initially, although it's not all that hard. To get them started, suggest to them that they carry out the following "thought experiment" (gedankenexperimente ??) :

- imagine the two cylinders are connected at only one end, and free to expand separately at the other end.

- imagine they are initially "cold", and at the same temperature (say 20°C, although this is arbitrary in the case of both cylinders being made from the same material!).

- imagine that the jacket is heated to a uniform temp of 80°C, and the shell is heated to a uniform temperature of 150 °C. What happens? ("Umm, the inner shell gets longer than the jacket" ..... "Brilliant!")

- now imagine that we force the jacket and shell back into alignment by attaching a rigid flange at the other end, so they are now of the same length. By considering the equilibrium of this flange, you should be able to extract from students that the flange exerts an axial tensile force on the jacket (to stretch it) and an equal and opposite compressive force on the inner shell (to shorten it)

Engineering Design Associates

Project Title : **Jacketed heat exchanger**

Date 07:4:99

Page 2 of 7

Ckd. erm

Since both jacket and shell have the same coefficient of thermal expansion (and Young's modulus), we may put:

δ_T = differential thermal expansion = $L \alpha \Delta\theta$

$= 20{,}000 \text{ mm} \times 10.3\times10^{-6} \frac{\text{mm}}{\text{mm.} ^\circ\text{C}} \times (150-80)\,^\circ\text{C}$

$= \underline{14.42 \text{ mm}}$ — δ_T

Let F_T = "thermal force", ie. the force imposed by the flange in our "thought experiment" to equalize lengths of jacket & shell.

Now $\delta_j = \frac{LF_T}{A_j E}$ = extension of jacket from its natural length

$\delta_s = \frac{-LF_T}{A_s E}$ = compression of shell " " " "

$\delta_T = \delta_j - \delta_s$ (see previous diagram)

$= \frac{LF_T}{A_j E} + \frac{LF_T}{A_s E} = \frac{LF_T}{E}\left(\frac{A_j + A_s}{A_j A_s}\right)$

Hence $F_T = \frac{\delta_T E}{L}\left(\frac{A_j A_s}{A_j + A_s}\right)$ — (1)

In general, $A = \frac{\pi}{4}(d_o^2 - d_i^2) = \frac{\pi}{4}(d_o+d_i)(d_o-d_i)$

$= \pi(d_i + t)\,t$ [since $d_o = d_i + 2t$]

Thus $A_j = \pi(260.35 + 6.35)\,6.35 = \underline{5{,}320 \text{ mm}^2}$ — A_j

$A_s = \pi(154.06 + 7.11)\,7.11 = \underline{3{,}600 \text{ mm}^2}$ — A_s

(hint: make sure your class comes to some common agreement on these area figures.)

Now, (1) → $F_T = \frac{14.42 \times 196{,}000}{20{,}000}\left(\frac{5320 \times 3600}{5320 + 3600}\right)\left(\frac{\text{N}}{\text{mm}^2}\right)\left(\frac{\text{mm}^4}{\text{mm}^2}\right)$ — F_T

$= \underline{303{,}428 \text{ N}}$

Note that the relative deformations from the "heated but unstressed" positions are:

$\delta_j = \frac{LF_T}{A_j E} = \frac{20\,000 \times 303\,428}{5320 \times 196\times10^3} = \underline{5.82 \text{ mm}}$ — δ_j

and by a similar process, $\delta_s = \underline{-8.60 \text{ mm}}$ — δ_s

Engineering Design Associates

Project Title : **Jacketed heat exchanger**

Date 07:4:99

Page 3 of 7

Ckd. erm

It is convenient to switch notation here, by identifying the "thermal force" F_T with the axial compressive force P which is tending to buckle the inner shell.

Column Buckling (i.e. analysis of inner shell with regard to ability to resist buckling)

We have identified the inner shell as the component subject to the possibility of buckling failure. In order to answer the final question of part 3(c) ("Are additional measures required in order to prevent buckling?"), students will need to carry out an analysis of the inner shell as a column — initially with no lateral restraints.

The crucial issues the students must address are:

(i) end conditions, and hence effective length
(ii) radius of gyration, and hence slenderness ratio
(iii) categorization of column as "LONG" (Euler) or "SHORT" (Johnson)
(iv) calculation of critical column stress (P_{cr}/A), and hence P_{cr}
(v) calculation of allowable column load P_{all} (using F_d)
(vi) comparison of P_{all} to P in order to assess structural adequacy

..... more or less in the above sequence.

(i) End Conditions : "FIXED/FIXED" (see note at end of assignment sheet, 4(b))

For this case, $m_{F/F} \left(\equiv \frac{L}{\ell}\right) = 2$

Thus the effective length, $\ell = \frac{L}{m_{F/F}} = \frac{20\,000}{2}$

$= \underline{10,000 \text{ mm}}$ ℓ

(ii) Slenderness Ratio : Students may be bleary on the definition of the radius of gyration, r_{xx}, or at least confused about which second moment of area to use in finding it. The simple formula they need to remember is:

$I_{xx} = A\,r_{xx}^2$, where I_{xx} is the 2nd Mo. of Area of a cross-section about an axis perpendicular to the plane in which buckling is to occur.

In this case (and usually for circular sections), all buckling planes are equivalent, and I_{xx} is just taken about a diameter, i.e.

$$I_{xx} = \frac{\pi}{64}\left(d_o^4 - d^4\right) = \frac{\pi}{64}\left[(168\cdot28)^4 - (154\cdot06)^4\right] = \underline{11,711,876 \text{ mm}^4}$$ I_{xx}

Engineering Design Associates

Project Title : **Jacketed heat exchanger**

Date 07:4:99

Page 4 of 7

Ckd. evm

Now $r_{xx} = \sqrt{\frac{I_{xx}}{A_s}} = \sqrt{\frac{11,711,876}{3,600}} = \underline{57.04 \text{ mm}}$

Hence $\left(\frac{l}{r_{xx}}\right) = \frac{10\,000}{57.04} = \underline{175.3}$

(iii) "Long" or "Short" Column :

The transitional slenderness ratio, $\left(\frac{l}{r}\right)^*$, is given by (note that this is a property of the material) — (2)

$$\left(\frac{l}{r}\right)^* = \sqrt{\frac{2\pi^2 E}{S_y}} = \sqrt{\frac{2\pi^2 \times 196\,000}{205}} = \underline{137.4}$$

The actual slenderness ratio (175.3) is greater than this, so we categorize the inner shell as a "long" column and proceed initially to use the Euler formula.

(iv) Critical Column Load : There are a number of permutations of the formula which students can use to obtain P_{cr}. They may obtain it directly by : $P_{cr} = \frac{\pi^2 E I_{xx}}{l^2}$ or $\frac{\pi^2 E A}{(l/r_{xx})^2}$ — (3)

It is usually more instructive to calculate the critical column stress, (P_{cr}/A), and then get P_{cr}.

$$\left(\frac{P_{cr}}{A}\right) = \frac{\pi^2 E}{(l/r_{xx})^2} = \frac{\pi^2 \times 196\,000}{(175.3)^2} = \underline{62.9 \text{ MPa}}$$ $\left(\frac{P_{cr}}{A}\right)$

Hence $P_{cr} = 62.9 \times 3600 \text{ mm}^2 = \underline{226,560 \text{ N}}$ P_{cr}

(v) Allowable Column Load :

$P_{all} = \frac{P_{cr}}{F_d} = \frac{226\,560}{2.0} = \underline{113,280 \text{ N}}$ P_{all}

(vi) Structural Adequacy : Recall $P = F_T$

$= \underline{303,428 \text{ N}}$ P

Hence $P > P_{all}$ ⇒ INADEQUATE.

The obvious conclusion students should draw is that the inner shell would buckle under present conditions (even without applying a factor of safety): and so additional measures must be taken to avoid this.

Engineering Design Associates

Project Title : **Jacketed heat exchanger**

Date 07:4:99

Page 5 of 7

Ckd. evm

Design of Lateral Restraints

This part of the assignment is very open-ended (intentionally) but I suspect there is only a limited range of possible designs which students can pursue. My solution, which is more-or-less what is done in practice, is based on the following geometry:

sets of restraining plates welded to shell only; four plates per set; loose sliding fit inside outer jacket.

relatively short, say ~100 mm

In this scheme, the outer jacket is regarded as a rigid foundation.

The restraints provide "pinned joint" support. The question which designers (& students) need to address is how many sets of restraints need to be provided, and where.

Say there are N sets of restraining plates:

L

L_{end} L_{mid} L_{mid} etc .... L_{end}

The span between restraing plates is L_{mid}, for which the end conditions are "PINNED/PINNED", so the ratio of actual length to effective length for this span is

$$m_{P/P} \left(\equiv \frac{L_{mid}}{l_{mid}} \right) = 1$$

The span between end flanges and endmost restraining plates is L_{end}, for which the endconditions are "FIXED/PINNED", so

$$m_{F/P} \equiv \left(\frac{L_{end}}{l_{end}} \right) = 0{\cdot}7$$

Thus we can write

$$\left.\begin{aligned} l_{mid} &= L_{mid} \\ \text{and} \quad l_{end} &= 0{\cdot}7\, L_{end} \end{aligned}\right\} \qquad (4)$$

Engineering Design Associates

Project Title : **Jacketed heat exchanger**

Date 07:4:99

Page 6 of 7

Ckd. erm

We require $P_{all} \geq P$ where $P_{all} = \frac{P_{cr}}{F_d}$

i.e. $P_{cr} \geq F_d . P = 2{\cdot}0 \times 303{,}428 = \underline{606{,}856 \text{ N}}$

It is not immediately obvious whether we shall need a "long" or a "short" column, so for completeness we begin by looking at the value of P_{cr} for a column whose slenderness ratio equals the transitional slenderness ratio, $\left(\frac{l}{r_{xx}}\right) = \left(\frac{l}{r}\right)^*$. We use the Euler formula.

For this $\left(\frac{P_{cr}}{A}\right)^* = \frac{\pi^2 E}{\left(\frac{l}{r_{xx}}\right)^2}$ but $\left(\frac{l}{r_{xx}}\right)^2 = \left[\left(\frac{l}{r}\right)^*\right]^2 = \frac{2\pi^2 E}{S_y}$

Hence $\left(\frac{P_{cr}}{A}\right)^* = \frac{S_y}{2} = 102{\cdot}5 \text{ MPa}$

$P_{cr}^* = 102{\cdot}5 \left(\frac{\text{N}}{\text{mm}^2}\right) \times 3600 \text{ mm}^2 = \underline{369{,}000 \text{ N}}$ (INSUFFICIENT)

It is therefore clear that we require a "short" column (i.e. that we use the Johnson formula). We now seek a value for the maximum allowable effective length which will give at least the required value of P_{cr}.

For this $\left(\frac{P_{cr}}{A}\right)_{req'd} \geq S_y \left\{1 - \frac{S_y}{4\pi^2 E}\left(\frac{l}{r_{xx}}\right)^2\right\}$

or $\left(\frac{P_{cr}}{A}\right)_{req'd} = S_y \left\{1 - \frac{S_y}{4\pi^2 E}\left(\frac{l_{MAX}}{r_{xx}}\right)^2\right\}$ ——— (5)

Rearranging: $\left(\frac{l_{MAX}}{r_{xx}}\right)^2 = \frac{4\pi^2 E}{S_y^2}\left[S_y - \left(\frac{P_{cr}}{A}\right)_{req'd}\right]$

$\Rightarrow \quad l_{MAX} = r_{xx} \frac{2\pi}{S_y}\sqrt{E\left[S_y - \left(\frac{P_{cr}}{A}\right)_{req'd}\right]}$ but $(P_{cr})_{req'd} = F_d . P$

$= r_{xx} \frac{2\pi}{S_y}\sqrt{E\left(S_y - \frac{F_d . P}{A}\right)}$ ——— (6)

$= 57{\cdot}04 \text{ mm} \times \frac{2\pi}{205}\sqrt{196000\left(205 - \frac{2{\cdot}0 \times 303{,}428}{3600}\right)}$

$= \underline{4\ 672 \text{ mm}}$ l_{MAX}

From (4) we obtain that $(L_{mid})_{MAX} = \underline{4672 \text{ mm}}$

and $(L_{end})_{MAX} = \left(\frac{4672}{0{\cdot}7}\right) = \underline{6\ 674 \text{ mm}}$

The only way to achieve this spacing is with N=3, i.e. three sets of restraing plates, spaced something like: N

6000 | 4000 | 4000 | 6000

i.e. $L_{end} = 6000$ mm, $L_{mid} = 4000$ mm L_{end} L_{mid}

Engineering Design Associates **Project Title :** ***Jacketed heat exchanger***	**Date** 07:4:99
	Page 7 of 7
	Ckd. evm

Note of course that other values of L_{mid} and L_{end} are satisfactory, as long as they remain smaller than their maximum allowable values of 4672 and 6674 mm respectively.

Fabrication : The jacketed pipe exchanger with restraining plates welded to the outside of the inner shell is certainly able to be fabricated, and I won't insult tutors' intelligence by spelling out the welding sequence. However students should be encouraged to be methodical about this, and answer in tabular or sketch form. My preference would be for them to use a sketch showing all welds schematically, with each weld given a sequence number. Students should not write little essays for this question!

(Note of clarification : "welding sequence" does not mean the fabrication procedure for each weld, it simply refers to the sequence in which components are welded together.)

Alternative Methods of Avoiding Buckling

- change the process so that $\Delta\theta$ is reduced (eg. by reducing θ_s or increasing θ_j)
- use a jacket material which has a larger coefficient of thermal expansion (eg. mild steel, which has a value of $\alpha = 11.5 \times 10^{-6} \left(\frac{mm}{mm\ ^\circ C}\right)$ and would make good economic sense)
- reduce L (ie. more exchangers bolted together)
- increase d (will increase I_{xx} — F_T also increases, but only slightly
- increase t (" " " — " " " " " "
- reduce D (will effectively reduce F_T, ie. P)
- reduce T (" " " " ")
- ...... [possibly other (bizarre?) alternatives] .

3.6 Pressure vessels: internal pressure

Pressure vessels resist fluid pressure. They consist, in generic terms, of an impermeable envelope, or *shell*, surrounding some hollow space, and their purpose is to keep fluid in or out of that space. The fluid, of course, may be liquid or gas or both, and is at an elevated pressure. The vessels may also have to contend with corrosive environments, high or low temperatures, and external loads.

Depending upon its contained volume and pressure rating, it is generally fair to assume that a vessel's design will be governed by one or more national or industry-standard codes for the design of *unfired pressure vessels*. These design codes include *ASME* (American Society of Mechanical Engineers) Boiler and Pressure Vessel Code, Section VIII: Pressure Vessels (United States), BS 5500 (United Kingdom), AS-1210 (Australia), and DIN[3.6] Handbook 403 Quality Standards, 3: Pressure Vessels & Pipelines (Europe). They frequently will have the force of statutory regulation, so the pressure vessel designer must be aware of their detailed requirements. In the words of the famous, anonymous, vessel designer, "*Don't try this at home!*"

3.6.1 Failure of pressure vessels

Vessels are pressurised either *internally* (with fluid pressure on the inner surface exceeding that on the outer surface), or *externally*. Externally pressurised vessels (sometimes referred to as *vacuum vessels*) generally fail due to the buckling of their shells under circumferential compression. This is a complex phenomenon of elastic instability, and we shall not discuss it further here.

Internally pressurised vessels are much more common in practice. Their main modes of failure are listed as follows.

- Vessels are primarily prone to yielding under pressure. This in itself is not so bad (although the stretched dimensions of the yielded vessel may be inconvenient), but it serves as a warning of an imminent and much more damaging failure.
- If unchecked, yielding can progress to eventual rupture (bursting). This is bad.
- Vessels that are subjected to fluctuating pressure can also experience fatigue fracture, which is very bad, since there is no warning beforehand. However a detailed treatment of fatigue in pressure vessels is beyond the scope of this book.
- Occasionally (rarely), the functional performance of a pressure vessel will be considered to have failed if it undergoes excessive pressure-induced deflection.
- Fluid leakage of a pressure vessel through one of its bolted seals is indeed a form of failure, but is considered separately in Chapter 4.2.

When vessels of even moderate size fail, the consequences are almost always serious. This is largely because vessels that contain internal pressure can easily store a large amount of energy. Consider an *average* industrial pressure vessel like the one sketched in Figure 3.63, say an *air receiver,* 5 metres long by 1 metre in diameter, containing air at 1 MPa (approximately 145 psi, a common enough rating for industrial compressed air).

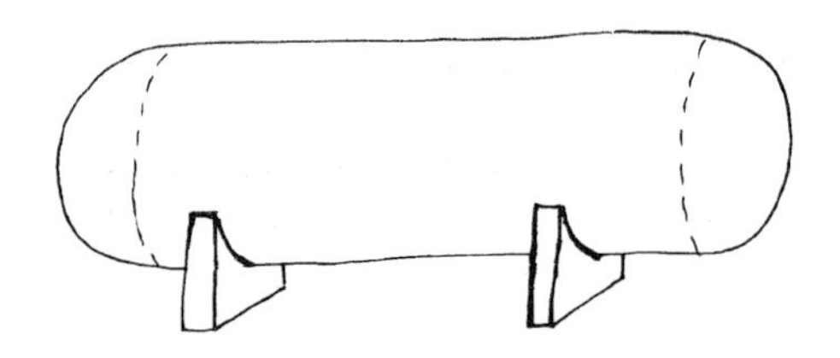

Figure 3.63 A typical industrial pressure vessel

Figure 3.64 High pressure glass polymer tank

3.6 Deutsche Industrie Normen (German Industrial Standards)

Since the elastic potential energy of the stored gas is essentially the product of its volume (3.7 m^3) and its pressure, we can calculate that the vessel contains approximately 3.7 MJ of high-grade energy. The sudden release of this energy, as in a vessel rupture, would have the same impact as a small bomb (approximately 0.6 kg of TNT). Refer to Figure 1.6 to see the results of an accident of this kind.

So it is important for designers to know about designing pressure vessels safely. In the following sections, we shall consider issues involved in designing the most common type of pressure vessel: cylindrical vessels subject to internal pressure.

Figure 3.64 shows an alternative design for a pressure vessel, a typical high pressure tank constructed from glass/polymer composite materials. Although pressure vessels are produced in a great variety of shapes and sizes, their design is almost invariably based on the simple approach presented in this text.

3.6.2 Identifying vessels

The topic of pressure vessels is very close to our heart. In fact, only a few centimetres from the adult human heart lies the most important pressure vessel in the human body. Called the *aorta* (approximated in Figure 3.65), it is our largest blood vessel, carrying all of the heart's pumped output of arterial blood from the left ventricle, around a steep bend in the upper chest and down into the abdomen.

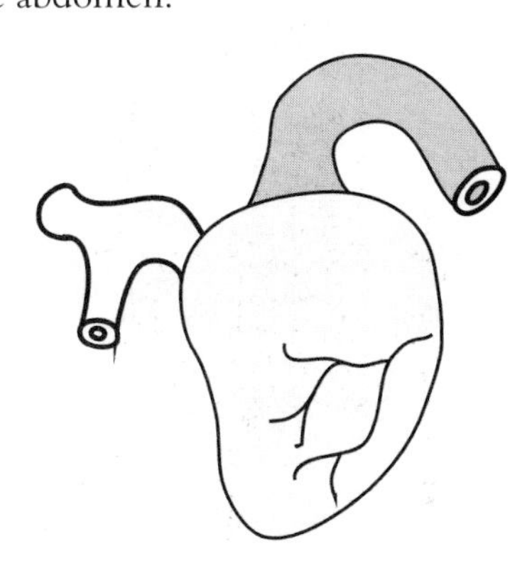

Figure 3.65 Heart and aorta (caution: do not perform cardiac surgery on the basis of this slightly inaccurate sketch!)

Like all blood vessels, it is a pipe for transporting pressurised fluid. But being adjacent to the heart/pump, it must contain fluid at its highest pressure anywhere in the body, and having the greatest internal diameter of any artery/pipe, it clearly faces the body's most demanding pressure-containing role. If the vessel walls should become thinned through disease or erosion, an *aortic aneurism* (localised ballooning of the aortic wall) can occur, often resulting in a ruptured pressure vessel of a most serious kind.

Pipes, of course, are evident in the technical world as well as the anatomical. Pipes are without doubt the commonest form of cylindrical pressure vessel confronted within industry. There they are most often steel and of welded construction, although lengths may be bolted together with end-connectors called *flanges*, as indicated in Figure 3.66. In the home, pipes are often made from copper *brazed* into a continuous network, and on the farm or in the garden pipes are commonly fabricated in plastic (mainly polyvinyl chloride, and polyethylene) joined by glue or compression fittings.

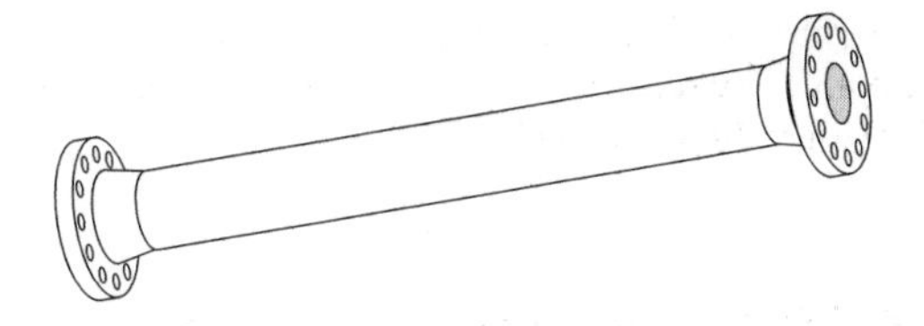

Figure 3.66 Pipe spool with flanges

Whereas we think of the purpose of a pipe as being the *conveyance* of a fluid, the notion of a pressure vessel also carries with it the function of *storage*. Shown in Figure 3.67 are some domestic examples of vessels used to store fluids at elevated internal pressures (and *to deliver them on demand!*) In the case of mass-produced containers like these, the challenge to the designer is primarily that of raw-materials optimisation. These superficially humble artefacts are in fact highly engineered products of great sophistication, and the skills required to design and manufacture them are closely guarded. Consider for a moment the 1.25 litre PET (polyethylene terephthalate) soft-drink bottle, the clear, pressure-containing component of which weighs about 40 grams. If it were thinned by just 1 gram per bottle, the worldwide saving in this important raw material would amount to something like 50,000 tonnes each year (not to mention the reductions in energy and emissions associated with its manufacture).

Figure 3.67 Soft drink bottle and aerosol can

The next two examples are not intended to store pressurised fluid for any appreciable time, but only to contain it momentarily. In the case of the hand-held bicycle pump of Figure 3.68, the cylinder wall acts as a pressure vessel while the piston compresses air inside it for delivery to the bicycle tyre (yet another pressure vessel). In resisting pressure, the vessel wall must also maintain a radial strain (relative increase in diameter) small enough so that its air seal with the piston is not jeopardised.

All reciprocating pumps, pneumatic and hydraulic cylinders (actuators), and piston engines possess components that act as pressure vessels in a similar fashion.

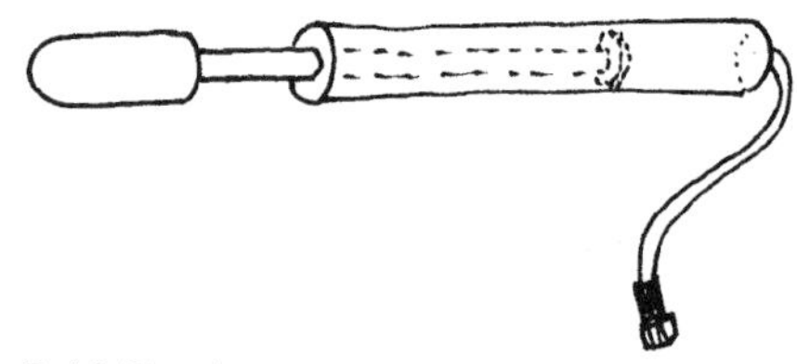

Figure 3.68 Bicycle pump

The emphasis of the cannon is also on *delivery rather than storage!* A cannon is a thick-walled pressure vessel open at one end, intended to contain the high-pressure and high-temperature gaseous combustion products of an explosive detonation while they accelerate a projectile through its muzzle. The successful use of such devices as weapons of warfare around the year 1320 required not only the discovery of suitable explosives like gunpowder, but also the development of reliable thick-walled vessels. Unfortunately, the mathematical theory of thick-walled pressure vessels was not articulated by Gabriel Lamé (1795–1870) until much later[3.7], so in the meantime the structural design of cannonry evolved by a painful, lethal process of trial and experimentation.

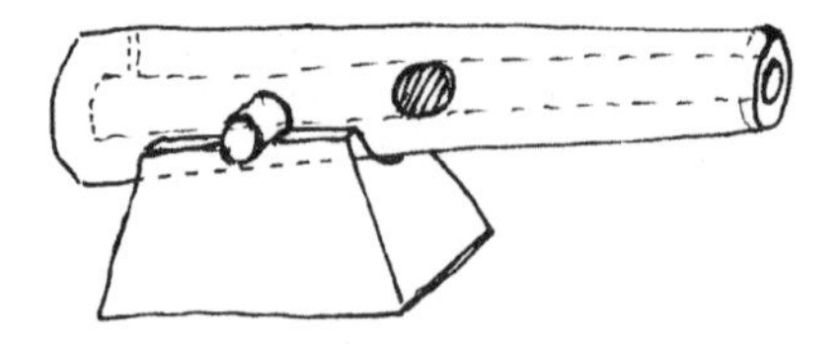

Figure 3.69 Old-fashioned cannon

Other instances of cylindrical vessels under internal pressure could be adduced: jet engine combustion chambers, aircraft fuselages, *sausage skins*, SCUBA[3.8] tanks, toothpaste tubes (during the squeeze), grain silos, water tanks, extruder barrels, boiler tubes, hydraulic jacks — the list is endless.

Figure 3.70 A plethora of pressure vessels in a petrochemical plant

In petrochemical plant (see Figure 3.70) and industrial equipment generally, the widespread incidence of pressurised fluid streams and reservoirs means that pressure vessels of all varieties are commonly encountered, not only cylindrical in shape but also with spherical, conical,

3.7 Lamé G. and Clapeyron, B.P.E. 1833, Mémoire sur l'équilibre intérieur des corps solides homogènes, *Mém. divers savans*, 4

3.8 Self Contained Underwater Breathing Apparatus

toroidal, ellipsoidal and non-axisymmetric surfaces. Any specific treatment of these would quickly exceed the introductory scope of this book, so we shall confine this discussion to cylindrical vessels only, looking first at simple, thin-walled theory and then progressing to a brief consideration of thick-walled vessels.

3.6.3 Thin-walled cylindrical pressure vessels

Cylindrical pressure vessels are characterised mainly by their internal diameter and their *shell thickness*. If the ratio of these two dimensions is greater than 20, the vessel is referred to as *thin-walled*, and a range of simplifications can be applied to its analysis.

1. Nomenclature

In this book we will use the following symbols and terminology in relation to pressure vessels.

- p = internal pressure used to design the vessel, or *design pressure* (expressed as a *gauge pressure*, or pressure above the ambient atmospheric pressure);
- t = shell thickness;
- D = shell internal diameter;
- D_o = shell external diameter: $D_o = D + 2t$;
- f = design stress intensity (a form of *allowable stress*);
- η = welded joint efficiency;
- c = total corrosion allowance (the sum of c_i and c_o).

The following are the symbols used for the three pressure-induced stresses:

- σ_H = hoop stress (often called *circumferential stress*, or *tangential stress*);
- σ_A = axial stress due to pressure (often called *longitudinal pressure stress*);
- σ_R = radial stress.

When discussing bending in Section 3.6.6, it will be useful to distinguish σ_A from the following:

- σ_B = bending stress magnitude (another stress in the axial, or *longitudinal*, direction);
- σ_L = total longitudinal stress, $\sigma_L = \sigma_A \pm \sigma_B$.

The symbols σ_1, σ_2 and σ_3 will be reserved for application to the ordered principal stresses, with $(\sigma_3 < \sigma_2 < \sigma_1)$, as usual. Some additional symbols will be introduced for use with thick-walled vessels. The nomenclature stated above is based on common parlance in the pressure vessel industry, and is illustrated in Figure 3.71. We note in passing one slightly confusing aspect of the terminology relating to principal stresses and welds in pressure vessel: *longitudinal welds* are subjected to *circumferential stress*, whereas *circumferential welds* are subjected to *longitudinal stress*.

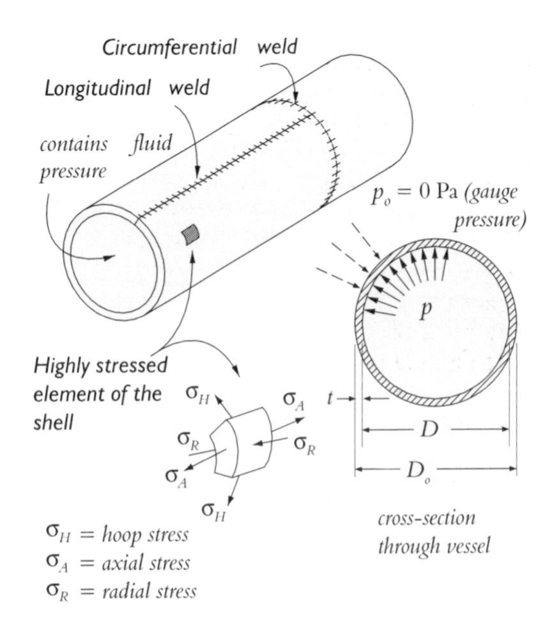

Figure 3.71 Nomenclature of thin-walled pressure vessels

As we shall soon discover, pressure vessels are *triaxially stressed* components, that is, they are subjected to appreciable stresses in all three spatial directions. The directionality of these stresses is important, since the stresses may be combined with additional, non-pressure stresses according to the direction in which they act. For this reason, it is imperative that the designer acquires a good spatial understanding of pressure vessels, and especially the orientations of the principal stresses in relation to a vessel's shell. The three principal stresses and their directions are indicated for one *highly stressed element of the shell* in Figure 3.71.

Ex. 3.18 Three-dimensional stresses — a 'spatial aside'

Spatial ability is an intriguing intellectual skill, that psychological studies have shown bears little correlation with standard intelligence. Some will find the *3D visualisation* recommended above to be trivially straightforward. Others will struggle to *see* the stress directions in their three-dimensional imagination, and for them perhaps playing with a physical model might be helpful.

Obtain a medium-sized plastic cylinder (maybe a suitably emptied beverage container?) that you can mistreat legally. Use a felt-tipped marker to annotate the vessel with the directions of the principal pressure stresses σ_H , σ_A and σ_R (pierce the vessel with a pencil, if necessary, to represent the direction of any stresses which are normal to the shell). Cut out a rectangular patch from the vessel shell, or curve some heavy paper to model such a patch. Using your fingers, grip the patch and impose the equivalent of each pressure stress on it one-by-one, naming the stresses as you proceed. *Become one with the pressure vessel.*

2. Design methodology

When approaching the design of a new pressure vessel, the designer will usually know (or at least have a *reasonable estimate* of) the following quantities or characteristics:

- volume of fluid to be contained;
- operating pressure;
- operating temperature;
- nature of fluid;
- major fluid connections;
- desired general shape (cylindrical/spherical/conical...; short/long) and orientation (horizontal/vertical);
- external loads (wind, piping, supports, attached structures, mechanical loads).

On the basis of the operating temperatures and pressures, and a preliminary decision about the likely material of fabrication of the vessel, the designer can then establish:

- design pressure, p;
- design temperature, T;
- material properties (S_U , S_Y , ...).

The designer is then in a position to begin the task of designing the vessel to ensure its structural integrity. But in which direction should he or she proceed from here? It is worth taking a mental *step back* at this point to review the designer's objectives by asking some key questions.

Q: What is the designer trying to determine?

A: The geometry of the vessel, or more specifically, D and t.

Q: What is the designer trying to avoid?

A: Failure of the vessel, or more specifically, yielding.

Q: How does the designer predict this failure?

A: By modelling the pressure stresses as a function of D and t to see when the stresses will cause yielding.

Q: What sort of stresses act on the vessel's shell?

A: Triaxial stresses.

Q: So how are these multiaxial stresses combined to predict yielding?

A: By using a *failure predictor* (*theory of failure*) like *MSFP* or *DEFP*.

Pressure vessel materials are usually ductile, so either of the *MSFP* or *DEFP* would be suitable. Because of its simplicity and conservatism, the *Maximum Shear Stress Failure Predictor* (*MSFP*) is generally employed in the design of pressure vessels. Recall that it predicts failure of the vessel shell when:

$$\tau_{\max fail} = \frac{S_Y}{2},$$

therefore we need to limit $\tau_{\max}$ within the vessel such that $\tau_{\max} \leq \tau_{\text{allowable}}$. The allowable value will be the *failure value*, $\tau_{\max fail}$, reduced by some factor of safety, hence:

$$\tau_{\max} \leq \frac{S_Y}{2F_d}. \tag{3.53}$$

It is common in various national codes for pressure vessel design to define (and tabulate values of) the *design stress intensity* as:

$$f \equiv \frac{S_Y}{F_d}. \tag{3.54}$$

At this point we note that the value of the safety factor, F_d, used here is mandated by pressure vessel design codes. The customary value is taken as $F_d = 1.6$ to resist yielding, or $F_d = 4.0$ to resist rupture. In effect, despite the simplified definition given in Equation *(3.54)*, the design stress intensity is actually the lesser of two values:

$$f = \min\left\{\frac{S_Y}{F_{d,\text{yield}}}, \frac{S_U}{F_{d,\text{rupture}}}\right\}$$

$$= \min\left\{\frac{S_Y}{1.6}, \frac{S_U}{4.0}\right\}.$$

The difference in the two values of safety factor arises purely from the different severity of consequences of the two modes of failure (refer to Section 2.3.5 for a discussion of factors of safety).

Recall that

$$\tau_{\max} = \frac{\sigma_1 - \sigma_3}{2},$$

so from Equations *(3.53)* and *(3.54)* we can write the requirement that:

$$\frac{\sigma_1 - \sigma_3}{2} \leq \frac{f}{2},$$

or alternatively,

$$(\sigma_1 - \sigma_3) \leq f. \qquad (3.55)$$

This is our design inequality for the pressure vessel. To develop a design rule, it merely remains to evaluate the principal stresses σ_1 and σ_3 in terms of design variables.

3. Pressure-induced stresses

For thin-walled vessels, the pressure stresses can be derived in a very simple fashion by considering the following three *free bodies* shown in Figures 3.72, 3.73 and 3.74. We use the equilibrium of forces in each of these free bodies to establish the desired stress relationships.

(i) Hoop stress

Hoop stress, σ_H, is usually the largest of the three principal stresses. We see from Figure 3.72 that in a thin-walled vessel it is given by:

$$\sigma_H = \frac{pD}{2t}. \qquad (3.64)$$

This is generally a large, positive (tensile) stress, since it is greater than the internal pressure by the ratio of

$$\frac{\text{shell radius}}{\text{shell thickness}},$$

which is always greater than 10 for thin-walled vessels.

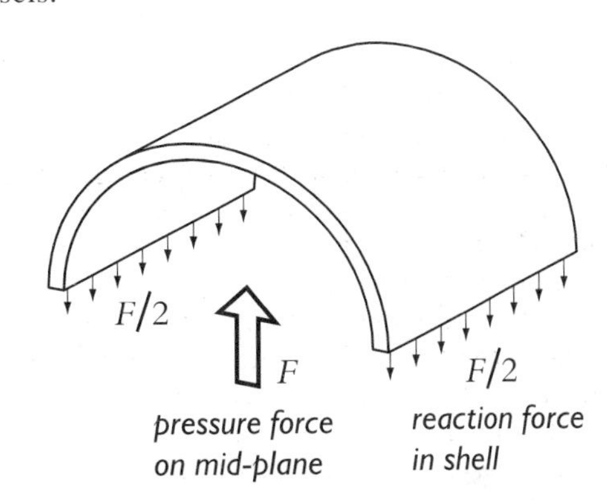

Figure 3.72(a) Free body for hoop stress, σ_H

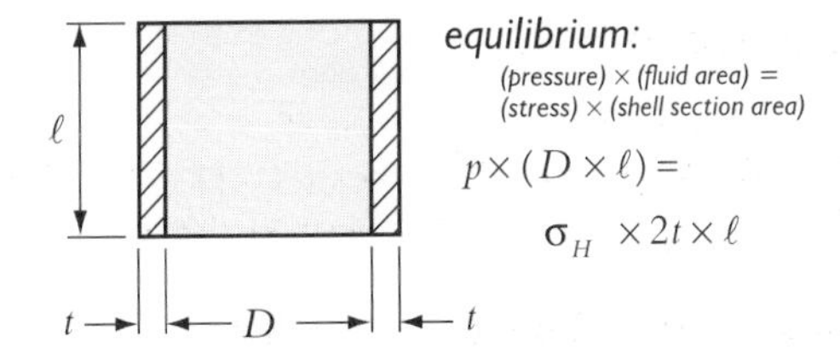

Figure 3.72(b) Equilibrium for hoop stress, σ_H

Ex. 3.19 Thinking about hoop stress

Atmospheric storage tanks are large cylindrical pressure vessels that rest on flat ground with their central axis vertical, and contain liquids at atmospheric pressure. Actually, only the top surface of the liquid is at atmospheric pressure; the liquid at every level below the surface experiences additional hydrostatic pressure due to the weight of the liquid above it, and exerts that hydrostatic pressure on the walls of the storage tank. The familiar relationship for hydrostatic pressure is $p_h = \rho g h$, where ρ is the liquid density (kg/m^3), g is the gravitational field constant (N/kg), and h is the depth (m) below the surface. Because there is no axial pressure acting on the vessel walls from contained liquid (since gravity restrains the fluid vertically), atmospheric storage tanks experience no pressure-induced axial stress — only hoop stress. Atmospheric storage tanks are used to store petroleum products in *tank farms*, water for drinking and fire-fighting purposes in remote or hazardous areas, wine in fermentation vats, and many other bulk liquids. A metal-walled, circular, above-ground swimming pool is a classic example of an atmospheric storage tank.

From your understanding of the analysis shown in Figure 3.72, predict the nature of the hoop stress in a circular, above-ground swimming pool of diameter 5 m, height 1.5 m, and wall thickness 1.6 mm. Plot the hoop stress as a function of water depth.

(ii) Axial stress

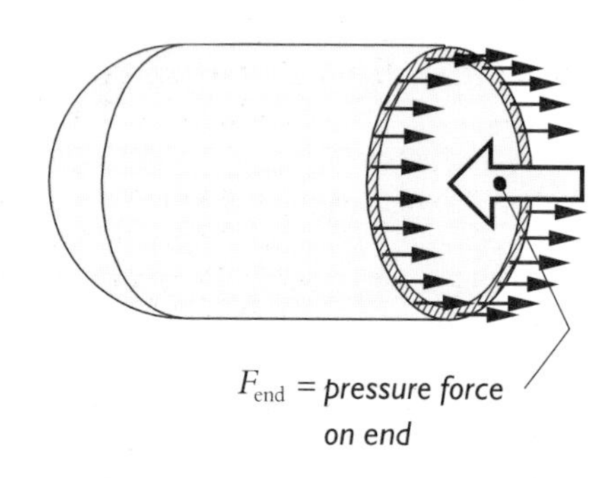

Figure 3.73(a) Free body for axial stress, σ_A

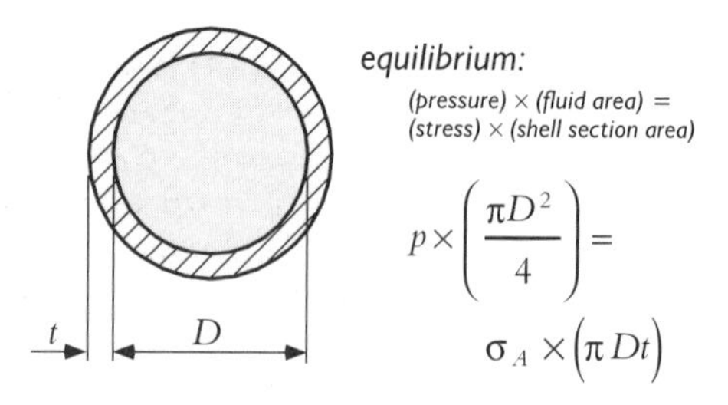

Figure 3.73(b) Equilibrium for axial stress, σ_A

From the above figure, the axial stress, σ_A, in a thin-walled vessel is given by:

$$\sigma_A = \frac{pD}{4t}, \qquad (3.65)$$

which is exactly half the magnitude of the hoop stress in Equation *(3.56)*. We also note that the sectioned view shown in Figure 3.73(b) is identical to the one that would be seen if we were to section a spherical pressure vessel along any of its central cutting planes. Indeed, the pressure-induced membrane stresses in a spherical pressure vessel are uniform in every direction, and are equal in magnitude to the axial stress given in Equation *(3.57)*. Therefore, for the same D and t, the maximum principal stress in a spherical vessel ($\sigma_1 = \sigma_2 = \sigma_A$) is half of the value in a cylindrical vessel ($\sigma_1 = \sigma_H$).

Ex. 3.20 Thinking about axial stress

When aloft, hot-air balloons are essentially spherical over the upper two-thirds of their envelope. The air inside is rendered buoyant by direct flame heating, so for practical reasons it is difficult to measure the air temperature and pressure inside a balloon directly. But for design purposes, it is important to have some means of estimating the membrane stress acting on the balloon material, and predicting its strength (which is temperature-dependent).

Imagine you are a hot-air balloon designer in the process of specifying the material to be used in a new balloon having diameter 30 m and carrying a gondola that will have a mass of 800 kg when full. Based on your understanding of the analysis shown in Figure 3.73, estimate the membrane stress in the spherical region of the balloon if the material thickness is to be 0.1 mm. Can you think of a way to estimate the average air temperature inside the balloon?

(iii) Radial stress

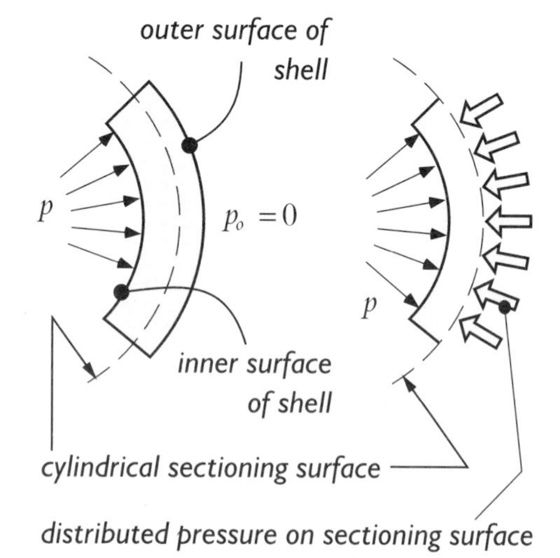

Figure 3.74 Free body for radial stress, σ_R

Radial stress, σ_R, is usually the smallest (least positive) of the three principal stresses. Considering horizontal force equilibrium, it can be asserted by inspection of the free body in Figure 3.74 that the radial stress varies from $-p$ (compressive) at the inner wall, to $p_o = 0$ at the outer wall. For thin-walled vessels, the variation is almost linear, so a good approximation to the *average radial stress* is:

$$\sigma_R = -\frac{p}{2}. \qquad (3.58)$$

4. Development of design rule

Summarising the stress results from the previous section, we have:

$$\sigma_1 = \sigma_H = \frac{pD}{2t} \qquad (3.59)$$

and

$$\sigma_3 = \sigma_R = -\frac{p}{2}. \qquad (3.60)$$

(Note that $\sigma_2 = \sigma_A$ is not used in τ_{max}.) Substitution of these expressions from Equations *(3.59)* and *(3.60)* into our *design inequality* of Equation *(3.55)* leads to the following simple design rule:

$$\frac{pD}{2t} + \frac{p}{2} \leq f,$$

and hence

$$t \geq \frac{pD}{2f - p}. \qquad (3.61)$$

Before it is ready to be used, the above inequation needs to be modified to take account of two practical issues — the effects of welding and corrosion.

5. Welded joints and inspection

Steel pressure vessels are made from initially flat plates (*strakes*) of material. After first being cut to size, their edges are prepared by careful grinding to form the required *edge profile* for welding. Then the plates are formed into their cylindrical shape by rolling through a specialised plate roller (or pressed into shape, in the case of hemispherical, torispherical or semi-ellipsoidal *dished ends*). The shaped plates are then welded together, first by *tacking* them at several points to align the parts correctly, and then by producing long, continuous *butt-welds* in several passes, to join the plates fully. These are often ground to remove stress concentrations at the welds.

Welds introduce residual thermal stresses into pressure vessels, so at one or more stages of their fabrication, vessels may be *stress-relieved*. This is usually performed by *soaking* in a large furnace at elevated temperatures for extended periods of time. Since this can lead to larger grain size in the steel's crystalline structure, resulting in consequent loss of ductility, *quench-annealing* (heating followed by rapid cooling) is sometimes used, especially with stainless steels. This heat-treatment process is different from that used to radically alter the grain structure in the heat-affected zone when welding high tensile (usually high carbon content, i.e. >0.15%) steels (refer to Chapter 4.4 Welded joints).

Welding is a well-developed fabrication process, used almost universally in the manufacture of modern steel pressure vessels. With care and skill, it is possible to produce a butt-weld in the shell of a pressure vessel which is at least as strong as the *parent plates*. However welding is more prone to variation in quality than the process by which the original *strakes* of the pressure vessel were produced. For this reason, some allowance may have to be made in case the welds are weaker than the parent metal.

The approach taken in the pressure vessel industry (and in the various regulatory design codes to which it adheres) is to apply a *welded joint efficiency*, η, to the allowable design stress intensity, f, of the shell material. The welded joint efficiency is a factor expressing the extent by which the parent metal can be presumed to have been weakened by the welding process. The value of η varies from 1.0 (for demonstrably perfect welds) to approximately 0.65, depending upon the quality of the manufacturing process (especially heat treatment) and the extent of *weld inspection*.

There are several non-destructive methods for testing welds. The most expensive, but effective, is radiography using X-rays. *Full radiography* means that 100% of the welds throughout a pressure vessel have been radiographed, thus qualifying the vessel to use a welded joint efficiency of $\eta = 1.0$. *Spot radiography* means that only a statistical sample (say 10%) of the welds has been radiographed, and this qualifies a vessel to use a lesser welded joint efficiency, say $\eta = 0.75$. Crack-testing of welds with *dye penetrant*, and ultrasonic testing, are also widely used. The governing pressure vessel design code must be consulted to establish the appropriate value of η in each case.

6. CORROSION ALLOWANCE

Corrosion is a deep subject, alluded to in Chapter 2.1.2. But the standard approach to corrosion taken by pressure vessel design practice is very simple-minded. Designers simply use a *corrosion allowance*, *c*, to increase the required shell thickness beyond that value needed to withstand pressure alone. The notion is that, during the design life of the vessel, no more than this amount of thickness will be lost to corrosion at any one location on the shell. Values of *c* are normally around 1.0 to 1.5 mm for each exposed surface (inner and outer) subject to corrosion, depending on the nature of the fluid and the intended service life of the vessel. Coated surfaces, or those exposed to non-oxidising fluids, may be allocated a zero corrosion allowance.

7. MODIFIED DESIGN RULE

With the two modifications referred to above, the design rule for thin-walled, cylindrical vessels given in Equation *(3.61)* can now be rewritten as:

$$t \geq \frac{pD}{2f\eta - p} + c \cdot \qquad (3.62)$$

3.6.4 Thick-walled cylindrical pressure vessels

1. NOMENCLATURE

The following symbols are introduced for thick-walled pressure vessels.

- a = shell inner radius ($a = D/2$);
- b = shell outer radius ($b = D_o/2$);
- p_i = internal pressure (when used in conjunction with p_o, otherwise the unsubscripted p will continue to be used for internal pressure);
- p_o = external pressure;

In addition, it is convenient to define the *diameter ratio* as:

$$\Phi \equiv \frac{D_o}{D} = \frac{b}{a} \cdot \qquad (3.63)$$

Note that this dimensionless ratio is always greater than unity, $\Phi > 1$. (The diameter ratio should not be confused with the *hollowness ratio* that was defined for shafts following Equation *(3.2)* as $\beta = d_i/d$, which is always less than unity. The reason for the differing definitions is that, for shafts, the outer diameter d is the key design variable, whereas for vessels, the inner diameter D is a more basic design variable since it contains the fluid.)

Recall our earlier definition of a *thick-walled vessel* as one for which (t/D)>(1/20). This can now be recast in terms of the diameter ratio:

$$(t/D) = \frac{D_o - D}{2D} = \frac{\Phi - 1}{2} \cdot$$

Hence, for thick-walled vessels, we require: $(\Phi-1)/2 > 1/20$, or

$$\Phi > 1.1 \qquad (3.64)$$

The terminology for stresses is the same as described in Section 3.6.3.

2. THICK-WALLED PRINCIPAL STRESSES (FULL EQUATIONS)

The equations describing the variation of principal stresses, σ_H, σ_A and σ_R, in thick-walled pressure vessels are not in themselves difficult, but their derivation is beyond the scope of this book. It is available in many standard texts on elasticity and the strengths of materials (for example, Timoshenko, 1956, Vol.II). Note that the expression for σ_A is sometimes not included in such texts, as it is a constant independent of the other two stresses and does not appear in their differential equation. Nevertheless it is easy to derive from a free-body such as the one presented earlier in Figure 3.73(a), on the assumption that the vessel is closed and cross-sections remain planar. We shall state the full form of the equations here for reference, including the effect of external pressure.

$$\sigma_H = \frac{a^2 p_i - b^2 p_o}{b^2 - a^2} + \frac{(p_i - p_o)a^2 b^2}{r^2(b^2 - a^2)} \qquad (3.65\text{-a})$$

$$\sigma_A = \frac{p_i a^2 - p_o b^2}{b^2 - a^2} \qquad (3.65\text{-b})$$

$$\sigma_R = \frac{a^2 p_i - b^2 p_o}{b^2 - a^2} - \frac{(p_i - p_o)a^2 b^2}{r^2(b^2 - a^2)} \qquad (3.65\text{-}c)$$

3. Thick-walled principal stresses (reduced equations)

In the common case of negligible p_0, the above equations reduce to the following.

$$\sigma_H = \frac{p}{\Phi^2 - 1} \cdot \left(1 + \frac{b^2}{r^2}\right) \qquad (3.66\text{-}a)$$

$$\sigma_A = \frac{p}{\Phi^2 - 1} \qquad (3.66\text{-}b)$$

$$\sigma_R = \frac{p}{\Phi^2 - 1} \cdot \left(1 - \frac{b^2}{r^2}\right) \qquad (3.66\text{-}c)$$

In this form it is very clear that all three stresses share a common term: $p/(\Phi^2-1)$, which is the constant axial stress. For the hoop and radial stresses, this term is multiplied by unity plus-or-minus an inverse-squared function of the radius, r. Graphs of the three stresses are shown to scale in Figure 3.75, superimposed upon a partial cross-section of a thick-walled vessel having $\Phi = 2.5$. Obviously, in thick-walled pressure vessels, the radial stress is no longer of small magnitude, and increases sharply to large compressive values near the inner wall of the vessel.

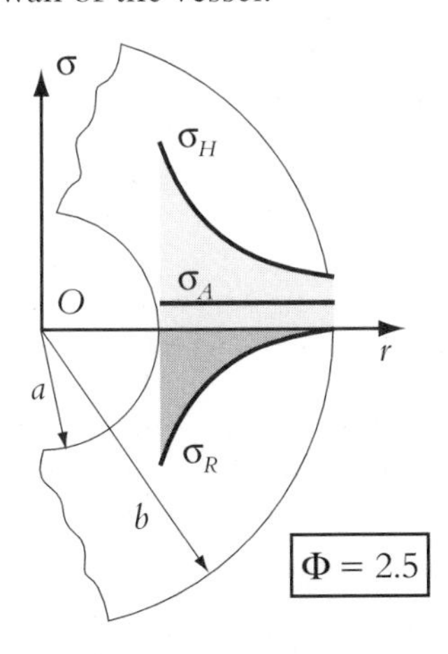

Figure 3.75 Stresses in a thick-walled vessel

It is also clear that the ordering of the three stresses as ($\sigma_R < \sigma_A < \sigma_H$) remains consistent throughout the full thickness of the vessel shell. In the absence of any added bending stresses (see Section 3.6.6), the identification of $\sigma_H = \sigma_1$, $\sigma_A = \sigma_2$ and $\sigma_R = \sigma_3$ can confidently be made, so the maximum shear stress at radius r in a thick-walled vessel can be expressed as:

$$\tau_{max} = \frac{\sigma_H - \sigma_R}{2} = \frac{p}{\Phi^2 - 1}\left(\frac{b^2}{r^2}\right). \qquad (3.67)$$

4. Peak stress values (thick-walled)

Designers will mainly be interested in the peak values of these stress functions, all of which occur at the inner and outer surfaces of the shell. Table 3.4 summarises these peak values.

Table 3.4 Peak values of stress

	inner wall ($r=a$)	*outer wall* ($r=b$)
HOOP	$\sigma_{H,i} = \frac{p}{\Phi^2 - 1}(1 + \Phi^2)$	$\sigma_{H,o} = \frac{p}{\Phi^2 - 1} \cdot (2)$
AXIAL	$\sigma_{A,i} = \frac{p}{\Phi^2 - 1}$	$\sigma_{A,o} = \frac{p}{\Phi^2 - 1}$
RADIAL	$\sigma_{R,i} = \frac{p}{\Phi^2 - 1}(1 - \Phi^2) = -p$	$\sigma_{R,o} = \frac{p}{\Phi^2 - 1} \cdot (0) = 0$
SHEAR	$\tau_{\max,i} = \frac{p\,\Phi^2}{\Phi^2 - 1}$	$\tau_{\max,o} = \frac{p}{\Phi^2 - 1}$

Incidentally, it is clear from the values in the above table that the thick-walled results for σ_H, σ_A and σ_R all asymptote to our previous results for thin-walled vessels (see Section 3.6.3) as Φ approaches 1. For example, consider the outer-wall hoop stress:

$$\sigma_{H_o} = \frac{p}{\Phi^2 - 1} \cdot (2) = \frac{2p}{(\Phi + 1)(\Phi - 1)}$$

$$\approx \frac{2p}{(2)(\Phi - 1)} \qquad \text{as } \Phi \to 1$$

$$= \frac{pD}{(D_o - D)} = \frac{pD}{2t},$$

which is the expression required by Equation 3.64. Similar results for axial and radial stresses are easily obtained.

3.6.5 Deformations in pressure vessels

In this brief discussion of pressure-induced deformations in vessels, we will consider only thin-walled cases. The approach to be taken with deformation in thick-walled vessels is entirely analogous, but a little too complicated for our purposes here.

Pressure vessels expand and contract slightly as they are pressurised and de-pressurised. If you have held a full soft-drink bottle in one hand while you have unscrewed its top with the other, you will have felt this phenomenon with your bare hands! Predicting the extent of diametral expansion in a pressure vessel is merely an application of some simple principles in elasticity.

The increase in a vessel's diameter, δ_{diam}, when it is pressurised is a consequence of its increase in circumference:

$$\delta_{\text{diam}} = \frac{\delta_{\text{circ}}}{\pi},$$

which in turn is a result of the circumferential strain (or *hoop strain*, ε_H) acting over the circumference of the shell:

$$\delta_{\text{circ}} = \pi D \varepsilon_H \cdot$$

Therefore the change in diameter is given simply by:

$$\delta_{\text{diam}} = D\varepsilon_H \cdot \qquad (3.68)$$

Since pressure vessel shells are triaxially stressed, the hoop strain would normally be evaluated using the full three-dimensional form of Hooke's Law:

$$\varepsilon_H = \varepsilon_1 = \frac{1}{E}\left[\sigma_1 - \mu\left(\sigma_2 + \sigma_3\right)\right] \cdot$$

However, we know that for thin-walled vessels, the radial stress ($\sigma_R = \sigma_3$) is dwarfed by the other principal stresses, so we shall use instead a simplified version of the above equation.

$$\varepsilon_H \approx \frac{1}{E}\left(\sigma_1 - \mu\sigma_2\right) \qquad (3.69)$$

$$\approx \frac{1}{E}\left(\sigma_H - \mu\sigma_A\right)$$

$$\varepsilon_H \approx \frac{pD}{4tE}\left(2 - \mu\right)$$

Substituting this result into Equation *(3.74)* gives the diametral expansion of a cylindrical vessel as:

$$\delta_{\text{diam}} \approx \frac{pD^2}{4tE}\left(2 - \mu\right) \cdot \qquad (3.70)$$

It is instructive to carry out a similar exercise for a spherical shell. The analysis is more-or-less identical up to Equation *(3.69)*, but then differs because for spherical vessels, $\sigma_1 = \sigma_2 = \left(pD/4t\right)$. This yields a diametral expansion of:

$$\delta_{\text{spherical diam}} \approx \frac{pD^2}{4tE}\left(1 - \mu\right) \cdot \qquad (3.71)$$

which is substantially smaller than the cylindrical result (the sphere expands by only about 40% of the expansion of a comparable cylinder).

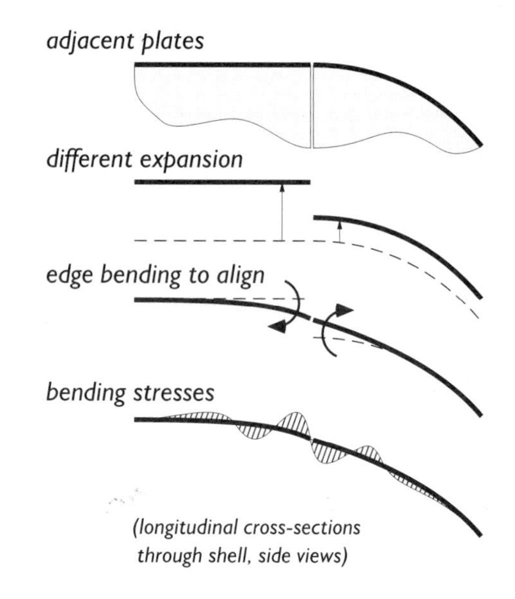

Figure 3.76 Discontinuity stresses

One important corollary of this is that differently shaped shells of revolution (cylinders, hemispheres, toroids, cones, *etc.*) of the same D and t will tend to expand by different amounts when pressurised. If they are welded together as adjacent parts of the same vessel (for example, where a cylindrical shell is joined to its hemispherical ends), the differential expansion will give rise to what is termed *discontinuity stresses*. These constitute a localised stress concentration consisting of *membrane bending stresses* within the shell near the join between dissimilar shell plates. Figure 3.76 illustrates the phenomenon. Fortunately, these stresses are self-relieving (that is, local yielding causes them to disappear).

Note that *membrane bending stresses* occur about a *membrane neutral axis* located at the mid-thickness of the shell. They are axisymmetric, and do not cause any overall bending of the vessel away from its central axis, just small, rippling deformations in the skin of the vessel. They are not to be confused with the *beam-bending stresses* discussed in the next section.

3.6.6 Stresses due to external bending

The shell of a pressure vessel can act as a beam as well as a pressure container. Cylindrical pressure vessels are often quite long relative to their diameters, and therefore the effects of transverse bending loads can be pronounced. These loads may be either distributed (such as self-weight or wind forces), or concentrated, as shown in Figure 3.77. They give rise to bending stresses that act about the *neutral plane* of the vessel in a manner by now familiar to all students of elementary beam theory. The bending moments are evaluated in just the same way as for any other beam.

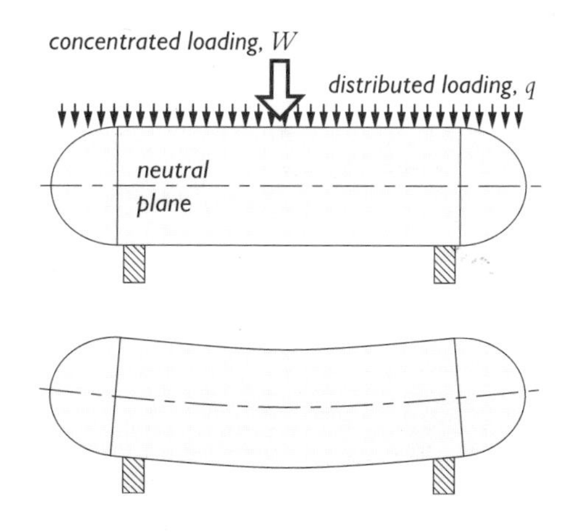

Figure 3.77 Vessel as a beam

Recall that for any beam, the maximum bending stress occurs at the extreme fibres of the section, and is given by $\sigma_{B,max} = M_{max}\, y_{max} / I_{zz}$. In checking the structural integrity of a vessel, it is prudent to use *corroded dimensions*, since they represents the worst case for stress. As a reminder of this, we will express the section properties of a pressure vessel with a hollow circular cross section as follows:

$$A = \frac{\pi}{4}\left(D_{oC}^2 - D_C^2\right) = \frac{\pi D_C^2}{4}\left(\Phi_C^2 - 1\right)$$

$$I_{zz} = \frac{\pi}{64}\left(D_{oC}^4 - D_C^4\right) = \frac{\pi D_C^4}{64}\left(\Phi_C^4 - 1\right)$$

where $D_C = D + 2c_i$, $D_{oC} = D_o - 2c_o$

and $\Phi_C = \left(D_{oC} / D_C\right)$.

Since $y_{max} = \pm D_{oC}/2$, we can write:

$$\sigma_{B,max} = \pm\frac{32 M_{max} D_{oC}}{\pi\left(D_{oC}^4 - D_C^4\right)}$$

or

$$\sigma_{B,max} = \pm\frac{32 M_{max}}{\pi D_C^3}\cdot\frac{\Phi_C}{\Phi_C^4 - 1} \qquad (3.72)$$

The above expressions are equivalent and exact. For the case of thin-walled vessels, the expressions for A, I_{zz} and $\sigma_{B,max}$ may be simplified slightly as follows:

$$A \approx \pi D_{ave}(t - c),$$

$$I_{zz} = \frac{\pi}{8} D_{ave}^3 (t - c),$$

$$\sigma_{B,max} = \pm\frac{4 M_{max}}{\pi D_{ave}^2 (t - c)},$$

where the average shell diameter is $D_{ave} = D + t$, and $(t{-}c)$ is the *corroded thickness*. In the approach that follows, we will use the symbol σ_B to represent the magnitude of $\sigma_{B,max}$.

The difference between pressure vessels and *ordinary* beams is that pressure vessels are already triaxially stressed with pressure-induced stresses, and their bending and pressure stresses must both be taken into account simultaneously. To achieve this, the bending stresses, σ_B, and the axial pressure stresses, σ_A, must be superimposed (since each is a stress acting in the same longitudinal direction). As mentioned in Section 3.6.3, it is helpful to distinguish σ_A and σ_B from the total longitudinal stress:

$$\sigma_L = \sigma_A \pm \sigma_B \,, \tag{3.73}$$

which is the new principal stress in the axial (longitudinal) direction. Since the axial principal stress has now been modified, it is possible that the structural integrity of the vessel has been affected. To check whether or not this is so, we must make reference to the failure predictor in Equation *(3.55)* — itself based on a consideration of the maximum shear stress in the vessel.

A good way to visualise the interaction between principal stresses and the maximum shear stress in a solid body is to use the Mohr's circle construction (outlined in Section 3.1.5 for the biaxial stress system in shafts). In three-dimensional stress fields, the Mohr's circle is adapted to include three circles, one for each of the three interacting pairs of principal stresses: $[\sigma_1,\sigma_2]$, $[\sigma_2,\sigma_3]$ and $[\sigma_3,\sigma_1]$. For a simple, thin-walled pressure vessel *not* subject to bending, the Mohr's circle construction is as shown in Figure 3.78. In this, the $[\sigma_H,\sigma_A]$ circle is shaded in white, the $[\sigma_A,\sigma_R]$ circle is shaded lightly, and the $[\sigma_R,\sigma_H]$ circle has dark shading. (For the later Mohr's circles in Figure 3.79, the same shading will be preserved, but circles will use σ_L instead of σ_A.)

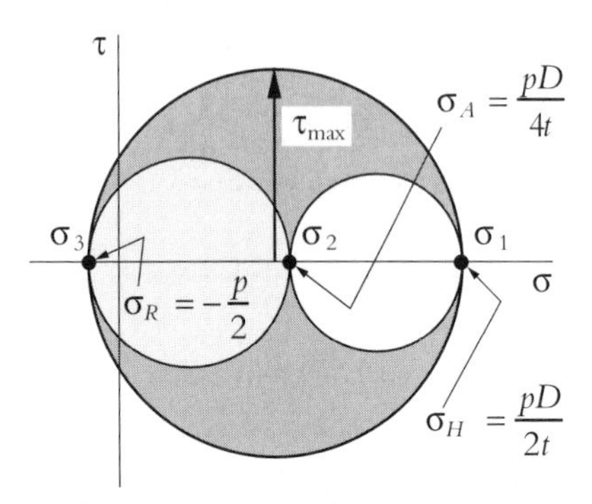

Figure 3.78 Mohr's circles for a pressure vessel (no bending)

Before addition of a bending stress, the axial stress does not enter into the calculation of maximum shear stress, since the expression for τ_{max} does not make reference to σ_2. In terms of Mohr's circle, τ_{max} is the radius of the outermost circle, and is determined only by σ_1 and σ_3.

The question now is whether the addition of $\pm\sigma_B$ to σ_A will make the total longitudinal stress large enough to be classified as σ_1, or small (negative) enough to be classified as σ_3. Three possibilities exist:

Case 0: Small bending stress

If σ_B is only small or moderate compared with σ_H (specifically, if σ_B is less than $\sigma_H/2$), then $\sigma_L = \sigma_A + \sigma_B$ will remain less than σ_H. In this case the outermost Mohr's circle will be unchanged, and the structural integrity of the vessel will not have been affected. Refer to Figure 3.79(a).

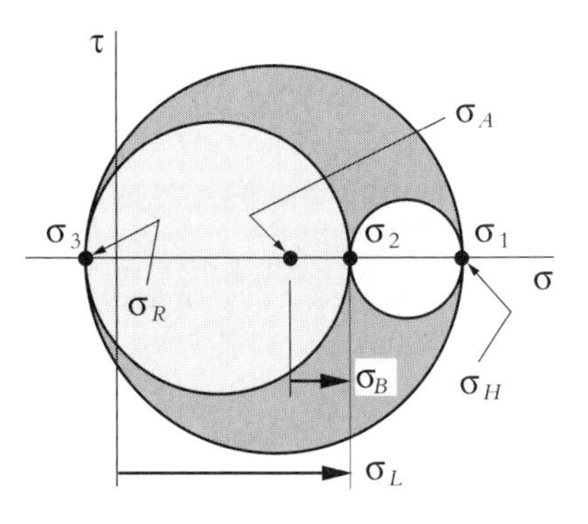

Figure 3.79(a) Small bending stress

Case 1: Large tensile bending stress

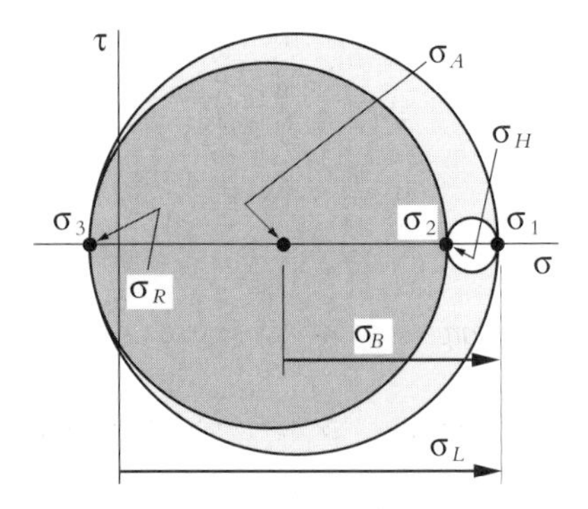

Figure 3.79(b) Large tensile bending stress

If σ_B is large enough (more than $pD/4t$), then $\sigma_L = \sigma_A + \sigma_B$ will be greater than σ_H, and so σ_L becomes the new σ_1. The outermost Mohr's circle will now be the one corresponding to [σ_L ,σ_R] interactions, and this will now be larger in radius than the old [σ_R ,σ_H] circle, so that τ_{max} will have increased. This is shown in Figure 3.79(b). The structural integrity of the vessel has been infringed. The vessel's *actual factor of safety* — defined in this case as $F_a = \tau_{max,fail} / \tau_{max}$ — will be reduced to some degree. If it falls below 1.0, the vessel will fail (at least, as predicted by the *MSFP*).

Case 2: Large compressive bending stress

If σ_B is large enough (more than $pD/2+p/2$), then $\sigma_L = \sigma_A - \sigma_B$ will be less (more negative) than σ_R, and so σ_L will become the new σ_3. The [σ_H ,σ_L] circle will now be the outermost Mohr's circle, and just as in the previous case, it will have a larger radius than the old[σ_R ,σ_H] circle. The same comments about τ_{max}, *actual factors of safety* and the vessel's structural integrity will again apply. Mohr's circle construction for this case is shown in Figure 3.79(c).

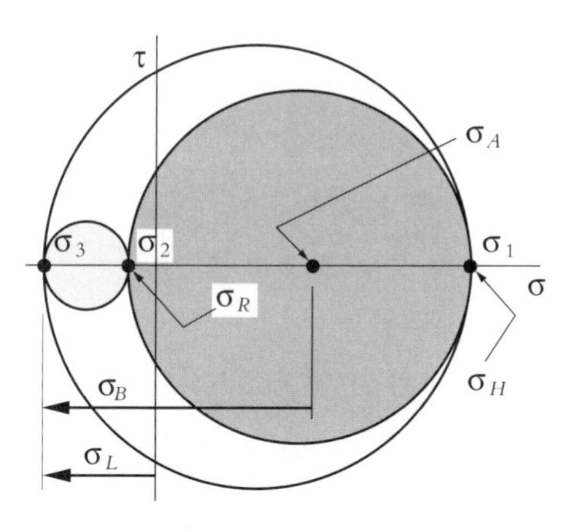

Figure 3.79(c) Large compressive bending stress

Ex. 3.21 Locating worst-case stresses

By means of a sketch, identify locations on the horizontal pressure vessel of Figure 3.77 that correspond to each of the three cases (Case 0, 1 and 2) described above.

3.6.7 Further complexities

Several categories of pressure vessel, or pressure vessel problem, lie beyond the scope of this text. They are listed briefly below, and illustrated schematically.

1. Non-cylindrical pressure vessels

Many pressure vessels, or parts of vessels, are not cylindrical. Rules for their design are usually included within pressure vessel design codes.

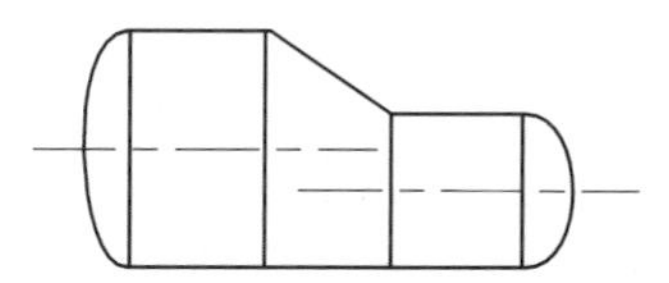

Figure 3.80(a) Asymmetric steam re-boiler

2. Pressure-bearing flat plates and ends

As a particular instance of non-cylindrical shells, pressure containers are occasionally designed having flat panels. This is not a good idea, and generally entails much ugly reinforcement and unsightly distortion. Stresses and deflections in flat plates are treated in Young (1989, Ch. 10).

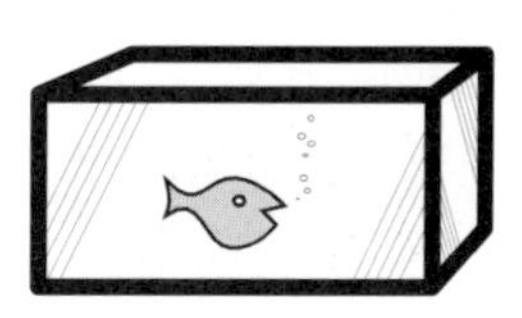

Figure 3.80(b) Aquarium

3. Vessels subject to external pressure

So-called *vacuum vessels* are encountered occasionally in industry. The extent of external pressure they must withstand is generally limited to well under an atmosphere (101.3 kPa).

Large pressure vessels and closed pipes can be accidentally subjected to internal vacuum if they are drained of liquid through a low outlet without first providing sufficient venting at the top so that air can fill the vacated volume. Many expensive accidents with large, roofed storage tanks have occurred in this manner.

Similarly, vessels can experience (atmospheric) external pressure if their liquid or gaseous contents contract upon cooling, and insufficient *breather vents* are provided. Tight-lidded saucepans on kitchen stoves sometimes suffer this effect.

Whereas the examples above are limited to relatively low (sub-atmospheric) values of external pressure, certain other types of pressure vessel can be subjected to very large external pressures:

Steam (or other hot, pressurised fluid) can be supplied to the annular outer compartment of *jacketed pressure vessels* for the purpose of heating their contents. If the contents can be at a lower pressure than the heating fluid, then the inner compartment must be designed for external pressure.

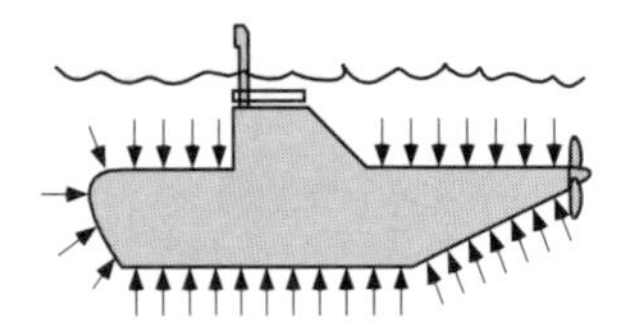

Figure 3.80(c) Submarine

Submersible ocean-going vessels, such as submarines and bathyscaphes, experience large hydrostatic hull pressures. At the deepest part of earth's ocean, the pressure reaches to about 100 MPa.

Interference fits and shrink fits are sometimes used in machinery (and in some pressure vessels) to create a tight join between an inner and an outer component. These exert a large external pressure on the inner component.

4. Thermal stresses in pressure vessels

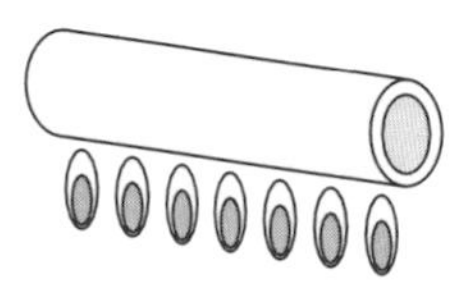

Figure 3.80(d) Gas-fired boiler tube

When vessels are heated, '*temperature gradients*' are unavoidably set up since different parts of the vessel will reach different temperatures. The gradients cause *differential thermal expansion* (meaning that different parts of the vessel will expand by different amounts), leading to thermal stresses as various parts of the vessel pull and push each other. Thermal stresses can be very high, but are always self-relieving. With large heat flux, the temperature drop across a single shell wall can be appreciable — this is especially pronounced in thick-walled shells.

Ex. 3.22

A pipe has internal diameter 100 mm and external diameter 110 mm. For each of the six combinations of location (inner surface, outer surface) and stress (hoop, axial, radial), calculate the error in stress values caused by assuming that the pipe is "thin walled". What are these errors if the outer diameter is 120 mm and 150 mm? Sketch a graph of error versus external diameter.

Ex. 3.23

Your company is preparing engineering specifications for a large gas pipeline between Rotterdam (on the coast of Holland) and Prague (capital of the Chech Republic), a distance of approximately 650 km. The operating pressure is 2.0 MPa and the internal diameter is 1000 mm. Due to international standards the following requirements have been imposed on the design:

- allowable design stress intensity $f = 120$ MPa;
- corrosion allowance = 1.5 mm.

Your company plans to use a plain carbon steel for this application and you need to determine if full weld radiography should be used.

- steel price is approximately $8/kg;
- full weld radiography costs $400 /metre of pipe length
- weld joint efficiency = 100% with full weld radiography, otherwise it is 75%;
- steel plates for this pipeline are available in thicknesses of 6, 8, 10, 12, 16, and 20 mm.

Ex. 3.24

(a) An aluminium (E=70 GPa) beverage can of diameter D= 75 mm and wall thickness t = 0.2 mm develops an internal gauge pressure of 100 kPa. Two rows of twelve cans are usually packed side-to-side in a box for shipment. Calculate the length and breadth of the shipping box needed for the pressurised cans.

(b) A hydraulic system uses 12 mm diameter steel (E = 210 GPa) tubing for interconnecting several hydraulic actuators. The length of tubing used is 10 m and you need to estimate the change in hydraulic volume as a function of pressure. Calculate the increase in the tube volume at 21 MPa (the maximum operating pressure). What is the volume increase when the tube is pressurised to a level where the hoop stress is 100 MPa? [The tubing is made of steel with 800 MPa yield strength and a design factor of safety of 1.8]

3.6.8 Summary of design rules for cylindrical vessels, internal pressure

(a) Thin-walled vessels

Pressure stress:

Model:

$$\sigma_H = \frac{pD}{2t}, \quad (3.56)$$

$$\sigma_A = \frac{pD}{4t}, \quad (3.57)$$

$$\sigma_R = -\frac{p}{2}. \quad (3.58)$$

Require: $(\sigma_1 - \sigma_3) \le f$. (3.55)

Rule: $t \ge \dfrac{pD}{2f\eta - p} + c$. (3.62)

(b) Diametral expansion:

Model (cylinder):

$$\delta_{\text{diam}} \approx \frac{pD^2}{4tE}(2-\mu). \quad (3.70)$$

Model (sphere):

$$\delta_{\text{spherical diam}} \approx \frac{pD^2}{4tE}(1-\mu). \quad (3.71)$$

Require: $\delta_{\text{diam}} \le \delta_{\text{diam,allowable}}$.

(c) Thick-walled vessels (pressure stresses)

Model:

$$\sigma_H = \frac{p}{\Phi^2 - 1}\cdot\left(1 + \frac{b^2}{r^2}\right), \quad (3.66\text{-}a)$$

$$\sigma_A = \frac{p}{\Phi^2 - 1}, \quad (3.66\text{-}b)$$

$$\sigma_R = \frac{p}{\Phi^2 - 1}\cdot\left(1 - \frac{b^2}{r^2}\right), \quad (3.66\text{-}c)$$

$$\tau_{max} = \frac{\sigma_H - \sigma_R}{2} = \frac{p}{\Phi^2 - 1}\left(\frac{b^2}{r^2}\right), \quad (3.67)$$

where $\Phi \equiv \dfrac{D_o}{D} = \dfrac{b}{a}$. (3.63)

Require: $(\sigma_1 - \sigma_3) \le f$. (3.55)

(d) External bending

Model:

$$\sigma_B = \frac{32 M_{\max}}{\pi D_C^3}\cdot\frac{\Phi_C}{\Phi_C^4 - 1}, \quad (3.72)$$

$$\sigma_L = \sigma_A \pm \sigma_B, \quad (3.73)$$

and use Mohr's circle construction to check whether $\sigma_L > \sigma_H$ or $\sigma_L < \sigma_R$ (see Section 3.6.6).

Require: $(\sigma_1 - \sigma_3) \le f$. (3.55)

3.7 Assignment on pressure vessels

3.7.1 Background

A large vertical pressure vessel is being designed for use in a petroleum refinery. It is to have an inner diameter of D = 2500 mm, and an outer diameter of $D_o = D + 2t$ (where t is the shell thickness). It will be welded from mild steel plates, and in service will store only gaseous contents. The general layout of the vessel is shown in Figure 3.81

Some concern exists over the fact that the refinery is located in a West African coastal region which is occasionally subject to cyclone activity. Under worst-case conditions, the free-stream wind velocity can reach 50 to 60 m/s. Because the vessel is so tall, the project manager wants you to carry out an initial check on its ability to withstand wind-induced bending stresses. (This will just be a quick check — formal design calculations to satisfy the requirements of the Wind Loading code will be carried out later in the project.)

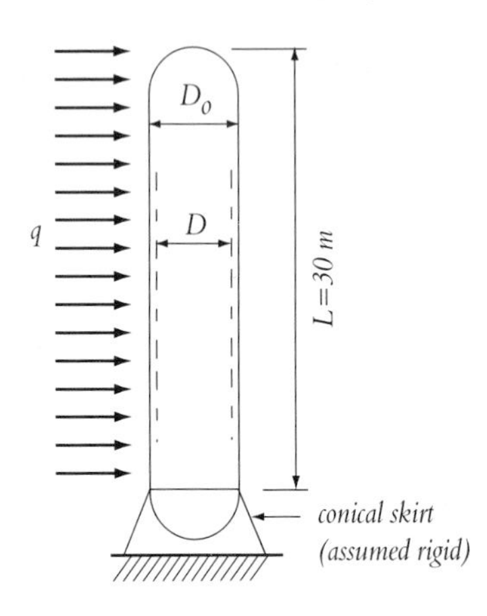

Figure 3.81 Vertical vessel

You have decided to approach these preliminary calculations for the vessel in two stages — firstly a design based on internal pressure alone in order to obtain a first estimate of the required wall thickness, then a check of the stresses in the vessel when both internal pressure and external wind loads are acting simultaneously.

3.7.2 The Design Problem

(a) *Design for internal pressure only:*
Using the design parameters given below, select a suitable plate size for the vessel. Available plate sizes are 6 mm, 8 mm, 10 mm, 12 mm, 16 mm, 20 mm, 25 mm, 32 mm, 40 mm, 50 mm, and 65 mm. At this stage, calculations should be on the basis of design to resist yielding or rupture due to internal pressure only. Design data can be summarized as:

- internal design pressure (gauge pressure), $p = 2.0$ MP a
- total corrosion allowance (internal + external), $c = 3.0$ mm
- welded joint efficiency, $\eta = 0.85$
- design factor of safety (yielding), $F_{dy} = 1.6$
- design factor of safety (rupture), $F_{dr} = 4.0$
- material yield stress, $S_y = 240$ MPa
- material tensile strength, $S_U = 620$ MPa

Note that the design stress intensity, f, is taken to be the minimum of S_y/F_{dy} and , S_U/F_{dr}. This is more-or-less the basis of design stresses used in several statutory pressure vessel design codes, notably *AS-1210* (Australia), *BS5500* (U.K.) and *ASME-VIII* (United States).

(b) *Design check for wind loading and internal pressure:*
Given the shell thickness selected above, what average free-stream wind velocity, V, would cause the stresses in the vessel to exceed their allowable values? Your analysis should consider static wind forces only (i.e. the effects of vortex shedding may be ignored here).

You may find it helpful to refer to the Mohr's stress circle when considering the stresses which act on this vessel when it is subject to both internal pressure and external wind loading simultaneously. In justifying your calculation you should use an appropriate failure predictor (*theory of failure*), together with the following assumptions:

- the stresses arising from wind loads and internal pressure can be evaluated separately and then combined using superposition;
- the horizontal wind load acting on the vessel is uniformly distributed along the vessel's vertical axis, and can be specified as a force per unit distance along that axis, q N/m. This distributed load is given by:

$$q = C_D\left(\frac{\rho V^2}{2}\right)D_o$$

(note: C_D is called the *drag coefficient*);

- $C_D = 0.4$ (assumed constant over the range of wind velocities, V, relevant to this problem);
- $\rho = 1.2$ kg/m^3 for air.

Engineering Design Associates

Project Title : **Wind-loaded pressure vessel**

Date 12:4:98

Page 1 of 6

Ckd. kne

VERTICAL, WIND-LOADED PRESSURE VESSEL

(a) PV design — internal pressure only

Approach 1

— use simple application of PV design equation (this approach is perfectly valid here).

— use $t \geq \frac{pD}{2f\eta - p} + C$

— yielding governs, since $\left(\frac{240}{1.6}\right) = 150$ (yield) which is less than $\left(\frac{620}{4.0}\right) = 155$ (rupture)

— $p = 2.0$ MPa $\quad \eta = 0.85 \quad f = 150$ MPa
$D = 2500$ mm $\quad C = 3.0$ mm

— $t \geq \left(\frac{2.0 \times 2500}{2 \times 150 \times 0.85 - 2.0}\right) + 3.0$ mm
$\geq 22.76 \Rightarrow$ choose $t = 25$ mm plate

Approach 2

— derive equ'ns from 1st principles (OK, but unnecessarily awkward in such a straightforward case. Also, prone to omission of η & C terms).

— $\sigma_1 = \frac{pD}{2t}$ (hoop stress)
$\sigma_2 = \frac{pD}{4t}$ (longitudinal stress)
$\sigma_3 \simeq -\frac{p}{2}$ (average radial stress)

— $\tau_{MAX} = \frac{\sigma_1 - \sigma_3}{2} = \left(\frac{pD}{2t} + \frac{p}{2}\right)/2 = \frac{pD + pt}{4t}$

— "Max. Shear Stress Theory of Failure" predicts failure when $\tau_{MAX} = \frac{S_y}{2}$:

Engineering Design Associates

Project Title : **Wind-loaded pressure vessel**

Date 12:4:98

Ckd. kne

– design inequality (not incl. welding effy) : $\tau_{MAX} \leq \frac{S_y}{2F_d}$

– but welded material has lower strength : $\boxed{\tau_{MAX} \leq \frac{S_y \eta}{2F_d}}$

– combining, obtain $\left(\frac{pD}{2t} + \frac{p}{2}\right)/2 \leq \frac{S_y \eta}{2F_d}$

$$\Rightarrow \quad \frac{pD}{2t} \leq \frac{S_y \eta}{F_d} - \frac{p}{2}$$

$$\Rightarrow \quad t \geq \frac{pD}{2\left(\frac{S_y}{F_d}\right)\eta - p}$$

– this is the "corroded thickness" — for actual min. thickness to resist pressure & corrosion, require :

$$\boxed{t \geq \frac{pD}{2\left(S_y/F_d\right)\eta - p} + C}$$ (as before, $f = S_y/F_d$)

– $t \geq 22.76$, choose 25 mm plate (as before).

(b) Cantilever design — distributed wind load (as well as pressure)

Given that $t = 25$ mm, calculate V to cause overstressing of vessel shell (at base of cantilever, with both p & q).

NOTE : – experienced designers would carry out the above analysis on the basis of the corroded dimensions, ie. $t_c = 22$ mm. Most students will omit this & so these calcs are based on $t = 25$mm. Results for $t_c = 22$ mm are included for comparison.

– uniformly distributed wind load, $q = C_D\left(\frac{\rho V^2}{2}\right)D_o$ N/m

– $C_D = 0.4$

$\rho = 1.2$ kg/m³

$D_o = D + 2t = 2500 + 2 \times 25 = 2550$ mm $= 2.55$ m

Engineering Design Associates

Project Title : **Wind-loaded pressure vessel**

Date 12:4:98

Page 3 of 6

Ckd. kne

– load, SF & BM diagram

V, q, L = 30 m, 'A', 'B', tens., comp., SFD, BMD, ⊖, q.L, $M_{MAX} = \left(\frac{qL^2}{2}\right)$

– bending stress due to wind loading :– $\sigma_B = \frac{My}{I}$

∴ max bending stress, $\sigma_{B,MAX} = \frac{M_{MAX}\, y_{MAX}}{I}$ (occurs at 'A' (tens.) & 'B' (comp))

– this is a longitudinal stress, & adds (or subtracts) to the longitudinal pressure stress $\sigma_{\ell,p} = \frac{pD}{4t}$

– three principal stresses are: $\sigma_c = \frac{pD}{2t}$

$$\sigma_\ell = \sigma_{\ell,p} + \sigma_{\ell,B}$$

$$= \frac{pD}{4t} + \frac{My}{I}$$

$$\sigma_r \simeq -\frac{p}{2}$$

– without bending, the max. princ. stress, σ_1, is σ_c.
– with sufficient levels of bending stress, the max. princ. stress can become σ_ℓ.
– best illustrated on a Mohr's circle diagram.

Engineering Design Associates

Project Title : **Wind-loaded pressure vessel**

Date 12:4:98

Page 4 of 6

Ckd. kne

'A'&'B' no bending, locus 'A' & 'B' the same

$$\left(\tau_{MAX} = \frac{\sigma_1 - \sigma_3}{2} = \left(\frac{pD}{2t} + \frac{p}{2}\right) \right)$$

τ_{MAX}, $-\frac{p}{2}$, $\sigma_\ell = \frac{pD}{4t}$, $\frac{pD}{2t}$

'A': τ_{MAX}, $+\sigma_B$, $-\frac{p}{2}$, $\sigma_\ell = \left(\frac{pD}{4t} + \sigma_B\right)$, $\frac{pD}{2t}$

'B': moderate bending (τ_{MAX} unaffected); τ_{MAX}, $-\sigma_B$, $-\frac{p}{2}$, $\sigma_\ell = \left(\frac{pD}{4t} - \sigma_B\right)$, $\frac{pD}{2t}$

'A': τ_{MAX}, $+\sigma_B$, $-\frac{p}{2}$, $\frac{pD}{2t}$, $\sigma_\ell = \left(\frac{pD}{4t} + \sigma_B\right)$

$$\tau_{MAX} = \left(\frac{pD}{4t} + \sigma_B + \frac{p}{2}\right)\Big/2$$

'B': large bending; τ_{MAX}, $-\sigma_B$, $-\frac{p}{2}$, $\sigma_\ell = \left(\frac{pD}{4t} - \sigma_B\right)$, $\frac{pD}{2t}$

$$\tau_{MAX} = \left(\frac{pD}{2t} - \left(\frac{pD}{4t} - \sigma_B\right)\right)\Big/2$$

$$= \left(\frac{pD}{4t} + \sigma_B\right)\Big/2$$

τ_{MAX} is larger on the tension side than on the compression side.

Engineering Design Associates

Project Title : **Wind-loaded pressure vessel**

Date 12:4:98

Page 5 of 6

Ckd. kne

– if $\sigma_{B,MAX} > \frac{pD}{4t}$, then longitudinal stress becomes the maximum principal stress, and τ_{MAX} has its worst-case value on the tension side of the cantilever; where:

$$\tau_{MAX} = \left(\frac{\sigma_\ell - \sigma_r}{2}\right) = \left(\frac{\sigma_{\ell,p} + \sigma_B - \sigma_r}{2}\right)$$

$$= \frac{\left(\frac{pD}{4t} + \frac{p}{2} + \frac{My_{max}}{I}\right)}{2}$$

τ, τ_{MAX}, σ_B, $\sigma_r = -\frac{p}{2}$, $\sigma_{\ell,p} = \frac{pD}{4t}$, $\sigma_c = \frac{pD}{2t}$, $\sigma_\ell = \left(\frac{pD}{4t} + \sigma_B\right)$

– limiting value of shear stress is $\tau_{all} = \frac{S_y}{2\,F_d}$

– design inequality is $\tau_{MAX} \leq \tau_{all}$

i.e. $$\frac{\left(\frac{pD}{4t} + \frac{p}{2} + \frac{My_{max}}{I}\right)}{2} \leq \frac{S_y}{2\,F_d}$$

$$\Rightarrow \boxed{\frac{M\,y_{max}}{I} \leq \frac{S_y}{F_d} - \left(\frac{pD}{4t} + \frac{p}{2}\right)}$$

– the L.H.S. represents the max. allowable bending stress due to wind loading, $\sigma_{B,MAX}$

$$\frac{My_{max}}{I} \leq \frac{240}{1.6} - \left(\frac{2.0 \times 2500}{4 \times 25} + \frac{2.0}{2}\right)$$
$$\leq 150 - (50.0 + 1.0)$$
$$\leq 99.0\ \text{MPa}$$

– $M \leq 99.0\,\text{MPa} \times \left(\frac{I}{y_{MAX}}\right)$

– $$\left(\frac{I}{y_{MAX}}\right) = Z = \frac{\pi}{64}\left(D_o^4 - D^4\right) \Big/ \left(\frac{D_o}{2}\right) = \frac{\frac{\pi}{64}\left(2550^4 - 2500^4\right)}{\left(\frac{2550}{2}\right)}$$

$$= \frac{0.15806 \times 10^{12}\ \text{mm}^4}{1275\ \text{mm}} = \boxed{123{,}970{,}000\ \text{mm}^3 = Z}$$

Engineering Design Associates

Project Title : **Wind-loaded pressure vessel**

Date 12:4:98

Page 6 of 6

Ckd. knw

— hence $M_{MAX} \leq 99.0 \text{ MPa} \left(\frac{N}{mm^2}\right) \times 123{,}970{,}000 \text{ mm}^3$

$\leq 1.227 \times 10^{10} \text{ N.mm}$

$\leq \boxed{12.27 \times 10^6 \text{ N.m}}$

— recall $M_{MAX} = \frac{qL^2}{2}$

$\Rightarrow q \leq M_{MAX} \frac{2}{L^2}$

$\leq 12.27 \times 10^6 \text{ N.m} \times \frac{2}{30^2 \text{ m}^2}$

$\leq \boxed{27{,}273 \text{ N/m}}$

but $q = C_D \left(\frac{\rho V^2}{2}\right) D_o$ N/m

$\Rightarrow V \leq \left(\frac{2q}{C_D \rho D_o}\right)^{\frac{1}{2}}$ m/s $= \left(\frac{4M_{MAX}}{C_D \rho D_o L^2}\right)^{\frac{1}{2}}$

$\leq \left(\frac{2 \times 27273 \text{ N/m} \left(\frac{kg.m}{s^2}\right)}{0.4 \times 1.2 \text{ kg/m}^3 \times 2.55 \text{ m}}\right)^{\frac{1}{2}}$

$\leq \left(44{,}564 \frac{m^2}{s^2}\right)^{\frac{1}{2}}$

$\Rightarrow \boxed{V \leq 211 \text{ m/s}}$

In primitive symbols : $M_{MAX} \leq \left[\frac{S_y}{F_d} - \left(\frac{pD}{4t} + \frac{p}{2}\right)\right] \frac{I}{y_{MAX}}$

$\therefore V \leq \left\{\frac{4}{C_D \rho D_o L^2} \left[\frac{S_y}{F_d} - \left(\frac{pD}{4t} + \frac{p}{2}\right)\right] \frac{I}{y_{MAX}}\right\}^{\frac{1}{2}}$

$\leq \left\{\frac{4}{0.4 \times 1.2 \times 2.547 \times 30^2} \left[\frac{240}{1.6} - \left(\frac{2.0 \times 2.503}{4 \times 0.022} + \frac{2.0}{2}\right)\right] \times 10^6 \times \ldots \right.$

$\left. \ldots \times \frac{\frac{\pi}{64}\left(2.547^4 - 2.503^4\right)}{(2.547/2)}\right\}^{\frac{1}{2}}$

corroded :—
$D_o = 2.547$ m
$D = 2.503$ m
$t = 22 \times 10^{-3}$ m $= 0.022$ m

≤ 191 m/s (corroded dimensions)

4

DESIGN OF MECHANICAL CONNECTIONS

He shall bring together every joint and member, and mould them into an immortal feature of loveliness and perfection.
Milton, Areopagitica, 1644

It is one thing to design a simple engineering component consisting of a single structural element; it is another to design a structural system of several cooperating components, where the structural and functional integrity of the whole is a complex amalgam of that of its parts. When dealing with such a system, four additional challenges are superimposed upon the designer's task:

1. to identify, by means of a process of *structural distillation*, the subsidiary structural elements that compose the system (we have discussed the notion of *structural distillation* extensively in Section 1.3 and elsewhere in this book.);
2. to define or delineate the nature of the *structural connections* between those subsidiary structural elements;
3. where necessary, to establish the functional requirements of such connections; and
4. to design the connections.

As was the case for the *generic structural components* examined in Chapter 3, the last of these challenges (that is, "*design the element*") is a succinct statement covering a multitude of responsibilities and decisions incumbent upon the designer. It involves choosing:

- materials,
- configuration,
- geometry and
- process requirements (such as details of fabrication, assembly and pre-loading)

for each of the connections in the system in order to meet the full range of objectives and constraints imposed upon it. It implies achieving a balance between the strength of the connections and the strength of the component parts being connected. Designing the *structural connections* in a system is at least as demanding a task as designing the structural elements they connect.

In the previous chapter, we developed design rules for a range of generic engineering components, following an essentially consistent methodology. In this chapter we concentrate much more on the structural connections between these components. We will discover that the connections themselves can be assigned structural and functional objectives, and that relatively simple design rules can be adduced for them using an approach based on mathematical modelling, design inequalities and factors of safety. Our intention is to develop this approach in a manner congruent with the one used for all the generic engineering components introduced so far in this book, thereby building confidence in the novice designer for the application of *design thinking*.

4.1 The nature of mechanical connections

We begin by exploring some philosophical issues around the topic of connections. The purpose of this brief section is to clarify our thinking and terminology in preparation for the more detailed discussion of specific connection types in Sections 4.2, 4.3 and 4.4.

4.1.1 'Mechanical' versus 'integral' connections

For our present purposes we will need to make a distinction between two main kinds of *structural connection*. In some cases, the *connection* between adjacent subsidiary structural elements will exist

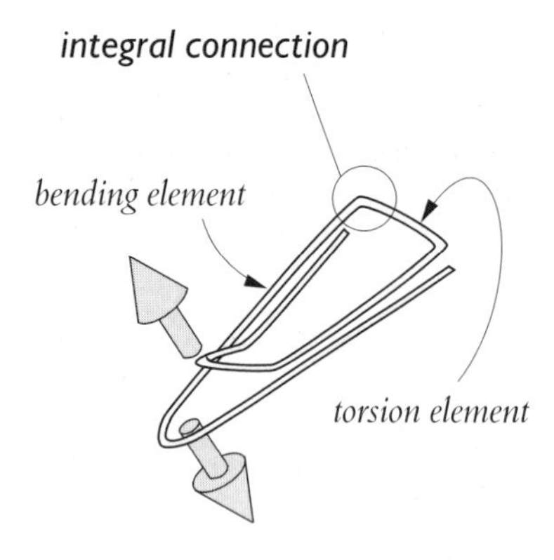

Figure 4.1(a) Integral connections in a simple part

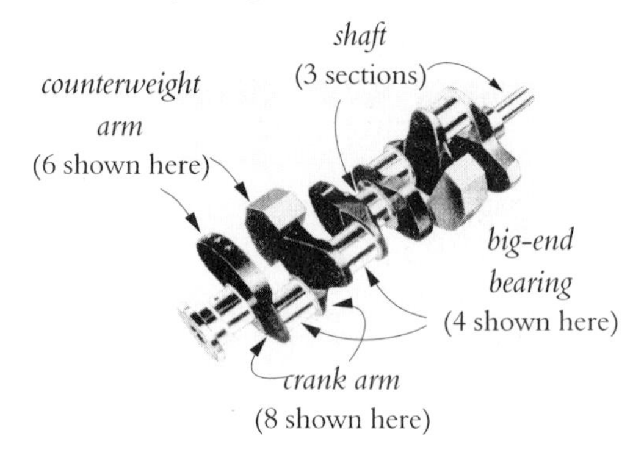

Figure 4.1(b) Integral connections in a more complex

only as a concept separating two regions of a contiguous load-carrying part. No one has to create a discrete connection in these cases, since the connection is already inherent in the body of the part. The part is formed *in one piece* from a single material, generally by manufacturing processes like extrusion, forging, casting, bending or pressing. Different regions within it fulfil different structural functions (and comprise distinct structural elements in the terms of a *structural distillation*) by virtue only of the shape of the part and the loads imposed on it. We will call the conceptual structural connections between these contiguous regions, *integral connections*. Figure 4.1 shows two examples (a paper clip and an automotive crankshaft) of parts possessing *integral connections*. This class of connection is *not* the subject of the present chapter.

In other cases, *a structural (that is, load-carrying) connection must be made between physically separate entities in a structural system.* The entities may be different *generic structural components* within the system (for example, the tree-limb, rope and tyre of the child's swing in Figure 4.2 are, respectively, a cantilever-beam, a tension link and a ring-beam). Alternatively, the connected entities may be sub-component parts of a single *generic structural component* that is too large to be made in one piece. For example, adjacent cylindrical plates in a large pressure vessel shell must be joined to form the vessel, and the long marine propeller shafts used in large ships are commonly built from several shaft sections (Figure 4.3). All these connections must be created (designed and fabricated) in a conscious effort separate from the creation of the parts that they join. We will refer to such connections collectively as *mechanical connections*.

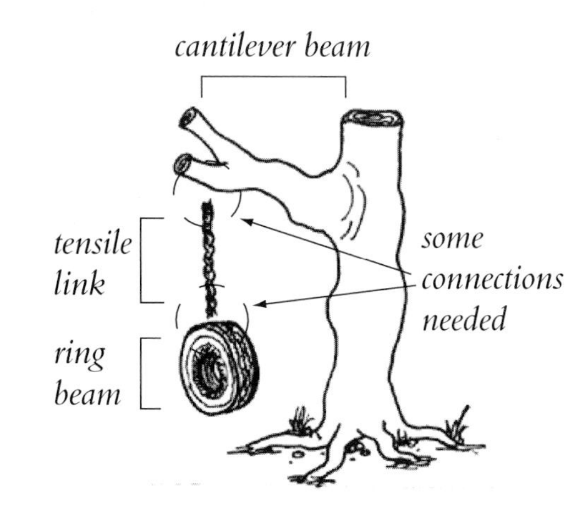

Figure 4.2 Mechanical connections between discrete structural components

4.1.2 Classes of mechanical connections

Mechanical connections come in a vast array of embodiments. It will be instructive to consider the range of mechanical connections used to join a familiar component of everyday life (namely, *paper*), and then to see whether some more abstract insights can be drawn concerning the nature of mechanical connections.

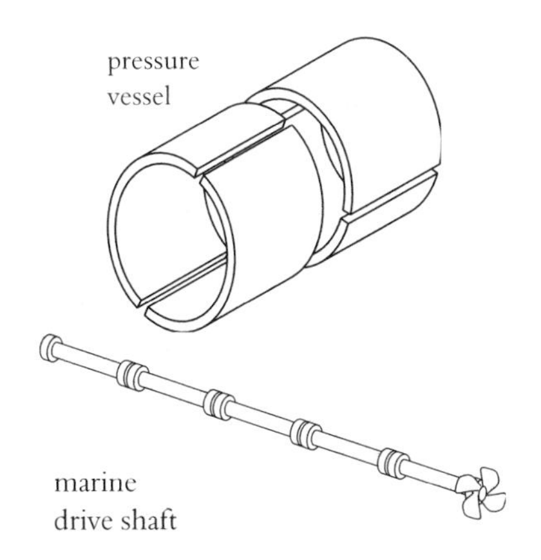

Figure 4.3 Mechanical connections within a single structural component

We define a *fastener* most generally as *any device that connects two or more solid objects to prevent relative motion between them to some degree.* Figures 4.4, 4.5 and 4.6 show a sample of the range of *fasteners* that have been developed to join paper. Some are ancient; others are very recent. Most will be familiar to readers (although some notes are provided below to explain those less common fastening methods whose sketches are marked with an asterisk).

Some of the examples shown would scarcely be called *fasteners* in general speech, yet they fulfil the requirements of our current definition. They each have different characteristics, and their own particular advantages and disadvantages. This is not surprising, since they were developed at various times to satisfy different functional objectives. Each paper fastener is an example of a *mechanical connection* (in the broadest sense of the term), and we will use their diversity to illustrate the various essential features of mechanical connections.

Explanatory notes:

Figure 4.4 All of these examples are clamped joints. They all use some mechanical component to establish a compressive force between adjacent sheets. This compression sets up frictional forces that prevent the paper sheets from sliding out.

Figure 4.4 (E) We do not suggest that woodworking clamps are often used to join paper! Rather, this sketch represents the wide array of mechanical clamps used to secure sheets of paper in printing, cutting and other paper processing machinery.

Figure 4.4 (H) The *glide-on spine* is an item of commercially produced stationery. Its main component is a plastic extrusion with an open, hollow, triangular cross-section. This is slid axially along the spine of a folder (or sometimes swivelled about a pivot at one end of the spine) in order to clamp outer covers between the arms of the cross-section.

Figure 4.4 (I) The *spring-back spine* is another item of commercially produced stationery. It is a length of light spring steel having an open, hollow, circular cross-section. It forms the spine of a folder that has rigid end covers. By forcing the covers backwards, the spine is opened to allow insertion of paper sheets. These become clamped when the spine is allowed to *spring back*.

Figure 4.4 (M/N) It could be argued that the wrapping in these examples serves not to *clamp* the sheets, but merely to *enclose* them. The distinction is semantic only.

Figure 4.5 All of these examples are shear links. They all pierce the paper sheets in some way, and interpose some kind of shear pin through the hole.

Figure 4.5 (A) Staples are perhaps the most commonly encountered mechanical connector for paper. If they are applied tightly enough, staples provide some clamping action that forces the paper sheets into firm frictional contact with each other. Loose staples function as *dual shear pins*, linking the paper in a simple shear joint.

Figure 4.5 (B) Many staplers are fitted with a reversible platen to allow the formation of *open* staples. These function much like a pin through the paper, and are more easily removed from it than standard *closed* staples.

Figure 4.5 (G) The screw-and-nut combination is an item of commercially produced stationery, often used to join the pages of stamp- or photo-albums. The screw tightens hard onto the end-face of the *nut*, which acts like a simple shear pin. Unless poorly sized, this device does not exert any compressive force onto the paper itself.

Figure 4.5 (J) Old-style book binding techniques involved both gluing and stitching the spine.

Figure 4.6 (A/B/C) These three examples are shear connectors like those in Figure 4.5, except that here the objective is to fasten paper to some surface rather than to other paper.

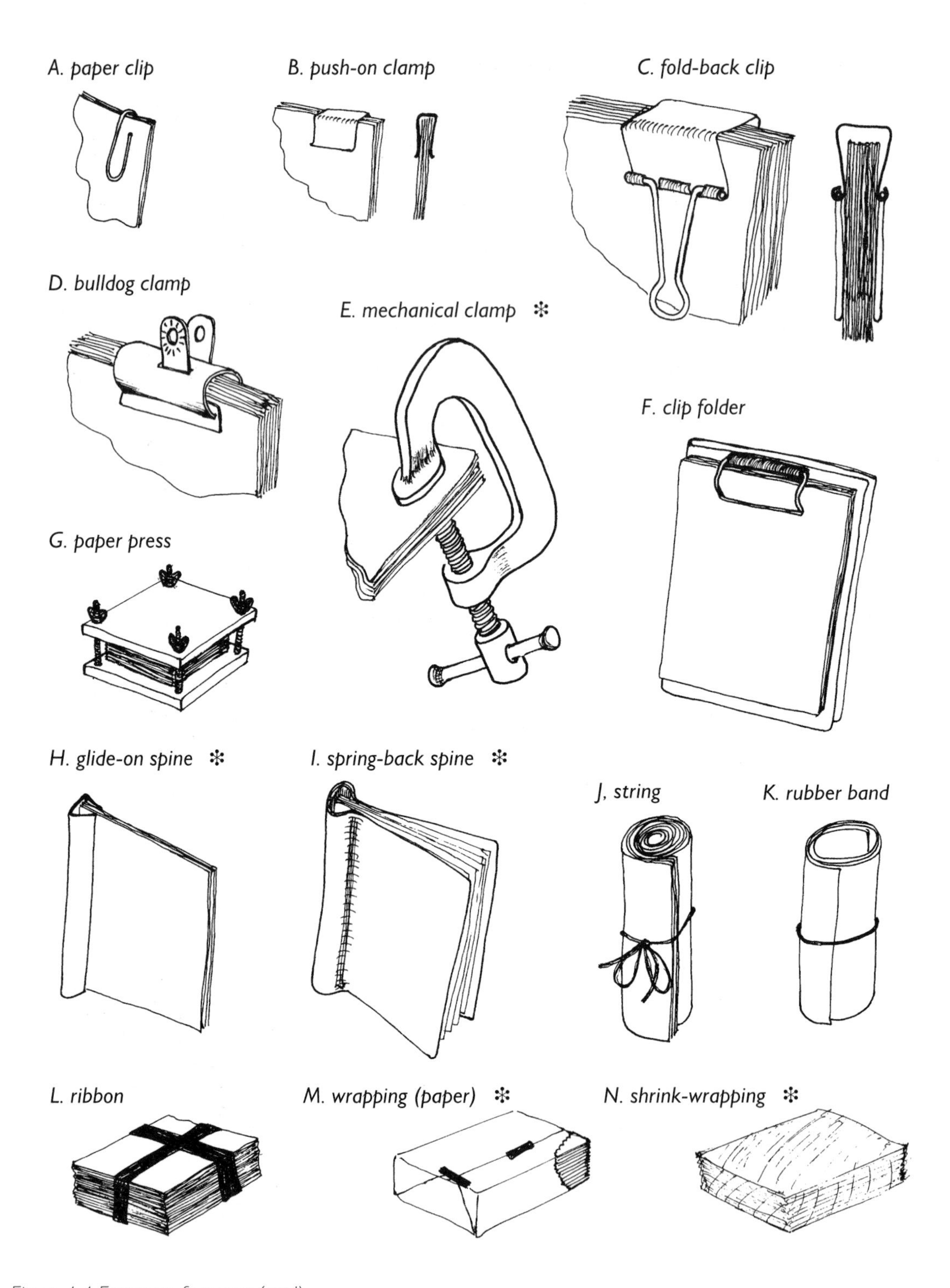

Figure 4.4 Fasteners for paper (set I)

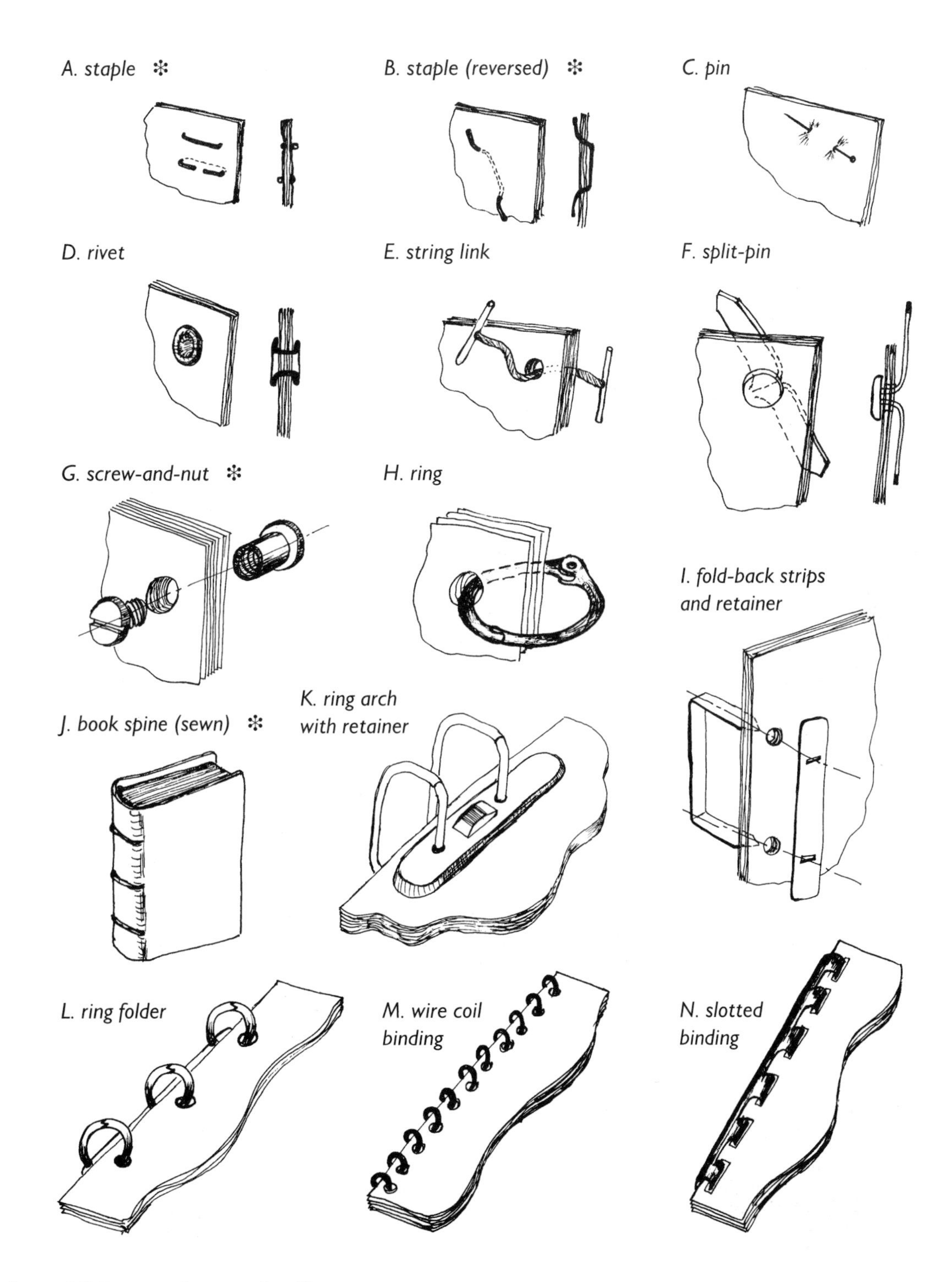

Figure 4.5 Fasteners for paper (set II)

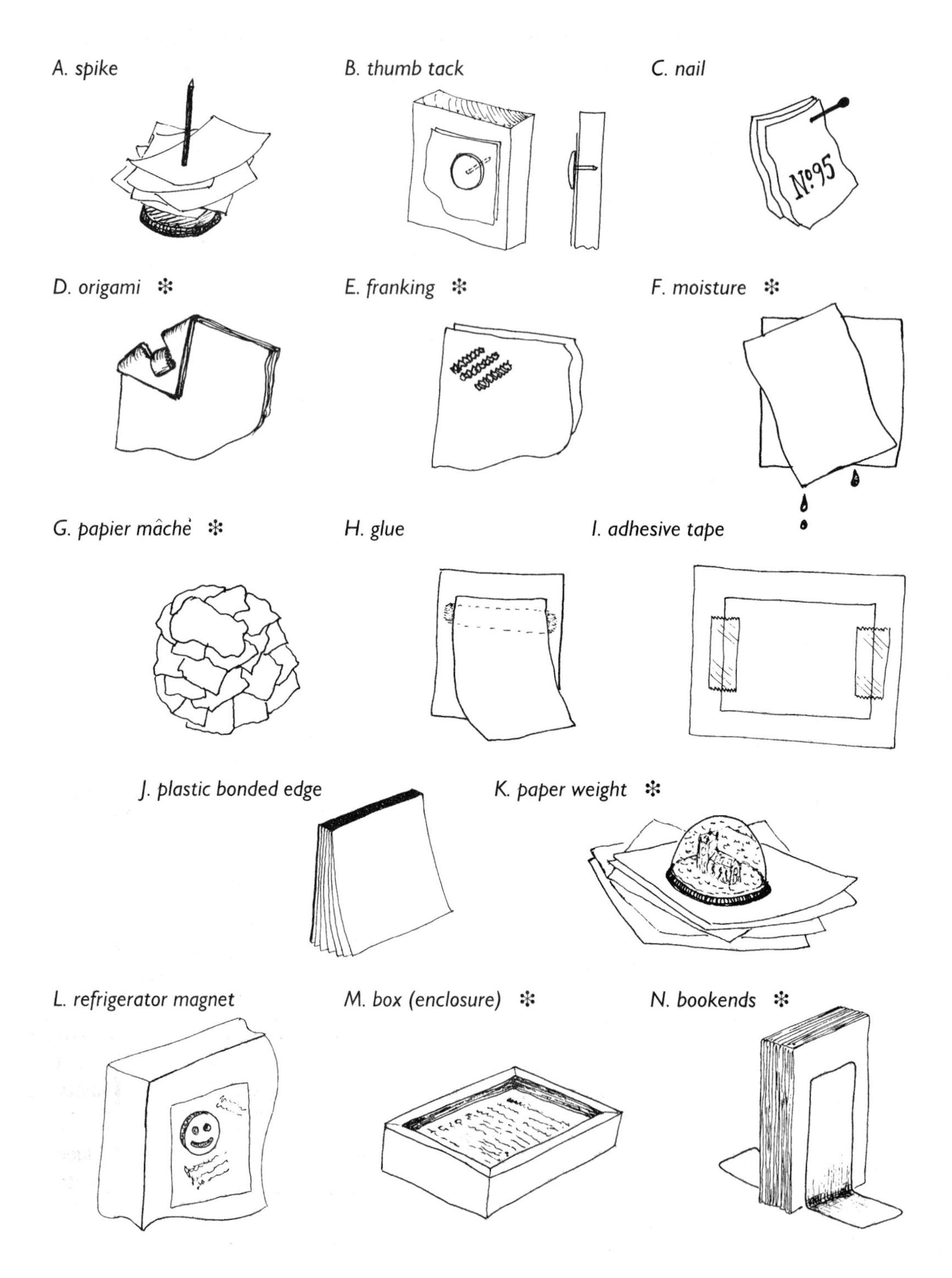

Figure 4.6 Fasteners for paper (set III)

Figure 4.6 (D) A corner-locking technique favoured by some students for assignment submissions, but universally loathed by those who must assess the documents.

Figure 4.6 (E–J) The tensile bonds available in examples (E, F, G, H, I, J) of Figure 4.6 do not rely on any compressive elements within the fastener.

Figure 4.6 (E) Franking is a little known paper-joining technique that uses a purpose-built machine to emboss rows of deep indentations into (usually only two) sheets of paper simultaneously. These indentations '*lock*' into those on the adjacent sheet, forming a weak joint.

Figure 4.6 (F) Moisture can act to decompose the surface of paper sheets slightly, so that it temporarily forms a thin layer of fibrous, starch suspension (not unlike the *pulp* from which paper is made). If allowed to dry while in contact with other similarly damp sheets, the paper can develop a weak bond.

Figure 4.6 (G) Papier mâché uses water and paste to soften paper into a slurry that can be moulded into shapes.

Figure 4.6 (K) A gravity-assisted paper fastener.

Figure 4.6 (M) The box represents any enclosure that acts as a constraint on the relative motion of paper sheets. In the open box shown, gravity (*self-weight*) provides an increasing clamping force on sheets lower down the stack (the lower the sheet, the greater the clamping force). Other such enclosures (not relying on gravity) include paper bags and plastic pockets.

Figure 4.6 (N) Bookends impart a small end-to-end clamping force to the paper, limited in magnitude by frictional forces between the bookends and their supporting surface.

In the forty-two examples above, we can identify the following categories of mechanical connection.

(A) Connections formed by linking the connected parts (shear link)

Examples are Figure 4.5 (A, B, C, D, E, F, G, H, I, J, K, L, M, N), and Figure 4.6 (A, B, C, D). These connections all pierce the connected elements in some way, and interpose some other linking element through the hole. The linking element is often called a *shear pin*, since it generally experiences shearing forces. In abbreviated terms, these are *pinned joints*.

(B) Connections formed by clamping the connected parts (compressive force exerted by an element in bending or tension, or by some other source)

Examples are Figure 4.4 (A, B, C, D, E, F, G, H, I, J, K, L, M, N), Figure 4.6 (K, L, M, N), and potentially Figure 4.5 (A, D, I, J, K). The existence of clamping forces in Figure 4.5 (A, D, I, J, K) depends on how tightly the joint has been formed. In most of these example (all those from Figures 4.4 and 4.5), an element can be identified that is in bending or tension, and which consequently imparts compressive contact forces to and between the connected elements. Again in abbreviated terms, these are *clamped joints*, of which *bolted joints* will form a subset.

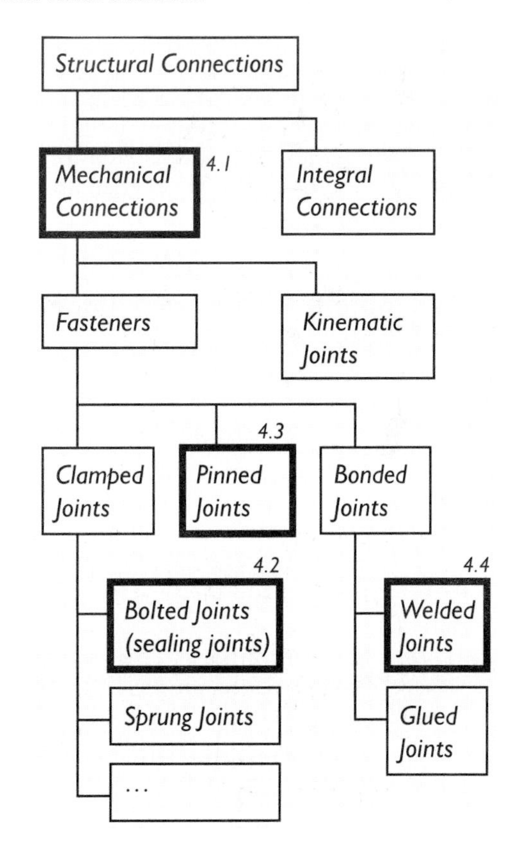

Figure 4.7 A simple taxonomy of structural connections

(c) Connections formed by bonding the connected parts (tensile bonding agent)

Examples are Figure 4.5 (J), and Figure 4.6 (G, H, I, J). These are *glued joints*.

(d) Connections formed by coalescing the material of the connected parts (establishes a tensile bond in the material itself)

Examples are Figure 4.6 (E, F, G). Very similar to the connections in the previous category, except that here the bond is formed within the connected material itself. These are *welded joints*.

On the basis of the discussion to date and the categorisation noted above, we propose Figure 4.7 as a *road map* of terms describing classes of structural connection referred to in this book. Figure 4.7 also highlights where the sections of the present chapter fit into the overall *road map*.

More will be said about the definitions of *bolted*, *pinned* and *welded* joints in the appropriate sections that follow. We note in passing that the topic of *structural connections* is a very wide one, and we make no attempt to deal with several large sub-topics. These include *kinematic joints* (joints designed to move), *glued joints*, and the motley assembly of *clamped joints* that do not use bolts as their source of compressive force.

4.1.3 Assignment on abstract thinking about fasteners

(a) Aims

This assignment deals with fasteners. It also highlights a crucial aspect of *design thinking*: the ability to move from abstract statements of problems and ideas to the development of solutions that specify or embody hardware. It emphasises:

- shifting one's thinking at will between abstract concepts and the physical hardware by which these concepts are realised;
- specifying and communicating essential constructional detail by sketches; and
- thinking about the wide range of fasteners used for joining even simple materials, and the functional objectives that those fasteners aim to fulfil (refer to Table 4.1).

Table 4.1 Fastener objectives

Objective	Criteria
Fastening strength	*Force required to dismantle fastened assembly*
Ease of actuation	*Force required to assemble fastener and objects being joined; time to make the connection*
Ease of unfastening	*Force required; time to unfasten objects*
Ease of access	*Extent to which the space available exceeds the size of hands, fingers, tools*
Minimum protrusion into the environment	*Height of protrusion above bounding surface; increase in aerodynamic drag*
Precision of relative location of objects	*Distance representing maximum mismatch of one object with respect to the other*

(b) Preamble

Engineering designers must develop a heightened sense of self-awareness, especially the ability to switch thinking at will from one level of abstraction to another. For example, a powerful advantage is provided to the vehicle designer who is able to progress from thinking about physical objects such as rubber tyres to the conceptualisation of general pictures of impact loads on flexible membranes stressed under internal pressure.

This exercise is concerned on the one hand with a general, abstract definition of a device called a *fastener*, and on the other hand with ways of physically realising such devices in engineering practice. We may start with knowledge of specific things like screws and nails used to fasten objects together, and then try to formulate a wide ranging and abstract definition, which may in turn suggest new and novel types of fastener. The general definition of a fastener is as previously proposed:

a device that connects two or more solid objects to prevent relative motion between them to some degree.

This general definition has geometrical implications in that there is presumably a common surface of contact between the two objects, usually planar, occasionally cylindrical, occasionally something else. Sometimes the two objects are oriented with respect to each other in their environment so that there is a front and back to the fastened assembly, sometimes this is not the case. It is also implied that there is a clamping force maintaining contact between the two objects. Often this force will be generated by the elasticity or *springiness* of some part of the fastener, and there are various ways of obtaining this spring action. Other possible sources of a clamping force are: wedges, magnetism, gravity, fluid pressure, plastic deformation and perhaps others.

Beyond mere geometry, we can also establish the desirable attributes of fasteners in terms of some or all of the objectives and criteria in Table 4.1. There may be additional objectives (and corresponding criteria) relevant to particular applications in which fasteners are used. Table 4.1 is not intended to be exhaustive.

(c) Thinking About: Fasteners for Paper

Many different methods for fastening sheets of paper were illustrated in Figure 4.4 to 4.6. Choose one of these three figures, and for each of the fastening methods shown in it, identify which of the above attributes of the fastener are important. If other attributes are important, record them as well. Your response should be concise and presented in a tabular fashion.

(d) Thinking About: Other Common Fasteners

- Make a list with illustrative sketches of about ten common fasteners used for various purposes in houses, apartments, cars, bicycles, sailboards, furniture, cabinets, buildings, bridges, structures, boats, aircraft, etc. If possible, include some examples of non-planar surfaces.
- For each fastener in your list, explain the nature of the clamping force and the source from which it is derived.
- In cases where the clamping force is derived from the elastic deformation of part of the fastener, sketch and describe the '*spring*' which provides this recoverable deformation. How would the stiffness of the spring be determined experimentally? (The stiffness of a spring is the force required to produce unit deflection. It is the k in the equation: $F=kx$).

(e) A General Symbolic Description of Fasteners

The final part of this exercise aims to encourage you into some fairly abstract thinking. It may seem somewhat unusual, but remember *Rule No. 1*: "Don't Panic!"

Develop, in whatever way you think appropriate, a general symbolic representation for fasteners. The system of representation should be largely (although not entirely) graphical/pictorial, and should contain standardised means for referring to:

- contact surfaces;
- degree of constraint (or its converse, residual relative motion);
- the source of clamping force;
- and perhaps other important characteristics of a fastener.

Relate your abstract approach to real-life examples by making a second (very schematic) sketch of each of the ten fasteners you included in your list in (**d**), above, and applying your system of symbolic representation to it.

4.2 Bolted joints and screws: design for clamping contact

It is commonplace to think of bolted joints in terms of one or more bolts *holding some components together*. However such a definition is too loose for our purposes. In this book we regard bolted joints to be a subset of *clamped joints* (see Figure 4.7 in the previous section), so the bolts must perform a more specific task than merely *linking* the connected components. A bolt that merely links adjacent components without ensuring any appreciable clamping force between them is properly classified as either a *shear connector* (*pinned joint*), or perhaps a *tensile link*, and may be designed according to the methods of Section 4.3 or 3.4.2, as appropriate.

In order to qualify as a *bolted joint* in the terms of this present section, the bolts must *maintain a minimum clamping contact force* between the parts being joined. Furthermore, they must continue to clamp the joint with at least this required contact force, when the external loads (those tending to separate the joint) vary through their full load cycle.

Not surprisingly, a bolted joint consists of a tensile element that we refer to as a *bolt*, and special *connection regions* on the two components being connected.

These *connection regions* come in various shapes, but always possess a *contact surface* (where the compressive contact, or *sealing force*, is transmitted against the partner component), an *anchor point* to the tensile element, and some stressed material between the *contact surface* and the *anchor point*. These rather abstract terms are illustrated in Figure 4.8. Note that the tensile elements of a bolted joint may be called *bolts*, *machine screws*, *studs*, *screws*, *tie rods*, and so forth depending on details of their shape; and the *connection region* is frequently called a *flange*. Also note that a relatively soft layer of jointing material, called a *gasket*, is often (but not always) inserted between the two contact surfaces to promote better sealing. The gasket's usual location is indicated in Figure 4.8(b).

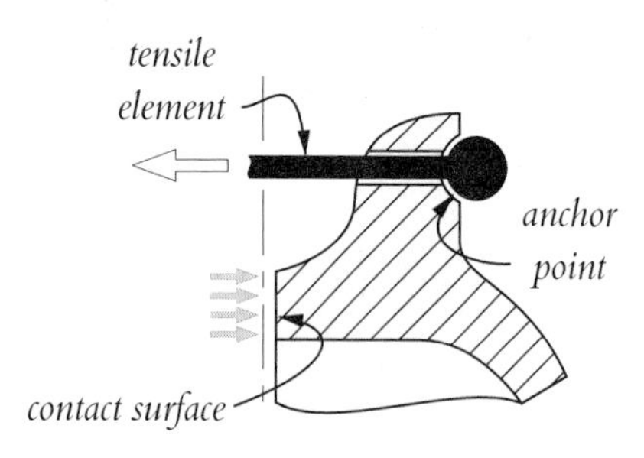

Figure 4.8(a) Essential items in half a bolted joint

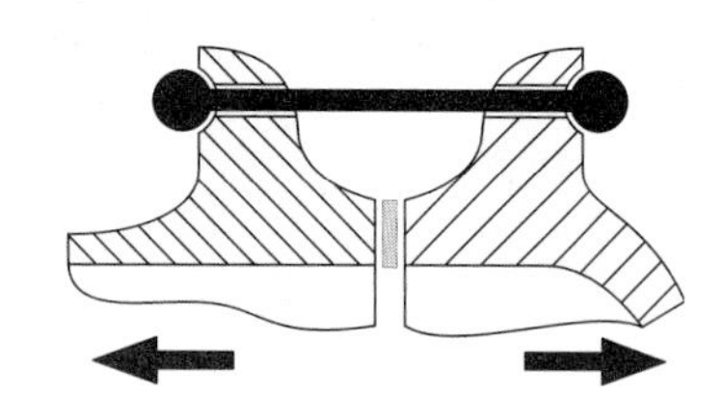

Figure 4.8(b) Two halves of a bolted joint, with gasket

4.2.1 Function of a bolted (clamped) joint

We consider cases in which both:

- compressive contact must be maintained (to at least some minimum value) between the joined components, and
- the external force (acting parallel to the bolt's axis) varies with time.

The need for continuous compressive contact may arise from several functional requirements, but principally these will be either:

(A) that the joint must provide a seal against fluid leakage; or

(B) that the joint must provide sufficient frictional resistance to prevent any transverse sliding of one component relative to the other.

Function (A) is common in pressure vessels, piping, internal combustion equipment, and pumps. Function (B) is commonly encountered in structural engineering, foundation and anchor bolts, and assemblies of machine parts. From a structural viewpoint, the approaches to designing bolted joints for these two functions (fluid sealing on the one hand, and no-slip friction on the other) are identical.

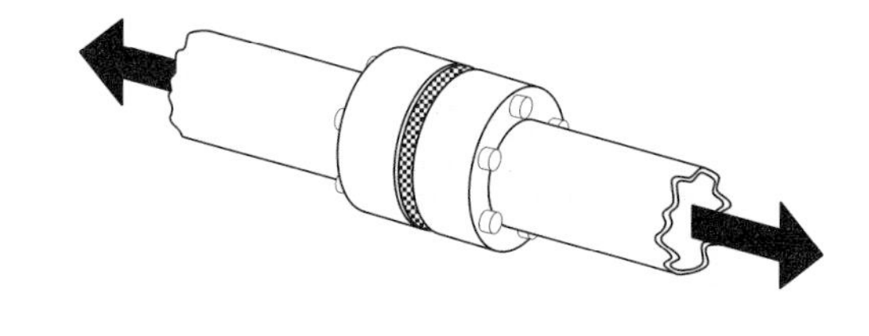

Figure 4.9 External force on a bolted joint

The *external force* on the joint (that which *tends to separate* the components, as illustrated in Figure 4.9) will generally arise from one of two possible causes:

- *pressure force* from contained fluid; and
- *traction* due to gravity, differential thermal expansion, and mechanical or structural loads.

Some of the many applications of bolted (*clamped*) joints are shown in the sketches of Figure 4.10.

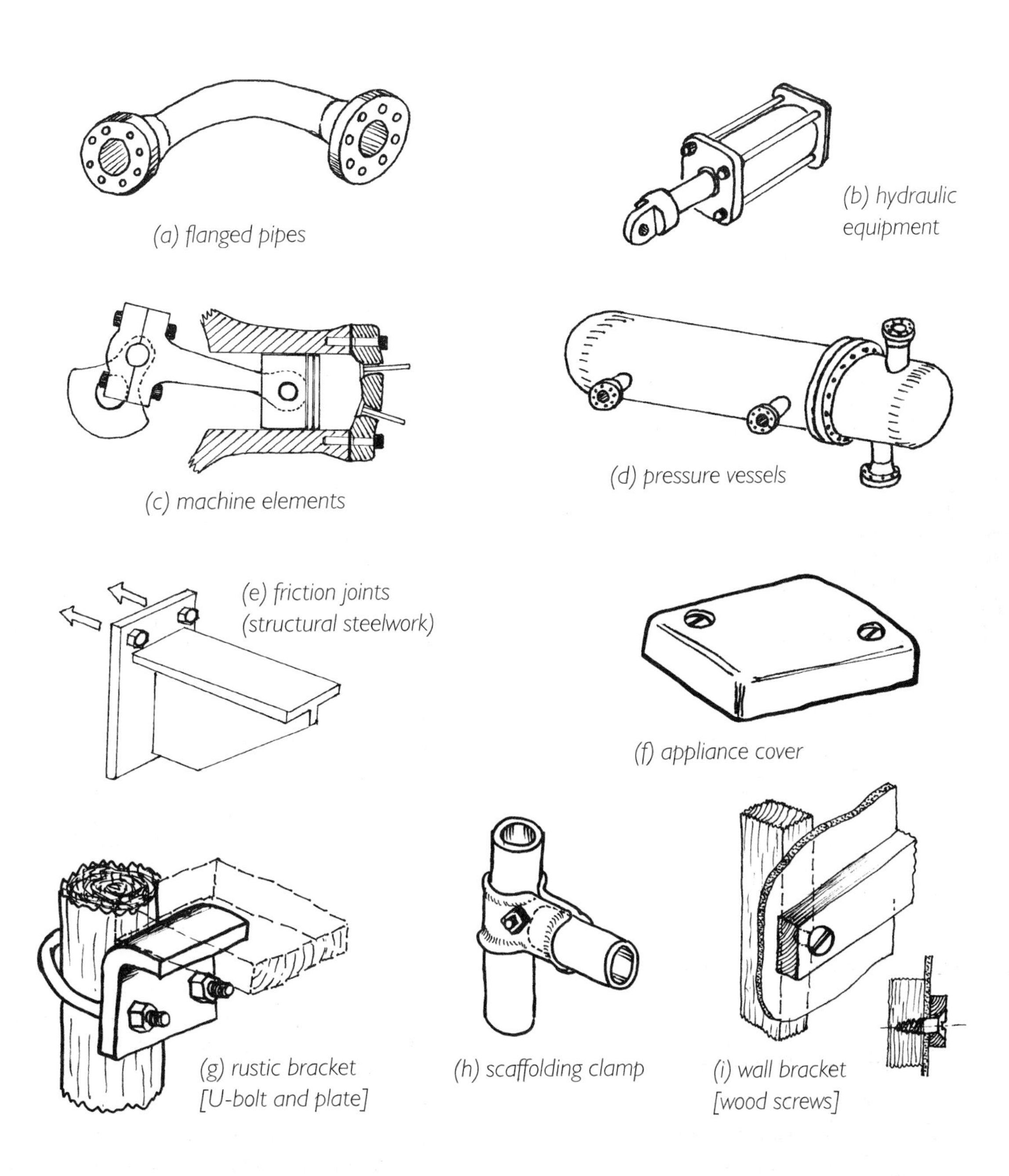

Figure 4.10 Bolted joint examples

Ex 4.1 Identifying the elements of bolted joints

For each of the examples of bolted joints shown in Figure 4.10, identify (i) the *tensile element*, or *bolt*, (ii) the *flange(s)* or their equivalent, (iii) the *anchor point* between the bolt and flange, (iv) the *contact surface* where the flanges meet, and (v) the *gasket* (if present). Illustrate your answers with neat sketches of the sectioned joints, after the fashion of Figure 4.8(b). Note that the example in Figure 4.10(c) includes two different types of bolted joint.

4.2.2 Bolted joint failure modes

The potential failure modes for a bolted (*clamped*) joint relate to its principal function (that is, *maintaining a clamping contact force* in order to ensure either sealing or no-slip friction), and also to the structural integrity of the joint's various elements (*bolt, contact surface, gasket, flange*). The two main failure modes considered in this text are:

(a) loss of adequate clamping contact force; and

(b) yielding (or fracture) of bolts in tension.

Loss of clamping force leads to potentially serious consequential failures: either leakage of the fluid, or transverse slippage of the joined components. Leaking joints may be hazardous if the fluid is itself hazardous (hot, caustic, toxic, asphyxiating, flammable, explosive); or lead to slow damage (for example, a leaking water pipe may cause ground subsidence and cracking of nearby buildings); or be unsightly (oil spots on driveways are usually a result of a leaking bolted joint); or simply wasteful. Slipped friction joints are dangerous, since they can result in bolts being loaded in shear as well as tension, possibly causing structural failure of the bolt. They can also cause misalignment of machinery parts.

When do bolted joints leak? The simple-minded response is that they leak when the joint contact force reduces to zero. However, in practice, most gaskets and jointing materials require some minimum, non-zero stress, called the *gasket seating stress*, to ensure that all potential leakage paths across the gasket-to-flange contact surfaces are closed up. So a minimum allowable joint contact force can be derived for a particular joint geometry and gasket material by multiplying the 'gasket seating stress' and the effective contact area of the gasket.

Similarly, for bolted friction joints, a minimum allowable joint contact force can be obtained by dividing the worst transverse *slip-inducing* force to be resisted, by the minimum likely static coefficient of friction between the gasket or bolt washer (if any) and the connecting surfaces.

Yielding or fracture of bolts in tension is clearly a serious mode of failure, since it can lead to sudden structural collapse or rupture of the entire system. Design to resist this mode of failure proceeds according to the general approach given in Section 3.4.2, modified if necessary by considerations of fatigue strength (see Section 2.4).

Other more complex modes of failure exist, including (c) crushing of the contact surface, (d) radial bursting or erosion of the gasket at the contact surface, (e) failure of the bolt's thread system through binding or stripping, and (f) yielding or cracking of the flange through excessive bending. These are all very real and must be reviewed in critical applications, but lie outside our detailed scope.

Ex 4.2 Failures in bolted joints

For each of the examples of bolted joints shown in Figure 4.10, state whether the structural objective of the joint is to provide sealing, or to prevent transverse slippage, or something else. In each case, discuss the implications of the joint failing to fulfil this structural objective. What consequences might arise from a broken bolt?

4.2.3 A simple mathematical model for bolted (clamped) joints

(a) The idealised bolted joint

Clearly bolted joints can be constructed in a diverse range of shapes and configurations. It will be helpful to make reference to a standard form of bolted joint for the purpose of developing our mathematical model. The intention is that all (or most) *real-life* bolted (*clamped*) joints can be associated with the standard form of this idealised bolted joint.

At this point, we should make explicit a colour convention (*in black-and-white!*) adopted throughout the present Section on bolted joints, and already used in Figures 4.8 and 4.9. External forces on a joint are coloured black, compressive forces exerted on the contact surface (for example, gasket forces) are shown shaded in grey, and tensile bolt forces are shown as white.

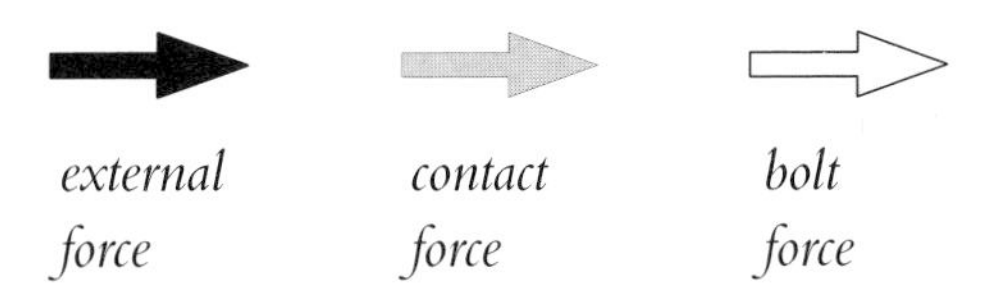

Figure 4.11 shows a bolted, flanged joint at one end of a simple system consisting of two components in connection: a flanged vessel and a cover plate. These components are subjected to a separating force of some kind — perhaps from pressure in the vessel as suggested by Figure 4.11(a), perhaps from traction as in Figure 4.11(b), it doesn't really matter.

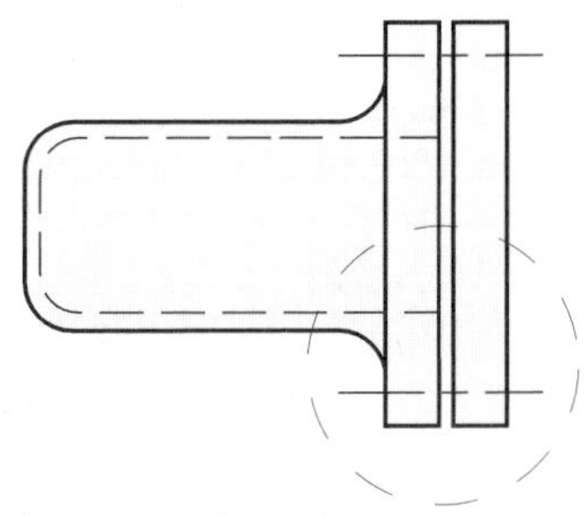

(a) external force from pressure:

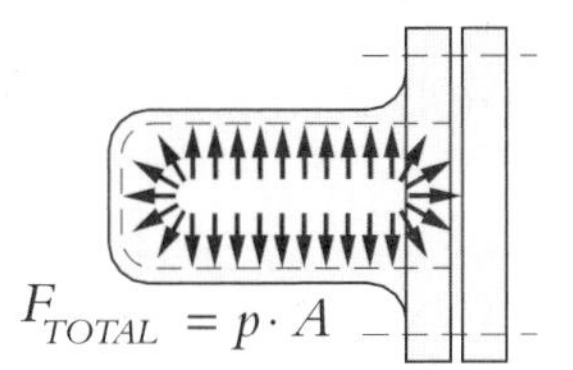

(b) external force from traction:

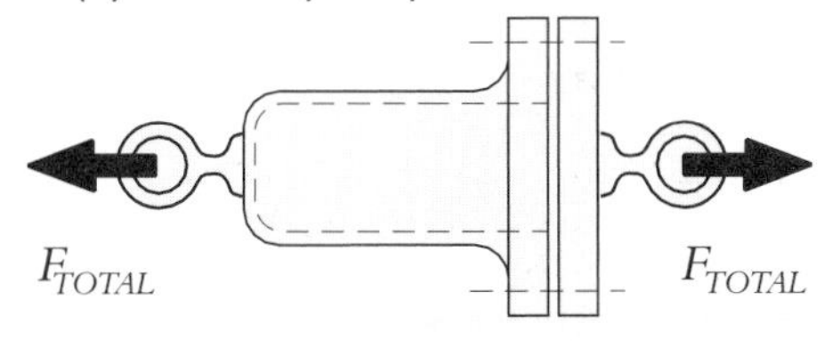

Figure 4.11 Components joined by a 'standard' bolted joint

Attention now focuses on just a part of the idealised bolted joint, namely, that portion associated with the single bolt shown in the standard partial view of Figure 4.12. Say there are n bolts in the overall bolted joint. Figure 4.12 is therefore a free-body diagram of the partial joint consisting of one bolt and a $(1/n)$th portion of the flanges and $(1/n)$th of the gasket. The symbol F denotes the external force exerted on this partial *one-bolt* joint.

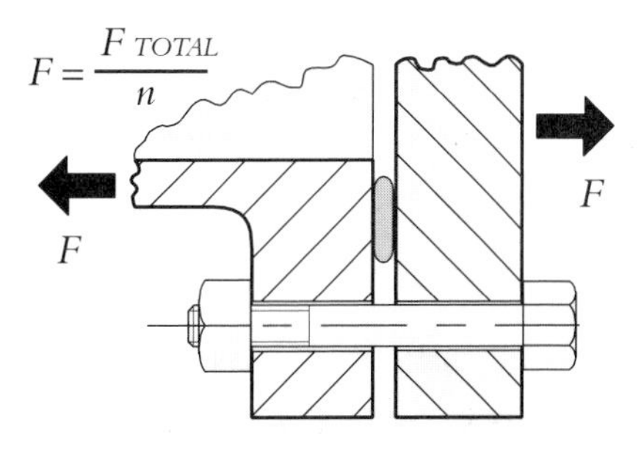

Figure 4.12 The idealised bolted joint (standard partial view)

(b) Free-body diagram for cover plate (three forces in equilibrium)

To introduce the bolt force and gasket contact force explicitly, we consider the forces on the partial cover plate (on the right of Figure 4.12). Our simple mathematical model for the behaviour of bolted joints is, in fact, based on the equilibrium of this component. The free-body diagram for the isolated (*partial*) cover, rotated through ninety degrees, is shown in Figure 4.13. It can be seen that three forces act on the cover:

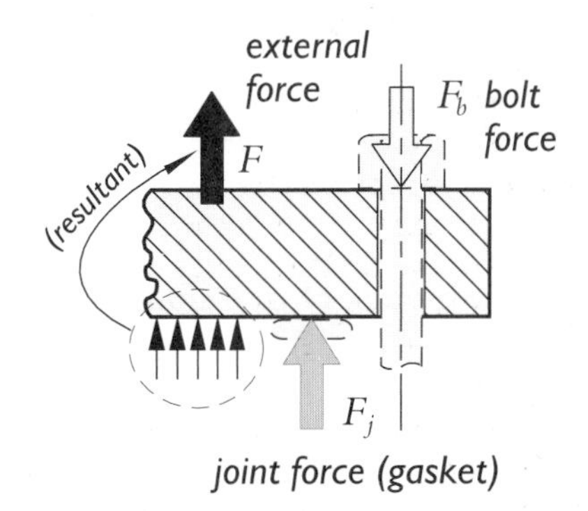

Figure 4.13 Three forces in equilibrium acting on the partial cover

- F = the external force on the partial joint (same as in Figure 4.12);
- F_b = the bolt force in the single bolt (*tensile shown as positive*); and

- F_j = the joint force (here acting through the gasket) on the partial contact surface of the joint (*compressive shown as positive*).

Several points merit emphasis:

- all forces are *per bolt*, therefore

$$F = \frac{F_{TOTAL}}{n},$$

and

$$F_j = \frac{\text{(total joint contact force)}}{n},$$

where n is the number of bolts used;

- despite the way they are drawn in Figure 4.13, all three forces are coaxial due to the axi-symmetry of the overall joint — therefore moment equilibrium is automatically satisfied;
- hence vertical force equilibrium is the only salient consideration, requiring the bolt force to be equal in magnitude to the sum of the external force and the joint contact force;
- note the sign conventions adopted: F is positive for forces tending to separate the joint, F_j is positive in compression, F_b is positive in tension;
- F_j is only ever a compressive force (gaskets cannot exert tensile forces);
- F_b is only ever a tensile force (in bolted joints, bolts cannot experience compressive forces).

(c) Elastic behaviour of bolt and joint

So much for statics. The remainder of the mathematical model derives from a consideration of the axial deflections occurring in the bolted joint. The bolt and gasket (or contact area, if no gasket is used) are assumed to behave in a linear, elastic fashion. This is a good assumption for the bolt — it is a relatively long component in simple tension, and usually stressed well below its elastic limit.

Gaskets, on the other hand, are thin (in other words, axially very short) elements in compression, and often manufactured from nonmetallic, fibrous, anisotropic materials with peculiar, non-linear, elastic properties. If metallic, they may be stressed to near, or some way beyond, their elastic limit. However these issues only have a significant effect on the stress-strain behaviour of the gasket while it is being *pre-loaded* from an unstressed condition to an initial state of high compression when the joint is first *bolted up*. Once in its pre-loaded state, perturbations of the joint contact force, F_j, about its initial value, F_{ji}, produce a fairly linear strain response in the gasket. In other words, the assumption of linear, elastic behaviour for a *pre-loaded gasket* is a reasonable one over the range of operating forces to which it is then subjected.

As a side issue, we note that gasket materials can also exhibit *viscoelastic* (time-dependent) behaviour. For critical applications (for example, in *head bolts* used to fasten the cylinder-head to an automotive engine block), special precautions are sometimes taken to pre-load the joint correctly. It may be necessary to initially *pre-tension* the bolts, run the machine for some time and several heat-up/cool-down cycles, and then *re-tension* the bolts to take up any relaxation in the gasket and bolt seating.

(d) Stiffness of bolt and joint elements

In what follows, we shall use the term *joint* to refer collectively to the three compressed elements of the bolted joint, namely the gasket and the two flanges. Since the bolt and joint are considered to be *elastic*, we can write for each of them that

$$\text{(total force)} = k \times \text{(deflection from rest)},$$

where the stiffness, k, is assumed to be constant. While the above relationship applies only approximately (especially for the gasket, for reasons discussed above), the following relationship for small changes in force and deflection applies much more reliably:

$$\Delta(\text{force}) = k \times \Delta(\text{deflection}),$$

where Δ denotes a change from the initial *pre-loaded* state.

The same free-body diagram as was presented in Figure 4.13 is shown again in Figure 4.14, but with the bolt and joint represented as linear springs, each having its own stiffness: k_b for the bolt, and k_j for the joint. The stiffness, k, of either the bolt or the joint can be derived from the simple stress-strain relationship, $\sigma = E\varepsilon$, which is just a special form of *Hooke's Law* for elastic behaviour.

For prismatic elements in uniaxial tension or compression (just like the bolt and joint of a bolted connection), it is easy to show that the ratio of force to deflection is

$$k = \frac{AE}{\ell}.$$

For our bolted connection, each of the terms (A, E and ℓ) must be evaluated for either the bolt or the joint (*flanges plus gasket*). Applying this to the bolt, we write:

$$k_b = \frac{A_b E_b}{\ell_b}, \qquad (4.1)$$

where the cross-sectional area, A_b , is based upon the nominal shank diameter of the bolt (since the threads have little influence on the axial stiffness of the bolt). The interpretation of terms A_b , E_b and ℓ_b is straightforward, since there is only one tensile element to consider.

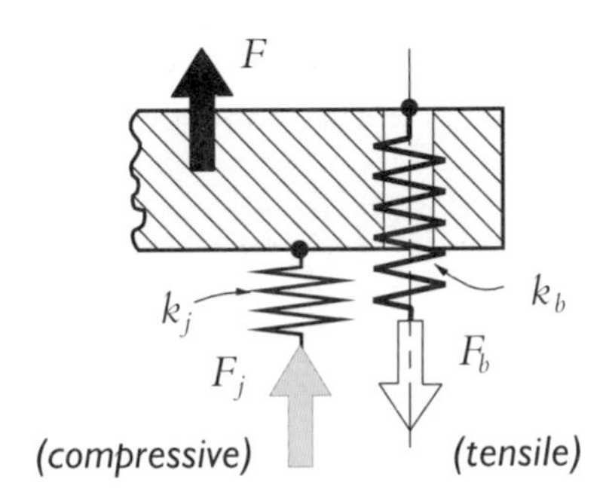

Figure 4.14 Elastic bolt and joint

For the joint, however, there are three distinct elements subject to compression: the lower flange, the gasket (if present), and the upper flange. Although we can write:

$$k_j = \frac{A_j E_j}{\ell_j}, \qquad (4.2)$$

the correct interpretation of the terms A_j , E_j and ℓ_j is not quite so apparent. Before simplifying matters, it is necessary to consider that the joint (and hence k_j) is made up of several different stiffnesses in series, as depicted in Figure 4.15.

The compound stiffness for these compressed elements in series (Chapter 3.3.2, Figure 3.31) is:

$$k_j = \left(\frac{1}{k_{f\,1}} + \frac{1}{k_g} + \frac{1}{k_{f\,2}} \right)^{-1} \qquad (4.3)$$

where each of the constituent stiffnesses can be separately evaluated in the usual way. For example:

$$k_{f\,1} = A_{f\,1} E_{f\,1} / \ell_{f\,1} \text{ and } k_g = A_g E_g / \ell_g .$$

In practice, if a gasket is used, its flexibility (that is, $1/k_g$) will usually dominate the right-hand-side of Equation *(4.3)* so that the overall joint stiffness can be approximated by the gasket stiffness, $k_j \approx k_g$ $(=A_g E_g/\ell_g)$. Alternatively, if no gasket is used, the joint stiffness may still be considerably simplified, but by using different area and length terms.

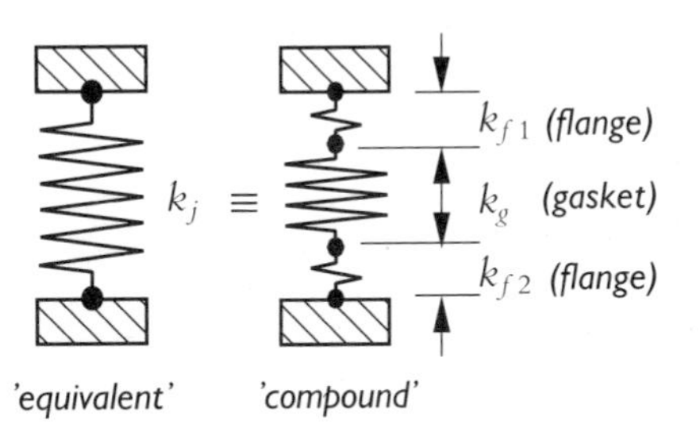

Figure 4.15 Series of compressed elements comprising the joint

Figures 4.16 and 4.17 depict the geometrical quantities (A and ℓ) influencing the stiffness of the bolt and joint, and give some indication of their relative magnitudes.

For the gasketed arrangement in Figure 4.16(a), D_j is the mean gasket diameter, and w_j is the nominal gasket width. A_b and A_j are, respectively, the bolt cross-sectional area and the gasket contact surface area, and are the areas to be used in the k_b and k_j expressions. Although actual areas would need to be calculated in each case, it can be seen from the sketch that, to a reasonable approximation, the joint area in this gasketed configuration is of the same order of magnitude as the bolt area. Therefore in gasketed joints, the ratio of bolt and joint stiffness will mainly be governed by the large difference in bolt and gasket lengths (see Figure 4.17(a)). If the gasket material is metallic so that E_b and E_g are similar, then the gasket will usually be much stiffer than the bolt, since $\ell_b >> \ell_g$. Of course, soft non-metallic gasket materials are commonly used, so the safest approach is to check the true values of k_b and k_j by calculation.

In the un-gasketed arrangement of Figure 4.17(b), the joint is established using metal-to-metal contact between the two flanges. Complex compressive stresses in the flanges are set up underneath and around the bolt heads. Following a crude but widely-used approximation, the compressed volume of flange material can be assumed to be equivalent in axial stiffness to a hollow cylindrical block extending the full thickness of the flanges (from bolt head to bolt head), having an inner diameter, d_b, and an outer diameter of approximately $3d_b$. From this it follows that, for un-gasketed joints, the joint area is around eight times the bolt area. Since the joint and bolt lengths are virtually identical in this case (see Figure 4.17(b), below), and their Young's moduli will be more-or-less equal, the resulting joint stiffness, k_j, for an un-gasketed joint will generally be much larger than that of its bolt, k_b.

Ex 4.3 Estimating stiffness

A large bolted joint is being designed to join two steel flanges, each of 80 mm thickness and 250 mm inside diameter. Sixteen M32 bolts will be used. These have the same Young's Modulus as that of the flanges — around 210 GPa. Three alternative gasket materials are being contemplated for use in this joint: aluminium (E_g =70 GPa), stainless steel (E_g =195 GPa), and a special *metal-jacketed, compressed-fibre-filled* gasket (E_g =1.9 GPa). In all cases, the gasket will have a mean diameter of approximately 280 mm, and a cross-section 20 mm wide and 3 mm thick. Estimate k_b and k_j for the three gasketed cases: aluminium, stainless steel and fibre-filled, and also for the case of no gasket.

(e) Deflection of bolt and joint: elastic elements in parallel

Our analysis so far has developed simple but useful representations of the *statics* of bolted joints (*equilibrium of forces on cover*) and the *elasticity* of their major elements (*bolt and joint stiffness*). It now remains to apply the requirements of *compatibility* (matching the deflections of structural elements so that their interfaces remain in contact) to complete our mathematical model.

Consider Figure 4.18. The *bolt* (tensile element) and *joint* (a lumped series of compression elements)

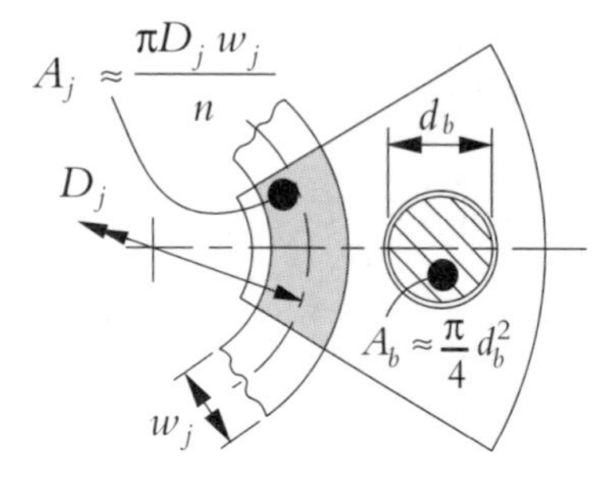

Figure 4.16(a) Relative areas of bolt and joint (gasketed)

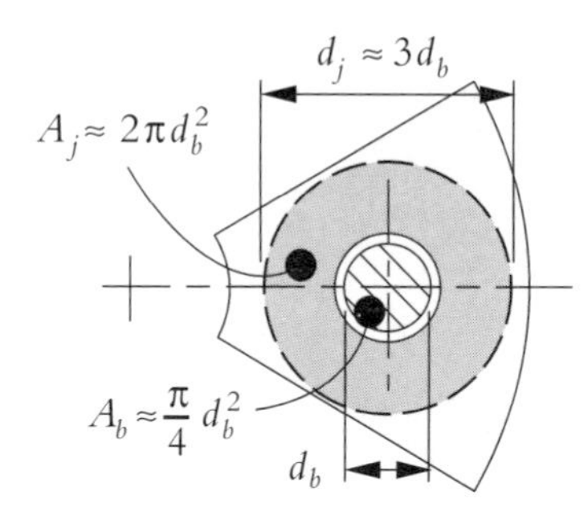

Figure 4.16(b) Relative areas of bolt and joint (un-gasketed)

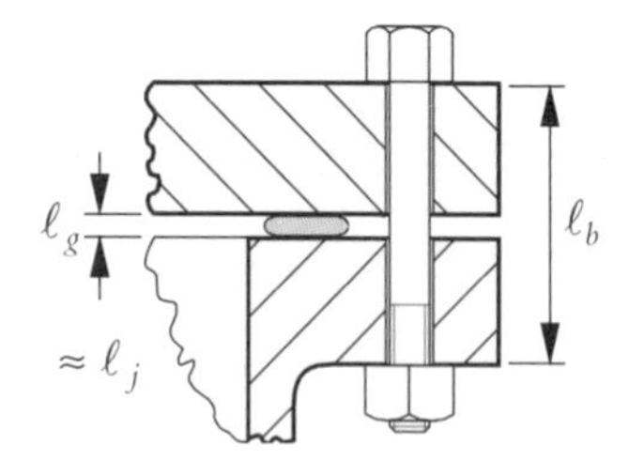

Figure 4.17(a) Relative lengths of bolt and joint (gasketed)

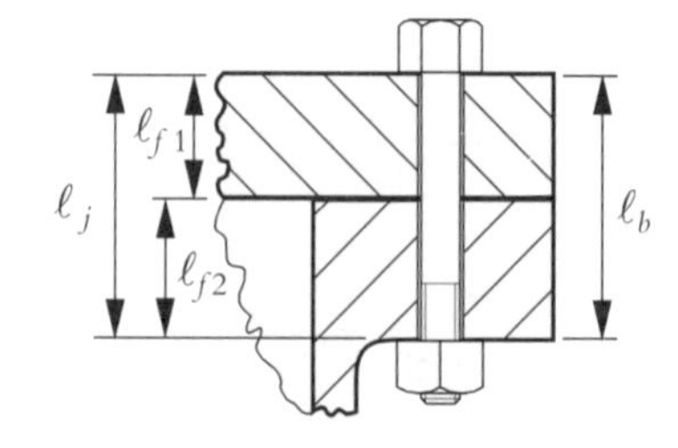

Figure 4.17(b) Relative lengths of bolt and joint (un-gasketed)

share a common interface at the *anchor points* where the bolt connects to the cover and flange of our idealised bolted joint. They therefore behave like two springs connected in parallel, since they are constrained to have the same deflection. In formal terms:

$$\Delta\delta_b = \Delta\delta_j \,, \tag{4.4}$$

where $\Delta\delta_j$ is the (*upwards positive*) extension of the joint from its initial (*pre-loaded*) position, and $\Delta\delta_b$ is likewise the (*upwards positive*) extension of the bolt from the same initial position, both extensions being due to the effect of some externally imposed force. In the above diagram the cover plate and lower flange are portrayed as being rigid, but note that any elastic deformation of the flanges is accounted for in the compound stiffness of the joint, k_j.

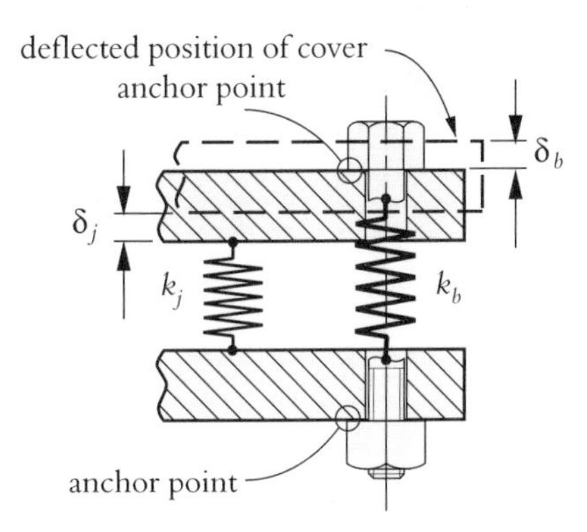

Figure 4.18 Compatibility of joint and bolt deflections

4.2.4 Bolted joint design equations

(a) Derivation of bolted joint design equations

Axial force equilibrium for the partial cover of our idealised bolted joint (Figure 4.13) requires that:

$$F_b = F + F_j \,. \tag{4.5}$$

In Equation *(4.5)*, F is the external force (per bolt) applied to the joint. We may regard it as the independent variable — the bolt force and joint contact force will both vary in response to changes in the external force. A special case exists when the external force is zero. In this situation, the bolt force and joint contact force are at their *initial* values (that is, the initial values established when the joint has been *pre-loaded*). By equating F to zero in Equation *(4.5)*, we obtain

$$F_{b\,i} = F_{j\,i} \,, \tag{4.6}$$

where $F_{b\,i}$ and $F_{j\,i}$ are both constants.

The *compatibility* requirement has already been expressed in Equation *(4.4)*. It referred to extensions of the bolt and joint from their initial (*pre-loaded*) position due to the imposition of the external force, F. These extensions, $\Delta\delta_b$ and $\Delta\delta_j$, can also be expressed in terms of the stiffness of, and change in force acting on, the bolt and joint. Using basic stiffness definitions, we can write:

$$\Delta\delta_b = \left(\Delta F_b / k_b\right) \tag{4.7-a}$$

and

$$\Delta\delta_j = -\left(\Delta F_j / k_j\right), \tag{4.7-b}$$

where $\Delta F_b = F_b - F_{b\,i}$ is the change in the bolt force from its initial value, $F_{b\,i}$ (at F=0), to its new value, F_b (at F=F). Similarly, $\Delta F_j = F_j - F_{j\,i}$ is the change in the joint contact force from its initial to its new value. (Recall that F_b and F_j are the bolt and joint force values obtained after application of the external force, F. They are the dependent variables we seek to express as functions of F.)

The negative sign in Equation *(4.7-b)* arises from the particular sign conventions adopted for F_j and $\Delta\delta_j$. Joint contact forces are defined as being positive in compression, so that a positive ΔF_j is an increase in compression. However $\Delta\delta_j$ is defined as being *positive for an extension of the joint*, not a compression — hence the need for a sign change.

When Equations *(4.7-a)* and *(4.7-b)* are combined with Equation *(4.4)*, we obtain

$$\frac{F_b - F_{b\,i}}{k_b} = -\left(\frac{F_j - F_{j\,i}}{k_j}\right),$$

and by using Equation *(4.6)* to eliminate $F_{j\,i}$ from this, we can write

$$\frac{F_b - F_{b\,i}}{k_b} = \frac{F_{b\,i} - F_j}{k_j}.$$

Finally, permutations of Equation *(4.5)* can be substituted successively into the above equation to eliminate either F_j or F_b and obtain both of the following results.

$$F_b = F_{bi} + \left(\frac{k_b}{k_b + k_j} \right) F \qquad (4.8)$$

$$F_j = F_{bi} - \left(\frac{k_j}{k_b + k_j} \right) F \qquad (4.9)$$

(b) Comments on equations for bolt force and joint force

Equations *(4.8)* and *(4.9)* are the *bolted joint performance equations*. The first, *(4.8)*, expresses how the bolt force, F_b, varies with external force. The equation has a positive sign in front of the F term, indicating that the bolt force increases from its initial value in linear proportion to F, so that it is always greater than or equal to F_{bi}. The slope of this linear increase is determined by the ratio $k_b/(k_b+k_j)$, which is termed the *relative bolt stiffness*, and is always less than unity.

The second equation, *(4.9)*, expresses how joint contact force, F_j, varies with external force. The equation has a negative sign in front of the F term, indicating that the joint contact force decreases from its initial value in linear proportion to F, so that it is always less than or equal to F_{ji} $(=F_{bi})$. The slope of this linear increase is determined by a different ratio, $k_j/(k_b+k_j)$, which is termed the *relative joint stiffness*, and is always less than unity.

A clear comprehension of these two relationships between the internal forces (F_b and F_j) of the bolted joint and the external force upon it, is the key to bolted joint design. The extent of increase in bolt force, F_b, with F, determines the structural integrity of the bolt. The extent of decrease in joint contact force, F_j, with F, determines the functional integrity of the joint as a seal or clamp. Note that the relative bolt and joint stiffnesses are dimensionless, and sum to unity.

(c) Graphical representation of bolt force versus external force

Graphs are an effective way to represent and visualise the relationships in Equations *(4.8)* and *(4.9)*. Figure 4.19(a) shows the most common form of graphical construction used for this purpose. It is a graph of bolt force, F_b, on the vertical axis against external force, F, along the horizontal axis.

A line of slope $k_b/(k_b+k_j)$ corresponds to bolt force. A second, dashed line of unity slope (corresponding to the equation $F_b=F$) is drawn at a 45° angle through the origin, so that the

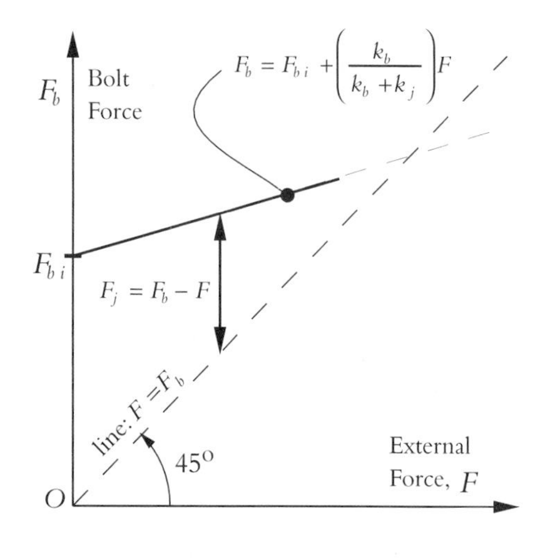

(a) General bolt force

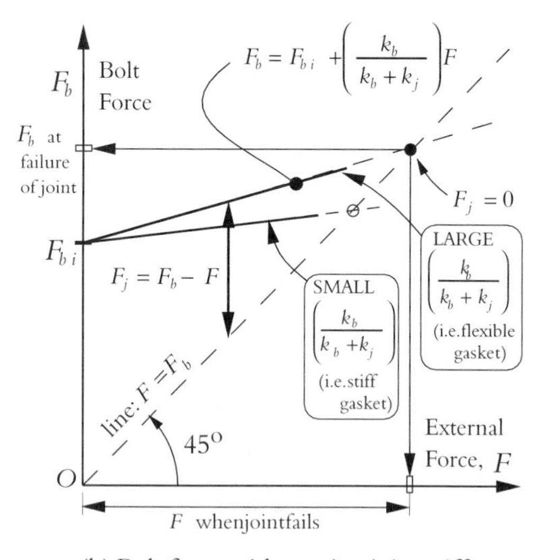

(b) Bolt force with varying joint stiffness

Figure 4.19 Bolt force versus external force

difference between it and the bolt-force line can be easily constructed. This difference corresponds to the joint contact force, since $F_j = F_b - F$. The point of intersection between the bolt-force line and the dotted external-force line indicates the point at which the joint force, F_j, becomes zero.

Figure 4.19(b) shows an extended version of the earlier graph, this time with a second bolt-force line of a lesser slope (corresponding perhaps to a *stiffer gasket*, or perhaps to a *more flexible bolt*, but in any case to a smaller value of relative bolt stiffness).

(d) Graphical representation of bolt force and joint force versus deflection

Figure 4.20(a) shows an alternative form of graphical construction used to represent the bolted joint performance equations *(4.8)* and *(4.9)*. This time we plot forces F_b and F_j on the vertical axis, against deflection, δ, along the horizontal axis. Because of the choice of axes, the slope of lines on this graph corresponds exactly to stiffnesses, k_b and k_j. A line of slope $+k_b$, representing the bolt force, F_b, is drawn first from the origin. After the joint *pre-load*, F_{bi} $(=F_{ji})$, is selected, a second line of slope $-k_j$ can be constructed, representing the joint contact force, F_j. Although constructed in the sequence described above, these lines are shown in Figure 4.20(a) proceeding from the undeflected starting points (before the joint is *pre-loaded*), upwards to the intersection point corresponding to the *pre-load* force, $F_{bi} = F_{ji}$.

In Figure 4.20(b), the above graph is extended by projecting vertical lines with deflection coordinates further and further to the right of the initial *pre-load* point. At any deflection, $\Delta\delta$, measured from this point, the corresponding external force, F, can be found by projecting a vertical line at the appropriate location, and constructing the difference in vertical ordinates between the bolt-force line and the joint-force line. This range of external forces is shown by the vertically shaded region of Figure 4.20(b).

(e) Failure criteria

Failure modes for bolted joints were discussed in Section 4.2.2. They are of two kinds — functional and structural. Both are potentially serious.

Functional failure of a bolted joint occurs when it can no longer maintain a pressure seal or prevent transverse slippage of the joined elements. These failures can be predicted to occur when the joint contact force, F_j, falls below certain minimum values.

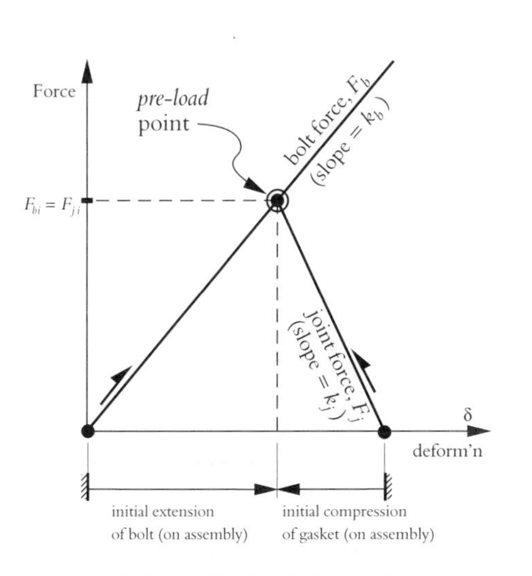

(a) General joint deformation

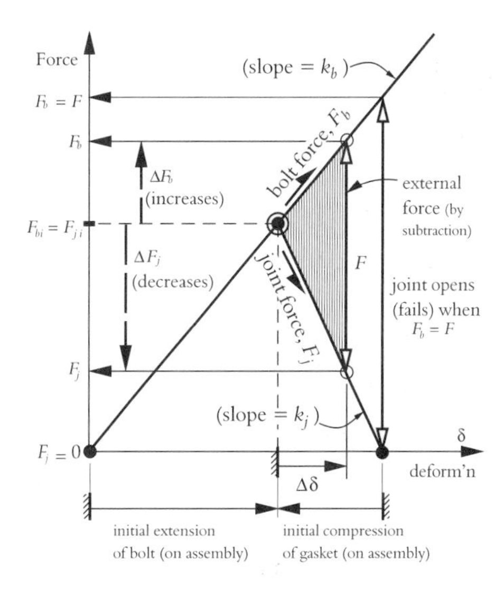

(b) Joint deformation at failure

Figure 4.20 Bolt and joint forces versus deflection

Ex 4.4 Graphical prediction of joint failure

How would you adapt a graph like the one shown in Figure 4.19(a) to determine the external force, F, that would cause joint failure at some minimum allowable value of joint force, $F_{j\ min}$?

Ex 4.5 More graphical prediction of joint failure

How would you adapt a graph like the one shown in Figure 4.20(b) to include a more realistic representation of functional (F_j) and structural (F_b) failure of the bolted joint?

(f) Pressure sealing joints

For pressure sealing joints, the minimum required joint contact force is that which will *seat* the gasket, and is estimated on the basis of an equation like *(4.10)*, below.

$$F_{j\ min} = \frac{\sigma_{\text{gasket seating}} A_{\text{gasket,effective}}}{n} \cdot F_d \qquad (4.10)$$

NOTES ON EQUATION *(4.10)*:

- $F_{j\ min}$ is a force *per bolt*, hence the n in the denominator.
- This equation seeks to hint, in a very simplified fashion, at the approach taken to gasket sealing design by the most commonly used design guide: the *ASME Code for Pressure Vessels*, Section VIII, Division 1, Appendix 2. The serious designer is referred to that august publication for study!
- Values for $\sigma_{\text{gasket seating}}$ are advised as y in Table 2-5-1 of the *ASME Code* (1980 edition).
- F_d is a factor of safety provided explicitly in this required joint force expression for demonstration purposes. However, the designer should ensure that such a safety factor is not already *built in* to the published values of $\sigma_{\text{gasket seating}}$. The *ASME Code* uses an m factor for this.
- Some types of gasket are notable for having a zero seating stress. Such gaskets possess what is called 'active sealing' characteristics, in that the gasket itself deforms under the action of the imposed fluid pressure. The higher the pressure to be sealed against, the tighter the seal. Elastomeric *O-rings* are the most common form of *active sealing* gasket.
- $A_{\text{gasket,effective}}$ is used here rather than some expression like $\pi D_j w_j$, since the *ASME design code* uses a notion of *effective gasket seating width* rather than actual gasket width when calculating gasket contact area. Again, life is complicated, and the excruciating details of gasket facing designs (that is, the patterns of raised nubbins, faces, grooves and scoring that are carefully machined onto the gasket contact area of flanges) can have a significant effect on sealing contact. The simple approach of Equation *(4.10)* presents a useful reduction of these complexities.

(g) Friction joints

Friction joints are used to clamp two components together so tightly in an axial direction, that friction prevents them from slipping against each other in a transverse direction. The *axial* direction referred to here follows the axis of the tension element (bolt or screw), and is the direction of the *normal force* that is always associated with any *frictional force*. The *transverse* direction is any direction perpendicular to the axial direction.

The frictional force able to be developed at the joint's contact surface will be limited to the product of the joint contact force, F_j, and some appropriate coefficient of friction, μ. Since one or more gaskets, washers, or other friction-modifying layers (including rust, grease and oil) might be present at the joint contact surface, conservative design philosophy dictates that we must use the slipperiest pair of surfaces as the basis of design. In other words, we select the minimum friction coefficient, μ_{min}, between any of these adjacent pairs of surfaces. Provided there is no slipping, we may write that

$$F_{\text{friction}} \leq \mu_{min} F_j$$

where F_{friction} is the frictional resisting force (per bolt) developed within the joint to resist slippage due to some external transverse force. Clearly, the upper limit of this is

$$F_{\text{friction, max}} = \mu_{min} F_j \,.$$

If external transverse forces exceed this resisting force, the joint will start to slip. If we denote the worst total transverse, slip-inducing force likely to be imposed on the joint as $F_{\text{transverse}}$, then we require

$$n\,F_{\text{friction, max}} \geq F_{\text{transverse}}\,.$$

For friction joints, the minimum required joint contact force is that which will just maintain static (non-slipping) frictional contact between the contact surfaces. It may be estimated on the basis of an equation like *(4.11)*, below.

$$F_{j\ \min} = \frac{F_{\text{transverse}}}{n\,\mu_{\min}} \cdot F_d \qquad (4.11)$$

NOTES ON EQUATION *(4.11)*:

- $F_{j\ \min}$ is a *force per bolt*, whereas $F_{\text{transverse}}$ is the total transverse force imposed externally on the joint, hence the *n* in the denominator.
- Again, F_d is a factor of safety provided explicitly in this required joint force expression for the purpose of demonstrating our design methodology.
- The right-hand side of Equation *(4.11)* includes the term $F_{\text{transverse}}$, so named to emphasise its direction of action in relation to the *clamping axis* of the joint. The transverse force is in fact an *external* force on the joint, but is not the same as the external force, *F*, that we have formerly mentioned as tending to *separate* the joint. *F* is an *axial* external force, and must not be confused with $F_{\text{transverse}}$. It is possible to have a friction joint for which *F* is always zero, although these are rather trivial to design, since they require no exercise of the bolted joint performance equations *(4.8)* and *(4.9)*. It is common for friction joints to have both *transverse* and *axial* forces acting on them simultaneously.

Ex 4.6 Transverse and axial external forces on friction joints

(*Warning!* Herein lies a *Deep Subtlety* attained by only the most enlightened.) For all those examples of friction joints in Figure 4.10, identify the source of the transverse force on the joint tending to cause it to slip. Then, separately, for all joints (both *friction* and *sealing*) in Figure 4.10, identify the source (if any) of an external axial force on the joint that would tend to separate the clamped contact surfaces. (Note that for friction joints, the axial force will often be a consequence of the transverse joint.) Show these forces in a neat, sectioned view of the joint.

(h) Structural failure of bolts:

Structural failure of a bolted joint occurs when the bolt is overstressed. These failures can be predicted to occur when the bolt force, F_b, rises above some specified maximum allowable value. This maximum allowable value will be estimated on the basis of an equation like *(4.12)*, below (but see Section 4.2.5 for a discussion of fatigue, which is a very important issue for bolted joints).

$$F_{b\ \max} = \frac{S_{fail}\,A_s}{F_d}\,. \qquad (4.12)$$

NOTES ON EQUATION *(4.12)*:

- $F_{b\ \max}$ is a *force per bolt*.
- F_d is a factor of safety, on this occasion employed in the familiar role of reducing the presumed strength of the bolt material from S_{fail} to some smaller allowable value.
- S_{fail} is an appropriate material strength property (*failure stress*) for the particular mode of structural failure being designed against. In general, it will be equal to the *proof stress*, S_P, (see Section 4.2.6(a) for a definition of proof stress), or the yield stress, S_y, or the ultimate tensile stress, S_U. Each of these will be associated with its own factor of safety, and the governing mode of failure will be the one having the smallest value of $\sigma_{\text{allowable}} = S_{fail} / F_d$. Typically, high-strength bolting materials have very large yield stress values in proportion to S_U (that is, *low ductility*), and it is not uncommon for them to be stressed in service up to 90% of their rated tensile strength.
- A_s is the *stress-area* of the bolt. It is always less than the shank cross-sectional area, $A_b = \pi d_b^2 / 4$, due to the stress-concentrating effect of the thread form. Numerous tests have established that the *stress-area* is the area of a circle whose diameter is the mean of the thread's *pitch diameter* and *minor diameter*. For metric and unified thread systems, these diameters are given by $[d_b\text{-}2(3/8)p_b\sqrt{3}/2] = (d_b\text{-}0.649519p_b)$, and $(d_b\text{–}1.226869p_b)$, respectively, where p_b is the thread pitch. For these thread systems, the diameter of the *stress-area* is therefore $(d_b\text{-}0.938194p_b)$.

(i) Design equations

Equations *(4.10)*, *(4.11)* and *(4.12)* correspond to bolted joints performing at the limit of functional and structural failure. They suggest very directly three inequalities that can be used as design rules for bolted (*clamped*) joints, each of which is essentially a statement that we require $F_j \geq F_{j\,min}$ or alternatively $F_b \leq F_{b\,max}$.

Require:

$$F_j \geq \frac{\sigma_{\text{gasket seating}} A_{\text{gasket,effective}}}{n} \cdot F_d . \quad (4.13)$$

Require: $$F_j \geq \frac{F_{\text{transverse}}}{n\,\mu_{\min}} \cdot F_d . \quad (4.14)$$

Require:

$$F_b \leq \frac{S_{fail} A_s}{F_d} ;\; S_{fail} \in \{S_P , S_y , S_U\}. \quad (4.15)$$

4.2.5 Fatigue of bolted joints

Equation *(4.15)* is not really adequate in situations involving fatigue. For many applications the external *joint-separating* force, F, is time dependent. This causes the force experienced by the bolt to fluctuate with time, and fatigue fracture becomes an important design consideration. We need to use an *A-M diagram* (refer Chapter 2.4.2) to design our bolted joint.

There is nothing remarkable about applying the *A-M* diagram approach to bolts, and we shall not dwell on the details that should be familiar to the reader. A few brief notes will suffice.

- The minimum load on the bolt is $F_{b\,i}$.
- The load amplitude is

$$\frac{1}{2}\left(\frac{k_b}{k_b + k_j}\right)F.$$

- More particularly,

$$\frac{\text{amplitude of bolt force}}{\text{amplitude of applied force}} = \left(\frac{k_b}{k_b + k_j}\right).$$

- If we can reduce k_b and increase k_j, we can effectively reduce the *fatigue effect* of the applied load. This is done in many applications where fatigue is of great concern.
- One elegant way to do this is to *thin down* the shank diameter of the bolt, as shown in Figure 4.21. Note that this is possible without harmfully increasing the tensile stress in the bolt only because standard *straight-shank* bolts all have some *extra meat* in them to withstand excess stresses. This is because the worst tensile stresses (due to *stress concentration* in the threads) occur within the uppermost few threads of the nut or flange, so the peak stresses along the bolt shank are correspondingly lower.

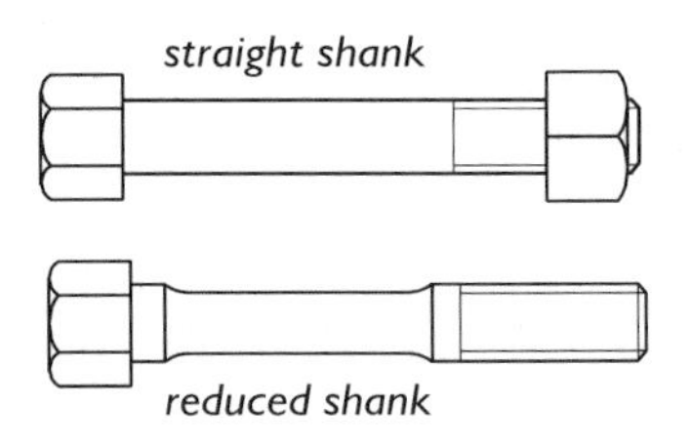

Figure 4.21 Reduced-shank bolt

Ex 4.7 Fatigue effects of shank thinning

The bolt shown in Figure 4.21 is M20X2.5 (coarse pitch), 150 mm in length. The threaded section extends 40 mm from the free end of the bolt. The bolt is one of several used to join two steel flanges. The flange faces are wide and flat, and no gasket is used. When installed, the end 30 mm of threads are engaged. The shank is reduced in diameter from 20 mm to 16 mm over a 100 mm length.

(a) Calculate the bolt stiffness, and the relative bolt stiffness, before and after thinning.

(b) If the ratio of external force, F, to bolt pre-load, $F_{b\,i}$, is 2:1, calculate the slope of the *load line*, $(\sigma_a/\sigma_m) = (F_{b,amplitude}/F_{b,mean})$, for these bolts before and after thinning.

(c) If the bolt material has $S_U = 1000$ MPa and $S_e = 300$ MPa, calculate the value of external force per bolt, F, which these bolts could sustain with a factor of safety of $F_d = 1.5$, before and after thinning. Assume that the peak tensile stresses in the bolt are unchanged by the thinning of the bolt shanks.

Ex 4.8 Costs and benefits of fatigue reduction

Discuss what is gained by thinning the bolt shanks in the previous exercise, and what might be the 'costs' associated with this measure. Suggest other measures that could be taken to improve the fatigue characteristics of these bolts.

Ex 4.9 Pre-loading for fatigue reduction

It is strongly recommended in bolting design literature that for *statically loaded* (that is, non-fatigue) joints, the bolts should be pre-tensioned to 90% of the *proof load* (or roughly 80% of the bolt's yield stress — see next section for a discussion of *proof stress*). However for *dynamically loaded* joints (that is, joints subject to fatigue — usually because the external force fluctuates from zero to some maximum value, F, with time), it is strongly recommended that the bolts should not be pre-tensioned to 90% of the *proof load*, but instead to some smaller value of $F_{b\,i}$ as determined by the designer to ensure safe fatigue life. Explain the stark difference between the two recommendations.

4.2.6 Threaded fasteners for use in bolted joints

(a) Tensile and proof strengths of bolting materials

In bolting literature, it is common to use the terms *proof strength* and *tensile strength* when describing bolt material characteristics. The *tensile strength*, S_U, is familiar enough. The *proof strength* of a bolt, S_p, however, is defined as the maximum force (divided by its tensile stress area, A_s) that a bolt can withstand without acquiring a permanent set. This

Table 4.2 Metric mechanical property classes for steel bolts, screws, and studs

Property class	sizes from	sizes to	S_p MPa	S_y MPa	S_U MPa	material (steel)	(†) heat treatment
⋆ 4.6	M5	M36	225	250	400	low- or medium-carbon	—
⋆ 4.8	M1.6	M16	310	340	420	low- or medium-carbon	—
5.8	M5	M24	380	395	520	low- or medium-carbon	—
8.8	M16	M36	600	635	830	medium-carbon	Q+T
9.8	M1.6	M16	650	710	900	medium-carbon	Q+T
10.9	M5	M36	830	895	1040	low-carbon martensitic	Q+T
12.9	M1.6	M36	970	1100	1220	alloy steel	Q+T

⋆ classes 4.6 and 4.8 are deprecated for use in structural joints

† $Q+T$ denotes *quenched and tempered*

is slightly different from — and slightly smaller than — the yield strength of the bolt material, Sy, which is defined on the basis of a 0.2% permanent set. (Across the various metric property classes for bolt materials, the proof stress is found to range from 88% to 96% of the yield stress.)

Representative strength data for metric property classes of bolts is given in Table 4.2. These metric property classes use a designation system of the form "*A.B*", where "*A*" is the nearest integer to S_U MPa/100 and "*B*" is the nearest single-digit decimal fraction corresponding to (S_y/S_U). For metric bolts, the property class "*A.B*" is generally stamped on the head of the bolt.

The Society of Automotive Engineers (*SAE*) also publishes a widely-used set of bolting specifications. The main grades of *SAE* bolting are listed in Table 4.3.

(b) Thread terminology

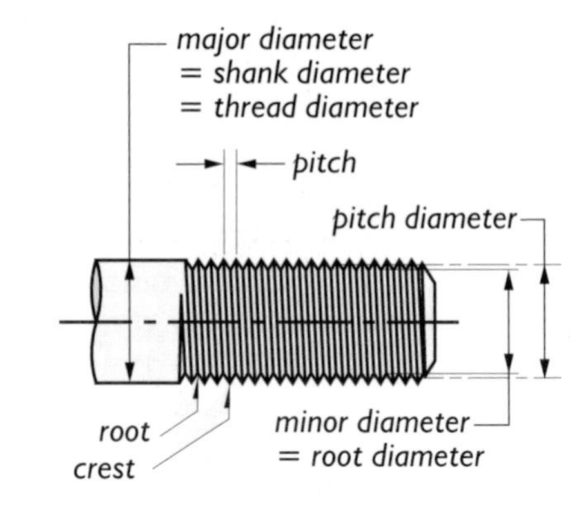

Figure 4.22 Terms used for thread systems

Table 4.3 Tensile properties of SAE bolting

SAE grade	sizes from	sizes to	min S_P MPa	min S_U MPa	max BHN (‡)	material (steel)	(†) heat treatment	head marking
1	1/4"	1-1/2"	228	379	207	low-carbon	—	
2	1/4"	1-1/2"	193 – 475	379 – 475	207 – 241	low-carbon	—	
3	1/4"	5/8"	552 – 586	689 – 758	269	medium-carbon	—	
5	1/4"	1-1/2"	510 – 586	723 – 827	285 – 302	medium-carbon	Q+T	
★ 6								
7	1/4"	1-1/2"	724	916	321	medium-carbon alloy steel	Q+T	
8	1/4"	1-1/2"	827	1034	352	medium-carbon alloy steel	Q+T	

‡ *BHN* is the *Brinell Hardness Number* — the best way of determining bolt strength in the field;
† *Q+T* denotes *quenched and tempered*; ★ grade 6 is not often used

Designers should be familiar with some of the jargon of thread systems.

- p_b = pitch of the threads (axial distance between adjacent threads);
- d_b = nominal bolt size; basic thread diameter; major diameter;
- d_r = root diameter; minor diameter;
- d_{pitch} = the pitch diameter is the diameter at which the axial distance measured through a thread (from leading flank to trailing flank) is exactly half the pitch;
- *root* = smallest-diameter region of thread form;
- *crest* = largest-diameter region of thread form.

The main terms are illustrated in Figure 4.22.

(c) Bolt and screw terminology

A panoply of threaded fastener types exist. They are described and differentiated by a large and sometimes confusing terminology. We will not attempt to cover the whole range here in detail, but rather provide a few key terms to help designers find their way.

Threaded fasteners have two main parts — a *head* and a *body*. Furthermore, the body of a threaded fastener has two main regions: the *threads* and (optionally) the *shank*. Thinking philosophically for a moment about shanks, threads and the roles of heads will help us to clarify the distinction between the two main classes of threaded fastener: *bolts* and *screws* (Refer Figure 4.23).

Broadly speaking, the fasteners that we would generically describe as *bolts* (this term includes *studs* and *cap screws* as well as *bolts*) have a shank region immediately beneath the head. The diameter of the shank is equal to the outer diameter of the threads, so it is physically impossible to screw a bolt all the way into a threaded hole — the threads would foul with the shank. Instead, a *bolt* is designed to be inserted through a clear hole or slot, and fastened with a *nut* (although on occasion it will be convenient to screw them part-way into a threaded hole). The main function of the head of a bolt is to provide an anchor point for fastening.

Conversely, the fasteners that we would classify as *screws* generally have no shank, but instead the threads extend along the full length of their body right up to the head (or run out into a reduced-diameter, unthreaded region called the *neck*, just under the head). They are designed to be screwed into threaded holes in machinery (although on occasion, it will be convenient to use them with a nut). The main function of the head of a screw, apart from providing an anchor point for clamping the fastened component, is to deliver torque to the body of the screw.

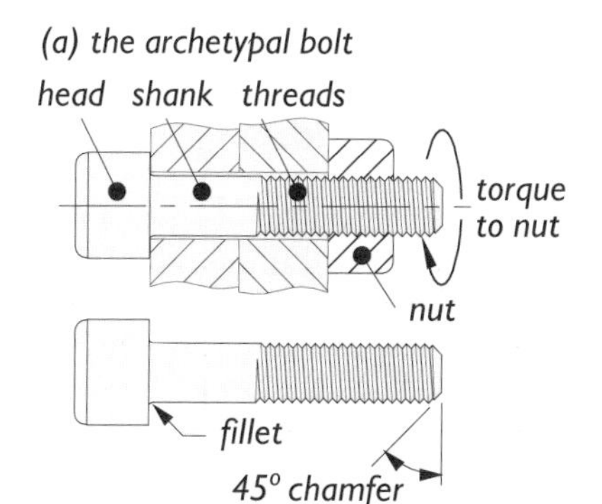

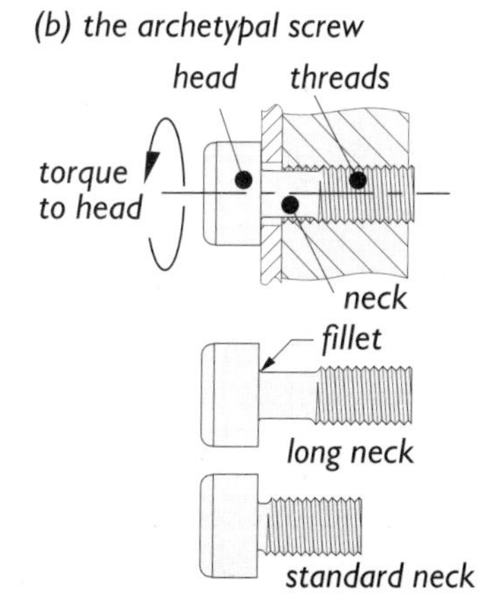

Figure 4.23 Bolt and screw archetypes

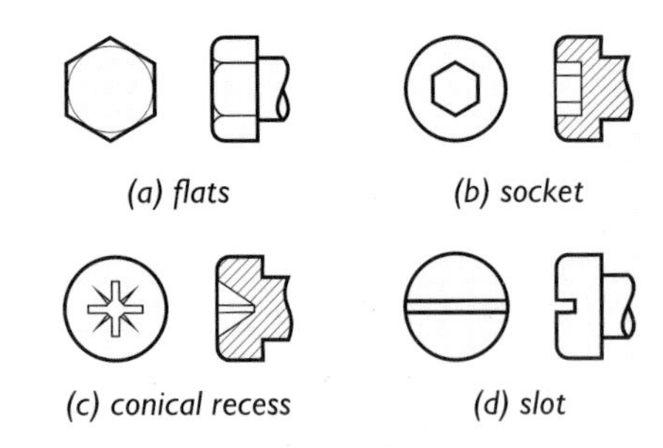

Figure 4.24 Torque-delivery methods

The head incorporates the means for delivering torque to the fastener, and usually has a larger diameter than the body (but not always — consider the case of *socket set screws* where the entire head lies inside the diameter of the body; or *studs*, which have no head at all). Torque delivery to the head is always by means of flat surfaces arrayed around it, providing contact regions for torsional driving forces imparted from a tool of some kind (spanner, screwdriver, hexagon key). The geometrical form of these contact surfaces is either:

- a set of external *flats* around the prismatic head, the transverse cross-section of which is usually hexagonal or square (Figure 4.24a);
- an internal prismatic *socket* (usually hexagonal) that can be driven by a prismatic key (Figure 4.24b);
- a conical-type recess with locking ridges arrayed around the central axis in a *star* or *cross* pattern (Phillips-head screw; Figure 4.24c);
- a *slot* (Figure 4.24d).

(d) Available types of threaded fasteners

Many of the distinctions between threaded fasteners are on the basis of the geometry of their heads. Some of these are shown in Figure 4.26 for bolts and nuts, and in Figure 4.27 for machine screws. No attempt is made to provide dimensions in these figures, although the sketches are drawn generally to scale. The list portrayed is not exhaustive, but sufficiently wide to provide designers with a brief overview of types and variations of available threaded fasteners. The sizes of all sketches have been scaled so that they share the same basic thread diameter.

4.2.7 Further details regarding threaded connections

(a) Nuts, dilation

The nut will be in compression due to the applied force F_b. In addition, the nut will dilate radially as a result of the threads attempting to *force the bolt and nut apart* (indicated schematically in Figure 4.25).

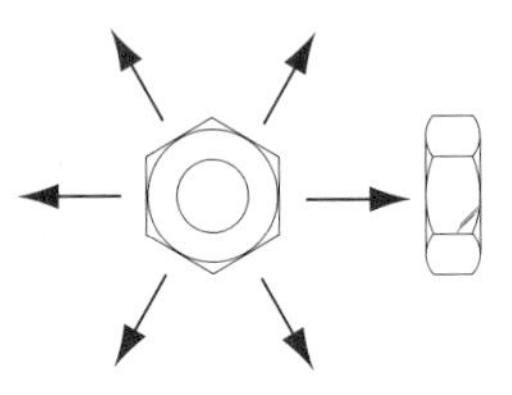

Figure 4.25 Nut dilation

(b) Threads and axial force distribution

In general a large proportion of the load in the nut is carried on the first thread. For example, in a 12mm bolt with fine pitch, the first thread experiences about 230% of the average thread stress (the ratio is 180% for coarse-pitch threads). In the light of this, the designer should be aware that:

- a softer nut will allow more even distribution of stresses on threads than a hard nut; and
- a coarse thread has better stress distribution than a fine thread, and also causes less severe stress concentrations.

(c) Estimation of pre-loading torque

Pre-loading bolted joints is essential if they are to work properly. Usually (but not always) this pre-loading is carried out by applying the *right* amount of torque to the bolt head or nut so the bolts reach their required initial tension. This torque is difficult to predict since it is largely friction dependent, but within 20% accuracy is given by

$$T \approx \frac{F_{b\,i}\, d}{5},$$

where d is the nominal thread diameter. For critical applications where large bolts are involved, the pre-loading is sometimes performed hydraulically

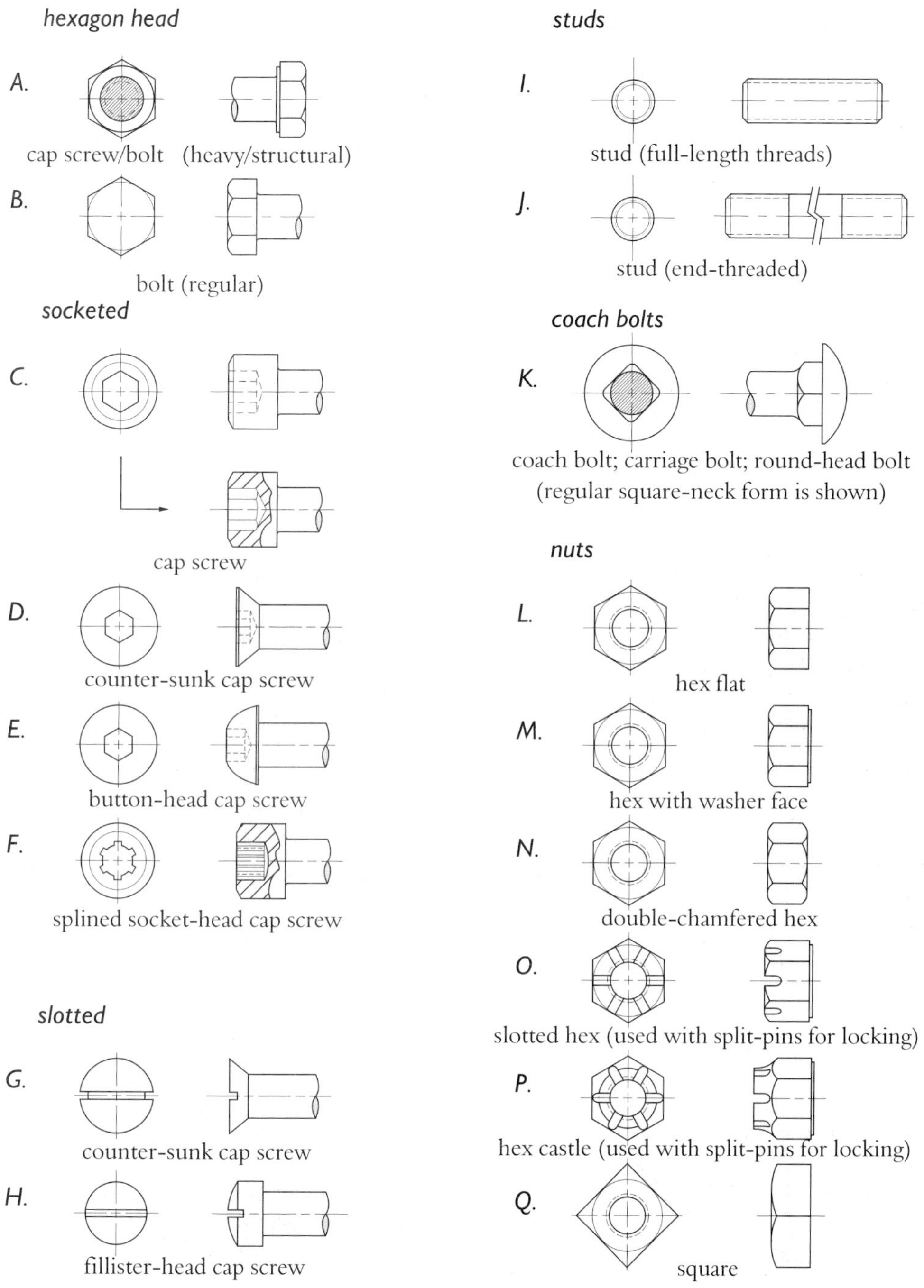

hexagon head

A. cap screw/bolt (heavy/structural)

B. bolt (regular)

socketed

C. cap screw

D. counter-sunk cap screw

E. button-head cap screw

F. splined socket-head cap screw

slotted

G. counter-sunk cap screw

H. fillister-head cap screw

studs

I. stud (full-length threads)

J. stud (end-threaded)

coach bolts

K. coach bolt; carriage bolt; round-head bolt (regular square-neck form is shown)

nuts

L. hex flat

M. hex with washer face

N. double-chamfered hex

O. slotted hex (used with split-pins for locking)

P. hex castle (used with split-pins for locking)

Q. square

Figure 4.26 Heads for threaded fasteners (bolts and nuts)

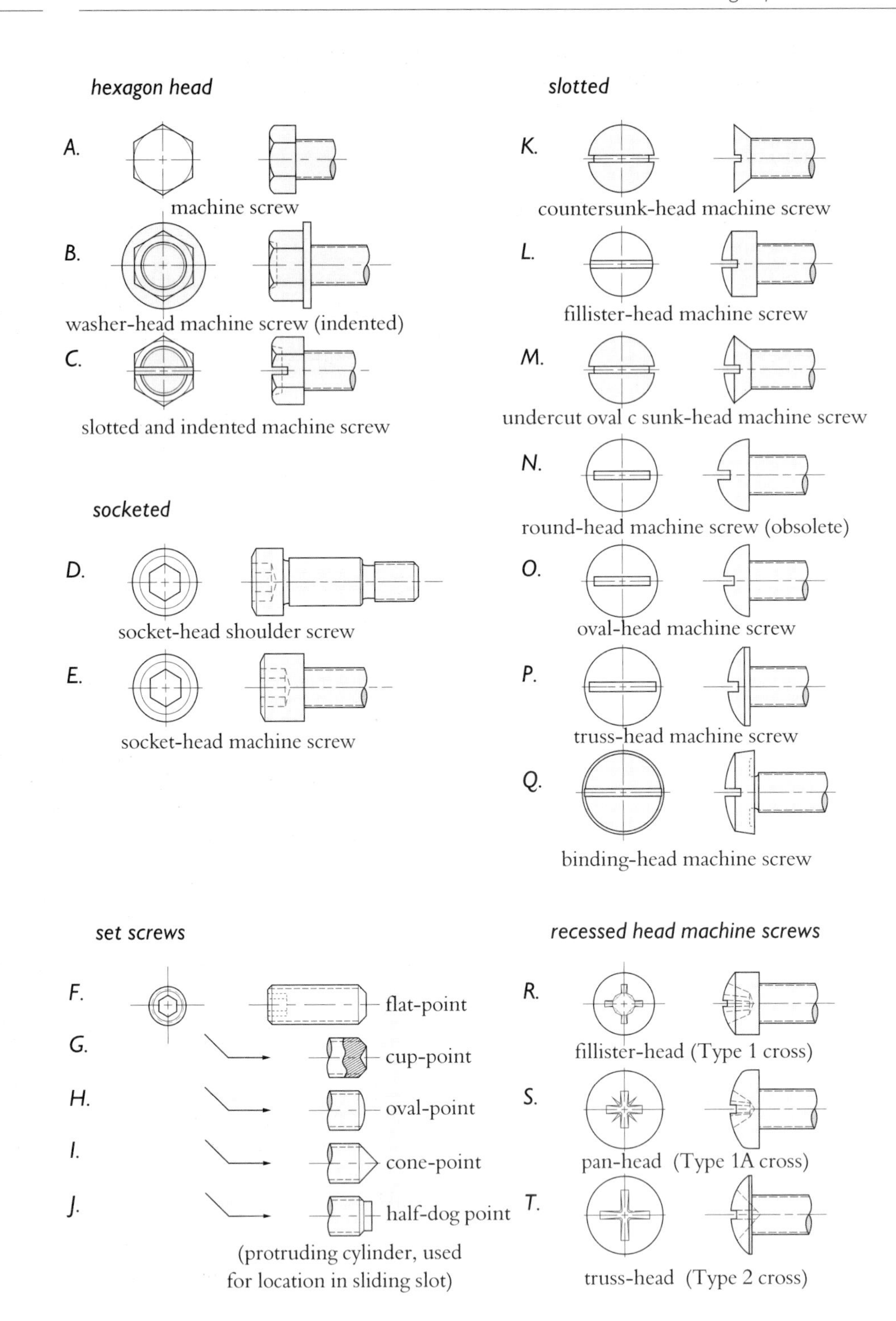
hexagon head
A.
machine screw
B.
washer-head machine screw (indented)
C.
slotted and indented machine screw
socketed
D.
socket-head shoulder screw
E.
socket-head machine screw
set screws
F.
flat-point
G.
cup-point
H.
oval-point
I.
cone-point
J.
half-dog point
(protruding cylinder, used for location in sliding slot)
slotted
K.
countersunk-head machine screw
L.
fillister-head machine screw
M.
undercut oval c sunk-head machine screw
N.
round-head machine screw (obsolete)
O.
oval-head machine screw
P.
truss-head machine screw
Q.
binding-head machine screw
recessed head machine screws
R.
fillister-head (Type 1 cross)
S.
pan-head (Type 1A cross)
T.
truss-head (Type 2 cross)

Figure 4.27 Heads for threaded fasteners (machine screws)

(pulling the bolt directly, with a more direct measurement of axial force). Alternatively the bolt tension can be estimated more accurately by measuring bolt extension.

4.2.8 Summary of design rules for bolted (clamped) joints

(a) Clamping integrity (joint force)

Model: $$F_j = F_{bi} - \left(\frac{k_j}{k_b + k_j}\right) F \quad (4.9)$$

where $$k_b = \frac{A_b E_b}{\ell_b}, \quad (4.1)$$

$$k_j = \frac{A_j E_j}{\ell_j}, \quad (4.2)$$

$$k_j = \left(\frac{1}{k_{f1}} + \frac{1}{k_g} + \frac{1}{k_{f2}}\right)^{-1}. \quad (4.3)$$

(b) Pressure sealing joints:
Require:

$$F_j \geq \frac{\sigma_{\text{gasket seating}} A_{\text{gasket,effective}}}{n} \cdot F_d \quad (4.13)$$

(c) Friction joints:

Require: $$F_j \geq \frac{F_{\text{transverse}}}{n \mu_{\min}} \cdot F_d . \quad (4.14)$$

(d) Bolt integrity (bolt force)

Model: $$F_b = F_{bi} + \left(\frac{k_b}{k_b + k_j}\right) F. \quad (4.8)$$

(e) Static loading:

Require: $$F_b \leq \frac{S_{fail} A_s}{F_d};$$

$$S_{fail} \in \{S_P, S_y, S_U\}. \quad (4.15)$$

(f) Dynamic loading:

Model: from *(4.8)*

$$F_{b\,\min} = F_{bi},$$

$$F_{b\,\max} = F_{bi} + \left(\frac{k_b}{k_b + k_j}\right) F.$$

Hence calculate mean and amplitude of fluctuation in F_b, and use *A-M* diagram to check for adequacy of bolts to resist fatigue. Adjust F_{bi}, k_b, k_j, n, to suit.

Require: $$\left(\frac{\sigma_a}{S_e} + \frac{\sigma_m}{S_U}\right) \leq \left(\frac{1}{F_d}\right),$$

which is the standard requirement of the *A–M* diagram approach.

4.3 Pinned joints and shear connectors

According to the proposed taxonomy of Figure 4.7, pinned joints are one of the *three main categories* of *fastener* that constitute *mechanical connections*. Pinned joints always require the connected elements being *pierced* in some way to provide a hole in one or both of them. The connection is formed when some other linking element is interposed through the hole. The linking element is often called a *shear-pin*, since it generally experiences shearing forces. Pinned joint connections are sometimes referred to as *shear-connectors* for this reason.

Fasteners are strictly distinct from kinematic joints. The former are designed to prevent movement, while the latter are designed to permit it. In fact, with pinned jointed structures, if the designer is not careful to constrain the structure properly, it can become finitely mobile. This feature of pin jointed structures is of particular concern to structural engineers. However, a shear pin provides an axis of rotation in a revolute kinematic joint. Hence, pinned joints are often identified within systems that are really simple mechanisms. Nevertheless, these still fit our broad definition of a *fastener* (*a device that connects two solid objects and prevents relative motion between them to some degree* — refer to the abstract-thinking assignment in Section 4.1.3), since such fasteners restrict relative motion in all but one degree of freedom.

Pinned joints consist of three elements: the two main connected parts bearing the holes, and the pin. In this section we consider structural failures arising from yielding or fracture along various planes and surfaces through these three elements, and simple design approaches for resisting such failures.

4.3.1 Typical pinned joints

(a) typical applications

Several examples of pinned joints were presented in Figure 4.5 (A to N), and Figure 4.6 (A to D) of Section 4.1.2, all in relation to the joining of sheets of paper. Some more typical *engineering* applications are shown in Figure 4.28.

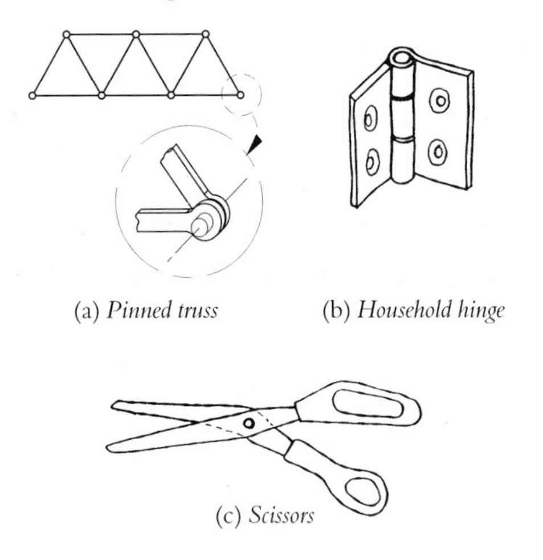

Figure 4.28 Engineering applications of pinned joints

The main features of such joints are that:

- the main connected members use plates or *eyes* aligned parallel to each other at the end connections;
- these *eye-plates* bear a hole for a shear-pin;
- the shear force may be exerted at a single shear plane through the pin, or it may be shared across several shear planes through the pin (as in the household hinge; Figure 4.28b), depending upon the number of *eye-plates* employed;
- there are no moments acting about the pin axis;
- shear forces act transversely to the pin, and parallel to the *eye-plates*;
- in *two-force* members such as those encountered in truss structures, forces act along members only (i.e. along the line joining the two points of application of force; Figure 4.28a);
- in *three-force* members (Figures 4.28b and 4.28c), a resultant direction must be found for the joint force, although it will always act transverse to the shear-pin.

(b) the representative pinned joint

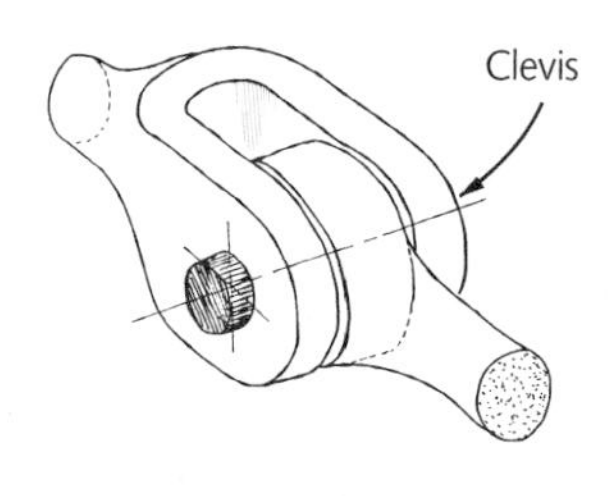

Figure 4.29 Representative pinned joint

Clearly, the geometrical variety of pinned joints is wide. It will be convenient for our discussion of common failure modes and design rules for such systems if we can adopt a representative configuration for pinned joints. The standard arrangement shown in Figure 4.29 will serve this purpose. This particular arrangement is a pinned joint connecting two tension rods of circular cross-section.

One rod terminates in a single *eye-plate*, which fits between the twin eye-plates of the *clevis* at the end of the other rod. The use of a clevis with twin eye-plates is not mandatory, but it provides the joint with symmetry about the vertical plane containing both rod centrelines, this being advantageous both for the uniform transmission of forces through the joint, and for simpler analysis. Other arrangements are possible, as shown schematically in Figure 4.30.

Note that asymmetric arrangements with single eye-plates, such as those in Figure 4.30(a), are common due to their ease of fabrication. Symmetric joints like 4.30(b) (which is the same as that of the representative pinned joint in Figure 4.29) and 4.30(d) are preferred since the main member forces are aligned. Asymmetric joints with multiple eye-plates as in 4.30(c) are rarely encountered.

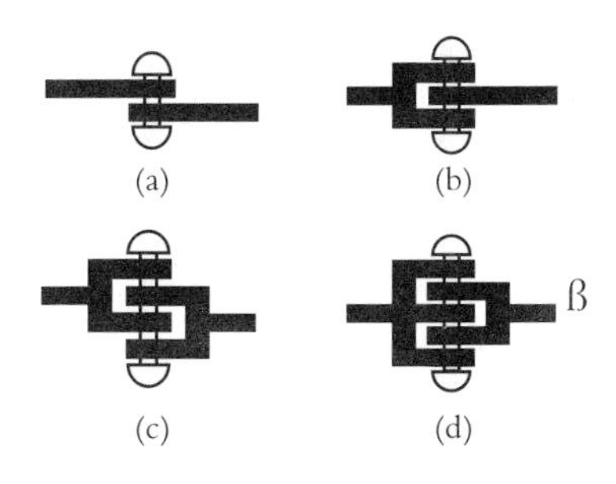

Figure 4.30 Alternative eye-plate arrangements

(c) key dimensions defined

The essential dimensions of our representative pinned joint in Figure 4.29 are defined in Figure 4.31. These dimensions merely serve to establish key sizes of components for the particular arrangement used in our representative pinned joint — they are not normative for all joint arrangements. For example, rods, eyes and even pins may be non-circular in form. The length of the clevis eye-plates is immaterial, and is not specified here.

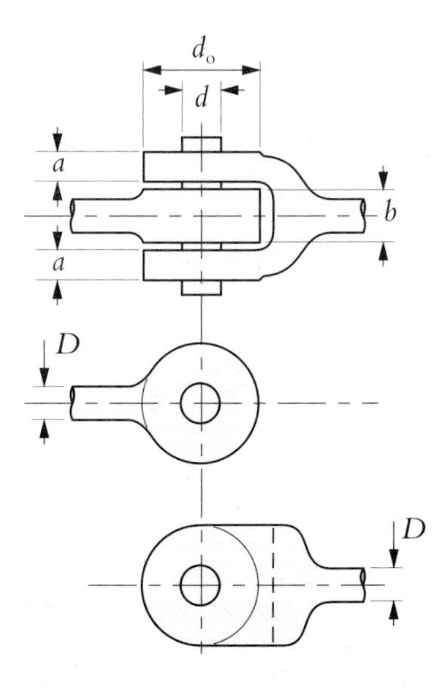

Figure 4.31 Key dimensions for the standard joint

Details of the *yoke* region of the clevis, whilst important in resisting bending stresses when the eye-plates transmit their forces back to the rod, are too specific to any one joint design to be shown here.

The main dimensional parameters are:

d = pin diameter (also inner eye diameter);

d_o = outer eye diameter;

D = rod diameter (gives cross-sectional area of main members);

a = outer eye-plate thickness (one of two plates used in clevis);

b = inner eye-plate thickness (one only).

4.3.2 Modes of failure

(a) general categories

We are concerned with the various *modes* of failure. The main modes are shown in Figure 4.32, and described in general terms as follows:

- tensile failure of the *rod* — Figure 4.32(a);
- transverse shearing of the pin — Figure 4.32(b);

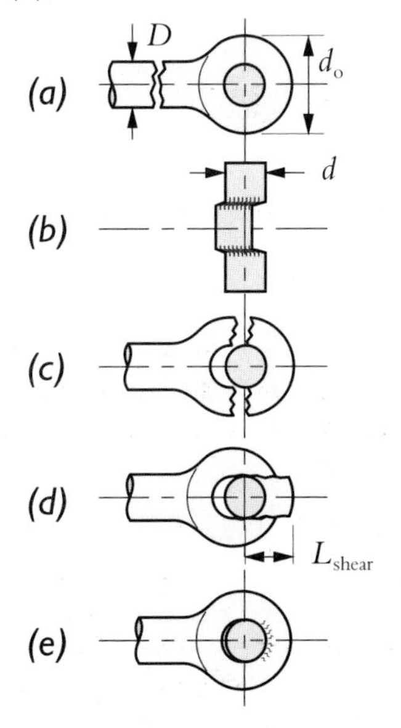

Figure 4.32 Tensile, shear and bearing failures

- tensile failure of an *eye* of the joint as shown in Figure 4.32(c);
- shear failure of an *eye* of the joint as shown in Figure 4.32(d);
- compressive *bearing* failure between the pin and eye — Figure 4.32(e).

In case (d), above, the shear length, L_{shear}, is *conservatively* estimated as

$$L_{\text{shear}} = \frac{d_o - d}{2}$$

where the two diameters are the outer and inner diameters of the eye respectively.

(b) illustrated planes of failure

The modes of failure listed above can be associated with eight distinct failure surfaces experiencing different kinds of stress. Grouping the failure modes according to stress type, these eight failure surfaces are shown in Figure 4.33 as a precursor to writing down the mathematical models that predict failure in each.

(c) summary of failure modes and locations

1. tensile failure of the rod;
2. tensile failure in the eye;
3. tensile failure in the net area of the clevis;
4. shear of the pin;
5. shear failure in the eye;
6. shear failure in the clevis due to *tear out*;
7. compressive failure in the eye due to excessive *bearing pressure* of the pin;
8. compressive failure in the clevis due to excessive *bearing pressure* of the pin.

4.3.3 Simple design inequalities

Having catalogued the major modes of structural failure of the representative pinned joint, it now remains to develop a mathematical model and design approach for predicting and controlling the various stresses leading to those structural failures. We will proceed to write down expressions for the stress on each of the eight failure surfaces. Then by a simple failure prediction model and the application of suitable safety factors, design inequalities will be presented to guide the designer in resisting all eight modes of failure. None of this is *deep*, nevertheless it provides a clear example of the way in which the design approach proposed in this book is applied.

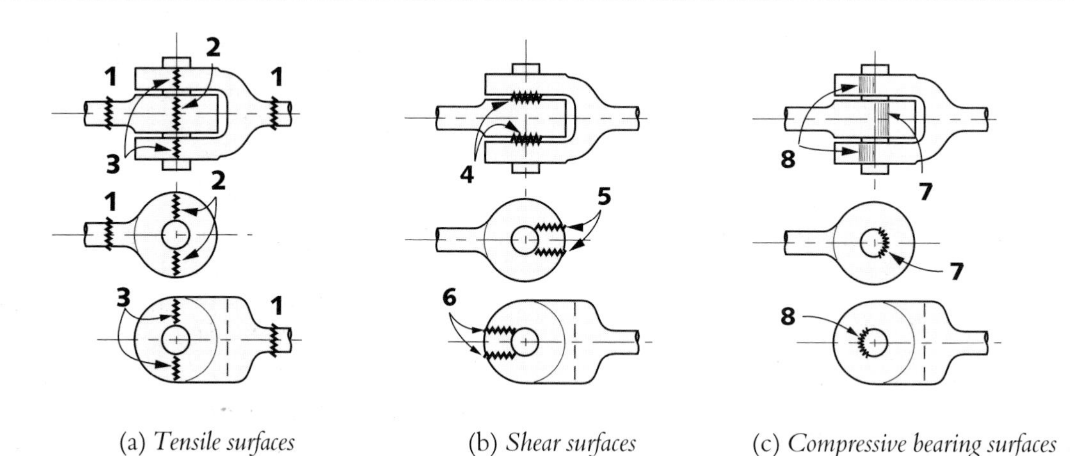

Figure 4.33 Potential failure surfaces

(a) sample derivation

The technique to be followed is the same in all cases. Therefore one sample design inequality shall be derived, and the rest left to readers as an exercise.

Consider mode of failure number 6: *shear failure* in the clevis due to *tear out*. Let the total axial force exerted upon the pinned joint be P. In this mode of failure, the force P is experienced as a shear force distributed over the failure surface denoted *6* in Figure 4.33. This surface divides into four equal sub-areas, and is illustrated in Figure 4.34, below.

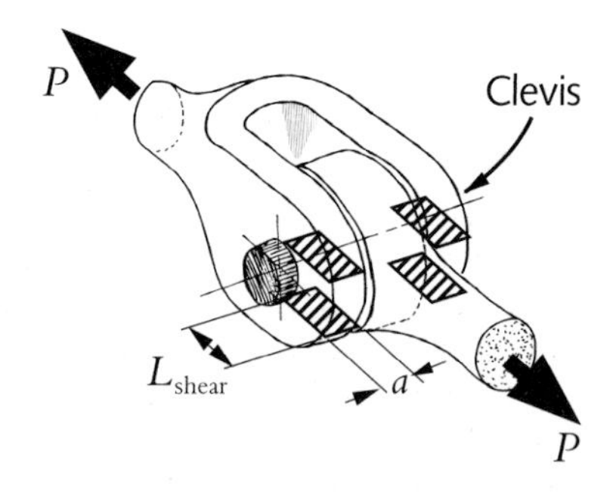

Figure 4.34 Shear plane within the clevis.

The total area of the shear stress surface is $A = 4 \times a \times L_{shear}$, where

$$L_{shear} = \frac{d_o - d}{2}.$$

Therefore the shear stress on this surface is given by:

$$\tau = \left(\frac{P}{A}\right) = \frac{P}{4aL_{shear}} = \frac{P}{2a\left(d_o - d\right)}.$$

This shear stress must be limited by the designer. It can be predicted that the joint will fail in shear when the maximum shear stress anywhere within it reaches value of $(S_y/2)$. Therefore the designer will want to restrict the shear stress to be less than this failure value by at least some acceptable factor of safety, F_d. In other words, we require

$$\tau \leq \tau_{allowable},$$

where

$$\tau_{allowable} = \frac{S_y}{2F_d}.$$

The designer will manipulate the available joint design variables — in this case, a, d and d_0 — to satisfy the above design inequality. Here, the design inequality can be rearranged to the following form:

$$a\left(d_o - d\right) \geq \frac{PF_d}{S_y}.$$

(b) stress modelling summary

All the failure modes examined above are summarised in Table 4.4.

Ex 4.10 Thinking conservatively

Why is the expression offered for potential failure case (d) in Figure 4.32 a *conservative* estimate? [Hint: you will need to examine the geometry of the eye.]

Ex 4.11 Stress derivations

By methods similar to that demonstrated above, derive the stress expression for each of the seven remaining failure surfaces shown and listed in Sections 4.3.2(c). Confirm the results summarised below in Table 4.4.

Ex 4.12 Design inequalities

Recast the results from Table 4.4 into the form of design inequalities.

Ex 4.13 Adapting for alternative geometries

To generalise the approach slightly for pinned joints with multiple eye-plates as in Figure 4.30, let us define N_A as the number of eye-plates on the 'eye-side' of the joint (plate thickness = a), and N_B as the number of eye-plates on the 'clevis-side' of the joint (plate thickness = b). How would the stress expressions in Table 4.4 need to be modified to account for N_A and N_B ?

Ex 4.14 Governing failure modes

Study the various expressions for actual and allowable stress in Table 4.4. Are there any failure modes that can be discounted as being redundant or automatically satisfied?

Ex 4.15 Geometrical proportions

By examining the expressions for actual and allowable stress in Table 4.4, see if you can devise any generalisations concerning suitable proportions in the joint (that is, desirable ratios between a, b, d, d_o and D).

(c) review of assumptions, approximations and simplifications

The approach outlined above is deceptively simple, yet involves several assumptions and approximations of which the designer should be aware.

- Pin contact force is assumed to be uniform. Lineal contact force between pin and eye-plates will be reasonably constant provided that the pin is sufficiently stiff to avoid appreciable transverse bending deflections. If the pin is too flexible, its bending deflections will lead to a highly non-uniform distribution of contact force against the eye-plate, as illustrated in Figure 4.35.
- Pin contact pressure is assumed to be uniform across the projected bearing area between pin and eye-plates. This is not strictly true. If there is some loose clearance between the diameters of the pin and the hole in the eye-plate, then contact between these two cylindrical surfaces is of the type leading to Hertzian contact stresses. These were discussed in Section 2.5 (refer to case 9 of Figure 2.40). Hertzian contact pressure is not uniform, but rather follows a semi-ellipsoidal distribution. Peak contact pressure occurs along the centreline of the projected area of contact, and can be found to be:

$$q_o = \sqrt{p_o \frac{E}{\pi\left(1-\mu^2\right)} \frac{\delta}{\left(1+\delta\right)}}$$

where $p_o = P/dw$ is the *projected area* average pressure ($w=2a$ or $w=b$ is the net thickness of the eye-plates), and δ is the relative diametral clearance, $\delta = (d_{hole} - d_{pin})/d_{pin}$. For large clearance and stiff material (E is large for steel), this peak value can be many times larger than p_o. However this result breaks down in the special but common case of zero clearance ($\delta = 0$), and the approximation that $\sigma_c \approx p_o = P/dw$, as quoted in Table 4.4, then leads to acceptable solutions.

- Stress concentrations are ignored, such that tensile and shear stresses exerted on imagined *potential failure planes* are assumed to be uniform over the whole area.

Table 4.4 Stresses on failure surfaces for pinned joints

Mode	Stress Type	Failure Location	Expression	Limiting Stress
1	*tensile*	*rod*	$\sigma_t = 4p/\pi D^2$	$\sigma_{t,allowable} = S_y/F_d$
2	*tensile*	*clevis*	$\sigma_t = P/b(d_0 - d)$	$\sigma_{t,allowable} = S_y/F_d$
3	*tensile*	*clevis*	$\sigma_t = P/2a(d_0 - d)$	$\sigma_{t,allowable} = S_y/F_d$
4	*shear*	*pin*	$\tau = 2P/\pi d^2$	$\tau_{allowable} = S_y/2F_d$
5	*shear*	*eye*	$\tau = P/b(d_0 - d)$	$\tau_{allowable} = S_y/2F_d$
6	*shear*	*clevis*	$\tau = P/2a(d_0 - d)$	$\tau_{allowable} = S_y/2F_d$
7	*compressive*	*eye*	$\sigma_c = P/db$	$\sigma_{c,allowable} = S_{yc}/F_d$
8	*compressive*	*clevis*	$\sigma_c = P/2da$	$\sigma_{c,allowable} = S_{yc}/F_d$

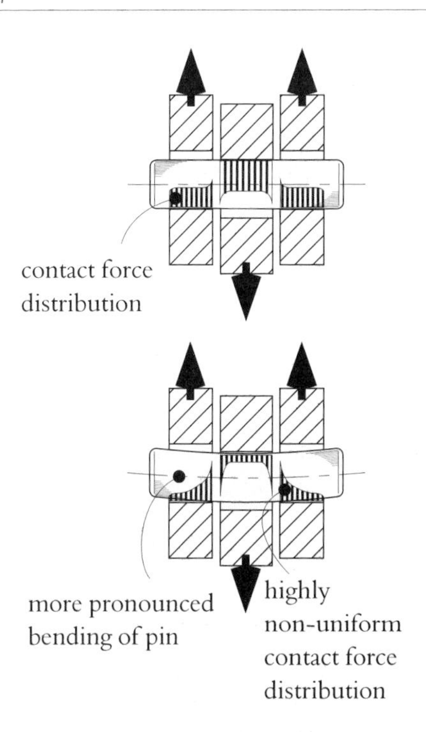

Figure 4.35 Effects of pin deflection

- Bending stresses in the pin are ignored. This is a reasonable assumption if there is a close tolerance fit of the pin in the eye-plate hole, and if the gap between adjacent eye-plates is small. But if either of these tolerances is large, then pin bending can be a significant issue.

4.3.4 Summary of design rules for pinned joints and shear connectors

The design rules for pinned joints have already been effectively summarised in Table 4.4, and the process of rearranging them into a convenient form is trivial. Rather than repeat the results here, it might prove more useful to summarise the design approach. In all cases, the approach has been:

- isolate the joint as a free body subjected to the external joint load;
- identify a possible mode of structural failure associated with yielding along a particular plane or surface within the joint;
- estimate the area of that potential failure plane or surface;
- identify the nature of the stress (tensile, shear, compressive, or bending) imposed on that potential failure plane by the joint load;
- estimate the value of that stress corresponding to failure (for example, $S_y/2$ for a shear stress), and so nominate an allowable limit to that stress by incorporating a safety factor (for example, $\tau_{all}=S_y/2F_d$ for a shear stress);
- estimate the stress on that failure plane or surface, assuming a uniform distribution of the applied load across the potential failure plane (or a linear distribution for bending stress);
- write down the design inequality by inserting the appropriate expressions into the inequation:

  (actual stress) $\leq$ *(allowable stress)* ; and
- select values of the geometrical design variables to satisfy all such design inequalities.

This approach is quite general, and may be extended to many simple cases in the design of machine elements for which no standard analysis is available.

4.4 Welded joints

Welding is a generic term describing a process for joining metal parts through the application of a combination of heat and pressure. Probably one of the oldest methods for joining metal components, it has evolved from the blacksmith's *impact welding*, through *flame welding* to the wide range of production welding processes used in industry today. The origins of impact welding date back to the first millennium, when sword manufacture relied on the impact welding of alternate layers of iron and brittle carburised steel, to form a tough blade. This process was certainly used by sword makers of Damascus, Syria, from whence the famed *Damascene* sword blades originate.

In the early 20th century flame welding was developed, using a combination of oxygen and acetylene gas to produce the necessary temperatures in the flame to melt iron welding rods. The procedure of flame welding relied heavily on the specialist skills of the welder and electric arc welding was developed to provide acceptable mass manufacturing consistency in the process. Originally arc welding, introduced during the 1914-1918 war for defence production, made use of bare welding rods, resulting in a brittle welded joint. The highly effective arc welding process, using shielded electrodes, was introduced during World War II, mainly for ship building. We have already made reference to the shipbuilding records set by the Liberty Ship makers (Section 1.1.2d). These manufacturing achievements were almost entirely due to the effective use of arc welding processes. *Resistance welding* which predates all other forms of electric welding processes, was invented in the late 19th century. This process was used for *spot* and *seam* welding and for joints in chain links.

In 1940, a new form of arc welding was introduced, using a non-consumable tungsten electrode and an inert gas shield to protect the weld bead from oxidation during production. This *tungsten-inert–gas* (*TIG*) welding process produced superior welds, and permitted the production welding of a wide range of materials including aluminium and stainless steel. In 1948 the *TIG* welding process was enhanced by the introduction of a consumable electrode for continuous machine welding. This relatively new *metal-inert-gas* (*MIG*) welding process is now in widespread use for all types of continuous arc welding in manufacturing. Its consistency of performance makes it the welding process of choice for many automated welding applications using robots. There are many varieties of welding processes in industrial use, including *spot welding*, *resistance welding*, *friction welding*, *flame welding*, *explosive welding* as well as manual and automated *electric arc welding*. In what follows we model the structural integrity of generic continuous electric arc welded joints. Modelling the structural behaviour of other forms of welding processes are beyond the scope of this introductory discourse.

Continuous electric arc welding in engineering structures employs the electric arc to achieve highly localised melting of the metal parts being joined. The melted parent metal, together with the molten metal from the electrode that has been sacrificed in producing the electric arc, coalesces in a small pool at the arc site, where it quickly *freezes* to form the finished weld *bead*. The weld bead is created continuously by slowly progressing the electrode tip and weld bead along the edge of the material to be joined, leaving a finished *welded joint* in its wake.

A byproduct of many arc welding processes is *slag*, a non metallic compound usually present as the coating of the welding electrode. The electrode coating is needed to facilitate the flow of material into the weld seam, generally preventing instantaneous oxidation of the bead surface, which would prevent the clean adhesion of weld to parent metal and weld beads to each other. The *slag* is expected to float on the surface of the weld bead, forming a brittle coating *chipped* off before the next weld pass. This work is performed by a skilled operator, or in some cases by automatic means, in such a way that the finished joint can be as strong as, or stronger than, the *parent metal* it joins together.

Designers of welded fabrications must consider issues relating to the materials of the joined plates and electrodes, and possibly to pre- and post-weld

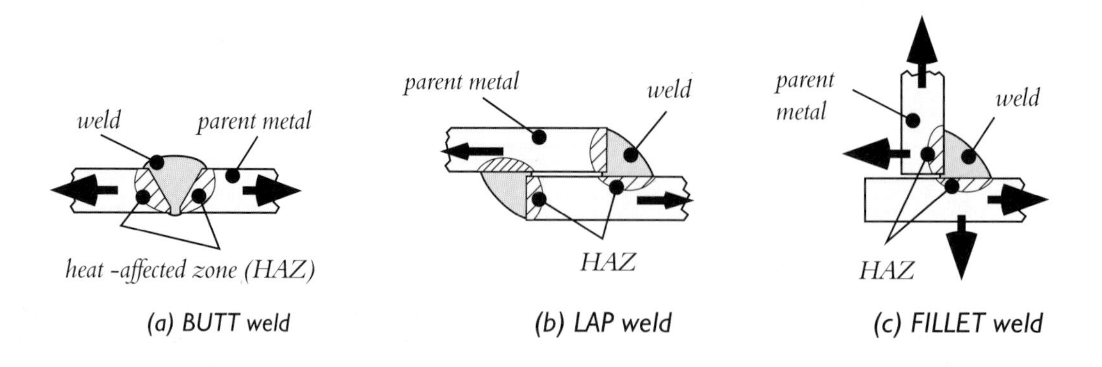

Figure 4.36 Weld types

heat treatment for ensuring adequate material properties. But primarily, the welded joint designer's task is to specify the required type and size of the weld.

4.4.1 Types of welded joint

Continuous welded joints are of three main types: *butt*, *lap*, and *fillet* joints (although one could argue that a lap joint is really comprised of fillet welds). Basic examples of each of these are shown in Figure 4.36.

Butt welds are usually continuous — they are generally intended to reproduce at least the strength of the parent plate. Fillet and lap welds may be made in a continuous line along the joint. Often, however, they are discontinuous — staggered at intervals along the joint.

This is done because:

(a) welds are expensive to make, and if the structural or sealing requirements of a joint do not necessitate a continuous weld, then discontinuous welds will provide a cheaper solution;

(b) the heat from welding tends to distort the components being joined. Discontinuous welds entail less heat, and so may cause less of a thermal distortion problem.

Small welds may be laid down in a single *pass* of the welding electrode. Large welds may require several passes in order to build up the required size, as indicated in Figure 4.37.

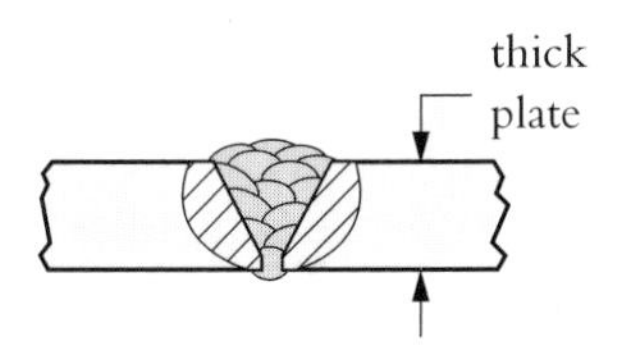

Figure 4.37 Multi-pass welding

In order to (a) provide a suitable region for containing the molten weld metal, (b) allow for good *penetration* of the weld through the parent plates, and (c) minimise the likelihood of *defects* (air gaps, or inclusions of nonmetallic *slag*) in the weld, the edges of the plates to be joined are carefully prepared by grinding/shaping to controlled profiles. Some of these are shown in Figure 4.38. All of the welds in Figure 4.38 are *full-penetration* welds, in that once-molten metal fills the joint cross-section in an unbroken region from one side of the joint to the other. Welds are not always of the *full-penetration* variety, as the extra strength is not always needed (and the extra cost is unwarranted).

The outer surface of a weld is often ground (manually, using an angle-grinder), partly to remove traces of the slag deposited by some welding techniques, partly for cosmetic reasons, but mainly to reduce the severity of stress concentrations set up at the sharp outer edges of the weld cross-section. Figure 4.39 illustrates this process.

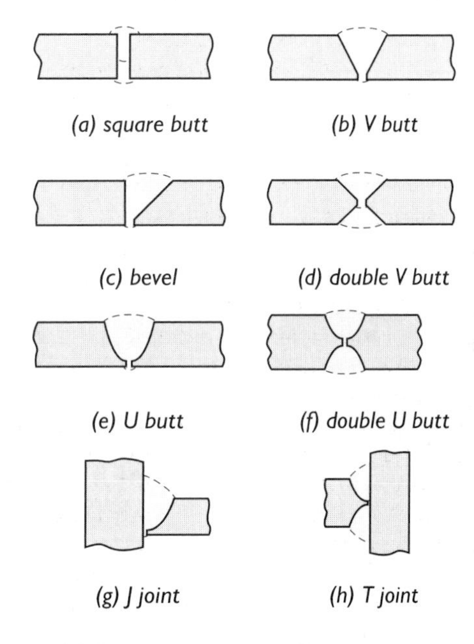

Figure 4.38 Edge preparation profiles

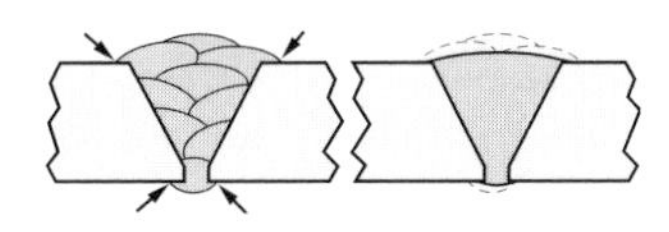

Figure 4.39 Weld surface ground smooth

Another commonly encountered weld type in structural systems involving thin sheets, or pressed components made of thin plates, is the *spot weld*. Spot welds are made by using two non-sacrificing electrodes, one on either side of the two sheets or components to be joined. The electrodes are clamped tightly against the work, making a small contact-area with it. Strong electrical current is then passed through the sheets from one electrode to the other, so that the electrically resistive properties of the material and contact region cause localised heating sufficient to melt the metal. By their nature, spot welds are discontinuous.

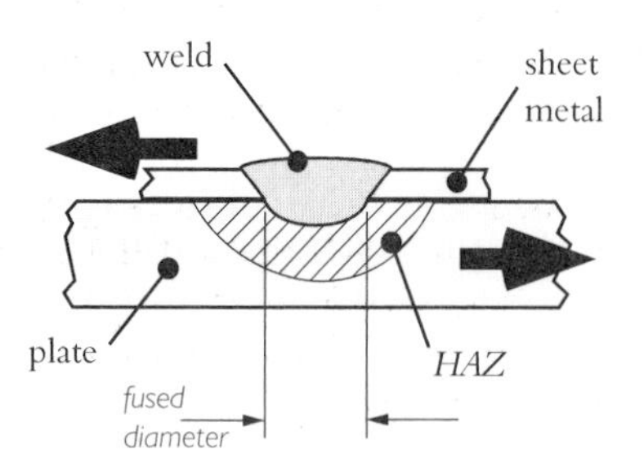

Figure 4.40 Spot weld between thin sheet and plate

Figure 4.41 Robotic welding in automotive manufacture

The designer must decide on their spacing and location to provide adequate attachment strength with the size of spot that can be produced by a given electrode. Figure 4.40 shows the general form of a spot weld. Strength of this weld in shear is based on the fused diameter of the spot, which may be significantly smaller than the diameter visible from outside the sheet. Figure 4.41 illustrates robotic welding processes on a car body assembly line. In the background, several spot welding machines can be seen joining pieces of the car frame. The development of a reliable spot welding process permitted the almost full automation of car body assemblies like this one at the General Motors Southgate plant in California.

4.4.2 Design of welded joints for strength

Strict guidelines for the design of welded joints are provided in national standards and codes of standard practice governing structural steel design, and sometimes also in standards for the design of specialised equipment such as piping, storage tanks and pressure vessels. It is wise for designers of equipment and systems involving welded joints to avail themselves of such documented advice, especially since the *advice* often has the force of statutory regulation. However it is also important for designers (whether of welded joints or anything else) to understand the simple structural concepts that underlie code-based rules. This understanding helps to de-mystify the rules (often turbidly and indirectly expressed), build confidence in the designer's approach, and provide insight into the focal design issues.

Perusal of a variety of structural welding design codes will reveal that, in general, the approach outlined here is common to them all:

- Welding electrodes of certain qualified compositions are assigned a representative set of structural properties (for example, S_U in various ranges from 430 to 620 MPa; S_y ranging from 350 to 420 MPa; and acceptable levels of ductility in the vicinity of 20% elongation).
- These material properties, in conjunction with specified factors of safety for particular cases, are used to assign an allowable stress value for the weld.
- Calculated estimates of actual stress values (direct or shear) in the weld metal under design conditions are compared to corresponding allowable stress values in order to determine the adequacy of the weld.
- In certain cases, the required weld size is determined instead by some minimum proportion to the size of the plates being joined, regardless of stress.

Stated in this way, the code-based rules are found to follow an identical approach to that adopted throughout this book. Of the steps described above, the key steps requiring special skill, insight, or judgement on the part of the designer are (a) the assigning of allowable stress values, and (b) the estimation of actual stress values in the weld.

(a) allowable weld stresses

The strength of a full-penetration butt weld is taken to equal or exceed that of the plates it joins. Therefore the maximum allowable direct stress on a butt weld is simply the minimum of the yield strength values of the two parent plates. No special factor of safety is applied to this allowable stress (except in the case of incomplete penetration butt welds, for which the allowable stress is reduced somewhat to account for the sharp stress concentration effects).

In fillet welds, the allowable shear stress is limited to $0.33 \times S_U$ (based on the strength of the electrode metal), which corresponds to a safety factor of approximately 1.5 for fracture. Depending on the welding electrode used, this implies a value of $\tau_{allowable}$ ranging from approximately 140 to 205 MPa.

Situations involving combined stresses (*direct stress combined with shear stress*) on the weld throat are discussed in Section 4.4.7.

The formulations given in welding design codes for the allowable loads on spot welds in shear similarly imply a safety factor of approximately 1.5 to 1.6 on the fracture stress.

If fatigue or low-temperature conditions apply, allowable stresses may be further reduced from those described above.

(b) calculation of actual weld stresses

Since properly tested butt welds are considered to be merely a continuation of the parent plate, separate calculation of weld stresses is not required — the weld stresses are the same as the plate stresses.

The shear stress in a spot weld is estimated merely as the parallel load per spot divided by the effective fused area of the spot ($\pi d_e^2/4$, where d_e is the effective fused diameter).

The geometry of fillet welds is much more complicated, however, and the approach in the following sections is recommended as a basis for estimating maximum actual shear stress values in fillet welds. For the purpose of this analysis, fillet welds are assumed to possess a simplified cross-sectional geometry like that shown in Figure 4.42. Note that the fillet surface is assumed to be at 45° to the parent plates, and any reinforcing thickness (curved weld surface above the fillet plane) is ignored. The height, or *leg length*, of the weld is denoted h, this being the dimension conventionally used to specify the size of the weld. The dimension t is often referred to as the *throat* of the weld, since in some (but not all) cases it is the width of the most critically stressed cross-section through the weld.

The objective of any stress-estimating exercise applied to fillet welds is to find the worst combination of applied shear load and cross-sectional area for imagined sectioning planes through the weld metal. These longitudinal sectioning planes are represented by the diagonal line at angle θ in Figure 4.42, and later by the shaded faces in Figures 4.44 and 4.45. Shear stress on such planes is found simply by dividing the shear force component acting parallel to the plane, by the area of the sectioned plane. The maximum shear stress in the fillet weld is then just the maximum value of this ratio.

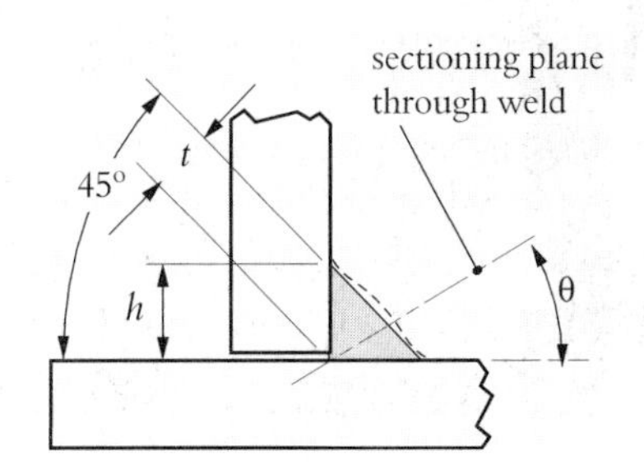

Figure 4.42 Idealised fillet weld

4.4.3 Welded fillet joints under parallel loading

Figure 4.43 illustrates the two main kinds of load that can be applied to fillet welds. A welded fillet joint of unit length subjected to *parallel loading* (that is, applied forces acting parallel to the long axis of the weld) is shown in Figure 4.43(a).

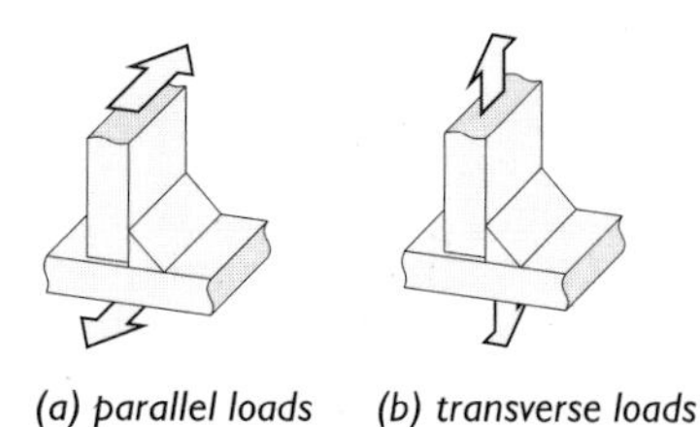

Figure 4.43 Loads on fillet welds

(a) mathematical model

Figure 4.44 is a free-body diagram of the parallel-loaded fillet joint in Figure 4.43(a), with the plane of maximum shear stress highlighted. By inspection, this plane corresponds to the weld throat ($\theta = 45°$), since this is clearly the longitudinal sectioning plane of minimum area, and all such planes carry the same shear load, W_P.

The expression for maximum (*actual*) shear stress in the parallel-loaded fillet weld can now be written down as:

$$\tau = \frac{W_P}{\left(h/\sqrt{2}\right)L_e} \tag{4.16}$$

where L_e is the effective length of the weld.

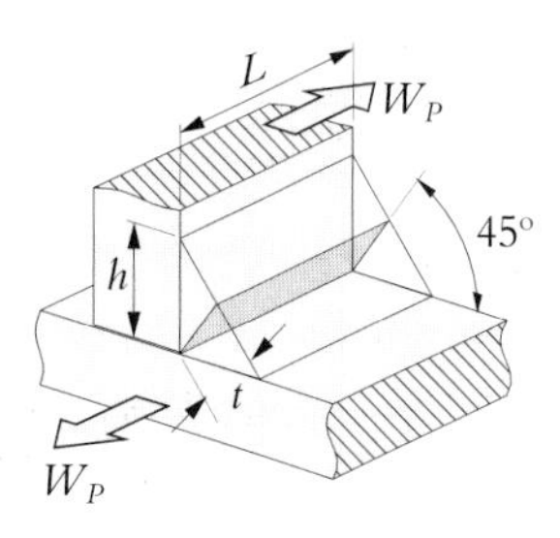

Figure 4.44 Parallel loads on a fillet weld

(b) allowable parallel load per unit length

If we now let:

$$w_P \equiv \left(\frac{W_P}{L_e}\right)_{\text{allowable}} \tag{4.17}$$

denote the allowable parallel load per unit effective length of fillet weld, then rearrangement of Equation *(4.16)* leads to the following result for parallel-loaded fillet welds:

$$w_P \equiv \left(\frac{h}{\sqrt{2}}\right)\tau_{\text{allowable}}. \tag{4.18}$$

The value of $\tau_{\text{allowable}}$ must be determined by reference to the governing welding code. If a conservative value of $\tau_{\text{allowable}} \approx 140$ MPa is used (at the small end of the range of values discussed in Section 4.4.2(a), above), Equation *(4.18)* gives

$$w_P \approx 100h \text{ (kN/m)} \tag{4.19}$$

as long as the size h is expressed in millimetres. For *intermittent welds*, L_e is the total length of full-size fillet weld, exclusive of any *end-effect* regions where the throat size might be smaller than in the main region of the weld.

4.4.4 Welded fillet joints under transverse loading

(a) mathematical model

The earlier Figure 4.43(b) showed a welded fillet joint of unit length subjected to *transverse loading* (that is, with the applied forces acting perpendicular to the long axis of the weld). Figure 4.45(a) is a free-body diagram of the transverse-loaded fillet joint in Figure 4.43(b), with an arbitrary longitudinal sectioning plane shown highlighted. The width of this plane is designated x.

In this case of transversely loaded fillet welds, the orientation of the plane of maximum shear stress in the weld is not immediately obvious, since all sectioning planes such as the one shown shaded experience both shear and normal stress components. These components are made explicit in the sectioned free-body diagram of Figure 4.45(b).

Working now from the sketch in Figure 4.45(c), it is evident that

$$x\cos\theta = h - x\sin\theta$$

due to the symmetry of the 45° fillet surface. Hence the width, x, of the sectioning plane is given by

$$x = \frac{h}{\cos\theta + \sin\theta}. \tag{4.20}$$

The shear stress on this arbitrary plane is the shear force component, $W_T \sin\theta$, divided by the area of the plane, $x \cdot L_e$, so

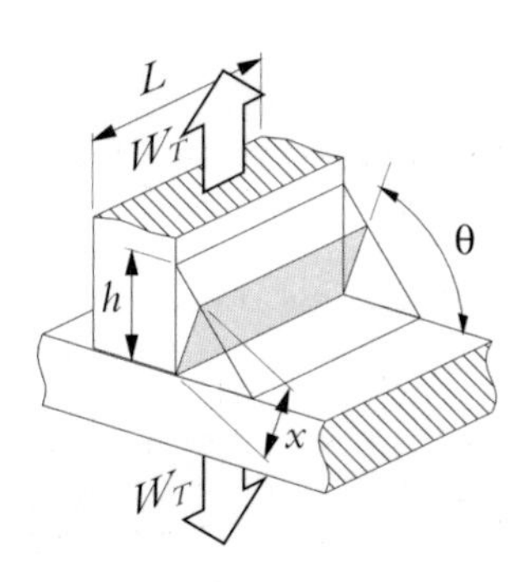

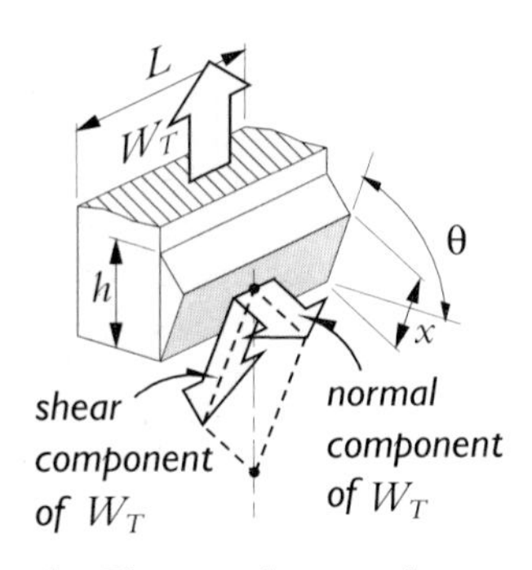

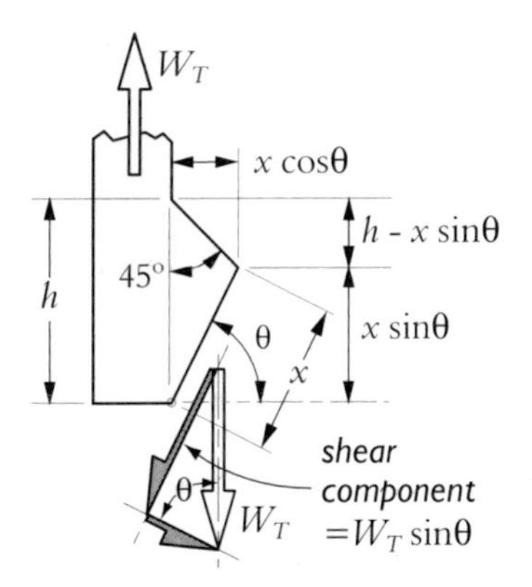

(a) loads on weld (b) Shear and normal components of loads (c) Geometry and force analysis

Figure 4.45 Geometry and force analysis of transversely loaded fillet weld

$$\tau = \frac{W_T \sin\theta}{x L_e},$$

where again L_e is the *effective length* of the weld. Therefore, by means of Equation *(4.20)*, we obtain

$$\tau = \frac{W_T}{h L_e}\left(\sin\theta\cos\theta + \sin^2\theta\right). \quad (4.21)$$

Differentiating *(4.21)* with respect to θ allows us to find the plane on which maximum τ acts.

$$\frac{d\tau}{d\theta} = \frac{W_T}{h L_e}\left(\cos^2\theta - \sin^2\theta + 2\sin\theta\cos\theta\right),$$

$$= \frac{W_T}{h L_e}\left(\cos 2\theta + \sin 2\theta\right),$$

which equals zero (*corresponding to a maximum shear stress*) when

$$\cos 2\theta + \sin 2\theta = 0. \quad (4.22)$$

Equation *(4.22)* is solved by

$$\theta = \frac{\text{artan}(-1)}{2} = 67.5°.$$

Substituting this value of θ back into *(4.20)*, we obtain the following expression for maximum (*actual*) shear stress in the transversely loaded fillet weld:

$$\tau_{\text{MAX}} = \frac{W_T}{h L_e}(1.21). \quad (4.23)$$

(b) allowable transverse load per unit length

Similar to the approach taken for parallel-loaded welds, we define the allowable transverse load per unit effective length of fillet weld as

$$w_T \equiv \left(\frac{W_T}{L_e}\right)_{\text{allowable}} \quad (4.24)$$

Rearranging *(4.23)* then leads to the following result for transversely-loaded fillet welds:

$$w_T = \left(\frac{h}{1.21}\right)\tau_{\text{allowable}}. \quad (4.25)$$

Setting $\tau_{\text{allowable}} \approx$ 140 MPa as before, *(4.25)* gives

$$w_T \approx 115h \text{ (kN/m)} \quad (4.26)$$

where h is again the *weld size* (*expressed in millimetres*). Comparison of Equations *(4.18)* and *(4.25)* shows clearly that *fillet welds are stronger (by about 17%) under transverse loading.*

(c) Examples of parallel & transverse welds

Figures 4.46(a) and 4.46(b) show typical examples of practical weld applications

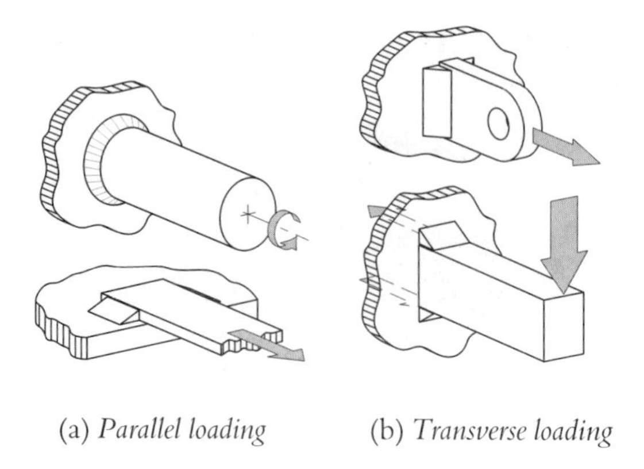

(a) *Parallel loading* (b) *Transverse loading*

Figure 4.46 Fillet weld applications

4.4.5 Welded joints subject to oblique loads

Sometimes fillet welds are subjected to a combination of transverse and parallel forces. This situation is known as *oblique loading*. Since the assumption of parallel loading has been found to lead to a more conservative result, the value of w_p given in Equation *(4.18)* is also applied to obliquely loaded joints.

4.4.6 Transverse welds subject to net bending moments

For the purpose of calculating stresses, transversely-loaded fillet welds have a somewhat awkward shape. We have seen that the plane of maximum shear stress is inclined at 67.5° for fillet welds subjected to uniform transverse loads. In the case where a net bending moment is exerted on a pattern of fillet welds in a joint, the transverse loads are non-uniform, and additional shear forces are present as well, so the situation is even more complex. Some simplification is in order.

One common simplifying approach is to treat the weld throat ($\theta = 45°$) as if it were in the plane of the joint. We imagine the fillet welds as being replaced by a *narrow rectangular stub*, as indicated in figure 4.47(c). This stub traces a similar layout in the plane of the joint as the original pattern of welding, and has a thickness equal to the throat thickness of the original fillet (that is, a thickness of t, rather than h). The net bending moment imposed on the joint is then regarded as being resisted by this thin section of equivalent weld, corresponding to the original weld throat. The estimation of direct (*bending*) and shear stresses on this *section* is straightforward since its planar section properties (area and second moment of area) are easily calculated. These stresses can then be used as conservative estimates of the normal and shear stresses acting on the weld throat. (Refer to the next Section for a discussion of how to apply a design procedure to such combined stresses.)

An example should help clarify this approach. Figure 4.47(a) shows a bar of circular cross-section welded to a supporting plate by a continuous fillet weld. In Figure 4.47(b) we see the actual longitudinal cross-section through this bar and fillet weld, and in Figure 4.47(c) this is replaced by the equivalent narrow section (*equivalent stub weld*) of thickness t. Figure 4.47(d) shows another view of the equivalent stub, emphasising that the pattern of the equivalent section in the joint plane follows the same path as the original fillet — in this example it is a circular annulus. In essence, the idealised stub section replacement for the weld *offers the same moment of resistance* to the applied load P as does the actual weld.

The bending and shear stresses in the equivalent stub weld are now calculated.

(a) Bending stress

The maximum bending stress in the weld is given by $\sigma_{B,max}=My_{max}/I_{zz}$, where $y_{max}= (d/2 + t)$ and

$$I_{zz} = \frac{\pi}{64}\left[\left(d+2t\right)^4 - d^4\right].$$

For $t<<d$, this is approximated by $I_{zz}\approx \pi d^2 t/8$, so we obtain

$$\sigma_{B,\max} = \frac{4PL}{\pi d^2 t} \quad \left(= f_n\right).$$

The symbol f_n stands for the maximum normal stress acting on the weld throat.

(b) Shear stress

The shear stress in the weld is easily found to be

$$\tau_{\max} = \frac{P}{A} = \frac{P}{\pi dt} \quad (= f_v).$$

The symbol f_v stands for the maximum shear stress acting on the weld throat.

4.4.7 Fillet welds subject to combined stresses

Fillet welds frequently occur in locations where both normal and shear stresses on the weld throat are appreciable. Welding codes recommend combining these as an equivalent stress given by

$$\sigma_{\text{combined}} = \sqrt{f_n^2 + 3f_v^2},$$

and limiting this combined stress to

$$\sqrt{3} \times 0.33 \times S_U,$$

(based on the strength of the electrode metal). This gives rise to the following design inequality:

$$\sqrt{f_n^2 + 3f_v^2} \le \sqrt{3} \times 0.33 \times S_{UW},$$

$$\Rightarrow \sqrt{f_n^2 + 3f_v^2} \le 0.57 S_{UW}, \qquad (4.27)$$

where S_{UW} is the nominal tensile strength of the weld electrode. Note that this value incorporates the same factor of safety as that introduced in Section 4.4.2(a).

Ex 4.16 Welded cantilever bar, rectangular cross-section

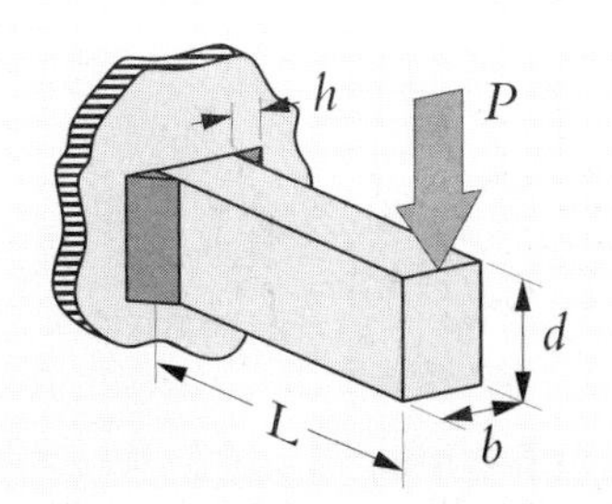

Figure 4.48 Bending of a welded rectangular bar

By a similar method to the one set out above, estimate the maximum normal and shear stresses (f_n and f_v) for the welded joint shown in Figure 4.48. Length of the bar is L = 300 mm, end load is P = 80 kN, depth of the section is d = 150 mm, and the weld size is h = 30 mm. Fillet welds are made down the vertical sides of the joint only.

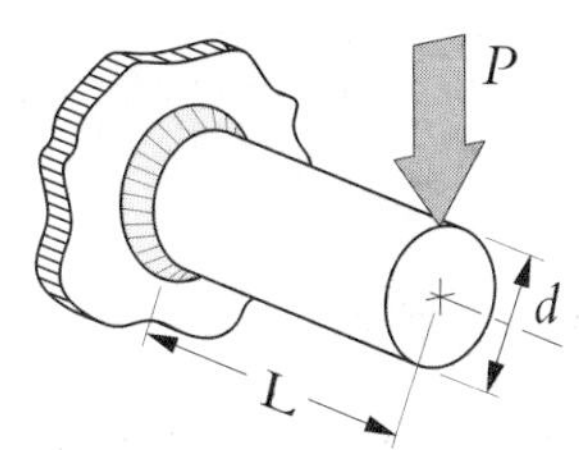

(a) *Loaded structure*

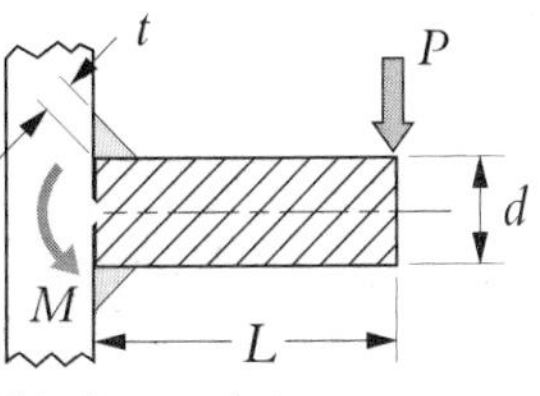

(b) *Sectional view*

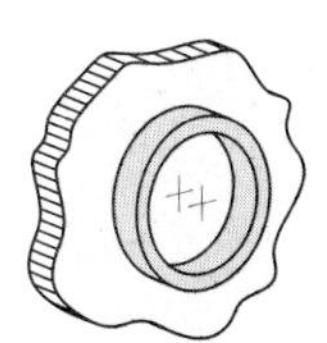

(c) *Weld replacement model*

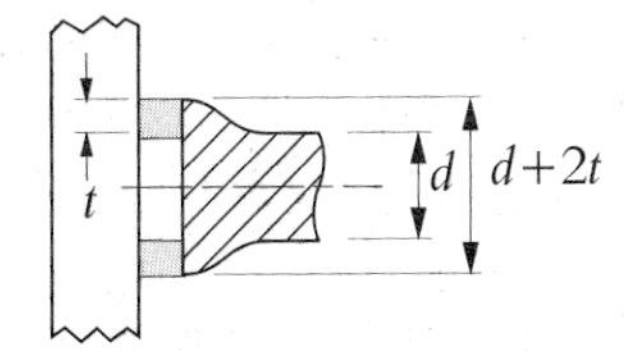

(d) *Idealised weld model*

Figure 4.47 Welded cantilever bar, circular cross-section

Ex 4.17 Failure predictor for combined stresses in welding

Show that the expression for σ_{combined} is based on the distortion energy failure predictor (*DEFP*).

For the circular cross-section cantilever of the previous Section, this becomes

$$\sqrt{\frac{16P^2L^2}{\pi^2d^4t^2}+\frac{3P^2}{\pi^2d^2t^2}} \le 0.57S_{UW},$$

or

$$\frac{P}{\pi dt}\sqrt{\frac{16L^2}{d^2}+3} \le 0.57S_{UW}.$$

Ex 4.18 Design of welded cantilever bars

(a) Derive the design inequality for a welded rectangular cantilever joint such as the one in Figure 4.48, above. Express the weld size, h, as a function of cantilever length L, end load P, weld length d, and allowable stress.

(b) Derive an expression for the weld size, h, required for a circular section bar of diameter D, length L, welded as indicated in Figure 4.46(a). Express h as a function of D and L.

4.4.8 Summary of design rules for welded joints

Butt welds

Allowable direct stress:

— use the minimum of the yield strength values of the two parent plates.

Actual direct stress:

— the weld stresses are the same as the plate stresses.

Fillet welds

Parallel loading:

Model: $$\tau=\frac{W_P}{\left(h/\sqrt{2}\right)L_e}. \qquad (4.16)$$

Require $$\frac{W_P}{L_e}\le w_P,$$

where $$w_P\equiv\left(\frac{h}{\sqrt{2}}\right)\tau_{\text{allowable}} \qquad (4.18)$$

and $$\tau_{\text{allowable}}=0.33\times S_{UW}.$$

Transverse loading:

Model $$\tau_{\text{MAX}}=\frac{W_T}{h\,L_e}(1.21). \qquad (4.23)$$

Require $$\frac{W_T}{L_e}\le w_T,$$

where $$w_T=\left(\frac{h}{1.21}\right)\tau_{\text{allowable}} \qquad (4.25)$$

and $$\tau_{\text{allowable}}=0.33\times S_{UW}.$$

Oblique loading:

— use parallel loading equations.

Bending of fillet welded joints, combined stresses:

— treat as equivalent weld section of thickness t (throat thickness);

— calculate f_n and f_v.

Require $$\sqrt{f_n^2+3f_v^2} \le 0.57S_{UW}. \qquad (4.27)$$

4.5 Assignment on joints

4.5.1 Background

Frequently in structural engineering, we require a bolted joint in which the load to be supported is a shear load (i.e. acting parallel to the plates being joined), but in which we would prefer that the bolt itself be subjected only to axial tension rather than to shear. This can be achieved by pre-tensioning the bolts sufficiently so that the friction forces between the joint members become high enough to prevent slip under the design load. This type of arrangement is called a *friction joint*.

More rarely, the situation arises in which a friction joint loaded in shear is also required to sustain an external tensile load (i.e. one acting perpendicular to the surfaces being joined, and tending to pull them apart). The resulting *tension-loaded friction joint* requires careful design because the applied tensile force will tend to reduce the joint clamping force established when the bolts were pre-tensioned. The joint clamping force must not be permitted to fall to a point where it no longer provides enough friction to support the shear load. Figure 4.49 illustrates exactly this kind of situation. It shows the arrangement of a bracket for supporting a large horizontal water pipe from a concrete wall in the basement of a multi-storey building which your company is designing.

There are a number of these brackets to be fabricated. They will be placed at intervals of 10.0 m along the length of the pipe, and fastened to the concrete wall by means of bolts fitted into expanding bolt anchors. The anchors do not perform well when the bolts are subject to lateral forces, hence the need for a friction joint. Details of the bolts and anchors are shown in Figure 4.50. Notice that a *fibre washer* is being proposed for insertion onto each bolt between the wall and the plate.

Notes

- As long as a reasonable amount of compression is maintained in the washer, it behaves in an approximately linear elastic fashion with a Young's Modulus of 550 MPa.
- Compressive stress in the fibre washer may be taken as being approximately uniform over the minimum of the two contact areas between the washer and its neighbours.
- The coefficient of friction between the compressed fibre material and steel is $\mu = 0.6$. Between the compressed fibre material and concrete the value is greater than this.
- The plate may be taken to be rigid.

4.5.2 The design problem

(a) Calculate the vertical and horizontal support reactions acting on this pipe bracket in service. Describe which part of the bracket (bolts, plate, washers) each of these external reactions acts on, and with what type of force (shear, tensile, bearing). Hence evaluate the external force imposed on each bolt of this bolted joint.

(b) Assuming that you require the vertical friction force which can be supported by the friction joint to be always greater than the vertical force which is imposed on it by a factor of at least 3.0, calculate the minimum allowable joint force, F_{jMIN} (on a *per bolt* basis).

(c) Develop an expression for the required initial bolt force, F_{bi}, on the basis of F_{jMIN}.

(d) By manipulating Equation *(4.8)* and the expression from (c) above, estimate the maximum tensile force experienced by the bolt, F_{bMAX}. Hence select an appropriate size for the bolt from the table below, using a factor of safety $F_d = 4.0$.

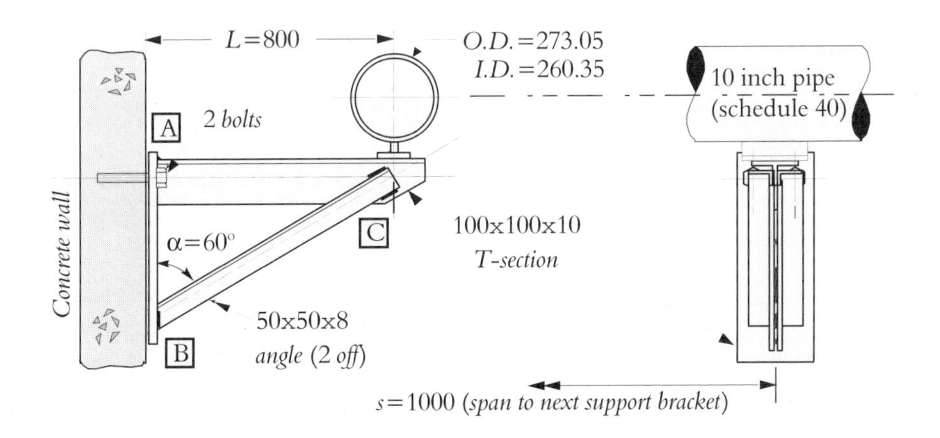

Figure 4.49 Sketch of Pipe Support Bracket, dimensions are mm (not to scale)

(e) Calculate the actual stiffnesses of the bolt and the gasket. Calculate the initial bolt force, F_{bi}. Comment on the effect of the relative magnitudes of k_j, and k_b. On the basis of these results, do you think the inclusion of washers was worthwhile?

(f) If this water pipe were to be frequently emptied and filled with water (say, several times a day), the fatigue life of the bolts might need to be considered. What measures could be used to reduce the susceptibility of these bolts to fatigue failure?

(g) Estimate (to within 20%) the torque required to pre-tighten the bolts. Refer to Section 4.2.7(c) for an appropriate method.

(h) Fillet welds are required in three different locations on the pipe bracket, as shown in Figure 4.49. Basing calculations on an arbitrary but reasonable size for these fillets of $h_A = h_B = h_C = 4$mm, decide on the required effective length of each weld, ℓ_{eA}, ℓ_{eB}, and ℓ_{eC}. Use the approach given in Section 4.4 and summarised in Section 4.4.8.

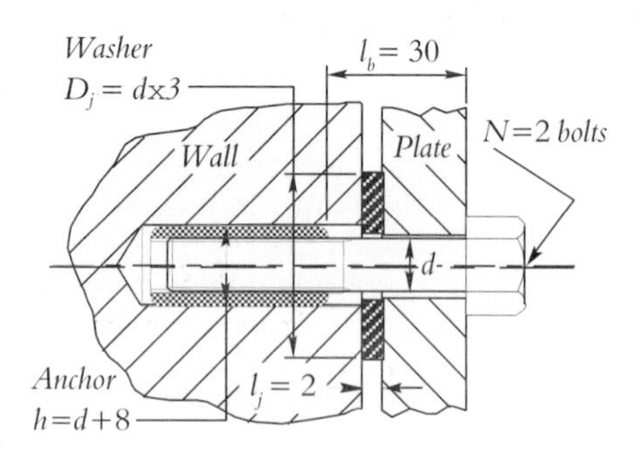

Figure 4.50 Detail of Bolted Joint (not to scale)

Table 4.5 Data for 'Unbrako' socket head Cap Screws

Thread Size	Pitch mm	B mm	A mm	Shank Area mm^2	Stress Area mm^2	Tensile Strength MPa	Tensile Strength kN	Yield Strength MPa	Yield Strength kN
M6	1.0	6.0	10.0	28.3	20.1	1240	24.9	1115	22.4
M8	1.25	8.0	13.0	50.3	36.6	1240	45.4	1115	40.8
M10	1.5	10.0	16.0	78.5	58.0	1240	71.9	1115	64.7
M12	1.75	12.0	18.0	113.0	84.3	1240	105	1115	94
M14	2.0	14.0	21.0	154.0	115.0	1240	143	1115	128
M16	2.0	16.0	24.0	201.0	157.0	1240	195	1115	175
M20	2.5	20.0	30.0	314.0	245.0	1240	304	1115	273
M24	3.0	24.0	36.0	452.0	353.0	1240	438	1115	394
M30	3.5	30.0	45.0	707.0	561.0	1240	696	1115	626
M36	4.0	36.0	54.0	1018.0	817.0	1240	1010	1115	910

B = shank diameter; A = Head diameter

Engineering Design Associates

Project Title : **Pipe support brackets**

Date 12:4:98

Page 1 of 6

Ckd. esc

PIPE SUPPORT BRACKET

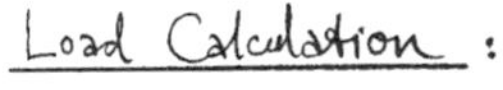

Load Calculation :

$A_W = \frac{\pi}{4} D_i^2$

$= \frac{\pi}{4}(260.35)^2$

$= 53{,}236.0 \text{ mm}^2$

$A_s = \frac{\pi}{4}(D_o^2 - D_i^2)$

$= \frac{\pi}{4}(273.05^2 - 260.35^2) = 5{,}320.4 \text{ mm}^2$

A_s (steel) $\rho = 7800 \left(\frac{kg}{m^3}\right)$

A_W (water) $\rho = 1000 \left(\frac{kg}{m^3}\right)$

$D_o = 273.05$ mm

$t = 6.35$ mm (= 1/4")

$D_i = 260.35$ mm

$W = W_s + W_W = (m_s + m_w)\,g$ (m : mass)

$= (V_s \rho_s + V_w \rho_w)\,g$ (V : volume)

$= (A_s \rho_s + A_w \rho_w)\,s.g$ (s : span)

Now $W_s = A_s \rho_s . s.g$

$= 5320.4 \text{ mm}^2 \times 10^{-6} \frac{m^2}{mm^2} \times 7800 \frac{kg}{m^3} \times 10 \text{ m} \times 9.81 \frac{N}{kg}$

$= 415.0 \text{ kg} \times 9.81 \text{ N/kg} = \boxed{4{,}071 \text{ N}}$

$W_W = A_w \rho_w . s.g$

$= 53\,236.0 \text{ mm}^2 \times 10^{-6} \frac{m^2}{mm^2} \times 1000 \frac{kg}{m^3} \times 10 \text{ m} \times 9.81 \frac{N}{kg}$

$= 532.4 \text{ kg} \times 9.81 \text{ N/kg} = \boxed{5{,}222 \text{ N}}$

$\Rightarrow W = W_s + W_w = \boxed{9{,}294 \text{ N}}$

W

Statics — Support Reactions :

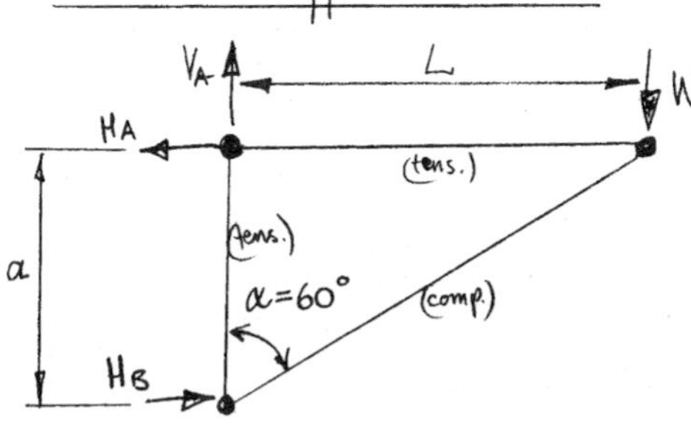

$\boxed{V_A = W}$

$H_B = \left(\frac{L}{a}\right) W$ (bearing force on wall)

$= W \tan\alpha$

$H_A = H_B$ (external force on joint)

$= W \tan\alpha$

$= 9294(\tan 60°) = \boxed{16{,}097 \text{ N}}$

Engineering Design Associates

Project Title : **Pipe support brackets**

Date 12:4:98

Page 2 of 6

Ckd. eac

(a) The three reactions are V_A, H_A and H_B.
Of these, — V_A is exerted as a shear force acting on the washers;
— H_A " " " " tensile " " " " bolts;
— H_B " " " " bearing " " " " plate.

Since the support reactions at the bolted joint are shared equally between 2 bolts, we have :-

(forces per bolt)

shear load (per washer) $= V_A/2 = W/2 = 4,647$ N		$= V$
tensile load (per bolt) $= H_A/2 = \frac{W \tan\alpha}{2} = 8,049$ N		$= F$

When discussing this with students, you may need to point out that the shear load (on the washers) will be used to determine the pretensioning force on the bolts, since it governs the minimum joint force F_{jMIN} compressing the washer. The tensile load (per bolt) is just the external tensile force on the bolted joint (denoted F in equations 13·4 & 13·5).

(b) Let us denote V_{SLIP} to be the value of V at which slipping begins (i.e. frictional resistance overcome).
We require $V_{SLIP} \geq 3.0 \times V$
At any instant, when the joint force per bolt is F_j then V_{SLIP} is limited to $V_{SLIP} = \mu F_j$
Hence

$$\mu \times F_j \geq 3.0 \times V$$

$$\Rightarrow F_j \geq \frac{3.0\, V}{\mu}$$

or $\boxed{F_j \geq F_{jMIN}}$ where $\boxed{F_{jMIN} = \frac{3.0 \times V}{\mu}}$ F_{jMIN}

Thus $F_{jMIN} = \left(\frac{3.0 \times 4647}{0.6}\right)$ N $= 23,235$ N (per bolt)

Engineering Design Associates

Project Title : **Pipe support brackets**

Date 12:4:98

Page 3 of 6

Ckd. esc

(c) Equ'n (4.9) is : $F_j = F_{bi} - F\left(\frac{k_j}{k_b + k_j}\right)$

The joint force varies from $F_{ji} = F_{bi}$ (when $F = 0$)

to $F_{jMIN} = F_{bi} - F\left(\frac{k_j}{k_b + k_j}\right)$ (when $F = F$).

$$\Rightarrow \boxed{F_{bi} = F_{jMIN} + F\left(\frac{k_j}{k_b + k_j}\right)} \quad \text{---- '}*\text{'}$$

(d) Equ'n (4.8) is : $F_b = F_{bi} + F\left(\frac{k_b}{k_b + k_j}\right)$

The bolt force varies from F_{bi} (when $F = 0$)

to $F_{bMAX} = F_{bi} + F\left(\frac{k_b}{k_b + k_j}\right)$ (when $F = F$).

Combining (4.8) and '*' (above), we obtain :

$$F_{bMAX} = F_{jMIN} + F\left[\left(\frac{k_j}{k_b + k_j}\right) + \left(\frac{k_b}{k_b + k_j}\right)\right]$$

$$= F_{jMIN} + F$$

We have previously evaluated $F_{jMIN} = 23{,}235$ N

and $F = 8{,}049$ N

$$\Rightarrow F_{bMAX} = (23{,}235 + 8{,}049) = \boxed{31{,}284 \text{ N}}$$

F_{bMAX}

Using a factor of safety, $F_s = 4.0$, we require a bolt which will not yield until

$$F_b = 4.0 \times F_{bMAX} = 4.0 \times 31{,}284 \text{ N} = \boxed{125{,}136 \text{ N}}$$

F_{yield}

From table we select M14 bolts, since they have a yield strength of 128 kN. ($\Rightarrow d \simeq 14$mm)

(e) Bolt & washer stiffnesses are given by :

$$k_b = \frac{A_b E_b}{l_b} \quad ; \quad k_j = \frac{A_j E_j}{l_j}$$

$$A_b = \frac{\pi d^2}{4} = 154.0 \text{ mm}^2 \text{ (from table)}$$

$$A_j = \frac{\pi}{4}\left(D_j^2 - h^2\right) = \frac{\pi}{4}\left((3\times 14)^2 - (14+8)^2\right) = 1{,}005.3 \text{ mm}^2$$

(see diagram, p.2)

Engineering Design Associates

Project Title : **Pipe support brackets**

Date 12:4:98

Page 4 of 6

Ckd. eac

Also, $E_b = 210 \times 10^3$ MPa ; $l_b = 30$ mm

$E_j = 550$ MPa ; $l_j = 2$ mm

$$\Rightarrow \quad k_b = \frac{154.0 \text{ mm}^2 \times 210 \times 10^3 \text{ N/mm}^2}{30 \text{ mm}} = 1.078 \times 10^6 \left(\frac{N}{mm}\right)$$

$$k_j = \frac{1005.3 \text{ mm}^2 \times 550 \text{ N/mm}^2}{2 \text{ mm}} = 0.276 \times 10^6 \left(\frac{N}{mm}\right)$$

F_{bi}

$$F_{bi} = F_{jMIN} + F\left(\frac{k_j}{k_b + k_j}\right)$$

$$= 23{,}235 + 8{,}049\left(\frac{0.276}{1.078 + 0.276}\right)$$

$$= 24\,878 \text{ N}$$

In this case we are using a relatively flexible joint material (the fibre washer), so that $k_j \ll k_b$

Here $\left(\frac{k_b}{k_j}\right) = \frac{1.078 \times 10^6}{0.276 \times 10^6} \simeq 3.9$. The implications of this are :

(i) <u>for bolt pre-tensioning</u> — since $F_{bi} = F_{jMIN} + F\left(\frac{k_j}{k_b + k_j}\right)$

$$= F_{jMIN} + F\left[\frac{1}{(k_b/k_j) + 1}\right],$$

F_{bi} is minimized by having a large $\left(\frac{k_b}{k_j}\right)$ ratio, ie. we don't have to pretension the bolts as much as would be required if k_j were greater.

(ii) <u>for bolt force amplitude</u> — since $F_b = F_{bi} + F\left(\frac{k_b}{k_b + k_j}\right)$

the amplitude of fluctuations in F_b as F changes from zero to F is

$$F_{b,amp} = \left(\frac{F}{2}\right) \cdot \left(\frac{k_b}{k_b + k_j}\right)$$

$$= \left(\frac{F}{2}\right) \cdot \left[\frac{(k_b/k_j)}{(k_b/k_j) + 1}\right],$$

which is maximized by having a large $\left(\frac{k_b}{k_j}\right)$ ratio.

If this were a fatigue-loading situation, this larger bolt force fluctuation would be undesirable.

(e) — (cont.)

(iii) for joint force amplitude — since $F_j = F_{bi} - F\left(\frac{k_j}{k_j + k_b}\right)$ the amplitude of fluctuations in F_j due to F are reduced by having a large $\left(\frac{k_b}{k_j}\right)$ ratio. This is helpful in minimizing the effects of hysterisis (i.e. loss of resilience) in the fibre washer.

In summary, it is probably not worth using the washers here since their only real benefit is to reduce the required pretensioning force, F_{bi}. The reduction is only slight: $F_{bi} = 24\,878$ N (with washers); otherwise $F_{bi}' \lessapprox F_{bMAX}$

$\lessapprox 31\,284$ N (no washers (i.e. $k_j \gg k_b$)).

The 25% reduction in F_{bi} might be of some use, but it's a marginal benefit (in this case).

(f) To improve fatigue resistance we could:

— use larger bolts (lower stresses — although this would also increase the amplitude of bolt force fluctuations)

— use a stiffer washer (see (ii) on previous page)

(g) From Lewis & Samuel, p. 206 : $T \simeq \frac{F_{bi} D'}{5}$

$\Rightarrow T \simeq 24\,878 \text{ N} \times 0.014 \text{ m} / 5$ (for M14 bolts)

$= \boxed{70 \text{ N.m}}$ (= required torque for pretensioning)

(h) Weld Sizing:

Adopt the terminology shown at right ...

Engineering Design Associates

Project Title : **Pipe support brackets**

Date 12:4:98

Page 6 of 6

Ckd. esc

All three are fillet welds. Weld Ⓐ is loaded transversely, welds Ⓒ are loaded in parallel, and the welds at Ⓑ are subject to oblique loading (for which we apply the same allowable load per unit length as for parallel loads, since this is the more conservative of the two). The approach used is to assign a size arbitrarily, and determine the corresponding required effective length of weld.

Weld Ⓐ: One weld line; max length = 150mm;

Transverse load: $F_T = 16097$ N ($= W_{tank}$)

Try 8mm fillet: $h_A = 8$ mm

$\Rightarrow f_t = \frac{h\,\tau_{ALL}}{1 \cdot 21} = (80 \times h)\ \frac{kN}{m} = 640$ kN/m (for $\tau_{ALL} = 96$ MPa)

$\Rightarrow l_{e,A} = F_T / f_t = \frac{16 \cdot 1 \text{ kN}}{640 \text{ kN/m}} = 0 \cdot 0251$ m (i.e. only 25 mm)

$\Rightarrow$ could use $h_A = 4$ mm; $l_{e,A} \geqslant 50$ mm

Weld Ⓑ: Four weld lines; Parallel load:

$F_P = \frac{W}{\cos\alpha} = \frac{9294}{\cos 60°} = 18588$ N

Try 4mm fillet: $h_B = 4$ mm

$\Rightarrow f_p = \frac{h\,\tau_{ALL}}{1 \cdot 41} = (68 \times h)\ \frac{kN}{m} = 272\ \frac{kN}{m}$ (for $\tau_{ALL} = 96$ MPa)

$\Rightarrow l_{e,B} = \frac{F_P}{f_P} = \frac{18 \cdot 6 \text{ kN}}{272 \text{ kN/m}} = 0 \cdot 068$ m (i.e. only 68 mm)

O.K., use $h_B = 4$ mm; $l_{e,B} \gtrsim 80$ mm (i.e. 4×20mm)

Weld Ⓒ: Two weld lines; Oblique load

$F_P = W = 9294$ N. Try 4mm fillet; $h_c = 4$ mm

$\rightarrow f_P = 272$ kN/m, as for case Ⓑ

$l_{e,c} = \frac{F_P}{f_P} = \left(\frac{9 \cdot 3}{272}\right) = 34$ mm

O.K., use $h_c = 4$ mm; $l_{e,c} \gtrsim 40$ mm (i.e. 2× 20 mm)

...

5

REVIEW: STRUCTURAL INTEGRITY OF ENGINEERING SYSTEMS

"Isn't it astonishing that all these secrets have been preserved for years, just so we could discover them." Wilbur Wright, June, 1903 [5.1]

5.1 Designing engineering systems: the 'big picture'

The ultimate objective of all design is to construct some system to repair some dissatisfaction with the status quo. As architect and design theorist, Horst Rittel (1987), suggests:

"Design takes place in the world of imagination, where one invents and manipulates ideas and concepts instead of the real thing..."

While Rittel is referring to system design in general terms, our concern here is to do with one specific aspect of systems, namely their total structural integrity. In the context of this general discussion we define the following terminology:

> *Generic engineering components* (*GEC*) are the basic building blocks of design for structural integrity. These are the components we deal with in Chapter 3. It is important to recognise that although we give several real-life examples of the application of design rules developed for *GEC*, the notion of the ideal *GEC* is that of an engineering abstraction from reality. We rarely (if ever) find in real life engineering structures the broadly idealised columns, shafts, beams and springs described in Chapter 3.
>
> *Structural elements* are the clearly identifiable *GEC* in any *component* or *assembly* of complex engineering structures. We find/identify these *GEC* by the process of *structural distillation* described in Section 1.3.
>
> A *component* of an engineering product or assembly is a single cohesive structure. A *component* may also contain a collection of structural elements, interconnected by one or other of the methods described in Chapter 4. A welded truss of a bridge, or a shell and tube heat exchanger in a chemical process plant are both *components* of assemblies. The wheel of a bicycle is an assembly of several interconnected parts that may be taken apart and re-assembled without damage to the connections.

As we have already noted, we apply engineering wisdom and judgement to make the following determinations:

1. What are the *key load-bearing components* of our structure? These are the structural elements that ensure proper functioning of our system.
2. What are the various *GEC* that might be used to adequately represent these load-bearing components? These result from a *structural distillation* of the structure.
3. What are the loads the structure and its *key components* will need to sustain in a *worst credible accident* ?

Design wisdom about structural integrity is best acquired through constant practice and through the careful study of design paradigms. Henry Petroski refers to John Roebling as a good case in point:

"Roebling achieved with uncommon success what other engineers had agreed could not be done. His principal advantage was not a more sophisticated theory or more careful calculations. Roebling achieved his success by concentrating his design judgement on how his bridge might fail. ... These were not merely design-theoretical speculations after the fact, for in his own writings, Roebling himself provided ample evidence that it was through studying failures

5.1 Quoted by Fred C. Kelly (Ed. 1996) in 'Miracle at Kitty Hawk: The letters of Wilbur and Orville Wright', New York: Da Capo

that he achieved success." Henry Petroski: "John Roebling as a Paradigmatic Designer", in Design Paradigms, 1994

Once the structural integrity of the individual structural components of the system have been determined, we need to work through a careful balancing of loads and stresses in the whole system to ensure that each component takes its proper share of the load.

American humorist, lawyer and physician, Oliver Wendell Holmes, clearly knew a thing or two about optimisation as evidenced by his famous 1858 work, *"The One-Hoss Shay"*, part of which will suffice for our discourse:

> *"Have you heard of the wonderful one-hoss shay,*
> *That was built in such a logical way*
> *It ran a hundred years to a day,*
> *And then, of a sudden, it — ah, but stay,*
> *I'll tell you what happened without delay,*
> *...*
> *How it went to pieces all at once, —*
> *All at once, and nothing first, —*
> *Just as bubbles do when they burst.*
> *End of the wonderful one-hoss shay,*
> *Logic is logic. That's all I say."*

It could be argued that, ideally, just like the *one-hoss shay*, we should design and build products in such a way that all of their components reach the end of their useful life at the same time. This approach would guarantee that the customer got best use out of every component. Yet the total result would be an economic disaster. This is the real point of Holmes' poem, originally intended as religious and political satire.

There seem to be two different quality measures embedded in Holmes' description of the magical *one hoss shay*.

(a) It lasted a hundred years, which may have been a desirable quality feature in the days of the horse and cart, but in times of rapid technological change, a life of this length is excessive by at least one order of magnitude. It must have been heavily over designed, with concomitant excessive cost of manufacture.

(b) The designer had achieved an absolutely uniform degree of structural integrity across every component in the system. While this may be a desirable condition, there may be no prior warning of the catastrophic failure the system will suffer at the end of its life.

In any event, both of these quality measures are associated with excessive design and manufacturing costs.

A more pragmatic approach to design demands that products have some components designed for long-term and some for limited-life. The limited-life components get replaced or reconditioned at maintenance intervals. These are the batteries, tyres, windscreen wiper blades and brake pads of automobiles, the seals on water valves, and light globes in most appliances. The longer-term components, such as the sealed compressor units on refrigeration equipment, are expected to last the *full life* of the product. In this context we have no clearly defined measure, or *benchmark* for the meaning of *full-life*. There is great variability of this measure between products.

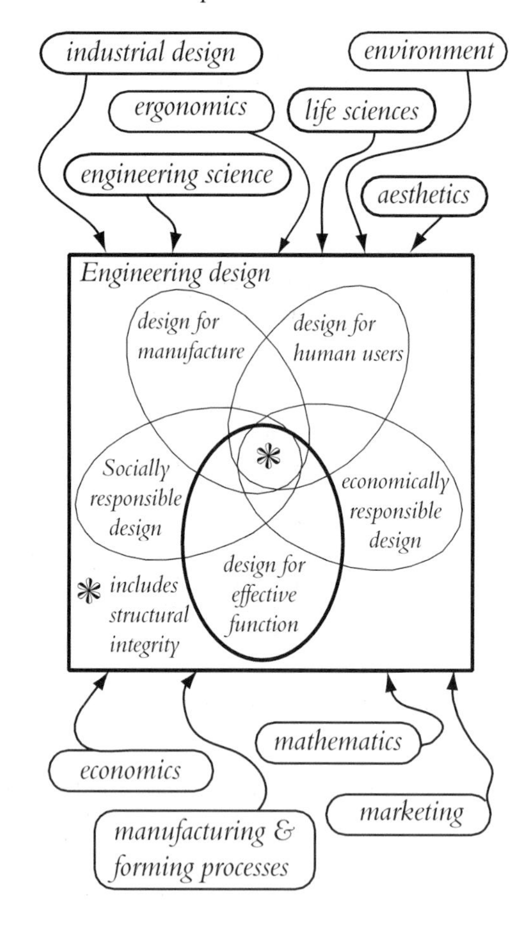

Figure 5.1 Structural integrity in the wider context of design

For most products, we regard *full-life* as the period for which the value of operating it exceeds the operating costs. In the case of automobiles, this period used to be of the order of decades, but as maintenance costs and insurance premiums have increased, the *full life* period of automobiles has become correspondingly much shorter. Furthermore, the life of a product is heavily influenced by the rate at which specific technologies change. For computing equipment *full life* is likely to be about three years, and for washing machines about ten years. Bearing these matters in mind, the style of design in which all components last the same length of time is economically quite unacceptable for most consumer products. After all, "who would be prepared to pay for a battery, a windscreen wiper blade, or a brake pad that all lasted as long as the engine of the automobile?" Alternately, "who would purchase an automobile that only lasted as long as the battery, windscreen wiper blade or brake pads?"

In contrast to component *life* in a product or system, its total structural integrity will benefit significantly from a *one-hoss shay* style of load sharing. While this approach is not always possible, it is certainly a desirable design goal to aim for.

Figure 5.1 places structural integrity into the wider context of design. A cynical observer might be tempted to suggest that, according to the apparently *pantheistic view* presented here, "everything that engineers do is design". We certainly do not embrace that representation of design. However, the process of formal planning and evaluation that constitutes design, touches all aspects of every product. Moreover, since designers are people who, *"by ambition imagine a desirable state of the world"*[5.2], their eclectic wisdom needs to draw on all the disciplines indicated in Figure 5.1.

5.1.1 The structural 'audit'

We have already established the special characteristics of some simple, generic engineering components, albeit in isolation. In presenting the discourse and problems of Chapters 3 and 4 we have taken some care to focus on these special characteristics in each case. Naturally, engineering designers are rarely, if ever, confronted with engineering structures in such a *compartmentalised* way. Experienced engineers will need to be able to dissect the structure into its elementary substructures *on the fly*. We have provided some simple examples of this dissection process in the section on *structural distillation*. In this section we examine some engineering systems, with the objective of identifying within them recognisable, simpler generic engineering components. The objective of this process is twofold.

1. If possible, we need to establish the generic components that constitute our engineering system. This is an essential first step in planning the structural analysis. Moreover, in those cases where the dissection indicates the presence of more complex components than those offered in our elementary list in Chapters 3 and 4, we need to make reference to standard texts on advanced structural analysis (see for example Young, 1989).

2. We need to build *engineering wisdom* through experience with familiar systems. This style of learning is largely associative[5.3]. The process of associating the *strange*, the newly acquired knowledge of generic engineering component behaviour, with the *familiar*, the engineering system of our personal experience, is a powerful reinforcement of learning.

This approach is entirely in tune with the process of acquiring engineering judgement through the study of paradigms[5.4]. We refer to the *wisdom* acquired through this process as *looking at systems through the engineer's eye*. We envisage that most of us have close and direct familiarity with some engineering systems. This familiarity is the result of experience with professional, household or recreational consumer products. By carefully examining the operation of such a product, we can perform its *structural audit*. We can *look at it through the engineer's eye* and find embedded in it generic engineering components and perhaps even estimate the loads on them resulting from the *worst credible accident*.

This process doesn't involve synthesis, but merely the building of experience. Ultimately, we might use the experience thus acquired, in synthesis through *association*, since all design synthesis is evolutionary and the result of our accumulated experience.

There is an urban legend surrounding Sir Henry Royce (1863-1933), co-founder of Rolls-Royce Ltd. Reputedly, he was able to *run* one of his engines in his imagination, examining all the components as they operated, and after some period *take it apart* mentally and examine the

5.2 Rittel, ibid.

5.3 Associative learning objectives in the ***'affective domain'*** have been identified by Kratwohl, Bloom and Masia (1964)

5.4 Petroski, op. cit.

components for wear. The Florentine sculptor and architect, Michelangelo [full name *Michelangelo Di Lodovico Buonarroti Simoni*] (1475-1564), just like many other sculptors, imagined the structure of a sculpture before he commenced his work on a block of marble. A symbolic representation of this process is evidenced in his famous unfinished sculpture *four slaves*, in the Galleria dell' Accademia in Florence. In this piece the human figures of the slaves appear to be struggling to emerge from the surrounding marble monolith.

What we are promoting here is a notion of the powerful human capacities to *visualise* and to *imagine* in the virtual world of the mind. The power of *gedankenexperiment*, or thought experiment, was used most effectively in abstract reasoning by Einstein and other physicists. In dealing with relativity Einstein imagined to himself what would happen if he were to sit on a light ray and travel with the speed of light– would time stand still?

"Would an observer stationary at a point on the earth's surface see a distant flash of lightning at the same instant as a second observer travelling in a train towards the place where the flash had occurred?" (Einstein, 1936; Einstein and Infeld, 1938)

Ex. 5.1 Some 'gedankenexperimente'

(a) Imagine you are in a train travelling towards a clock set at exactly 12 noon. If the train moves at the speed of light, how would you expect to see the hands of the clock moving? Would your observation of time on the clock change? If your train were travelling at the same speed away from the clock instead of towards it how would you see the clock hands move?

(b) Imagine that you are in a spacecraft in orbit around the earth. You light a candle and observe the flame. What shape is it?

In what follows, we make extensive use of *gedankenexperimente* in the structural audit of systems.

Figure 5.2 shows a hydraulic cylinder. This is a device for producing axial force (and motion) along the line AB, when pressurized fluid is introduced through ports P and Q on either side of the piston (shown dashed). We can identify various generic engineering components (for example, *beam*, *shaft*...) which are embodied within this device during its normal operation:

- the cylinder, within which the piston operates, is clearly a type of *pressure vessel*;
- the piston-rod, used to drive the load attached to its end at A, can operate either as a *tensile member*, when pulling on the load (contracting), or as a *column*, when pushing (extending);
- the end plates of the cylinder may be conservatively simplified to *beams*, although, in reality, they are more complex structural elements, *flat plates*, beyond the scope of our discussion here;
- the piston is a *uniformly loaded circular disc*. This is not a simple structural element considered in Chapters 3 or 4, but it can be easily analysed by reference to Young (1989);
- the four bolts holding the end-plates on the cylinder, act both as simple *tensile members* and also, in conjunction with the end plates and cylinder, as a *bolted sealing joint*;
- the connections for the load at A and B can be analysed as *shear connectors*.

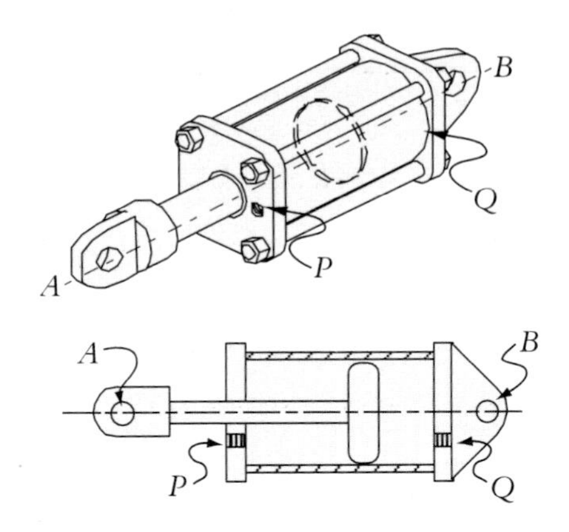

Figure 5.2 Hydraulic cylinder

We now examine two more simple engineering systems, and tabulate the *structural audit* in Table 5.1.

Example 1: Rotary clothes dryers

Figure 5.3 shows a *rotary clothes dryer* (an icon of Australian suburban life). It is a device for supporting damp items of woven material after washing, for drying in the wind and sun. To aid in the *pegging out* operation, a handle (B) is provided for winding the central support (C) down and up by means of a rack-and-pinion gear (not shown). The upper frame (D) of the dryer can rotate within the base (A).

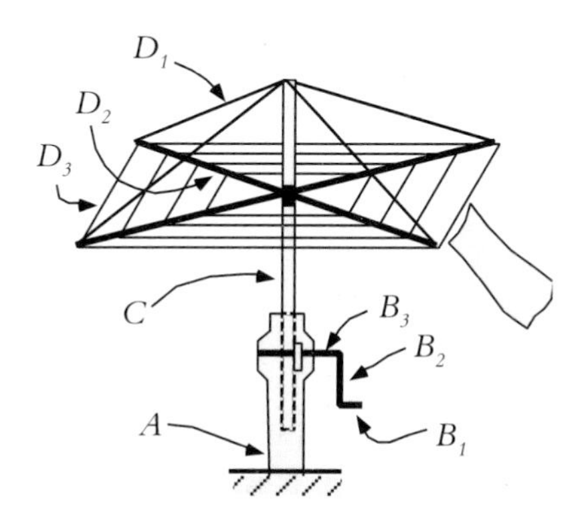

Figure 5.3 Rotary clothes dryer

In one embodiment of this design, the upper frame is made up of four steel tubes (D_2), held in a crosspiece mounted on the central support (C). Guy-wires (D_1) support the tubes in the vertical plane. In the horizontal plane the frame is stiffened by the wires (D_3) used for hanging clothes. Since the upper frame of the dryer is raised during operation, the dryer is colloquially referred to as a *clothes hoist*.

Example 2: Rotary mixer blades

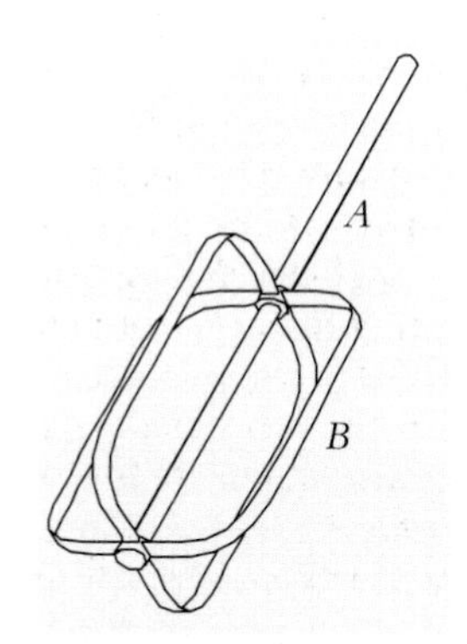

Figure 5.4 Rotary mixer blade

Table 5.1 Structural audit for rotary dryer and mixer blades

Nature of loading	Generic Component [class of element]	Rotary clothes dryer element								Mixer blade	
		A	B_1	B_2	B_3	C	D_1	D_2	D_3	A	B
Fatigue											
Impact		X				X					
Contact											
Bending	Beam	X	X	X		X		X		X	X
Tension	Tensile [rod]						X		X		
Buckling	Column					X		X			
Torsion [+Bending]	Shaft				X					X	
Pressure [+Bending]	Vessel										
Shear [+ Contact]	Shear connector [Pinned joint]										
Shear or Tension	Welded joint										
	Other										

Figure 5.4 shows one of two inserts (mixer blades) for an electric hand mixer (a common kitchen appliance). During normal use, each insert is rotated at high speed about its long axis while immersed in various kinds of viscous foodstuff. This simple device still qualifies as a *system*, since it is made up of several co-operating elements, combined by welding, into a single cohesive structure.

Table 5.1 lists all the generic engineering components and loading characteristics considered in Chapters 2, 3 and 4. We call these components and loading conditions the *structural varieties* of our engineering system. When a specific *structural variety* is active in our engineering system, its presence is identified by an **x** in the relational matrix of Table 5.1.

For the two simple examples shown, some of the *structural varieties* identified in Table 5.1 do not appear to be active. This is as it should be, and for most systems considered for structural audit, only a reduced set of *structural varieties* will be active. However, in all structural audits, the complete set of *structural varieties* of Table 5.1 need to be considered. This process will act as a *memory jogger* in the structural audit of more complex systems.

Ex 5.2 Thought-experiments in structural auditing

In performing a *structural audit* of a system, it is often useful to *imagine* the system being composed of some compliant material (say rubber) and imagine how the system might deform under the various loads imposed on it. In this way, the various elements of the structure that carry the imposed loads (the ones that suffer the most deformation in our *thought-experiment*), are quickly identified. Consequently, the subsequent identification of the *structural varieties* in the system becomes simplified.

Figure 5.5 shows a series of ten artefacts. You may be closely familiar with some of them. For the purposes of *structural auditing* we will regard these systems as engineering systems. Complete a structural audit of all these systems by drawing up a table similar to that of Table 5.1 for each system.

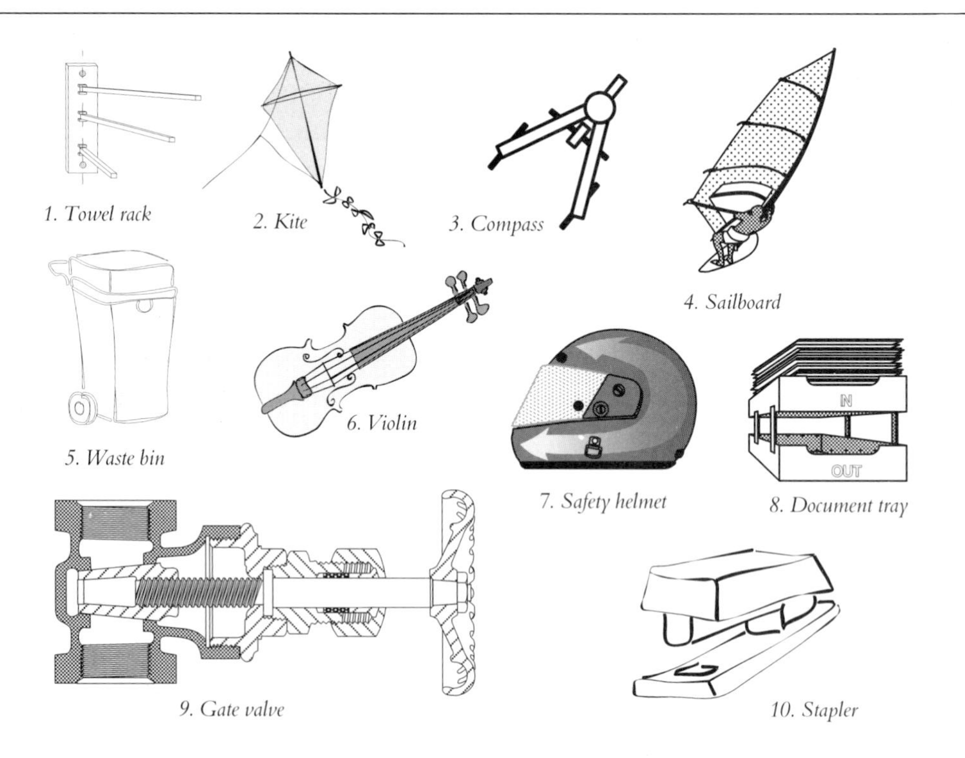

Figure 5.5 Artefacts for structural audit

In each case identify the various generic engineering components of the system, as we have done for the rotary dryer and the mixing blade examples above.

5.1.2 Some simple insights on structural analysis

The following set of examples require some understanding of the basic principles involved in structural analysis. These examples are offered in the form of a series of searching questions. Some of them may be answered by direct application of ideas discussed in the relevant sections of Chapters 1 to 4, while others require some thoughtful (*insightful*) extensions of these ideas. However, in all cases the answers to these questions rely on understanding rather than on simple recall of facts or figures.

1. Why does a piece of chalk fail along a 45° helix when twisted in pure torsion ? What material characteristic does this distinctive failure of chalk indicate?

2. Why do sausage skins split along their length (refer to Figure 5.6) when heated during cooking?

Figure 5.6 Sausage 'sizzle'

3. Why is the Sydney Harbour Bridge (refer to Figure 5.7) shaped in such a distinctive way? Please consider structural issues only.

Figure 5.7 Schematic shape of the Sydney Harbour Bridge

4. What is the purpose of an outrigger on a canoe? (Refer to Figure 5.8 to remind you of what an outrigger looks like.)

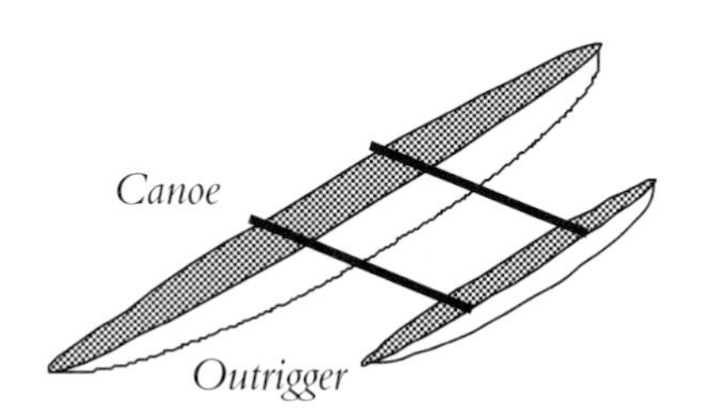

Figure 5.8 Outrigger on a canoe

5. Why is the Eiffel Tower shaped in such a distinctive way? Please consider structural issues only. Refer to Figure 5.9 to remind you what the Eiffel Tower looks like.

   This renowned Paris landmark is a metal lattice structure, designed and built for the Centennial Exhibition of Paris in 1889, to commemorate the French Revolution. Until 1930, at 300 m height, it was the tallest building in the world.

Figure 5.9 Eiffel tower

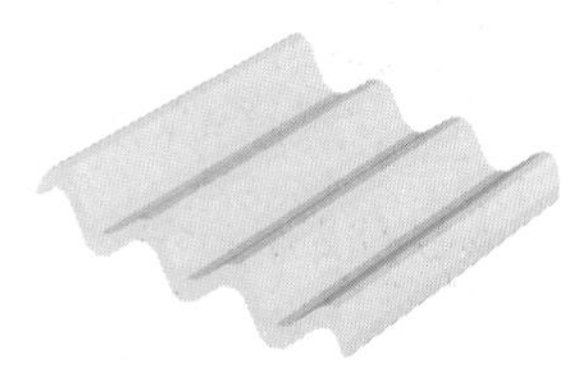

Figure 5.10 Corrugated iron

6. Corrugated sheets of iron are commonly used as roof covering. Why is corrugated iron corrugated? Please offer a structural explanation. (Refer to Figure 5.10 to remind you of what corrugated sheet iron looks like.)

7. A pump is discharging simultaneously into a cylindrical (A) and a spherical (B) pressure vessel, as shown schematically in Figure 5.11. Both vessels have the same diameter D and shell thickness. If the pressure is allowed to increase gradually but without limit, which vessel will rupture first, and why?

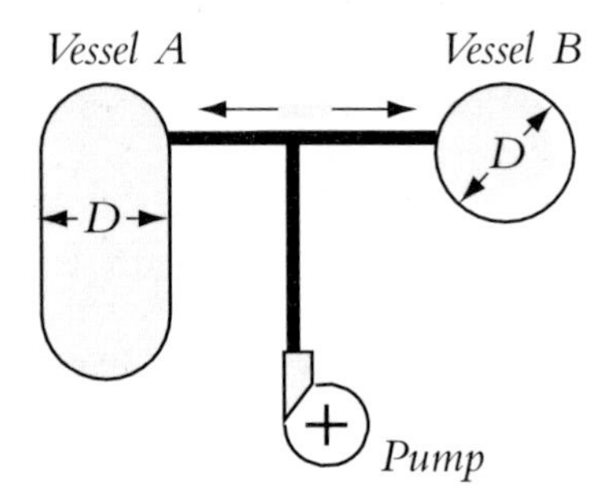

Figure 5.11 Pressure vessels pumped up simultaneously (schematic)

8. A gear drive is a transmission device for gaining either mechanical advantage (speed reduction set) or velocity ratio (speed *step-up* set). The nature of the transmission is determined by the choice of input-output direction. Figure 5.12 shows a reduction gear set, with the smaller gear the *input* and the larger the *output*.

   Why is the shaft driving the input gear smaller in diameter than the shaft driven by the output gear?

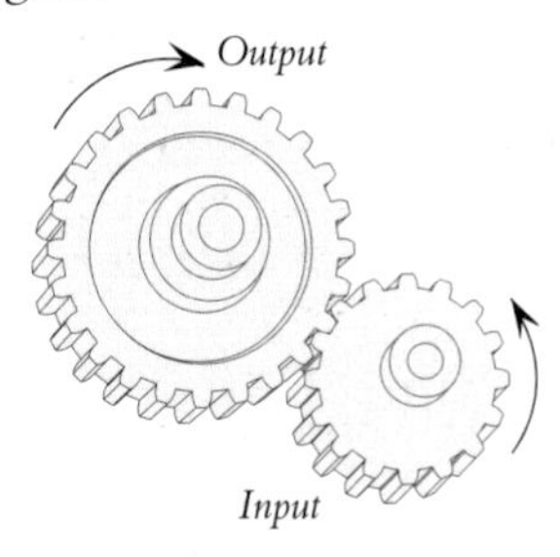

Figure 5.12 Reduction gear set

9. Figure 5.13 shows a type of pressure vessel colloquially referred to as a *bullet*. Discuss where the maximum shear stress occurs in the shell of a horizontal pressure vessel such as this. Sketch the bending moment diagram for this vessel, assuming that it contains a liquid of specific gravity 1.0 under pressure.

   For the purposes of this simple exercise, you may assume that the supports act as *simple supports*, and therefore they do not impose any horizontal forces on the bullet. Long vessels on two supports always have one sliding support, to allow for thermal expansion. In this case the sliding support is indicated schematically by the rollers under one support.

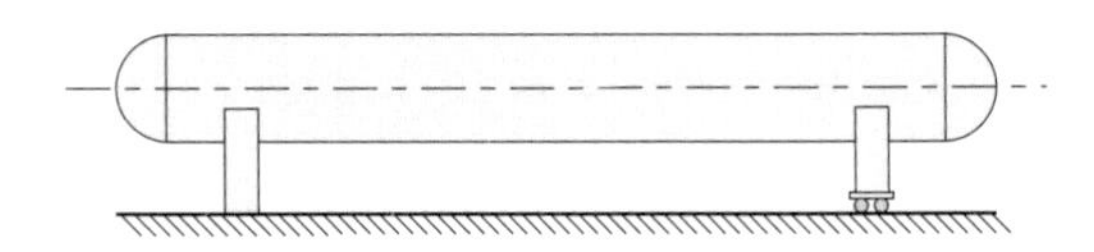

Figure 5.13 Bullet

Answers

1. The answer to this question is given in Chapter 2, Ex. 2.2.
2. As the sausage is heated, the contents expand under pressure generated by the vapourised fluid components. The sausage skin acts as a pressure vessel and eventually fails in tension as indicated, since the *hoop stress* in the circumferential direction is the largest stress acting.
3. The Sydney Harbour bridge acts as a *beam* carrying a uniformly distributed load. The shape of the truss reflects the natural bending moment diagram for such a beam.
4. The outrigger of the canoe acts to provide a restoring moment against *rolling* of the canoe about its longitudinal axis.
5. The Eiffel Tower is essentially a *cantilever* beam with a vertical axis. Its shape also reflects the bending moments acting on it as a result of wind loads (*worst credible accident*).
6. The corrugations of iron sheet displace material away from the neutral plane, thereby

increasing its second moment of area about the mean plane of the corrugations.

7. The circumferential stress, or *hoop stress*, in the cylindrical vessel *A* is always larger, than the longitudinal stress in the spherical vessel *B*. Hence the cylindrical vessel *A* fails first.

8. The power into the gear set is approximately the same as its power output. There will be some minor losses due to gear transmission inefficiency. Since power transmitted is the product of torque and angular velocity, the slower gear (*output*) will have a larger torque acting on it, and its output shaft will need to be correspondingly larger than the input shaft in order to cope with the larger stress.

9. Figure 5.14 shows the bending-moment diagram for the bullet.

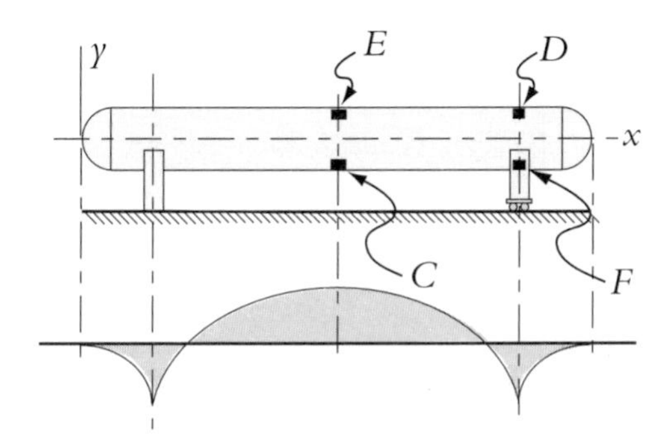

Figure 5.14 Bending moment diagram for bullet

Clearly, the largest shear stress acting on the bullet is a combination of the largest and smallest principal stresses according to

$$\tau_{max} = \frac{\sigma_1 - \sigma_3}{2},$$

The pressure stresses are:

$$\sigma_{Hoop} = \frac{pD}{2t}; \ \sigma_{axial} = \frac{pD}{4t}; \ \sigma_{Radial} = -\frac{p}{2};$$

where p, D and t are the internal pressure, the diameter and skin thickness of the vessel respectively.

Without the bending loads, the maximum principal stress, σ_1, is the circumferential *hoop* stress, $\sigma_1 = \sigma_{Hoop}$, and the minimum principal stress, σ_3, is the radial stress across the skin of the vessel, $\sigma_3 = \sigma_{Radial}$. The axial pressure stress, σ_{axial}, does not enter into the expression for τ_{max}. However, the bending stress, $\pm\sigma_B$, generated by the bending loads in the skin of the vessel, needs to be added to the axial stress, σ_{axial}.

The three possible cases depending on the relative magnitude of the bending stress, σ_B, are explored thoroughly, using Mohr's circle diagrams in Section 3.4.6. The worst-case locations for σ_B are either at mid-span (locations *E* and *C*) or at the supports (locations *D* and *F*). If σ_B is relatively small (Case 0, $\sigma_B < \sigma_{axial}$), then τ_{max} will be unaffected by σ_B, and will be *uniform everywhere* on the cylindrical shell of the vessel.

If, however, σ_B, is relatively large (Case 2, $\sigma_{acxial} - \sigma_B < \sigma_{Radial}$; which subsumes Case 1, $\sigma_{axial} + \sigma_B > \sigma_{Hoop}$), then the worst-case location for τ_{max} will be at locations C or D, wherever the *tensile* bending stress is greatest.

With these relatively simple examples of structural insight we have tried to illustrate the nature of design thinking. None of these examples relies on any deep appreciation of the mathematical theory of elasticity or mechanics. They simply reflect the practical application of the modelling ideas explored in Chapters 2, 3 and 4.

In most institutions where engineering design is taught, assessment of the understanding of the discipline is commonly evaluated by guided design exercises and *open book* examinations. These examinations permit reference to texts and notes during the examination. Unfortunately, the process is still severely time constrained, resulting in some unexpected responses from students.

It has been conjectured that these responses are the result of a mental regression from abstract thinking to a more concrete mode of thinking, brought on by the threat of the examination environment (see for example Samuel, 1984). Alternatively, it is possible that the unexpected answers listed in the following section result from the nature of competitive training experienced in schools, where rapid, rather than thoughtful, responses are rewarded.

5.2 Some 'radical new theories' in engineering design

"2 + 2 = 5, but only for very large values of 2". Chris Dowlen, Southbank University, UK, quoted at ICED99, Munich

In an examination students were asked to evaluate a thermodynamic formula due to Nusselt. The following creative response from one student indicates a mind perhaps better suited to art than science:

"The difference between Nusselt's expression for vapour condensing on horizontal tubes and the correct expression can be accounted for when examining Nusselt's home life.

He was not a happy man. Often he would step out in the cold morning air to go to his lab and curse the day he ever started analyzing[sic] and experimenting with heat transfer. He would start to think of his wife, Mrs. Nusselt, and their unhappy marriage, that was doomed to drag on for the rest of their days. He remembered the early days when passion was sparking and the heat transfer between their two young warm bodies was a function of the log mean temperature.

It was at their last night together when in frustration at his life he uttered that famous expression

$$h = 0.728\sqrt{\frac{g\rho(\rho-\rho_v)k^3 h'_{fo}}{hD\mu\Delta T}},$$

thus symbolising that life was not perfect but just an approximation of existence.".

The following answers are categorised according to the question to which they are *radically unusual* responses. In each case, apart from minor editorialising, the text is presented as written in the examination script book.

Why do sausage skins split along their length when heated ?

"because sausages can be thought of as a thin column. We know that max. stress for longitudinal direct is half of centipedal[sic]: i.e. $\sigma_L = q/2L; \sigma_C = q/L\,(q = \text{pressure})$ *. I don't know if these are the exact equations, but the same theory applies."*

"because the plasticity of the skin has already yielded as subjected to heat"

"skin is brittle: material fails in mode of maximum tension."

"column failure - buckles along length"

"it will split in the length, because the skin is being 'stretched' to hold the meat. Thus its 'grains' are longitudinal so it will split in length"

"long thin molecules have less resistance to shear than direct strain"

"skin property: cannot resist radial expansion"

"if the sausage is considered as a column, the reason for the longitudinal failure can be more easily explained.

The reason for the skin splitting at all is because it does not expand at a similar rate as the sausage meat under heat conditions. Hence, the sausage meat inside is effectively being compressed because the outside skin is resisting its expansion.

Hence the 'column' will fail in compression, under which conditions columns fail by buckling along its[sic] length."

"we know that the expansion is proportional in each side: i.e. if we have a cube it will be expanded in the same distance in all directions after being heated up. Since the sausage is sort of a rod, one side is longer than the other. Thus the longer side will expand more relatively than the shorter side. Owing to this it will split the sausage skin in the direction of its length."

Why is corrugated iron corrugated ?

"Corrugated iron is corrugated to distribute the force applied to it over a greater surface area"

"to take the loads the corrugated iron may be subject to."

"because it is able to reflect the sun-ray when used for roofing"

"so it can be used for channelling water in rain."

"to strengthen the material. The corrugations are in fact induced stresses in the iron. Therefore low sttresses put on the sheet will have no effect, unless they are larger than the induced stress."

"because corrugated iron has a better strength since the corrugation refers to dislocations of the structure of iron."

"obviously to increase its strength in some way or form, otherwise all the extra material required would be a waste of money."

"so that it can disperse the sunlight, because too much heat can cause oxidation and so this leads to corrosion of the iron."

"so that the force exerted on the sheet is divided and runs down the sides of the corrugations."

"to relieve stress concentrations."

"to prevent [water] dripping at nail holes."

"to allow expansion when heated."

"because cold work strengthens the steel."

"because the corrugations act like nodes in a long column and therefore prevent buckling of the sheet."

"since iron is brittle, we would want iron to have excellent wear resistance as well as high vibration damping. By corrugating iron the iron keeps its strength, but becomes more wear resistant and becomes a bit more ductile to withstand more (bending) stresses."

"so that it may be used in situations where the corrugations benefit the task for which the material is to be used, and iron and corrugation complement each other, since iron can be relatively easily corrugated."

"the material of corrugated iron is ductile and so it will fail in shear before it fails in tension. When the iron is loaded in tension or compression, the maximum shear stress will occur on a plane inclined at 45° to the direction of the applied stress. By making the iron corrugated more material is placed along these planes of maximum shear stress. As a result the sheet has a greater resistance to shear failure."

"because it will enable it to elongate and contract before failing due to tensile or compressive load."

"because it can give a strong bonding to cement when used in reinforcement."

"because it is not un-corrugated iron."

Which vessel will rupture first ?

"since the cylinder has a larger volume, the pressure can vary and distribute itself. The sphere will fracture first."

This answer, one of many similar ones, is representative of the main style of misunderstanding demonstrated by students when answering this question.

Why is the output shaft of a reduction gearbox thicker than the input shaft ?

"as it is a much larger shaft (i.e. has longer bearing span)-than the input. Diameter must be increased with length so that the chance of failure due to buckling, lateral deflections, excessive slope at bearing and excessive vibration is not increased."

"to reduce deflection: better for precise work."

"This may be explained by considering Torque = $I\alpha$ *(*I *= inertia and* α *= angular acceleration). Torque is constant through both shafts (input and output), but angular acceleration varies due to speed reduction. Therefore the output shaft with smaller acceleration, but subject to the same torque, requires larger inertia, hence larger diameter."*

" since output shaft holds a much bigger gear than the input shaft. Therefore the output shaft must be larger so that fracture of the output gear will not happen."

"input shaft suffers varying stress from different torque applied from bending, if critical speed is reached and vibration occurs. The output shaft encounters a constant rotational stress (from torque) and shear stress at fixed point."

"because on the input side the shaft has high speed and during speed changes the sudden changes could harm the shaft. Therefore it is made thin to absorb the sudden torque by twisting to some extent, which it could not do if it was thick."

"Since the shafts are connected by some sort of cog system, the distance travelled by each shaft would have to be the same, but in order to allow the output shaft to rotate more slowly, its circumference would need to be greater." [this concept is known to us from many such answers as the *law of conservation of distance*, or *LCD*]

"if the input shaft is smaller than the output shaft, the input is greater while the output is smaller. This is because the greater the radius, the slower the speed." [another version of *LCD*]

"the fatigue strength of a thinner specimen is better than a larger specimen. Since the input shaft experiences greater fatigue stresses we need a thinner shaft."

"The output shaft has a greater area and therefore requires more time to turn one revolution than the input shaft. Because it requires more time, it travels slower, therefore reducing the speed. Therefore the input shaft is thinner."

"The whole idea is to reduce the speed. Therefore the speed of the output shaft will be slower than the input shaft by reducing the diameter of the input shaft."

"Consider a bicycle: pedals do not have to go round nearly as fast as the rear wheels travel. i.e. Low input to high output. In the gearbox we have the opposite situation. So the input should be larger than the output."

"because the torque at the output produces greater power out."

Bending moment diagrams for the bullet

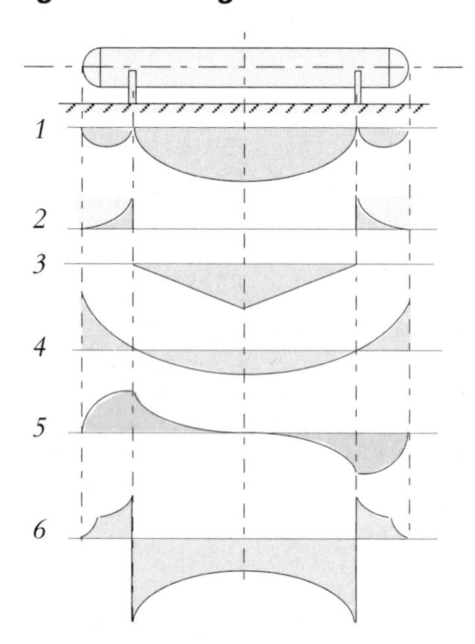

Figure 5.15 Some of the more 'radical' bending-moment diagrams offered by students.

Figure 5.15 shows only some of the many variations in answer offered by students to this problem.

The responses quoted in this section are offered by their authors with considerable conviction. They are never prefaced by a statement of uncertainty. It seems clear, that for many students presenting for examination, the answer *"I don't know"* is totally unacceptable. These *amusing* responses have been offered as a cautionary tale for both students and educators in engineering design.

In science we have come to accept the *law of parsimony* or *Occam's razor* for explaining physical phenomena. Occam's razor is a notion attributed to the theologian and scholar, William of Occam (1285-1349), who recommended that in the explanation of physical phenomena *entities are not to be multiplied beyond necessity*. Modern science interprets this notion by ensuring that wherever possible the explanations of physical phenomena are based on the simplest, or *least fanciful*, conjectures.

Students of engineering easily adapt to the learning environments of the engineering sciences. These doctrines, rich in formal procedures, have only limited opportunities for creative synthesis or *lateral thinking*. In sharp contrast to this experience, young engineers find adapting to the design discipline a difficult, and occasionally painful, process. When confronted with the almost alien concept of applying knowledge in unfamiliar circumstances, behaviour patterns tend to vary. Well disciplined students easily recognise that all the rules and *doctrines* of the sciences still apply to the physical world of design. They will formulate appropriate mathematical models even when faced with new and unfamiliar problems. Others, less disciplined, regard design as an opportunity for breaking away from the narrow confines of science, and some interpret creativity as *anything goes*.

In essence, design practitioners must share the blame for some of this lack of discipline. Those involved in the doctrines of science are well experienced in the procedures needed to seek solutions to problems. Moreover, they all have a common discoursive language of communication. No one has any doubts about the meanings of such terms as *melting temperature*, *heat transfer coefficient*, *electro-motive force* or *principal stress*. In design there are no clear procedures that offer a direct route to solving problems, and designers are yet to develop a discoursive *common language* of design.

5.3 Review questions

5.3.1 Engineering Estimations

The world's richest person has an estimated personal wealth of \$US 600 x 10^9 (at one time this was J. P. Getty, then it used to be the Sultan of Brunei, but in recent times Bill Gates has assumed this position). Using a well set-out engineering calculation, and making any necessary assumptions, estimate how many shipping containers it would take to contain the wealth of the world's richest person, if that wealth were all converted to \$US1 bills.

5.3.2 Failure

1. An accident assessor is carrying out an investigation into the structural failure of a piece of office furniture. The assessor's report contains the following statement:

> *"It is likely that the rear legs of the chair became permanently deformed when the occupant rocked backwards on the chair, causing the bending stress in the extreme fibres to reach the yield stress."*

(a) What is the mode of failure being discussed in this part of the report?

(b) What is the suggested cause of failure?

(c) Is a *failure predictor* (i.e. a theory of failure) needed when assessing the bending stress in the legs of the chair? Explain your answer.

2. Why do engineers use failure predictors (theories of failure)?

3. A generalised element of some engineering component has principal stresses $\sigma_1, \sigma_2, \sigma_3$. According to the Maximum Shear Stress Failure Predictor (*MSFP*), when this element is at the point of ductile failure, τ_{max} is taken to be $S_y/2$. Use a Mohr's circle construction to explain why this is so.

4. Figure 5.16 shows a schematic A-M diagram. The axes and various lines on the diagram need to be labelled. Identify the quantities you would plot on the axes and also identify the four design parameters A, B, C, D, which determine the specific shape of the diagram.

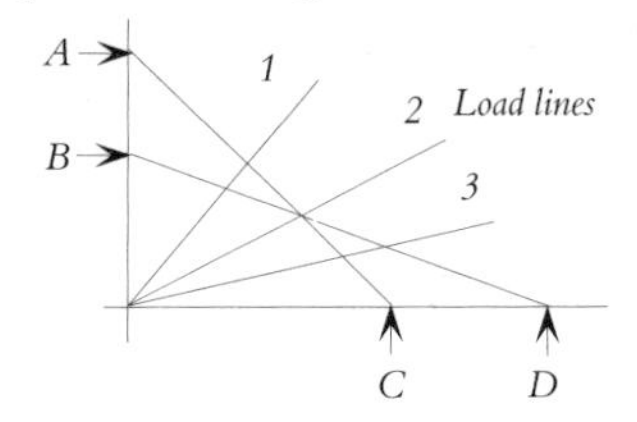

Figure 5.16 A-M diagram (schematic)

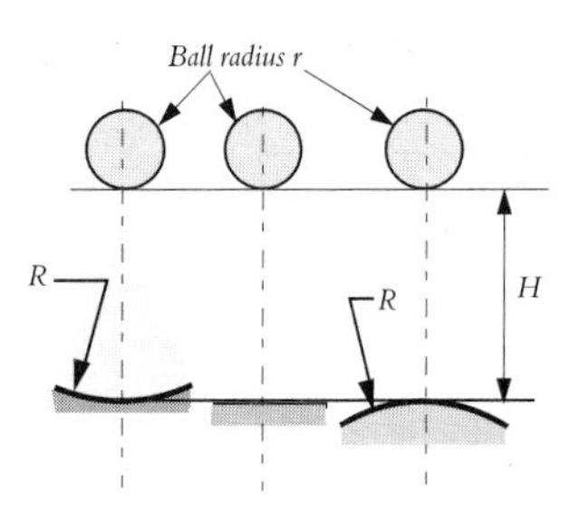

Figure 5.17 Ball dropping onto three surfaces

5. A ball bearing drops from a height H onto the three surfaces shown in Figure 5.17. The three surfaces are made of the same material as the ball bearing, and each is a surface of revolution which is symmetrical about a vertical axis through the ball. In which case will the contact stresses be greatest, and why?

5.3.3 Shafts

1. Shaft A and shaft B are identical. At a particular location, shaft A successfully withstands a 1000 Nm bending-moment while transmitting a 500 Nm torque. At a corresponding location shaft B successfully transmits a 1000 Nm torque, while subject to a 500 Nm bending moment. Which shaft has the larger factor of safety according to the Distortion Energy Failure Predictor (*DEFP*)? Briefly explain why.

2. Figure 5.18 shows the plan view and side elevation of a boom gate which is driven by a chain

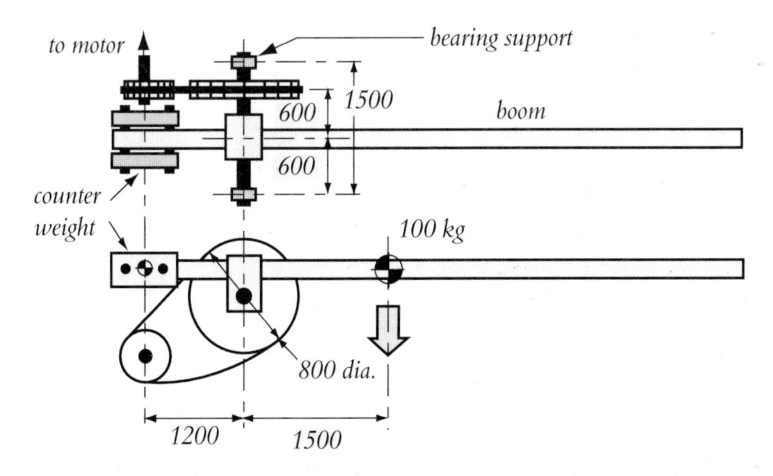

Figure 5.18 Boom gate (schematic, not to scale)

drive connected to a gearbox and an electric motor (not shown). The boom gate is to be raised from the horizontal position through 90° in 20 seconds. The mass of the boom is 100 kg, with a similar mass for the counterweight. The centres of mass for the boom and counterweight lie at 1500 mm and 1200 mm from the pivot respectively. The drive chain is slack on its return side. Calculate the design loads (moments and torques) on the driven shaft of the boom.

5.3.4 Mechanical springs

Two identical helical compression springs have coil diameter $D = 50$ mm, wire diameter $d = 4$ mm, and $n = 20$ coils. The springs are made from steel with a shear modulus of $G = 90$ GPa. Evaluate the combined stiffness when the two springs are connected in parallel (*a*) and in series (*b*) indicated in Figure 5.19.

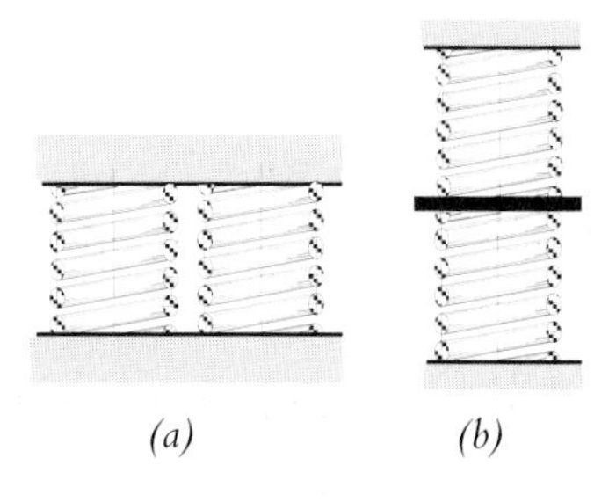

Figure 5.19 Springs in parallel (a) and in series (b)

5.3.5 Columns

1. A single slender column of length L and of hollow circular cross section is able to carry a given critical load. If the single column is replaced by two geometrically similar columns, acting in parallel, having the same total mass as the single column, will the load carrying capacity of the two columns be less or more than that of the single column they replaced ?
2. A pipe has a circular cross-section with second moment of area, $I_{zz} = 7 \times 10^6$ mm^4. The pipe is 100 m in length, and subjected to an axial compressive force of 80 kN. In addition to the simple supports at each end, how many intermediate simple supports are needed along the pipe to prevent it from buckling with a safety factor of $F_d = 1$? (Assume $E = 210$ GPa, and ignore gravity.)

5.3.6 Pressure vessels

Radiographic examination is a common technique for ensuring the quality of welded joints in critical items of equipment like pressure vessels. X-ray photographs are taken along the weld lines to search for any faulty welding, so that it can be ground out and repaired. The cost of this process is proportional to the length of welds examined. *Full radiography* examines 100% of welded joints in this manner, and so is very expensive. Describe the circumstances in which full radiography can nevertheless *reduce* the overall cost of a pressure vessel.

5.3.7 Bolted joints

1. The cylinder head of a high pressure gas compressor is held down by 10 equispaced bolts. Figure 5.20 illustrates the arrangement.

The pressure varies between 0 and 22 MPa sinusoidally in the cylinder, which has a cross sectional area of $5\text{x}10^{-3}$ m^2. The bolts are all pre-loaded to 10 kN.

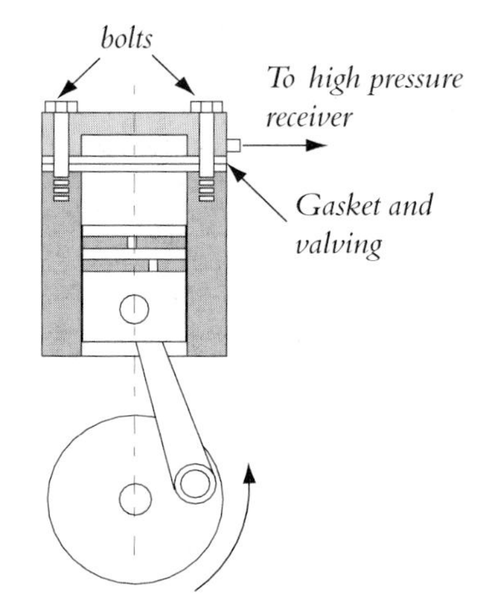

Figure 5.20 Gas compressor (schematic)

(a) Graph the bolt forces as a function of time when the bolt stiffness is the same as the cylinder head joint stiffness, and

(b) graph the bolt forces as a function of time when the cylinder head joint is ten times stiffer than the bolt.

(c) Will the joint fail (i.e. leak gas) in either case?

2. A bolted joint is pre-tightened to F_{bi} and is subjected to an external force F which fluctuates repeatedly from zero to F_{max}. The ratio of bolt stiffness to joint stiffness for this joint is $k_b/k_j = 2$. Sketch carefully a graph of force versus time, showing the variation of external force (F), bolt force (F_b) and joint force (F_j).

Evaluate the ratio $\dfrac{amplitude\ of\ bolt\ force}{mean\ bolt\ force}$.

3. A high pressure steam pipe (3.2 x 10^4 mm^2 internal area) in a refinery is connected to a pressure vessel by a flange with 16 high tensile bolts. The pressure in the vessel fluctuates sinusoidally from zero to 2 MPa. The bolts are pre-tensioned to 3.0 kN. Show on a diagram the way in which the load varies in one of the bolts for both:

(a) the case when the flange joint is very much stiffer than the bolt, and

(b) the case when the flange joint stiffness is twice that for the bolt.

Will the joint "leak" in either case?

5.3.8 Welded joints

1. Figure 5.21 shows a welded bracket subject to a vertical force W. There are two sets of welds on this bracket, each set consisting of four separate welds. All eight welds are identical in length, size, material and quality of fabrication. As W is gradually increased without limit, which set of welds will fail first, and why?

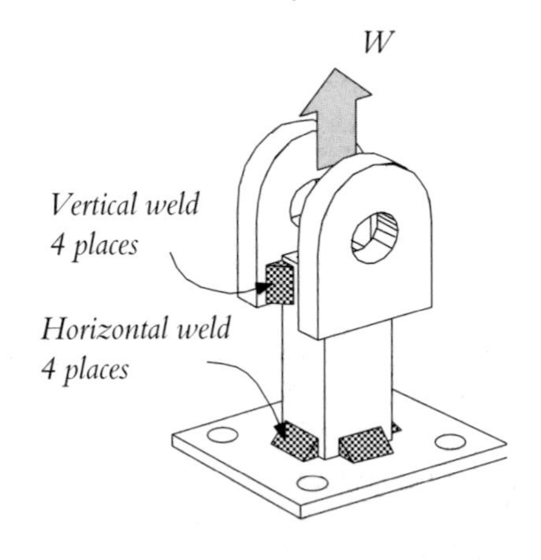

Figure 5.21 Welded bracket

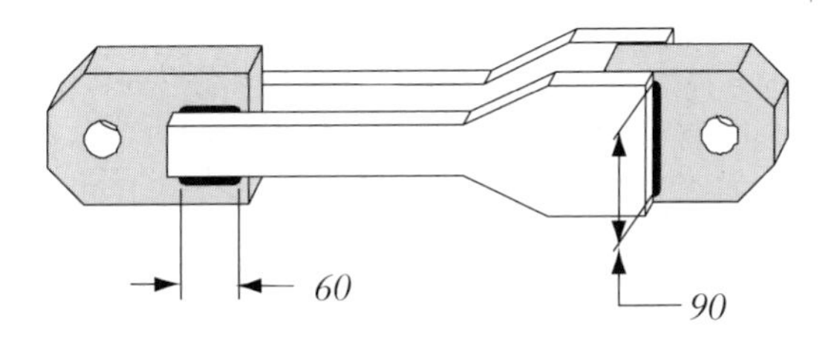

Figure 5.22 Welded tensile member

2. Figure 5.22 shows a fabricated tensile member intended for use on a piece of mining equipment. Assuming the allowable shear stress in the weld material is 140 MPa, and the fillet size of all welds on the component is 10 mm, calculate the maximum allowable tensile force which the strut can sustain, based on the strength of the welded joints. (The dimensions shown are for the *effective* lengths of each weld. Both faces of the component are identical.)

5.4 Design assignments

5.4.1 Optimum design of tripod

This assignment is in two separate but related parts, both concerned with optimum design — that is, obtaining the best results from given, finite resources. In the first part the emphasis is on mathematical modelling and calculations, while the second lays stress on practical insights, experiments and skills of construction.

PART I - DESIGN OF SIMPLE STRUCTURE (THEORETICAL)

This part of the project is concerned with the design of a structure of *minimum mass*. You are required to design a tripod capable of transmitting to the ground a force P of 50 kN from a point 3 m above ground level. The tripod must allow the force P to be applied in any direction. It has already been decided that the structure will be constructed of aluminium alloy, density = 2,770 kilogram per cubic metre. There is no restriction on the shape of the cross-section of the members.

The axial stress in tension members is not to exceed 85 MPa, while that in compression members is not to exceed 40 MPa. You may assume the weight of the tripod is small compared to the applied load, and that the ground is sufficiently stable to provide reactions for pinned supports. Figure 5.23 is a schematic view of the tripod.

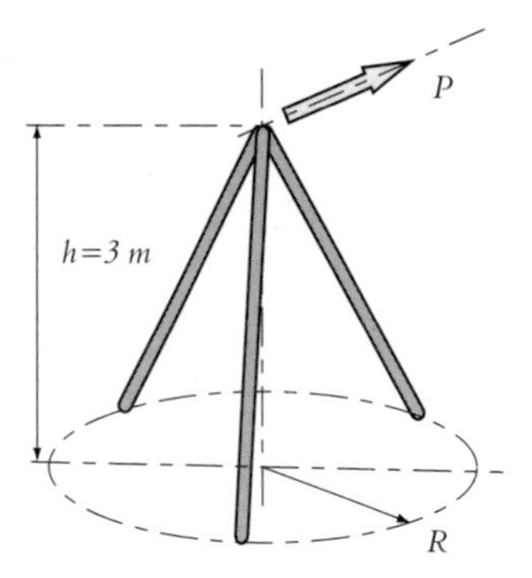

Figure 5.23 Tripod (schematic)

Instructions

(a) *Design* the tripod for the case where the attachments at the base lie on a circle of 1.2 metres diameter.

(b) If there is no limitation on base diameter, *determine* the optimum value of the base circle diameter for the structure of minimum mass. (A graph of tripod mass versus base circle diameter would be helpful.)

(c) *Prepare* a report setting out clearly the reasoning leading to your results. In your report, use a flow chart to *show* the major steps in the design. *List* all the assumptions used in your design calculations and critically review them. If you think that any of the assumptions lead to over-simplified calculations which could be seriously in error, make this point clear in your report.

Notes

(a) The allowable stress in compression members is less than that in tension members to try to allow for column action (the tendency to buckle elastically). In this case an arbitrary reduction has been made; briefly discuss whether the decision to make this reduction is justified.

(b) The design of this simple structure is an introductory exercise to illustrate features common to all design problems. The important ideas of *allowable stress* and *optimization* are introduced.

(c) Beware of getting bogged down in algebraic manipulations. Wherever possible, use a qualitative geometric evaluation, to prevent being enmeshed in mathematical detail.

Part 2 - Design of simple structure (Practical)

General Instructions

Design and build a structure which rests on a flat, horizontal surface (e.g. table or floor) and is capable of supporting a copy of this text book. No part of the base of the structure in contact with the supporting surface may intrude into a circular space 20 cm in diameter and 3 cm high directly below the book being supported.

Conditions to be Satisfied

The structure is to be made exclusively from one double page of a broadsheet newspaper, such as the Melbourne Age, London Guardian, or the New York Times (each page is about 85 cm by 60 cm), and no more than one metre of adhesive tape (Scotch Tape or equivalent, 12 mm wide). The best design will be that which supports the book at the greatest height, h above the supporting surface.

The book is to be horizontal, the maximum difference between the heights of any two points on its upper surface is not to exceed 2 cm. The book may not be stuck to the structure. The structure is to be sufficiently stable so that when the book is given a light tap it does not overbalance. A *light tap* is defined as the impact load experienced when the lower end of a 30 cm scale, which is held vertically and suspended from its upper end, is withdrawn a horizontal distance of 8 cm and released to hit the side of the book.

Idea Logs

Prepare idea logs containing sketches illustrating ideas for a wide variety of structures, drawn neatly and approximately to scale, together with brief notes of the pros and cons of the alternative designs considered. A fully dimensioned sketch of your chosen design should be included.

For the purpose of this exercise, *design* means *determine the relative position of the three members making up the structure, their lengths and cross-sectional areas.* Joints are not to be designed.

5.4.2 Optimum design of octahedral platform

This project is concerned with the design of a structure of *minimum mass*. You are required to design an *octahedral platform* capable of transmitting to the ground a force P of 50 kN from a point O_B at some height h above ground level. The *octahedral platform* is formed by inscribing an equilateral triangle ($A_1 A_2 A_3$) on a horizontal base circle of radius a centred on the origin O_A in the ground, and an opposed equilateral triangle ($B_1 B_2 B_3$) on a horizontal circle of radius b centred on O_B vertically above O_A, with $B_1 B_2 B_3$ and O_B lying in a rigid plate (the platform). The vertices of these base and upper triangles are connected with six (6) uniform legs as shown in Figure 5.24.

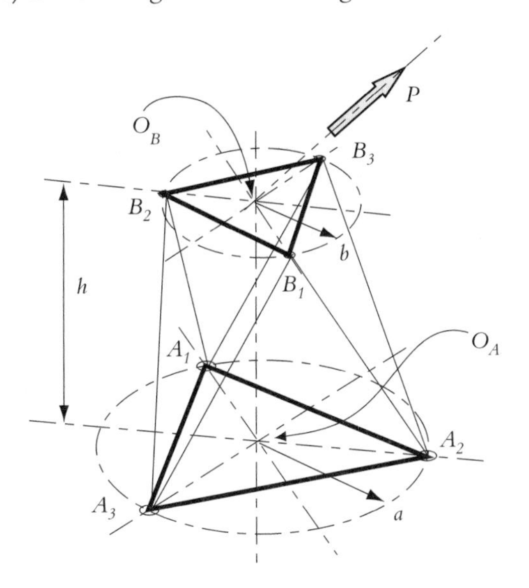

Figure 5.24 Octahedral platform

The structure must allow the force P to be applied in any direction. It has already been decided that the structure will be constructed of aluminium alloy, density = 2,770 kilogram per cubic metre. There is no restriction on the shape of the cross-section of the members.

The axial stress in tension members is not to exceed 85 MPa, while that in compression members is not to exceed 40 MPa. You may assume the weight of the tripod is small compared to the applied load, and that the ground is sufficiently stable to provide reactions for pinned supports.

INSTRUCTIONS

(a) *Design* the structure for the case where the load application point O_B is 3 m above the central base point O_A at ground level (i.e. h = 3 m), and a = 2.5 m and b = 1.0 m. .

(b) If there is no limitation on height h, *determine* the optimum value of h for the structure of minimum mass. (A graph of structure mass versus base height would be helpful.)

(c) *Prepare* a report setting out clearly the reasoning leading to your results. In your report, use a flow chart to *show* the major steps in the design. *List* all the assumptions used in your design calculations and critically review them. If you think that any of the assumptions lead to over-simplified calculations which could be seriously in error, make this point clear in your report.

NOTES

(a) The octahedral structure you are designing in this exercise is very closely related to an important class of robotic mechanism known as a *Stewart-Gough Platform* (Stewart, 1965). These are used widely in flight simulator applications, and have also been employed as manipulators in machining centres. In the *Stewart-Gough Platform*, the six legs which support the upper plate are all hydraulic or ball-screw actuators. When these are extended or retracted, they alter the position and orientation of the supported plate in a controlled fashion.

(b) For the purpose of this exercise, *design* means *"determine the relative position of the six members making up the legs of the platform, their lengths and cross-sectional areas"*. Joints are not to be designed, nor is the upper plate.

5.4.3 Determination of factors of safety

A tent-like awning is being designed to cover the concourse of a large automobile service station. The roof of the *tent* is to be fabricated from steel-reinforced woven polypropylene fabric, and will be constructed as several large bays, two of which are shown in Figure 5.25. In order to secure the edges of the roof material on adjacent bays, a large number of hold-down ties are required. Their purpose is to maintain tension in the roof material, and to anchor the roof to the ground.

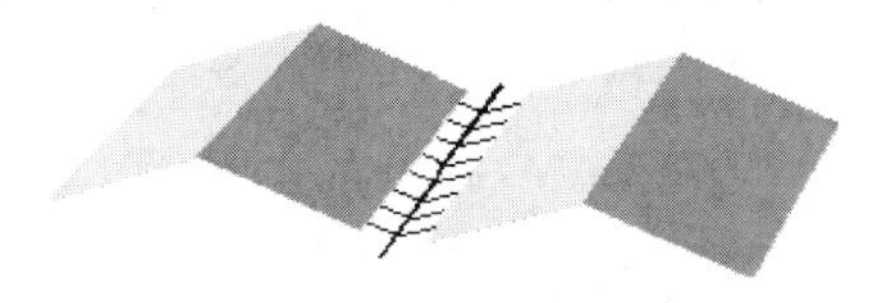

Figure 5.25 Roof structure for proposed service station (schematic)

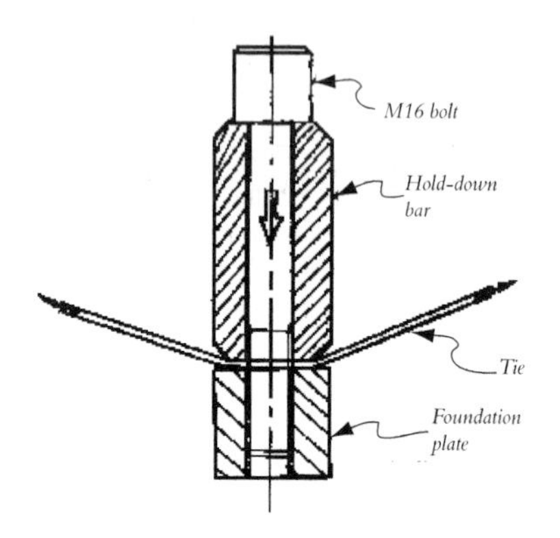

Figure 5.27 Hold-down bar arrangement

Each hold-down tie consists of a long strap with a cylindrical eye welded to each end, as shown in Figure 5.26. The ties are attached to a hem in the edge of the roof material by means of a continuous rod which is threaded through both hem and eye.

The manner of their attachment to the foundations is shown in Figure 5.27 — each tie is retained between two M16 high-tensile anchor bolts by a continuous clamping bar. The bolts are carefully tightened to an axial force of 35 kN (corresponding to approximately 20% of their yield strength). The design value of fabric tension in the assembled roof is 50 kN/m. The ties are made from stainless steel alloy 304, for which uniaxial tension tests have revealed the following material properties:

$S_y = 240$ MPa; $S_U = 585$ MPa; $E = 210$ GPa.

THE DESIGN PROBLEM

(a) List the possible modes of failure for the major elements of the hold-down system.

(b) Why is a *failure predictor* required in order to predict failure of the ties by yielding?

(c) On the basis of reasoned argument, decide upon a suitable factor of safety to be used in the design of the tie to resist yielding in combined tension and compression. Tabulate your calculation. Does the resulting factor of safety seem reasonable?

(d) Use any necessary assumptions to estimate the contact stress beneath the clamping bar.

(e) Using the Maximum Shear Stress Failure Predictor (*MSFP*), design the tie to resist yielding by a safety factor of 5.0. Ignore details of the end connections.

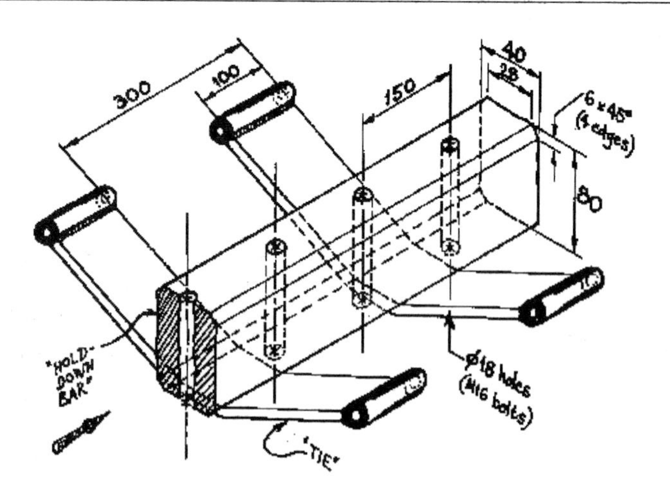

Figure 5.26 Detail of service station roof material securing arrangements

(f) Compare the *actual* factor of safety achieved in your design with the values which would arise if other failure predictors were used. Comment on the significance of this comparison.

When using the Distortion Energy Failure Predictor (*DEFP*), it is convenient to express the stress-state of the component in terms of the *von Mises* stress, σ_{vM}, where:

$$\sigma_{vM} = \sqrt{\sigma_1^2 + \sigma_2^2 + \sigma_3^2 - \sigma_1\sigma_2 - \sigma_2\sigma_3 - \sigma_3\sigma_1}\,.$$

The distortion energy per unit volume is then given by

$$u_{dist} = \frac{1+\mu}{3E}\sigma_{vM}^2\,.$$

5.4.4 Assignments on fatigue

(a) Designing a better clothes peg

The *Snappy Peg Company* manufactures a brand of clothes peg which relies on a helical torsion spring to hold the two wooden body halves together and to provide the closing force which clamps clothes to a line. The company has been receiving reports that some of these springs have fractured after extended periods of service. Although the consequences of these failures were minor, *Snappy* is concerned about its product's reputation for quality, and has called upon your services as a consultant to diagnose the problem.

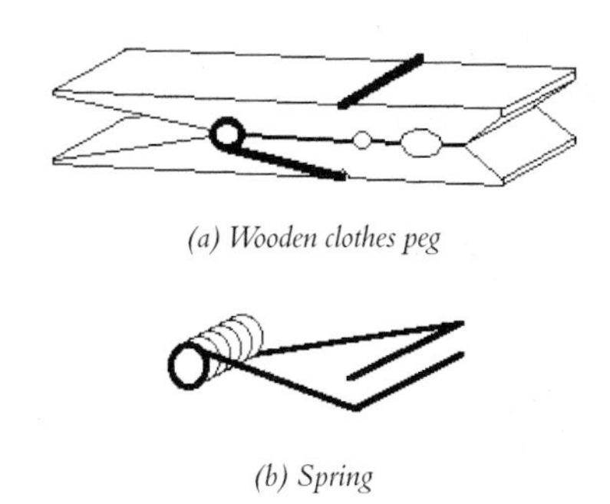

(a) Wooden clothes peg

(b) Spring

Figure 5.28 Wooden clothes peg and spring

The general form of the peg and spring is shown in Figure 5.28. Figures 5.29 and 5.30 respectively show the spring geometry and deflections imposed on the spring by the peg.

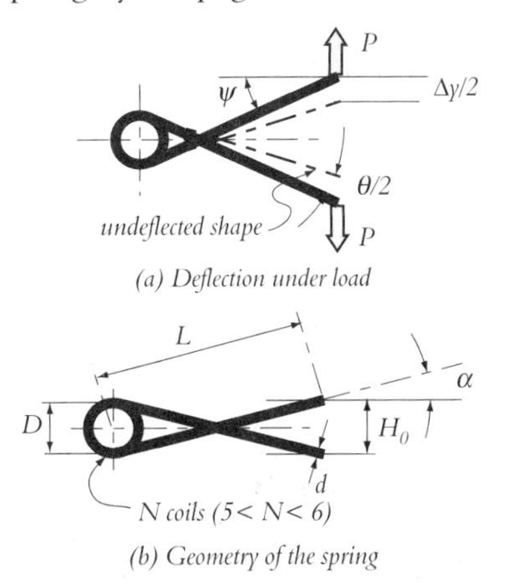

Figure 5.29 Spring geometry; The angle θ is defined in Figure 5.32; L= 20 mm; D=6 mm; d=1.6 mm; H_0=5 mm; α=15.6°.

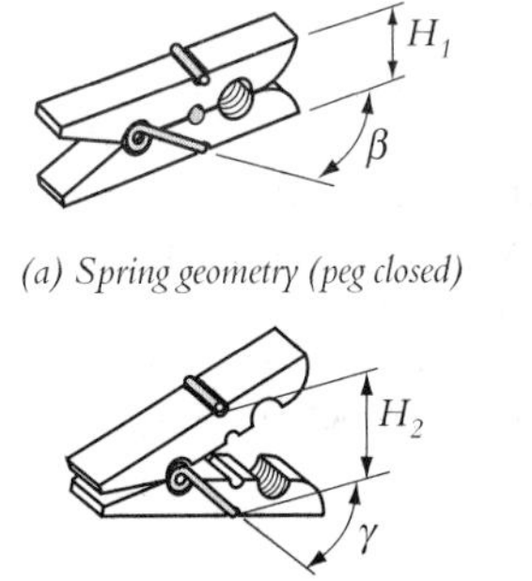

Figure 5.30 Spring geometry; H_1=12 mm; β=25.8° ; H_2=18 mm; γ=35.0°.

THE DESIGN PROBLEM

Your task is to check the existing spring design to ascertain whether it is adequate to provide an infinite fatigue life. In doing this, you should:

(a) Show that the vertical force P and separation of the spring tips, Δy , are related by :

$$P = \frac{E\,I\,\Delta y}{L^2 \cos^2\psi\left(L_{coil} + \dfrac{2L}{3}\right)}.$$

Definitions and some guidance for this proof are given in the notes at the end of this assignment.

(b) *Sketch* a graph of the stress cycle against time.

(c) *Evaluate* the mean and amplitude of the stress variation experienced by the spring during normal operation.

(d) Firstly, *calculate* an allowable stress in reversed axial tension, S_{all}, for the material. Then *establish* allowable limits for stress amplitude (and/or mean stress) to ensure an infinite fatigue life for this spring. [*Hint*: You will need to construct a suitable *A-M* diagram in order to show how you arrived at these limits.] A factor of safety of F_d=1.0 should be used in this part of your calculations — why?

(e) *Interpret* the results of (c) and (d) to predict the infinite fatigue resistance of these springs.

(f) If necessary, *briefly outline* the types of design changes which could be made to improve the spring's fatigue life. What other effects on the performance of the peg might these changes have? (Note that you are not required to determine the new value of any design parameter.)

ADDITIONAL DATA

- The spring is pre-tensioned (i.e. it exerts some force on the peg even when the peg is closed), as can be seen from the spring deflection between Figures 5.30(a) and 5.30(b).
- The spring is fabricated from round wire having a diameter of d = 1.6 mm.
- The coiled section of the spring consists of five (5) full turns and a partial turn of wire wound around the central axis of the coil at a mean diameter of D = 6.0 mm.
- The wire material is a high carbon spring steel which has been quenched and tempered to have the following properties: S_U =1700 MPa; S_y = 1530 MPa; S_e' = 880 MPa; E = 200 × 10^3 MPa.
- Coiled torsion springs such as this have a non-linear distribution of bending stress across the wire diameter from the inside to the outside of the coil. The fibres on the inside surface of the coil are most highly stressed. This is usually accounted for by multiplying the standard maximum bending stress for a straight rod of the same diameter, $\sigma_{B,max} = Md/2I$, by a stress concentration factor, K_f, where K_f is derived from K_t and the notch sensitivity factor, q, in the usual way. For this geometry, the theoretical stress concentration factor is approximated by:

$$K_t = \frac{4C-1}{4C-4}, \text{ where } C=D/d.$$

- Taking notch size to be equal to the coil radius of the spring, the notch sensitivity factor, q for this material may be taken from Figure 2.32 (Chapter 2).
- The surface finish factor, K_s for this cold-drawn material is given in Figure 2.34 (Chapter2).

NOTES

(a) A helical torsion spring is really like a tightly curved beam in bending. The wire of the spring is subject to bending, not torsion. It is called a *torsion* spring because the applied loads tend to twist it about the central *torsional-axis* of the coil. By implication, the symbol I here refers to the second moment of area of the wire about a bending axis:

$$I = \frac{\pi d^4}{64}.$$

(b) The spring shown in Figures 5.28(b) and 5.29(b) can be considered to be a system of three springs in series, each of which contributes to the vertical displacement of the tips:

- each straight end forms a cantilever spring [refer to Figure 5.31] ;

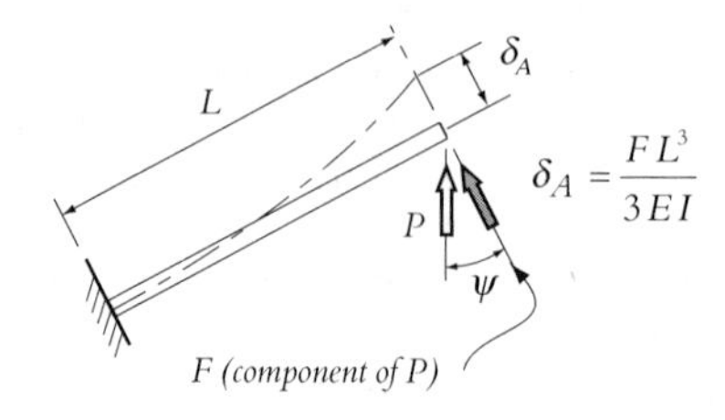

Figure 5.31 Deflection of a straight cantilever spring

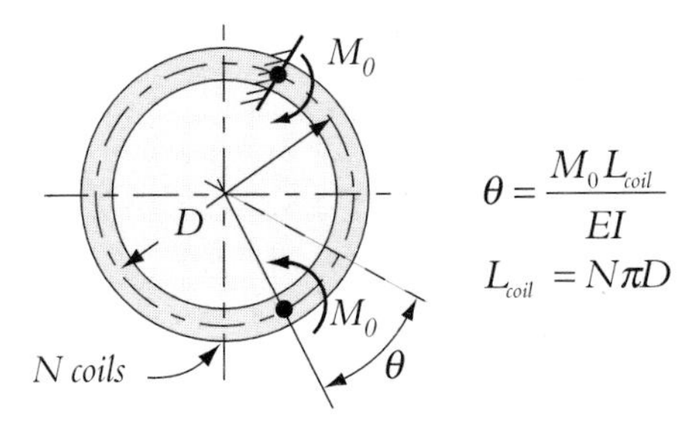

Figure 5.32 Deflection of a coiled torsion spring

- the coil forms one torsion spring [refer to Figure 5.32].

(c) The formula in Figure 5.32 is based on the simplifying assumptions that the bending moment in the coil is constant, and that the effects of tight curvature can be ignored in the calculation of strain energy. These are both reasonable approximations in this problem.

(b) Detent mechanism

Leaf springs are useful in situations requiring simple, cheap and compact spring components with relatively small deflections. One such application is the *detent mechanism* for a new model of *Dummo*™ *labelling gun*. These guns imprint alphanumeric characters onto coloured adhesive tape for labelling purposes. Characters are selected from a range of fifty letters, digits and punctuation marks arrayed around the periphery of a plastic print wheel, which is spun by the thumb and forefinger in order to select a desired character. The detent mechanism serves to locate the selected character centrally over the tape before imprinting. A general arrangement of the leaf spring in contact with some of the 50 detent teeth of the plastic print wheel is shown in Figure 5.33.

Because the detent mechanism is actuated fifty times for each rotation of print wheel, and because the print wheel is rotated many times to produce an average label, it is necessary for the leaf spring to withstand a large number of stress cycles during its service life.

Your task is to design the leaf spring for infinite fatigue life. The dimensions of the spring are up to you to decide, except that:

(i) the angle α has been already fixed at approximately 10°, giving a deflection of 0.5 mm each stress cycle.

(ii) for space reasons, it is suggested that the half-length, ℓ, should be about 10 mm.

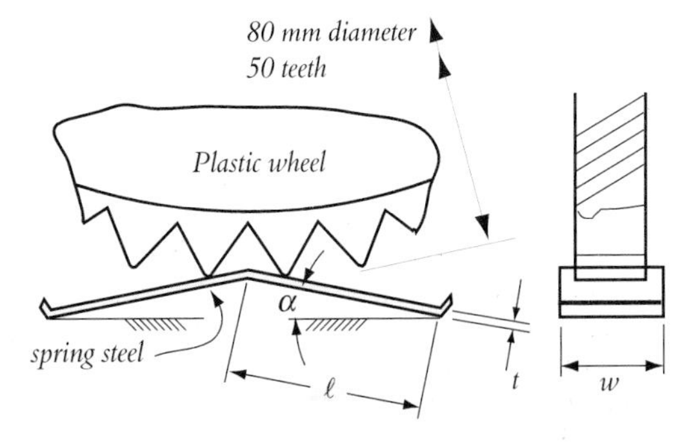

Figure 5.33 Detent mechanism

THE DESIGN PROBLEM

(a) Show that for a centrally loaded beam such as that in Figure 5.34, the relationship between force and deflection is given by:

$$P = \frac{\delta E}{2\cos^2\alpha}\,\frac{w t^3}{\ell^3},$$

where E = Young's Modulus, and other symbols are defined in Figures 5.33 and 5.34. Note that standard formulae pertaining to cantilever beams and rectangular cross-sections are given in the Appendix.

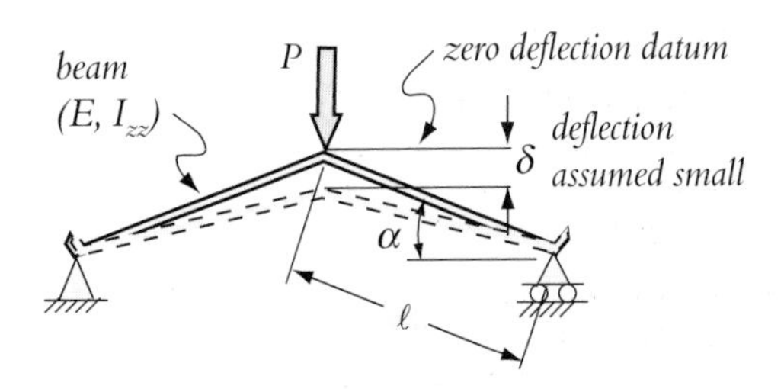

Figure 5.34 Leaf spring geometry under load (exaggerated)

(b) On the basis of the previous expression for P (see above), write down an expression for the maximum bending stress in the spring. Assume that when the plastic wheel is in the orientation shown in Figure 5.33, the spring and wheel are just lightly touching each other (i.e. assume this corresponds to the *zero deflection datum* position of the spring).

(c) Sketch the graph of stress versus time for the spring when in use. Identify the mean stress and the stress amplitude on this graph.

(d) Calculate an allowable stress in reversed axial tension, S_{all}, for the material. The spring is to be made from heat-treated medium-carbon steel having $S_e' = 420$ MPa and $S_U = 820$ MPa. The surface finish corresponds to that for a ground component. The bend in the centre of the spring is to have a radius of $r = 1$ mm, on which basis a theoretical stress concentration factor, (uncorrected for notch sensitivity of the material), can be estimated from the brief table supplied[5.5]. A factor of safety of $F_d = 1.5$ may be used. Briefly justify , if possible, this low value for F_d.

(e) Using the intermediate results obtained so far, develop a design *inequality* for the thickness, t, of the leaf spring to ensure infinite fatigue life. This *inequation* should relate t to other constants and parameters including, E, ℓ, δ and α.

[*Hint*: you will need to consider the *A-M Diagram*.]

(f) Calculate a suitable thickness for the leaf spring, using the suggested value for ℓ. Available sheet-metal thicknesses (based on preferred metric sizes) are listed in Table 5.3.

(g) It can be shown that in order to rotate the print wheel, a tangential force of approximately

$$F_t = P\left(\tan\alpha + \mu\right)$$

must be exerted on its rim. The manufacturer has specified that F_t must lie between 0.5 N and 1.5 N, so that the dangers of repetitive strain injury are minimized. Given a value of $\mu = 0.1$ for the coefficient of friction between the steel and plastic surfaces, select an appropriate width, w, for the leaf spring. If necessary, you may vary your previously selected values of t and ℓ, as long as all design inequalities remain satisfied.

(h) Summarize your final spring dimensions in a table or diagram.

Table 5.2 Stress concentration factors

r/c	K_t
1.5	1.9
2	1.5
2.5	1.4
3.0	1.3

Table 5.3 Available ferrous sheet

Thickness mm		Thickness mm	
First Choice (R10)	First Choice (R20)	First Choice (R10)	First Choice (R20)
0.1	0.1		
	0.112		1.12
0.125	0.125	1.25	1.25
	0.140		1.40
	0.160		1.60
	0.180		1.80
0.200	0.200	2.00	2.00
	0.224		2.24
0.250	0.250	2.50	2.50
	0.280		2.80
0.315	0.315	3.15	3.15
	0.355		3.55
0.400	0.400	4.00	4.00
	0.450		4.50
0.500	0.500	5.00	5.00
	0.560		5.60
0.630	0.630	6.30	6.30
	0.710		7.10
0.800	0.800	8.00	8.00
	0.900		9.00
1.000	1.000	10.00	10.00

5.5 The figures in Table 5.2 are based on data from Peterson (1974)

Table 5.2 is based on stress concentration factor data for curved (r = radius of curvature) rectangular cross section beams (beam thickness =2c in the plane of curve), subject to bending in the plane of the curve. Table 5.3 is based on *preferred number series* (refer to Appendix B for an explanation of *preferred number series*).

5.4.5 Assignment on columns

A farmer intends to fabricate a large number of steel-framed gates for a rural property. The general layout of the gate is shown in Figure 5.35(a). F rom prior experience, it is believed that such a gate experiences its worst-case loading (*worst credible accident*) when a heavy human operator stands on the end of the frame to swing it open. For this reason the design is to be based on the loading configuration shown in Figure 5.35(b). The gates will be welded from rectangular RHS (*rolled hollow section*) chosen from Table 5.4. F or aesthetic reasons, all seven structural members of the gate frame will be made from the same sized section, and oriented to have the larger cross-sectional dimension (D) in the plane of the gate as shown in Figure 5.34(c).

(a) By evaluating the axial forces in this truss system, identify which structural members of the gate frame are subject to axial compression and show that the axial force in sections QS and SU is given by $-W/\tan\,\theta$, which also equals $-WL/2H$.

(b) Demonstrate why the critical buckling condition (i.e. the *most likely buckling failure*) is for member QU to buckle normal to the plane of the gate.

(c) Design the gate (i.e. *select a suitable steel section*) to resist such buckling with a factor of safety of F_d = 5.0. State any major assumptions, including those concerning end fixity, plane of buckling, and effective length. Justify your choice of equation for calculating the buckling resistance of the relevant members.

ADDITIONAL INFORMATION

For the steel to be used in these gates:

- Young's Modulus, $E = 210 \times 10^3$ MPa;
- material yield stress, S_y = 240 MPa.

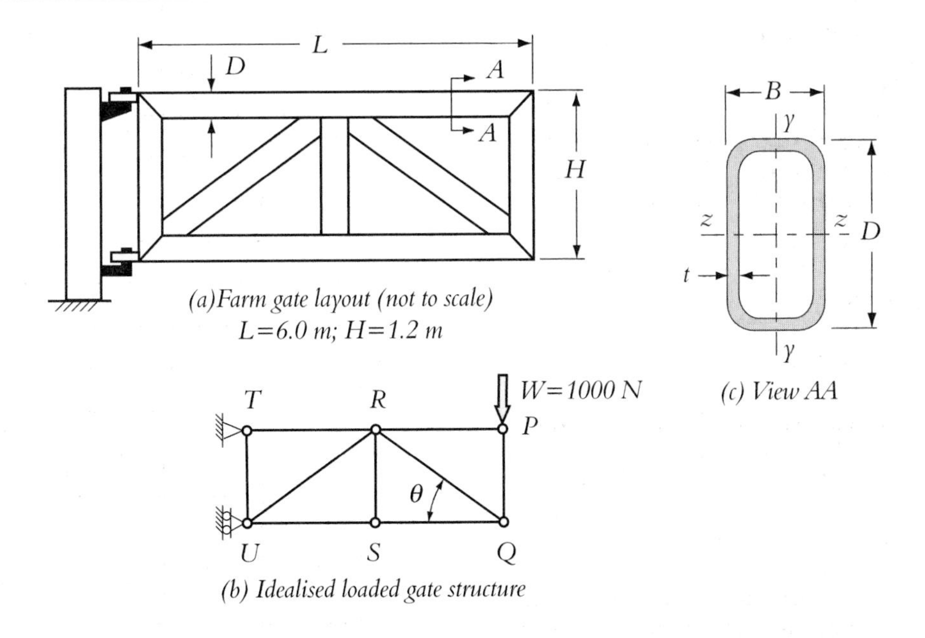

Figure 5.35 Farm gate

5.4.6 Assignment on bolted joints

A piston / conn-rod assembly has failed in service. The broken components are given to a consultant for inspection who finds that:

- the conn-rod is bent, the piston is broken, and the big end bolts have been snapped;
- upon close examination of the failure surface, one of the big end bolts appears to have failed in fatigue and the other in direct tension.

The general form of the reciprocating mechanism is shown in Figure 5.36. In addition, the following information is available:

Masses:

Piston = 1.0 kg; Conn-rod = 1.1 kg.

Small end and big end bearing shells = 0.035 kg each.

Two bolts and washers = 0.065 kg total.

The big end bolts are metric property class 10.9, having S_y = 830 MPa and S_U = 1040 MPa. They are *M10* hexagonal head bolts with a nominal shank diameter of 10.0 mm and head size of 16.0 mm across flats. They have rolled threads, with a pitch of P = 1.5 mm.

The bolts were pre-loaded to a torque of 61 N.m and it may be assumed that this represents a 32 kN bolt load when the system is stationary.

Table 5.4 Sectional properties of rolled rectangular hollow sections

Nominal Size Designation	Depth of Section	Breadth of Section	Wall Thick-ness	Area of Section	Second Moment of Area (z-z)	Second Moment of Area (y-y)
	D	B	t	A	Izz	Iyy
	mm	mm	mm	mm^2	$mm^4 \times 10^6$	$mm^4 \times 10^6$
50 x 25 x 2.5	50.8	25.4	2.6	357	0.108	0.0358
50 x 25 x 3	50.8	25.4	3.2	426	0.124	0.0405
65 x 40 x 3	63.5	38.1	3.2	591	0.299	0.134
65 x 40 x 4	63.5	38.1	4.0	717	0.348	0.154
75 x 40 x 3	76.2	38.1	3.2	674	0.475	0.159
75 x 40 x 4	76.2	38.1	4.0	820	0.588	0.184
75 x 50 x 3	76.2	50.8	3.2	756	0.585	0.311
75 x 50 x 4	76.2	50.8	4.0	924	0.692	0.366
75 x 50 x 5	76.2	50.8	4.9	1,080	0.785	0.413
75 x 50 x 6	76.2	50.8	6.3	1,350	0.919	0.479
90 x 40 x 3	88.9	38.1	3.2	756	0.705	0.184
90 x 40 x 4	88.9	38.1	4.0	924	0.833	0.214
100 x 50 x 3	101.6	50.8	3.2	921	1.20	0.404
100 x 50 x 4	101.6	50.8	4.0	1,130	1.43	0.479
100 x 50 x 5	101.6	50.8	4.9	1,330	1.64	0.544
100 x 50 x 6	101.6	50.8	6.3	1,670	1.95	0.640

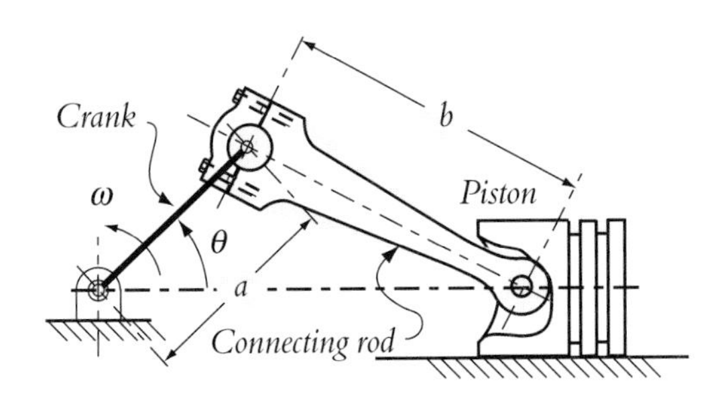

Figure 5.36 Reciprocating mechanism consisting of Piston, Connecting rod (Conn-rod) and Crank; a = 52 mm, b=137 mm, ω=314 rad/s

The Design Problem

(a) The linear acceleration of the piston, f_p, varies with crank angle (and therefore with time) according to (Bevan, 1955)

$$f_p \approx -a\omega^2\left(Cos\theta + \frac{aCos2\theta}{b}\right).$$

It is worthwhile deriving this simple equation from first principles, since it clearly implies that for long conn-rods (i.e. $b>>a$), the piston acceleration may be closely approximated by the horizontal component of the big end's centripetal acceleration,

$$f_{p,b>>a} \approx -a\omega^2 Cos\theta .$$

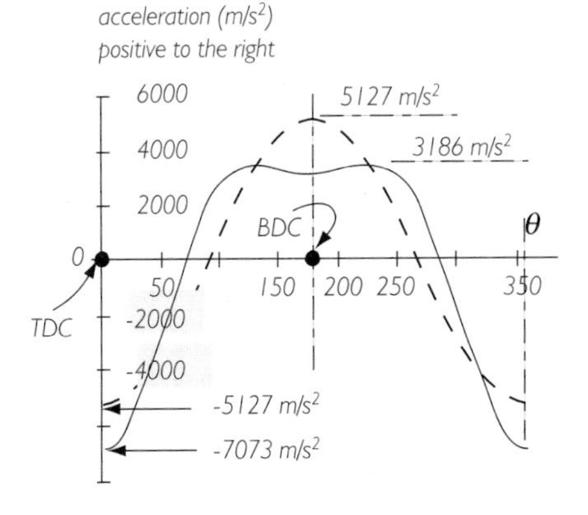

Figure 5.37 linear acceleration of the piston versus crank angle. The horizontal component of the big end's centripetal acceleration is also shown (dashed) for comparison.

What would be the percentage error in making such a simplifying assumption in this case (refer to Figure 5.37)?

Evaluate the maximum external force, F, experienced by each bolt during engine operation at this speed? (Ignore pressure effects in the cylinder.) By sketching an appropriate graph, *show* the manner in which F varies with time.

By making suitable estimates and approximations concerning the relative stiffness of the bolt and the clamped big end of the conn-rod, *calculate* the variation in bolt force, F_b .

(c) *State* the likelihood of this failure being caused by fatigue. What further information would you need to come to a firm decision regarding the failure mode?

(d) Describe the effect on the fatigue resistance of the bolts of:

- using a relatively soft gasket material (such as copper or bronze) in the bolted joint;
- using bolts which have the same stress area as the existing bolts (see below), but have their shank area reduced as shown here in Figure 5.38.

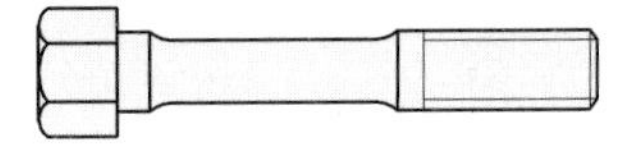

Figure 5.38 Reduced-shank bolt

Additional Information

- For metric bolts, the effective *stress area* (i.e. the minimum cross-sectional area used for calculating worst-case tensile stress at the threads) is calculated to be :

$$A_s = \frac{\pi}{4}\left(d - 0.9382P\right)^2,$$

- whereas the cross-sectional area of the shank is simply given by :

$$A_b = \frac{\pi}{4}d^2.$$

- For steel bolts and conn-rod : $E = 210$ GPa.
- The following equations describe the effect of a variable externally applied force F on the two other major forces in a bolted joint:

— Bolt force variation:

$$F_b = F_{bi} + F\left(\frac{k_b}{k_b + k_j}\right); \quad \text{[equn. (4.8)]}$$

— Joint force variation:

$$F_j = F_{bi} - F\left(\frac{k_j}{k_b + k_j}\right). \quad \text{[equn. (4.9)]}$$

NOTES

(a) The maximum acceleration along the axis of the conn-rod is found to be -7073 m/s^2, which occurs when the piston is at top dead centre (crank angle $\theta = 0^\circ$).

(b) You may use the simplifying assumption that 2/3 of the conn-rod mass is included in the total reciprocating mass.

(c) Use suitable assumptions to estimate the value of S_e for the bolt .

5.5 Some cautionary notes

5.5.1 A cautionary note about sign conventions

Throughout this book we have used consistent sign conventions for bending moments, shear forces and shear stresses. In Chapter 2, where we have first introduced these sign conventions, there was no indication that *our* conventions are anything but widely accepted procedures throughout the profession. Consequently, the unwary, inexperienced, user of simple structural models based on bending moment and shear force diagrams might become concerned when confronted with other texts or other disciplines, where our sign conventions are no longer obeyed. The purpose of this brief note is to explain the rationale behind our sign conventions and to point out that the structural world is not governed by unassailable absolute coordinate systems and sign conventions. As the name suggests the notion of a convention is that of a consistent rather than some absolute frame of reference.

(a) Bending moments and shear forces.

The long axial direction along a beam's neutral axis is chosen as the x-axis. The y- and z-axis directions are transverse to the x-axis and their orientation is determined by the *right-hand screw* rule.

In the x-y plane, centrally loaded, simply supported horizontal beams will deflect in the direction of the load, independent of the positive axial direction chosen for the load.

The governing moment-deflection equation

$$\frac{M}{EI_{zz}} = \pm\frac{1}{R} = \pm\frac{d^2y}{dx^2},$$

has the sign of the right hand side (*RHS*) of the equation determined by the y-axis direction chosen for the elastic curve. When the positive y-axis is chosen in the same direction as the radius of curvature R of the elastic deflection curve, the *RHS* of the moment equation becomes negative. This was the sign convention chosen in some earlier structural analysis texts (see for example Timoshenko, 1955). Conversely, when the positive y-axis is chosen opposite to the radius of curvature, R, the *RHS* of the moment equation becomes positive. This is the sign convention chosen by many current texts on structural analysis (see for example, Young, 1989; Benham et al., 1997). We have adopted this sign convention in our book, as seen in Section 1.3.2.

The relationship between shear force, V, and bending moment, M, is chosen to be

$$V = \frac{dM}{dx}; \quad M = \int V dx.$$

Once we have chosen a sign convention for bending moment, the sign convention used for shear force is consistent with the above relationship.

The colloquial terminology adopted in some texts is that of positive moments for *sagging beams*, where the elastic deflection curve has a negative radius of curvature, R, and negative moments for *hogging beams*, where the elastic deflection curve has a positive radius of curvature, R.

The sign conventions we have adopted for bending moment and shear force are illustrated in Figures 1.25 and 1.26 in Chapter 1.

(b) Shear stresses

Sign conventions adopted for shear stresses are governed by several geometric constraints. In adopting the conventions described above for shear force, we have chosen to recognise *anti-clockwise shear forces as positive*. Correspondingly we might intuitively expect to find anticlockwise acting shear stresses on an element (as seen in Figure 2.9b) as positive. That is indeed our choice of convention, but not for the reason of consistency with shear forces. Our choice of convention is based on the notion that shear stresses acting in the positive direction of an axis are positive.

Consider the element, illustrated in Figure 2.9 in Chapter 2, and a face of this element *parallel* to the *yz*-plane, with its *outward pointing normal* in the *positive x* direction. A shear stress acting on such a face in the positive *y*-axis direction is the positive τ_{xy}. This is the convention we have used for describing shear stresses.

In some texts (see for example Shigley, 1986) the convention for shear stress is taken to be *positive when acting clockwise*. The rationale for this choice of convention rests with the Mohr's circle construction. With this convention, the angle of rotation needed to transform the combined-stress axes into the principal axes (zero shear stress) is in the same sense as on the Mohr's circle figure as it is in the physical element. In our convention, the sense of this rotation on the Mohr's circle figure is in the opposite direction to that in the element. The effect of our choice of convention for shear stress is illustrated in Figure 2.10. Again we hasten to caution readers that this convention is just as much a matter of convenience, as long as consistency is maintained, as it is for bending moments and shear forces.

5.5.2 A cautionary note about material selection

We have discussed material selection briefly in Chapter 2. Novice designers might easily form a misconception based on our brief discussion and/ or any of the myriad material selection references available to modern engineers. This misconception is that the science of new materials has already made and will continue to make great impact on the performance and cost of engineered products. We feel compelled to influence this perception by some brief quotations from Professor J. E. Gordon's wonderful book, *Structures or Why Things Don't Fall Down* (Gordon, 1976,1979) .

In the section of the book dealing with *"Materials, Fuel and Energy"*, Gordon explores the notions of energy efficient materials for a great variety of applications. He writes:

"...Many attempts are on foot at present to collect energy for technology from diffuse sources, such as the sun, the wind or the sea. Many of these are likely to fail because the energy investment which will be needed, using conventional collecting structures built of steel or concrete, can not yield an economic return. A quite different approach to the whole concept of 'efficiency' will be needed. Nature seems to look at these problems in terms of her 'metabolic investment', and we may have to do something of the same kind.

It is not only that metals and concrete require a great deal of energy, per ton, to manufacture... , but also that, for the diffuse or lightly loaded structures which are usually needed for systems of low energy intensity, the actual weight of devices made from steel and concrete is likely to be very many times higher than it would be if we used more sensible and more civilised materials."

Gordon goes on to explore material behaviour using the parameter of *structural efficiency*. For many high technology structures, such as those used in the aerospace industry, this parameter of *structural efficiency* translates to the *specific Young's modulus*, E/ρ. For columns it converts to $(E/\rho)^{1/2}$ and for flat panels, to $(E/\rho)^{1/3}$. He writes about this aspect of material performance:

"...It happens that for the majority of traditional materials, molybdenum, steel, titanium, aluminium, magnesium and wood, the value of E/ρ *is sensibly constant. It is for this reason that, over the last fifteen or twenty years, governments have spent large sums of money in developing new materials based on exotic fibres such as boron, carbon and silicon carbide.*

Fibres of this sort may or may not be effective in aerospace; but what seems to be certain is that not only are they expensive but they also need large amounts of energy to make them. For this reason their future use is likely to be rather limited, and, in my own view, they are not likely to become the 'people's materials' of the foreseeable future."

One might argue that this last part of Gordon's quote is a little anachronistic in the light of kevlar climbing ropes, carbon fibre fishing rods and golf

Table 5.5 Comparative materil performace (After Gordon, 1976)

Material	Energy to manufacture GJ/tonne	Oil equivalent tonne/tonne	E/ρ	$E^{1/2}/\rho$	$E^{1/3}/\rho$	Stiffness energy	Strength energy
Steel (mild)	60	1.5	25,000	190	7.5	1	1
Titanium	800	20	25,000	240	11.0	13	9
Aluminium	250	6	25,000	310	15.0	4	2
Magnesium			24,000	380	20.5		
Glass	24	0.6	25,000	360	17.5		
Brick	6	0.15	7,000	150	9.0	0.4	0.1
Concrete	4	0.1	6,000	160	10.0	0.3	0.05
Carbon fibre composite	4,000	100	100,000	700	29.0	17	17
Wood (spruce)	1	0.025	25,000	750	48.0	.02	.002
Polyethylene	45	1.1					

Note: Stiffness energy = the energy required to ensure a given stiffness in the structure as a whole;
Strength energy = the energy required to ensure a given strength in the structure as a whole.
The figures in the last two columns are based on mild steel being unity and they are very approximate.

clubs, and large segments of automobile bodies making use of fibre-impregnated composites. Nevertheless we should be very cautious in dismissing Gordon's views, particularly when we consider his comparative figures offered in Table5.5.

Professor Gordon cautions readers of his book that the figures in the first two columns of Table 5.5 are either close approximations or informed guesses. Our objective here is to point out that even if these figures are of the right order of magnitude, and they are certainly at least that reliable, there is an opportunity to make not only more effective, but environmentally more responsible choices in materials of construction.

PART 2

PROBLEM SOLVING STRATEGIES: AN ENGINEERING CULTURE

6

THE EVOLUTION OF DESIGN PROBLEMS

For when the One Great Scorer comes to mark against your name,
He writes – not that you won or lost – but how you played the Game.
Grantland Rice, American sports writer in 'Alumnus Football'(1941)

We now focus on some general issues about the whole process of designing. In doing so, our presentation needs to be placed in a context of the designer's task mix. Almost all design is done by humans, who take on various roles during the management of the whole design task. This part of our discussion makes specific reference to those aspects of the design activity that are (currently) still the exclusive domain of the human mind.

We are being somewhat pedantic in implying that not all design is performed by human beings. In the context of *artificial intelligence* (*AI*) there is a substantial school of thought that deals with *automated design*. It has long been the dream of AI practitioners that once the requirements and design boundaries have been determined, a computer programme can intelligently resolve most, if not all, conflicts in a design. In the introductory chapter to *Intelligent Design and Manufacturing* (Kusiak, 1992), Kannapan and Marshek write:

"...substantial research effort in the area of engineering design has been directed at significantly improving the productivity of the design process."

This statement clearly identifies the process of design as requiring some improvement in support tools so that the human designer may be more productive in the delivery of design results. Some of these tools are referred to as computer aided design (*CAD*) or decision support systems (*DSS*). *CAD* is often interpreted as the process of preparing virtual design embodiments by solid geometric modelling programmes (for example *AutoCAD*, *Pro/ENGINEER*, *UnigraphicsCAD*, *CATIA* and *SolidWorks*, to name just a few of the geometric modelling products available). However, computer support has a far wider influence on design than that encompassed by mere geometric modelling. Kinematic and dynamic modelling of engineering systems, thermo-fluid modelling of complex thermal and flow processes, and even process modelling of robot based manufacturing offer substantial support for the designer beyond the embodiment activity. Sophisticated search and logic programmes (sometimes referred to as *search engines* and *inference engines*) provide help with resolving complex interactive design requirements. Yet in spite of the available, impressive, array of tools for computer based design, they can not be expected to replace the most important *design-front-end* activity of the human designer. What is invariably the commencement point of any design is the *recognition* that a *need for design* exists. Once the *need* has been identified, many *conflicts* must be *negotiated* through to some comprehensive statement of *objectives* and *requirements* of the ultimate design (the result of the design process), and the way we might measure the succes of the result (*design criteria*). It is this critically important human-directed aspect of designing that ultimately determines the majority of the features of the design result.

Chapter 6 briefly introduces the various elements of the design process, and Chapter 7 presents some useful decision making tools. We introduce the notion that design is *not a science*, but a *discipline*, where there are *no easily refutable conjectures*. As a direct result of this notion, we can not suggest that the design approach we propose is mandatory for best results. On the other hand, in what follows we try to develop a disciplined way of thinking about the general process of design, specifically useful for inexperienced designers. While our approach in Part 1 followed a slightly *deconstructionist* approach to design for structural integrity(*structural distillation*), Part 2 follows a strongly *constructionist approach* to design. We still need to *see problems through the designer's eye*. However, here we need to see the *whole* as it becomes *constructed from the several parts* rather than the other way around.

6.1 Cultural development of the designer

"Without confidence in its own culture, a nation does not seriously exist." Donald Horne[6.1]

Author and academic, Professor Donald Horne, was referring to a country's cultural awakening, but his comment may equally well be understood in terms of a growing interest in the design and development of high quality engineered products. Almost half a century ago, media guru, Marshall McLuhan, wrote (McLuhan and Fiore, 1967):

"Ours is a brand-new world of allatonceness. 'Time' has ceased, 'space' has vanished. We now live in a global village...a simultaneous happening."

McLuhan was referring to the psychological effects of globalised media. Even he could not have foreseen how globalised media would impact on manufacturing competitiveness. Manufacturers can directly access suppliers' databases at the *"flick of a keyboard"* anywhere in the world, day or night. Consumers are constantly exposed to internationally demanded standards of product quality. Many large manufacturers can now source manufacturing technology, and have direct access to product quality management tools, from anywhere in the world. These factors are largely responsible for such products as the MDX helicopter (see Figure 6.2), fuselage made in Australia, engine in England, gearbox in Israel. Also, many manufacturers of consumer products supply parts and even whole assemblies to a wide range of market outlets. The term for some of these products is *"badge engineering"*. Fashion houses like Calvin Klein and Gucci don't manufacture watch movements, yet they market watches.

In such an agressive atmosphere of marketing no manufacturer can now afford to spend five years in cradle-to-launch activities (typical time period for new models of automobiles). Where products involve computer technology, there is an even greater urgency to deliver to market. Most computing chips have a useful technology advantage for relatively brief periods of the order of one or two years.

With many consumer products, the manufacturer's competitive advantage (*edge*) results from *agility* (the capacity to rapidly respond to market demands). For products that rely on significant specialist engineering expertise, such as hydraulic turbines, turbo-alternators, or aerospace products, the competitive *edge* results from *flexibility* (the capacity to respond to a wide range of alternative design requirements).

Designers have significant influence in both aspects of manufacturing, *agility* and *flexibility*. In addition, quite apart from its marketing significance, a well designed and manufactured product can act as a source of national pride and a driving force for developing a strong and vibrant *design culture*. In this culture we find that *the disciplines of engineering design* and *industrial design* appear to have developed separately but in parallel.

The Encyclopaedia Britannica describes industrial design as:

"...design of products made by large scale industry for mass distribution. Designing such products means, first, planning their structure, operation, and appearance, then planning these to fit efficient production, distribution, and selling processes. Clearly, appearance is but one factor in such a complex process. Nevertheless, in consumer goods especially, appearance design is widely accepted as the principal virtue of industrial design; it is that portion of the whole least subject to rational analysis and, like craft secrets of the past, most advantageous in commercial competition. On the other hand, design of equipment for production, for services, and for sports is expected to demonstrate utility, but in these products, too, appearance design is increasingly important."

This is a highly simplified description of what industrial designers do, and, in essence, it is not all that different from what we might expect that engineering designers do. Perhaps the best way to describe the difference and the commonality between the two disciplines, industrial and engineering design, is to focus on the considerable specialist and organisational skills each discipline needs to encompass. The industrial designer's specialist skills might be seen to lie in the visual, ergonomic and material selection aspect of design. The engineering designer specialises in technical analysis, mathematical modelling and structural integrity issues. In many design situations there is significant overlap in the task mix of both types of designers.

6.1 Australia's Best:A Digest of Australian Achievement, 1988, The Australian Design Council.

6.1.1 Engineering design as wisdom

Engineering designers have been struggling with the definition of the discipline since 1966, when the first conference on engineering design was held in England (Gregory, 1966). The discipline is supported by a substantial body of literature (see the *Annotated bibliography of engineering design* in our reference list), yet most institutions teaching engineering design do not have a prescribed text for supporting the presentation of general design procedures. Many institutions feel compelled to offer their own versions of design to undergraduates, including a combination of guided design exercises and lecture material. It is commonly hoped that the guidance in the design exercises will convey and instil some elements of the general design process.[6.2] Our experience is that application of the general design principles is heavily infuenced by the technology context within which the project is set (Samuel and Weir, 1991). This process of building contextual or domain-specific experience has been referred to by French as *received wisdom* (French, 1988).

Gaining *wisdom* in engineering design, whether domain-specific or general, is a process that is low in procedural content and is often difficult to articulate. We quote McLuhan again[6.3], whose highly appropriate description of media is equally applicable to the formal design process as we proceed from *need* to *design objectives*:

"Environments are invisible. Their ground rules, pervasive structure, and overall patterns elude easy perception."

This is the portion of the whole design experience *least subject to rational analysis.* Broadly speaking, the engineering design discipline is an embodiment of current engineering culture. In that sense, it encompasses the current practices, ethics and linguistic character of all that exists and all that has gone before it in engineering. The essential objective of this book is to reveal and highlight the key features of that culture to young engineers.

6.1.2 Engineering design as discipline

First of all we note that engineering design is a discipline, with disciples who follow and ultimately disseminate the discipline to others. Consequently, the discipline is a lifelong learning experience and engineering design is as much about the practice of engineering as it is about the education of engineers to deliver that practice in the most effective way.

The engineering design discipline is often sharply contrasted with the engineering sciences, which are doctrines, rather than disciplines. These doctrines commonly presume some model of reality to enable them to deliver their predictions of behaviour along the lines of their current orthodoxies. Engineering science develops along the rigorous paths of the well recognised scientific method, involving model building and experimental verification. Engineering science looks for easily refutable (testable) rules for predicting model behaviour. Unless a rule can be tested, and refuted if false, it has no place in engineering science.

Engineering design is a problem finding and problem solving procedure. While it is strongly based in information management, it is not science. Nevertheless, we make consistent and substantive use of engineering science models for problem solving. Moreover, our problem solving methods follow closely the procedures used for scientific problem solving. These procedures have at various times been identified as *the design method.* Using a *design method*, or a prescriptive procedure for managing design information does not constitute *design science*. There is a sharp difference between our design discipline and the doctrinaire sciences. While we often use the design method for problem solving, it represents a *sufficient*, though *not necessary*, procedure. There is no way of verifying (or refuting) the proposition that *only the design method leads to useful design solutions.* What, in fact, can we identify as *useful design solutions?* What are the dimensions of a *good design?* These and associated issues are the subject of this second part of our book.

Both engineering science and engineering design are living organisms that respond to changes in technology and philosophy. However, the lead times for changes in the design discipline are, of necessity, shorter than the lead times for doctrine changes in science. As an example, the change from Newtonian mechanics to relativistic mechanics took several hundred years, while the change from horse-drawn vehicles to the automobile took less than 50 years. In printing the change from manual typesetting has changed to electronic typesetting in about 20 years.

6.2 A strong reminder of the older style of master-apprentice form of engineering training

6.3 McLuhan and Fiore (1967), loc. cit.

The culture of engineering science is rich in procedure. Thermo-fluid science relies heavily on the Navier-Stokes and Bernoulli's equations, and the various *laws* of thermodynamics. Mechanics of solids makes consistent use of equations based on energy conservation. While we have no such well established rules in engineering design, there are some *design-support* procedures that result in credible and effective design problem management. Clearly, the design discipline has its origins in an artisan culture. Great engineering designers of the last century such as Stephenson, Watt, and Brunel had no formal training in design, but rather acquired their considerable design skills and engineering wisdom amidst the apprenticed artisan culture of the day. Nowadays, maintenance of an informal artisan character in design would serve to relegate the discipline to one seriously lacking in scholarship. However, the benefits of acquiring experience and judgement through extended association with a "master practitioner" should not be dismissed.

It is instructive to examine the experience of Michael Faraday (1791-1867), widely regarded as the *father of electricity*. Of necessity, Faraday left school at age 13. At age 22 he was taken on as Sir Humphrey Davy's bottle washer and commenced his own experiments at age 29. Davy (1778-1829), a chemist with a considerably more affluent background than Faraday, was by this time the director of the Royal Institution. Consequently, Faraday was exposed to the substantial wisdom of Davy for seven years before commencing his own experiments. Faraday's genius could be easily mistaken for the triumph of *nature over nurture*, thereby ignoring the beneficial impact of formal training.

The Wright brothers had no more formal schooling than did Faraday. Yet they were able to triumph in the design and construction of flying machines, when many formally trained engineers and scientists failed before them (Kelly, 1996). In fact, they conducted carefully planned experiments and made extensive use of the *received wisdom* of their trade art in bicycle manufacture.

Perhaps the most significant issue in the development of a credible design culture is the recognition of the need for formal training in design *discipline*. Here we refer not to the practice of design, but to the carefully controlled application of its procedures.

6.2 Design problem solving

In his book, *Invention and evolution*, Michael French (1988) describes the design requirements of a humble clothes peg. In 1988 he claimed this to be an unsolved problem, but his description of the design detail in this apparently simple device we all take for granted, is a classic example of the richness of design studies. French in his description of the design problem did not suggest that there were no *useful embodiments* for clothes pegs, but that all these embodiments are the result of manufacturing compromises that leave *many of the original design needs unfulfilled*.
Another example of difficult consumer product design is the bayonet fitting on household electric lights. The connection between the light globe and the socket is made via two, low melting point, metal contacts on the rear of the globe. Unfortunately the heat from the incandescent globe softens these contacts, and the spring loaded socket contacts embed themselves in the soft metal. When the globe burns out, it becomes almost impossible to remove. To make matters worse, on the light fitting the *handedness* of the screw securing the socket to the fitting is invariably the same as that of the bayonet fitting. Hence, when trying unsuccessfully to unscrew burnt out globes, one inevitably unscrews the whole socket assembly (refer to Figure 6.1).

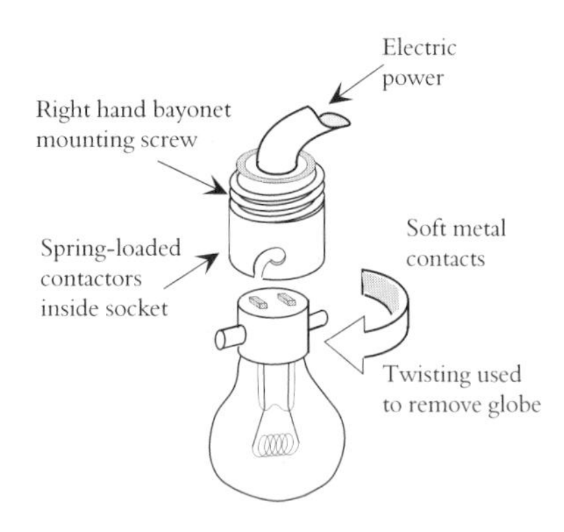

Figure 6.1 Household electric light globe fitting

Design involves both the intellectual and emotional substance of the designer. Intellectual because designing requires considerable analytical skill. Emotional because the designer is a human, designing for other humans, and this process brings with it a vast *grab-bag* of issues the designer must take into account. We refer to the environmental, social, aesthetic and economic nature of design. The designer must be *aware* of these matters to allow modification of the designed product accordingly. A key feature of a good designer is the ability to consider many points of view and to arbitrate between them to the overall benefit of the designed product. Another key feature of good designers is their capacity for coping with change (*flexibility*). In executing design, the designer must progress through an ever changing tableau of ideas and conflicts that bear on the ultimate outcome. This is true even in such simple, familiar products as the clothes peg and the light fitting.

Petroski (1996) presents the design process through the experiences of the inventor, identifying no less than 10 distinctly different patents for the simplest of engineering artefacts, the paper clip. As French (1988) has noted, *"design is a rich study"*.

As a first step in the development of design skills we need to characterise the substance and extent of design. There are at least three components to this characterisation. The first of these is *the designer*. We recognise that the designer is mainly human, although, as noted earlier, there is considerable and rapidly growing interest in artificial intelligence as a means of automating design (Brown et al., 1992). Hence we can argue that, at least in some situations, the designer may be *significantly aided by a machine*. Furthermore, the designer is often a *group*. When group responsibility is taken for a given design, the dynamics of team interaction come into play. Under these circumstances the team leader must be not only aware of the nature of these dynamic interactions, but capable of handling *conflict negotiations*.

Evaluation of psycho-social interactions involved in team dynamics is outside the scope of this text. Nevertheless, we need to recognise that engineering product design is rarely the domain of the individual. Perhaps the most challenging feature of participating in a design team is the individual ambition to add value to the ultimate result. The task of the effective team leader is to extract the best possible value from each individual in the team.

The second, equally important, component of design is the *procedure adopted for design*. Although this procedure is not rigidly defined, many of its steps are repeated in every design. In a formal way this is referred to as *the design method*. We also refer, among design practitoners, to the design method as an essentially *transportable property* of all design experiences. When reviewing case examples in design, it may be very tempting simply to adapt the procedures used for designing solutions to a problem *similar to our problem*. This is a false interpretation of the objectives of case examples, whose intent is to focus not on the *destination* but on the *journey*.

The third component of design is *the product* we are designing. Designers are called upon to address a vast range of products, but at this early stage of our discussion we do not focus on any specific product. Suffice to say that not all products are designable by all designers. Each product must be seen in its own context of application, since there is a considerable range of ideas and conflicts which are peculiar to the operating environment of a specific product (*domain-specific*). Moreover, experience and growing expertise in the design of specific products brings with it a kind of wisdom not easily transported to other domains of product design. Typically, specialist expertise in the design of rock-crushing machinery or steam engines has little relevance or transportability to jet engine or electric alternator design.

6.2.1 The designer as a person

Any discussion of the human designer must necessarily identify the skills possessed by the designer and how these skills are to be developed. We describe the behaviour of the human engineering designer in the context of design. We seek to distinguish those characteristics of the designer that influence the process of designing. This will enable us to recognise characteristics we need to develop in becoming good designers.

Engineering design is a relatively young discipline. James Watt (1736-1819), the inventor of the condenser for steam engines, was a land surveyor, colliery superintendent and businessman. Eli Whitney(1765-1825), the inventor of the cotton gin, was responsible for the modern concepts of mass production. Isambard Kingdom Brunel (1806-1859) designed and built railways, bridges and ships, in addition to working as an architect. Watt, Whitney and Brunel, among other great engineers, clearly performed tasks that are currently regarded the domain of engineering design. Yet these activities in the late 18th and in the 19th centuries were simply included into the broad general role of engineering. With the evolution of a great diversity in manufacturing technology, the artisan-engineer evolved into the manufacturing specialist and design specialist.

Coupled with the evolution of diversity in manufactured goods there has been a consistent evolution and development of specialist experimental and analytical tools in engineering, demanding some highly specialised technical skills. The engineering designer can not be expected to be equally conversant with all of these various specialties. Nevertheless, we must be sufficiently informed about them to identify those aspects of the design where they apply and to link the proper technology to these aspects of the design.

As we will demonstrate, the most significant skill of the engineering designer is the capacity for processing large quantities of diverse information in a well ordered way. This skill becomes especially useful when evolving technical problems from loosely identified needs. Often, many of the requirements of a design are unarticulated by the designer's clients. In general, the designer must identify the various functions which fulfil the requisite need with minimum compromise. Judgement in identifying the best (or at least a satisfactory) set of compromises, is part of this collection of design skills.

We add to these attributes a type of wisdom that regards all design problems, from the apparently trivial to the complex, as proper challenges for the designer's skills. This wisdom is related to the associative skills of *receiving*, *responding* and *valuing*. Designers face many challenges in their lives, and responding successfully to problem solving is a confidence-building experience. By receiving, responding to, and valuing each design challenge without *fear or favour* , the designer will invariably develop this confidence.

It is hardly accidental that this introduction to the *designer as a person* has the flavour of a motivational message. Much of the emotional makeup of the designer will develop in professional life, sometimes years after graduation. Our intention in dealing with it at this early stage is to set the scene for all the learning experiences that follow.

6.2.2 What constitutes design?

Before a discussion on how to go about designing products or systems, we need to explore what designers do. In other professions, such as the law or medicine, we can easily describe the activities of the participants. Doctors heal and lawyers litigate. Designers on the other hand find it singularly difficult to articulate the boundaries of their domain of influence. A few examples might help to illustrate, as well as deal with, this difficulty.

(a) Designing a better helicopter

Figure 6.2 is an artist's impression of the McDonnell-Douglas MDX twin turbine helicopter. In order to achieve a substantial *marketing advantage*, the design goal for this passenger transport aircraft called for nearly twice the payload-to-power ratio available from conventional helicopters. A wise combination of several conceptual design factors contributed to the achievement of this ambitious design goal. The main rotor blades have a flexible polymer mounting at the rotor hub. This arrangement, resulting from a French patent, significantly reduces the mechanical components used in the blade angle adjustment mechanism.

Figure 6.2 McDonnell-Douglas Explorer helicopter

The cabin and empennage (stabilising arrangement of the tail section) were to be constructed in carbon-fibre reinforced polymers. In aircraft construction, combining polymers and conventional aluminium panels is fraught with considerable risk in manufacturing and airworthiness testing. However, the weight saving was seen as a challenging goal, but achievable given the company's experience in the design of military aircraft that had already been built using this material combination.

The aircraft was designed without a tail-rotor. This feature is referred to as the *notar* design. The torque on the main rotor blades imposes an equal and opposite torque on the helicopter cabin. To overcome this problem, most conventional helicopters are fitted with a tail-rotor whose plane of rotation is approximately perpendicular to the rotation plane of the main rotor. Appropriate adjustment of the blade-angles on the tail rotor balances out the torque component imposed on the cabin by the main rotor.

In the *notar* design, the conventional tail rotor is replaced by a fluid dynamic behaviour called the *Coanda effect*, named after its discoverer, Henry Coanda (1886-1972), a Romanian aviator. This effect exploits the capacity of an air jet to *cling* to a curved wall. Figure 6.3 shows the operation of the *Coanda effect* schematically. Air jets are blown out on the side of the tail boom, resulting in a local low pressure region and net thrust on the boom.

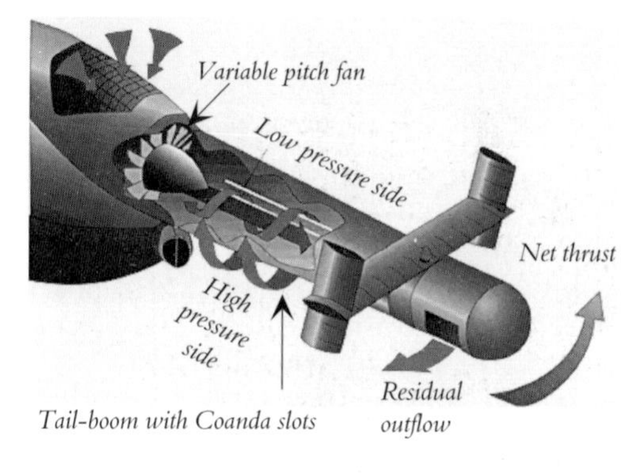

Figure 6.3 Coanda effect on Notar helicopter tail boom. (Courtesy of Kulikov Aircraft Co.)

The *notar* design allows the aircraft to approach buildings and people more closely than is possible with conventional helicopters. Additionally, the weight saving resulting from the absence of the tail-rotor drive and blading is substantial.

The MDX is clearly the result of substantial *know-how (domain-specific wisdom)* about helicopter design. However, marketing staff at McDonnell - Douglas had to establish a *need*, or a market opportunity for this design. Once etablished, the need was serviced by a creative cooperation of experience and design skill. Experience with composite manufacture, polymer mounted main rotors, and the *notar* design, were essential elements of this design synergy. The final result was also critically influenced by the broad range of design skills available to McDonnell-Douglas.

In summary then, this design problem involved:

- identification of a *need* to be serviced, namely the need for a more effective civilian helicopter service: this was the *design goal*;
- evolution of *technical design objectives* and the technical design limits, or *design requirements* within which the design embodiment should be achieved (examples here are : maximum cruise speed at sea level ≥ 278 km/hr; maximum endurance ≥ 4 hr);
- creation of *solution alternatives* from which a final design embodiment would be chosen;
- focused *decision making* on the final design choice;
- delivering clear and unambiguous *design specifications* to the manufacturer: this would include, among other things, drawings, airworthiness testing procedures and material testing specifications.

(b) Designing a better mousetrap

Ralph Waldo Emerson (1803-1832) American philosopher and poet, is credited with the following quote:

"If a man write a better book, preach a better sermon, or make a better mouse-trap than his neighbour, tho' he build his house in the woods, the world will make a beaten path to his door."

The *standard* mouse-trap has been in use for a considerable time. One current form of its embodiment is shown in Figure 6.4. At first sight, it may be surprising that anyone would want to improve on such a simple and apparently effective device. Yet, there are several design problems with the standard mouse-trap. It was designed to be manufactured with simple wire-forming technology. While the manufacture is completely automated, there are nine separate components to be manufactured and assembled (refer to Figure 6.4). Setting of the trap is fraught with some degree of risk to the fingers. If the spring is too strong, the trap can almost decapitate the mouse, and cleaning the trap subsequently can be unpleasant and often results in simply discarding the trap together with its victim.

Figure 6.5 shows a new design for the mouse-trap. Made mainly from moulded polymer, it has only four components. It is easy to set, with a single motion. It is also easy to clean, if necessary, by placing it in a dishwasher. While we can not vouch for its capturing effectiveness, it is, at least from the manufacturing and ergonomic points of view, a *better mouse-trap*. Of course, there are many other possible embodiments for catching mice. Our intention in this design demonstration was to compare these two traps with similar *design intent*.

The summary design procedure for the mouse-trap involved:

- identification of a *need*: recognising that there were some manufacturing and ergonomic problems with the original design, leading directly to the *design goal* of creating a better mousetrap;
- setting *technical objectives*, based on experience with manufacturing processes and some wisdom in ergonomics and kinematics: fewer components, easier manufacturing procedure, ergonomic improvements;
- identifying *requirements* within which the embodiment design can survive successfully: material properties and total cost are examples;
- creating *solution alternatives* for the embodiment;

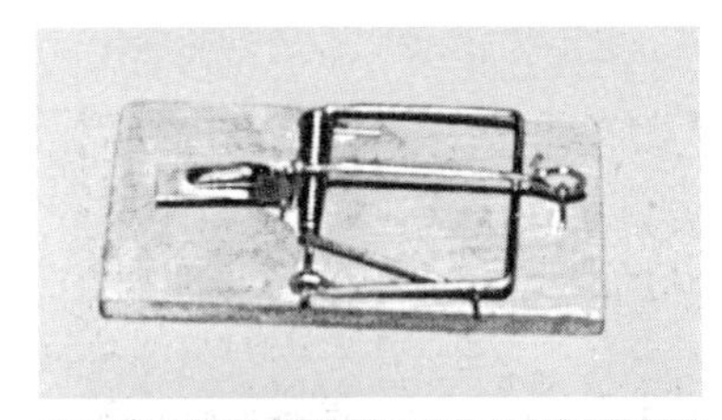
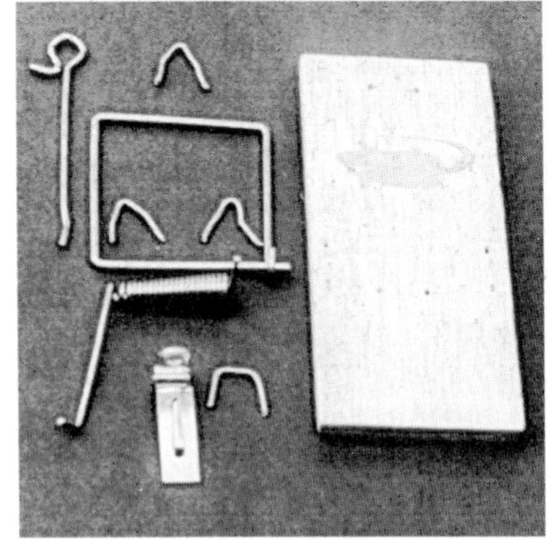

Figure 6.4 Traditional mouse-trap and its component parts

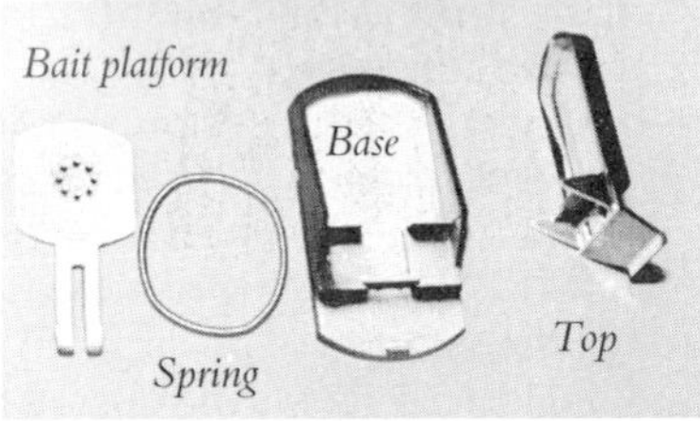

Figure 6.5 New design for mouse trap

- *making decisions* about choice of embodiment: based on evaluating materials, appearance, ease of use, relative cost;
- delivering *design specifications*: including drawings, and material identification and testing.

These two examples of design, one technically complex, the other simple, illustrate the generic design procedure that governs all designs.

Figure 6.6 shows six further examples of consumer product designs, all of which represent a design response to some need. All of them were evolved through the generic process described for the MDX helicopter and the mouse trap. Each is the result of considerable experience, or wisdom, associated with some specific domain of knowledge. Most of the designs in Figure 6.6 are of the type we refer to as *evolutionary*. In these designs, there is still considerable innovative content, but it is usually the result of past experience evolving into some future improvement of an already accepted product.

By contrast, the Bishop variable ratio steering gear, now almost commonplace in many commercial automobiles, represents a different type of innovation– see Figure 6.6(f). It is also the result of a design response to some need. However, the unusual, and completely new, technical content makes this design *inventive* beyond the mere

(a) Office chair

(b) Luggage trolleys at an international airport

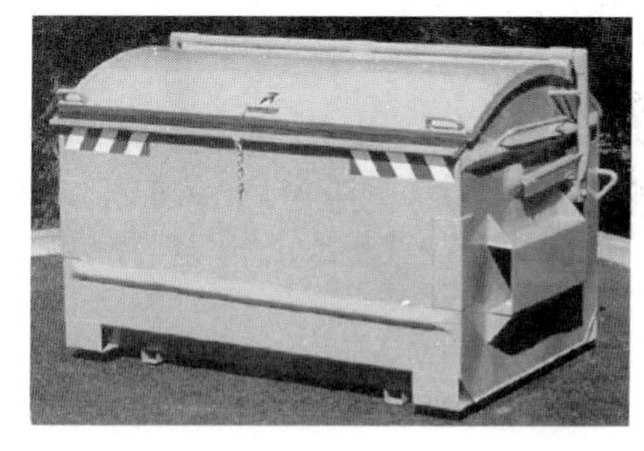

(c) Light fitting

(d) Industrial waste container

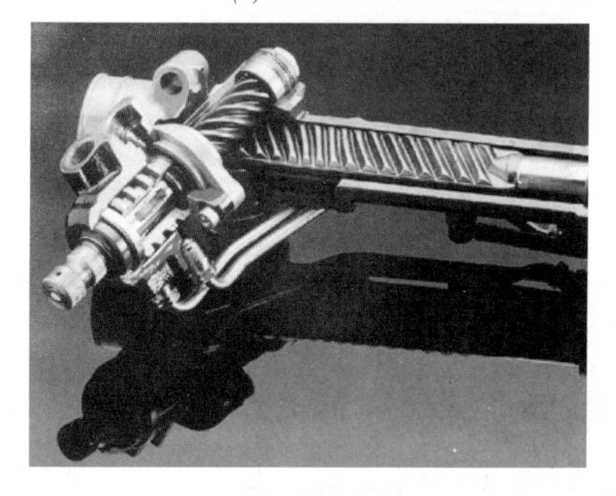

(e) Anamorphic printing

(f) Arthur Bishop's variable ratio steering gear

Figure 6.6 Some further examples of consumer product designs

improvement of an established product. This steering gear renders the steering more direct and responsive at *centre lock*, while driving on a straight path. The steering ratio (turns per degree of steering angle) increases with increasing steering angle, allowing easier parking and manouvering around hard corners.

Figure 6.6 (e)shows a type of printing for flat surfaces that transform images into proper features when the flat surface is deformed into its final container shape. The process relies on a geometric transformation known as *anamorphic* transformation, well known to painters and artists for some centuries (Leeman, 1975). Its application by the container industry is the result of the evolution of new materials and inventive new application of an ancient art. We again see the synergy of experience and wisdom playing key roles in creative design.

6.3 Evolution and enformulation of design problems

6.3.1 The nature of information and problems

In all the examples considered so far, we have dealt with *hard* information. In each case we had some easily identified technical objectives to meet. In all cases the designer may have retained significant creative freedom in the choice of embodiment, but a somewhat restricted freedom to choose a design concept in response to a need. In the case of the *mouse-trap* we didn't consider chemical or electronic or perhaps ultrasonic alternatives to the *real need* : namely a cheap and effective eradication of rodents. We refer to the process of establishing a suitable design concept and *establishing technical objectives* from a specific *need* as *enformulation*.

Enformulation is the result of substantial technical information processing. It brings into play the designers domain-specific know-how (*received wisdom*), as well as a broad awareness of currently available technology, and economic and manufacturing imperatives. Scientists solve problems by operating on abstract models of the problem. Designers can benefit significantly from *enformulating problems initially at an abstract level*. Rodenacker (1970) notes that the most significant creative impact on a design occurs at the highest level of abstraction. A simple and easy-to-understand *need* requiring design enformulation is illustrated by the following example.

At the entrance to some public buildings the presence of birds (mostly pigeons) on the parapet over the entrance can cause a minor hazard to health and clothing (refer to Figure 6.7). This situation represents a *problem*, which is a state wherein we can perceive a *goal* but not the means of attaining it.

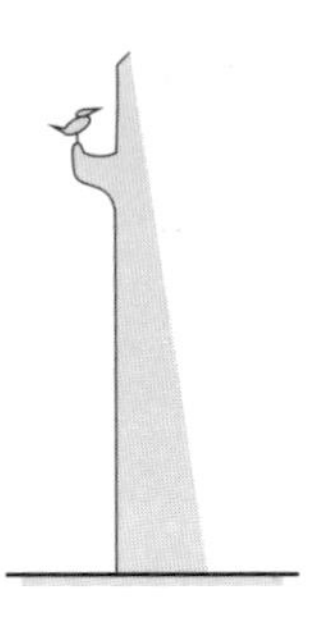

Figure 6.7 Bird hazard at building entrance

Designers (in this case we can think of them as *problem solvers*) respond to problems. Designers (problem solvers) make plans and act on them in ways which they predict will achieve the desired goal.

We identify two categories of problem-solving activity:

(1) planning actions to achieve a goal: we call this *design*;

(2) correcting deviations from desired performance: we call this *trouble-shooting*.

Some engineering texts treat both of these problem solving procedures under the heading of design. Both involve technical problem solving, however, trouble-shooting is more focused on correcting specific, and well-defined, technical problems. Design on the other hand, retains a large element of problem evolution. In what follows, we focus mostly on design problem solving.

As we have noted, the design goals for the products described in the previous section were the result of *hard* information. In contrast, the design goals we might develop for the *bird hazard* would be based on *soft* information. We have no data to tell us how serious the *bird-hazard* is in terms of its impact on eye injuries or dry-cleaning bills. If we had this type of data, we could estimate the comparative value of design effort that could be effectively devoted to solving this problem. After all, there are many, possibly more worthy, design problems making calls on our design capacity. Part of the design *enformulation* involves the *hardening of information*.

The following are some examples of problem statements. Some contain *soft* and others contain *hard* information. Unfortunately, designers are occasionally also confronted with what appear to be problem statements, but contain no information at all. One of the many skills that designers need to develop is the capacity to distinguish between these types of problem statements.

"There are too many road deaths" – *soft* information

"The road deaths per passenger kilometre on Victorian roads is 5% higher than in all other states" – *hard* information;

"Students need large desks" – *soft* information;

"The average student desk top measures 1m x 2m" – *hard* information;

"Our cities are ugly" – *content-free,* contains no useful information;

"Our public transport system is inefficient" – *content-free,* contains no useful information;

"Nobody thinks about the hardships a paraplegic endures" – *content-free,* contains no useful information;

"Clean air is essential for modern society" – *content-free,* contains no useful information.

Motherhood statements was the old-fashioned way of referring to *content-free* problem statements. We prefer to use the more expressive, gender-neutral, version to indicate that these type of problem statements carry no useful information and in general they are to be avoided.

6.3.2 Problem evolution

In an ideal world we could imagine design proceeding, in some nice serial fashion, from *problem* to *solution*. We could also idealise the role of the designer as some *external agent* to the *problem*, who, once advised of the *problem,* could be relied upon to generate the *solution*, almost without interference from the originator of the *problem*. This is the model of problem solving to which we have become accustomed in solving scientific problems. Additionally we have come to expect that some problems, specifically those in mathematics, have unique solutions.

Real design proceeds through a sometimes protracted series of negotiations between designer and client. This is also the case when the designer and client (originator of the problem) are the same person, perhaps *wearing different hats*. Clearly, when the designer and client reside in the same body, these negotiations may take place with less articulation than otherwise. Nevertheless, the negotiations remains an integral part of the design process.

The first negotiation to be managed is the clear identification of a *design goal*. Most commonly, problems in design are stated in a *solution-oriented* way. Examples are: *build a better mouse-trap*; *design a new refrigerator*.

In these example *problem-statements* a specific style of outcome is embedded by the world-experience of the problem's originator. She/he has already experienced mouse-traps and refrigerators, but found some disagreeable features that might be overcome by a *new design solution*. We will allude to this type of mental delimitation of the problem later in the chapter, but it is important to realise that this type of delimitation imposes the most severe restrictions on our problem solving process.

There are many clichés and aphorisms in the design repertoire that arise from experience with this type of delimited problem statement :

" we need to think outside the square" - a variant on de Bono's notion of lateral thinking;

"polishing the bullock-cart will never lead to the automobile or aeroplane";

"what you want is not necessarily what you need";

"designers are always asked to 'fix it, but don't change a thing'*"*.

A useful way of thinking about problems is to regard them as *need-directed*, or *goal-directed*. A *need-directed* problem description represents a vague, general, dissatisfaction with some aspect of the current world. Additionally, in general, a need-directed problem description has goals which are implicit, or *unarticulated*. The design process resulting from a need-directed problem description is, in general, *open-ended*. The designer and problem originator must evolve, through some form of negotiated procedure, clearly articulated design goals. No design work can proceed without these design goals.

In contrast to need-directed problem descriptions, the goal-directed problem statement has a specific goal (or goals) which is (are) more clearly articulated. Both types of problem statement are valid, and have different advantages. While we strongly advise avoiding problem descriptions bound up with some imagined solution embodiment, we recognise that these types of problem descriptions are common. Consequently, the process of negotiation from problem description to manageable design goals can be regarded as either a *forward-chaining* or *backward-chaining* problem solving procedure.

Figure 6.8 shows a generic flow-diagram for evolving design goals from need-directed problem descriptions. We can negotiate with our client to develop acceptable and achievable alternative design goals. This is indicated by the *Level 1* boxes in Figure 6.8. Each design goal will evolve its own set of technical objectives, indicated by the *Level 2* boxes. The sets of technical objectives will lead on to some set of embodiments at *Level 3* of our problem solving process. We could move on to *Level 4* (not shown), representing detailed design specifications, eventually followed by manufacturing specifications, and so on.

A simple, familiar, example will illustrate the procedure. Suppose that we are concerned with purchasing an automobile. The two choices under consideration are:

- a two seater convertible sports coupé: for speed and personal image;
- a medium size, four seater sedan: for conservative urban transport.

At this stage of the decision making process, we have already *forward-chained* to the embodiment stage of our problem solving. We began with a problem description of the form:

my transport arrangements are unsatisfactory.

After some negotiation and evaluation of the various transport alternatives, we have identified the *design goal* as:

I need to select an automobile best suited to my needs.

Our *technical objectives* may have included some specification of the maximum cost, or the minimum passenger accommodation. It is unlikely that we would have considered such matters as minimum rate of acceleration, fuel economy or other such technical performance characteristics. In the current climate of automobile marketing these performance characteristics are very similar for automobiles in similar price brackets.

Hence, we come to the choice of *embodiment*. Our decision problem is embedded in our perception of what each alternative choice might offer as *excitement quality* to our purchase. The *sports coupé* will provide excitement and a special personal image, while the *sedan* will provide conservative urban transport to an eventual family. If we now submit our whole problem evolution process to a *design-review* and ask ourselves:

'what is it I really want out of this purchase?: what is the real essence of the problem?',

then we are essentially *backward chaining* in the problem evolution process.

The result of the *design-review* might lead to our recognising that our *strategic needs* are best served by the *conservative sedan*, and we can still get the *image and excitement* aspects by *joining a sports racing club*. Naturally there may be other *design* alternatives. Whatever the outcome, however, the

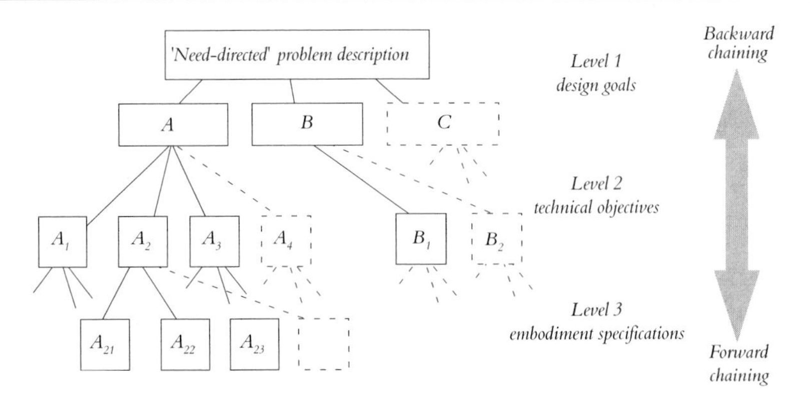

Figure 6.8 Evolving the design problem

essence of this *backward-chaining* process is that *design reviews* are absolutely essential to the overall health of the problem evolution process.

Before moving on to discuss the enformulation process, we note that problem descriptions can cover a spectrum of styles, ranging from the completely *need-directed* to the completely focused *goal-directed* form. Table 6.1 lists some examples of problem descriptions.

Ex 6.1 The 'bug-list'

An exercise for design teams of two to six people.

In every walk of life we find problem situations that become irksome and eventually grate on our consciousness. We might hear people referring to such situations as:

"this X really bugs me!",

where X represents some irksome problem. Typical examples are:

"I hate to get up on cold mornings";

"there are some really bad smells in my refrigerator";

"there is never enough space on notice boards - nobody clears off old notices";

"I hate looking for parking space at the supermarket".

Historically, the term *it bugs me* is probably derived from the technical vocabulary suggesting *there is a bug in the system*. Radio and other communication systems, relying on early forms of electromagnetic relays, would sufer loss of contact, particularly in the tropics, when small mites, *bugs*, became lodged between the relay contacts. In these cases there really were *bugs* in the system, preventing it from functioning. Nowadays, *"there's a bug in my system"* has become associated with any difficult-to-identify problem that makes a system malfunction. Hence, we have such terms as the *"Y2k bug"*, a reference to malfunctions expected when computer clocks showed two zeroes in the date at the end of the 9th decade of the 20th century.

(a) Make a list of twenty problems that *bug* you. When completed, present your list to your design team, perhaps expanding on each item in your *bug-list*. In essence, you will be expected to *defend* your *bug-list* within your peer-group of designers.

(b) The design team should jointly agree on a suitable *problem-statement*, selected from the several *bug-lists*, to be developed into a sustainable *design goal*. The team should then negotiate their way through to some reasonable *technical objectives*. In each case, the owner/proposer of the chosen *problem* will act as *design client* to the rest of the design team.

(c) Prepare a poster presentation of your chosen *problem*, to be viewed by other design teams. The poster must identify the *problem-statement* and the *bug* from which it was derived, and the resulting *design goal*, in the form of a series of sketches. Your design team should plan the poster presentation to be, as far as possible, self-explanatory.

Notes:

1. In generating the *bug list*, and its distilled *design goal,* your team should keep a record of ideas explored during the design negotiation process, as well as notes and sketches. (We call this an *idea-log*: refer to McKim, 1972. In general, the *idea-log* represents a series of *snap-shots* of the designer's

Table 6.1 Some typical problem descriptions

Problem type	Problem description
Completely need-directed	*– emergency vehicle access in the central business district needs to be improved*
Order of increasing specific goal focus ↓	*– people suffering from arthritis find it difficult to open packaged food: a solution is needed*
	– design specifications are needed for an apple harvester
	– design specifications are needed for a domestic washing machine
	– design a horn for an automobile
Specific goal-directed	*– design an antenna for a geo-stationary communication satellite*

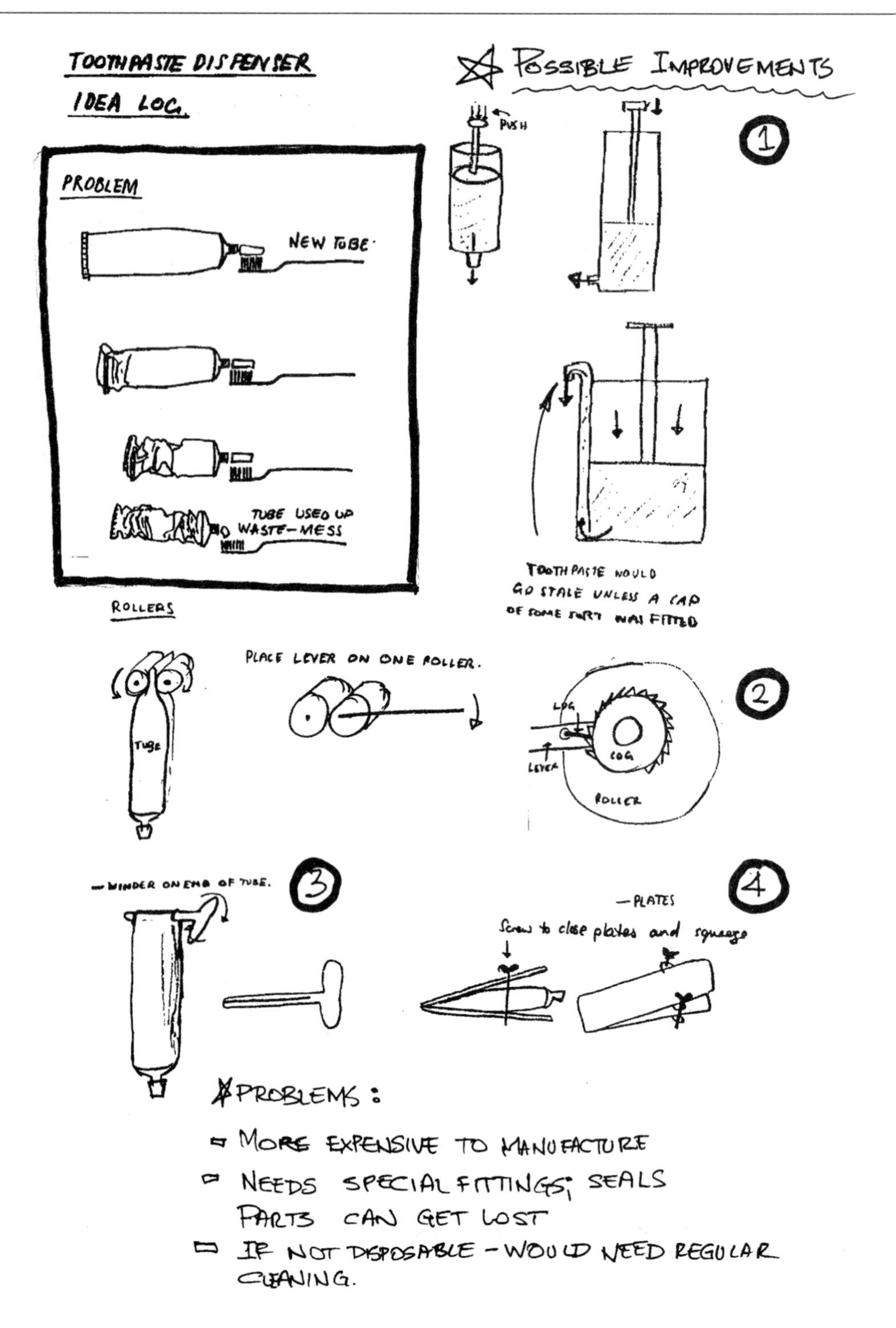

Figure 6.9 Typical idea-log: a problem and some solution alternatives

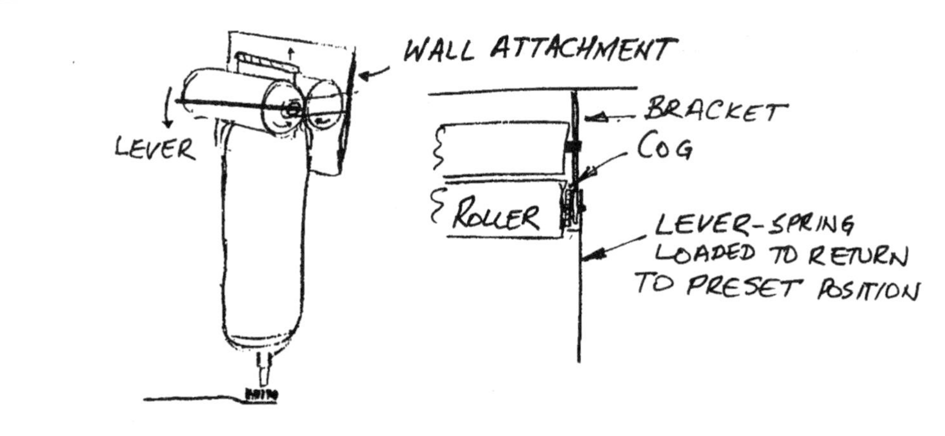

TUBE HANGS FREELY FROM ROLLERS & GETS DRAWN UP WITH USE

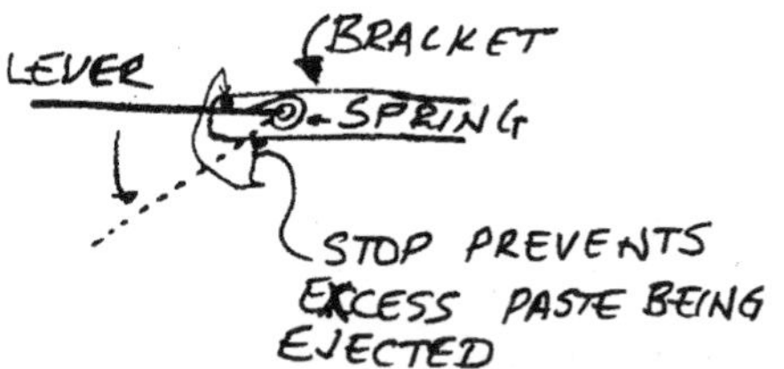

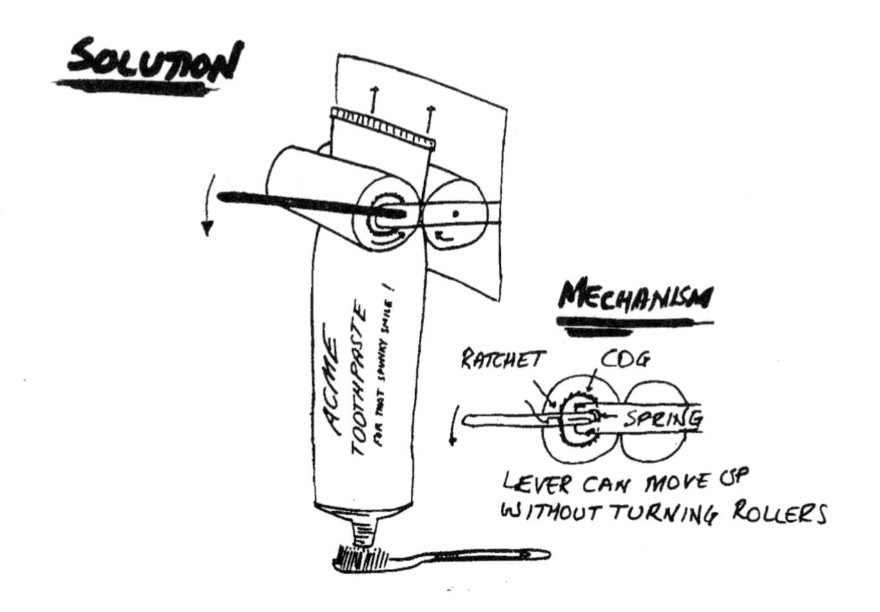

DEVELOPING THE SOLUTION

Figure 6.9 (continued) Typical idea-log: developing the solution

thinking process during problem evolution. Figure 6.9 shows a typical idea-log.)

2. To help you generate your *bug-list*, we offer a list of problem areas you may wish to explore. The list is intended only as a memory-jogger, and you need not adhere to problem areas offered. In each category we note some examples of design goals generated in past bug exercises:

Food *(production, preparation, distribution): keeping hospital food warm; making buttered toast; improved handling of eggs (to reduce breakage);*

Shelter *(mass produced housing or household items):keyless homes; toothpaste dispenser; improved bathroom fixtures; quiet toilet cistern; auto dog food dispenser;*

Clothing *(cleaning devices, disposable items): home dry-cleaner; auto ironing system;*

Transportation *(public or private transport, highways, freeways, rapid transit, congestion): campus/ urban congestion; car escape system; braking-rate indicator; quick windshield de-icer; improved car jack; improved spark plug remover;*

Communications *(written, spoken, telephone, electronic, radio television): improved notice board management; braille communication aid; device to awaken the hearing-impaired; television advertising avoider (to skip over advertising during entertainment broadcasts);*

Recreation *(children's toys, games, sports items, camping/hiking gear): spherical rolling toy (roll in any direction); anti-loss golf ball; improved design for harp tuning;*

Education *(lecture theatres, multi-media, libraries): distributed libraries and learning aids; quiet lecture theatres (means of controlling noise in large classes);*

Environmental *(deforestation, pollution, waterways, urban environment): reduced urban noise; portable pollution meter; trip planner (means of avoiding congested traffic areas);*

Human support *(aids for aged and disabled, access to buildings, hospitals, health and emergency services): anti-damage thermometer case; optical support for micro-surgery; therapeutic exerciser for weakened muscles; special aids for arthritic people (opening packages, turning on faucets etc.);*

3. For reasons that remain a mystery, we find that some student groups engaged in the bug-list exercise become obsessed with scatological concerns – typically: loud toilet flush, cold toilet seats, smelly toilet, seat too low, seat too high and similar. These may be valid design concerns, but try not to limit your thinking in this way.

Ex. 6.2

Figure 6.6 shows several products that had significant design input. In each case, write down a *problem-statement* to which the various design embodiments, shown in Figure 6.6, could be the design responses. Also write down what you consider to be appropriate *design goal*, that might follow from each suggested *problem-statement*. Do your design goals match the design outcomes shown in Figure 6.6? Do you think some other *design goals* might have better served your problem statements?

Note: The review procedure suggested by this exercise is a preliminary *design audit*. We will refer to *design audits* later in this chapter.

6.3.3 Problem enformulation

Part of the designer's enformulation task is to process *soft* information into *harder* information, by identifying *islands of certainty* associated with the problem. Figure 6.10 shows an abstracted model of the design enformulation process.

In the model of the design enformulation shown in Figure 6.10, we see the vaguely stated *problem* mapped onto several sets of technical objectives. As a matter of course, there will be considerable overlap between these sets of technical objectives. Nevertheless, each of these sets of objectives commonly represent distinctly different design alternatives. It is in the nature of design that there is no unique one-to-one mapping between *problem* and *technical objectives*. We have avoided calling these sets of technical objectives *solutions* to the problem since, in general, a solution implies some form of embodiment. Considerable creative design work is involved in moving from sets of technical objectives to the final embodiment. Each of the samples of product design shown in Figure 6.6 represents only one of many possible embodiments that will all satisfy some given set of technical objectives.

We call the process of mapping from *problem* to technical objectives *enformulation*. Enformulation imposes a *structure* on a problem. Additionally, in the enformulation/structuring process we identify *design parameters* that, in general, represent the value measures we use for evaluating the *goodness of fit* of the design solution to our problem. In Figure 6.10 we have used a *curved boundary* to represent the initial *problem-statement*, indicating its relative *softness*. The angular boundaries of the *islands of certainty* within the *soft* boundary of the original *problem-statement* represent established, *hard*, information about the problem. These may be limitations, or *requirements* that bound the problem, or *constraints* imposed by some statutory authority, or even some special *window of opportunity*, that delimits the time-scale or completion time of the enformulation process.

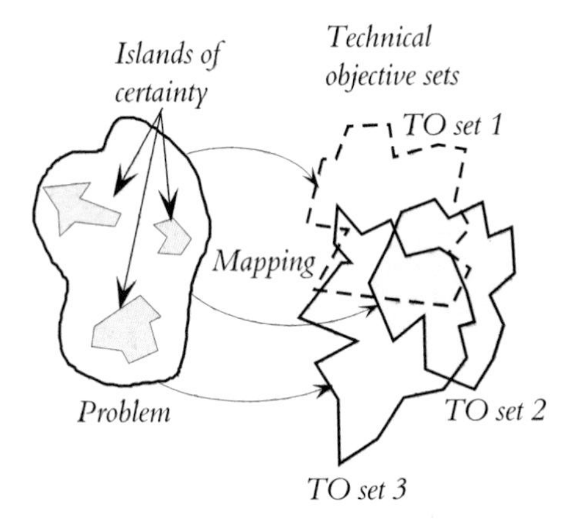

Figure 6.10 Abstraced model of the enformulation process

An example will clarify our terminology.

1. Problem statement (*soft*) : *"Too many bicycles are stolen from the university campus"*
2. *Islands of certainty* : The University of Smallville has 30,000 students, of whom 15% travel to the campus by bicycle. Of these, approximately 4,500 cyclists, two third (3,000) rely on some form of outdoor security for their cycles. The average value of these cycles is $ 500, and approximately 600 to 700 are stolen during the spring term. The total annual loss is approximately $ 325,000. This cost, spread over the cycling population, represents $ 73 annually. If we compare this cost to the $ 400 for car parking annually, or other forms of transport costs, we can see that charging a fee in the vicinity of $ 73 for guaranteed secure parking of cycles is quite reasonable. Moreover, we can consider a three year amortisation of the cost of installation, which is normal business practice for small capital expenditure. Hence we can afford to spend approximately $ 975,000 on installing secure parking for bicycles on the Smallville campus.
3. The layout of the campus is being reviewed, and the buildings administration is preparing a brief for consideration by the finance committee of the university. Plans for minor works on campus will be considered during August this year. This presents a *window of opportunity* for the submission of design recommendations for the provision of secure bicycle parking on campus. A suitable design evaluation must be completed by the end of July.
4. The *requirements* on the design of secure bicycle parking system are:

- it must accommodate at least 1000 cycles – an associated desirable design objective is to deliver a modular design, so that, should the need arise, the ultimate solution may be extended to accomodate up to 4000 cycles;
- it must not cost more than $325,000 per 1000 cycle module to build and install;
- the design plan must match the spread of cycle usage on campus (to be established by survey);
- the design must meet the requirement o f the *Metropolis Standard on Bicycle Parking Facilities*. This a mandatory *constraint* on the design.

CASE EXAMPLE 6.1: LEVEL CROSSING PROBLEM

Level crossings are used where two forms of transport systems, usually road and rail, intersect each other. In sparsely populated Australia, level crossings between road and rail are much more common than viaducts or tunnels. In the early 80's there were several traffic incidents involving property damage, injury, and even deaths at level crossings. Figure 6.11 is an extract from a news article that appeared in the Melbourne Herald in 1981. There was a public outcry for an inquiry to be conducted. What follows is the enformulation

TRAIN DRIVERS STARE DEATH IN THE FACE

By GEOFF EASDOWN

Surgeon Eric Wrigglesworth can still remember these words of a troubled train driver . . .

"When you see the looks on the faces of the women and kids staring up at you, knowing they've just missed out, you cannot forget . . .

"It makes you sick . . . sometimes physically."

On the level — it's hell

Train driver Leo O'Reilly . . . a brief delay and the pressure doubles.

HAUNTED BY THOSE FACES

GUIDE TO A CITY'S DEATH TRAPS

'It's there . . . and you hit it'

Jack Vernon

By LAWRENCE MONEY, Herald Transport Writer

Train driver Jack Vernon stared in horror as the white Holden buckled then twisted under his cabin.

Through the window he could see two children being tossed about as the train bulldozed the car down the track.

"I just froze," Mr Vernon said today about the level crossing accident that drove him out of the railways for good.

Melbourne Herald - 1981

Figure 6.11 Extracts from the Melbourne Herald in 1981

of this problem, together with the identification of its *design parameters.*

'NEED-DIRECTED' PROBLEM STATEMENT :

"There are too many deaths and injuries at level crossings."

Note : Major concern was expressed in the press about this problem in 1981. This had the effect of raising public awareness, which in itself may have reduced the problem.

EVOLVING THE PROBLEM : QUANTIFYING ISLANDS OF CERTAINTY:

- how serious is this problem?
- how much is it costing us?
- how much can we spend fixing it?
- where do we put our design effort (and hence expenditure) most effectively?

Table 6.2 shows accident data collected for the years 1982 to 1985.

There are three different types of level crossings, identified by the style of barrier or warning signal used to divide rail traffic from road traffic. Gates and booms are movable barriers, operated either by railway staff, or by automated signals. Flashing lights and bells are used with level crossings without barriers. They provide warning for road traffic users about approaching trains.

The majority of level crossings have no warning signals or barriers, as shown in Table 6.3. However, since these types of crossings are mostly in sparsely populated rural areas, we need to find an appropriate measure for the *seriousness of the problem* in terms of its social or economic impact.

Table 6.2 Consolidated accident data

	Type of crossing	Incidents
Type of loss: Deaths (Summary 1982-1985)	Gates/Booms	5
	Flashing lights/Bells	8
	No automatic signals	0
	Total	13
Type of loss: Property damage (Summary 1982-1985)	Gates/Booms	58
	Flashing lights/Bells	16
	No automatic signals	11
	Total	85

Table 6.3 Level crossings in service, by type

Type of crossing	No. in service
Gates/Booms	181
Flashing lights/Bells	416
No automatic signals	2,581
Total	3,178

The measure, or *parameter,* we have chosen, to compare the accident performance of these different types of level crossing, is the number of accidents incurred per 1000 crossings per annum. This data is shown in Table 6.4

Table 6.4 Degree of seriousness of problem

Rate of incidents/1000 crossings/year		
Type of crossing	Deaths	Property damage
Gates/Booms	6.9	80.1
Flashing lights/Bells	4.8	9.6
No automatic signals	0	1.1

Figure 6.12 shows the accident data over the period 1978 to 1985. Possibly, the overwhelming publicity given to this prioblem in 1981 has significantly influenced the results.

PROBLEM ELEMENTS FOR EXPLORATION

This is a very complex problem and it deserves further exploration before settling on some specific design goals. The following range of problem elements could be further explored by the designer/design team:

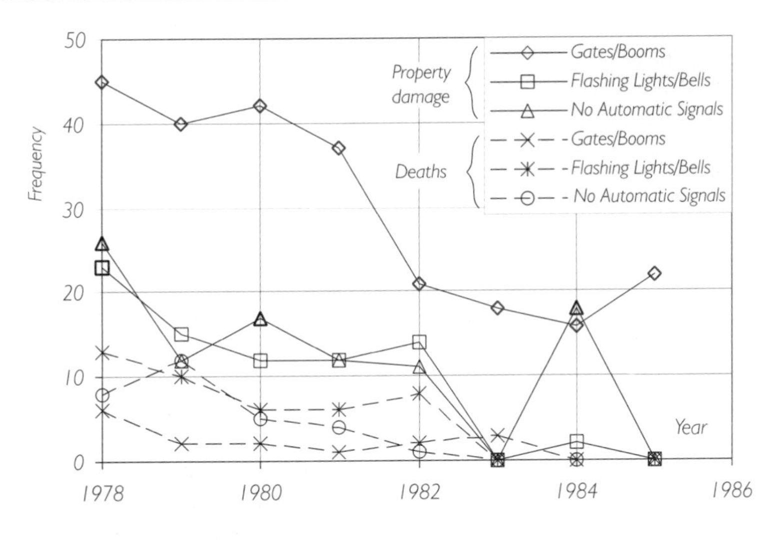

Figure 6.12 Level crossing accident data

Factors concerning the *car driver*

- possible over-familiarity with crossing – particularly in the case of local users?
- concentration/arousal – long straight bitumen roads?
- predicted response to signals:
  visual – passive, active;
  tactile – for example via the road surface;
  auditory.

Factors concerning the *crossing*

- presence/influence of side roads near crossing?
- influence of sunlight in morning/evening?
- illumination – of crossings, trains, cars.

Other factors

- health of train drivers?
- institutional barriers to concerted action – need to involve the service providers and local authorities.

RESOLUTION AND POSSIBLE DESIGN GOALS

1. advertising/public education campaign.
2. separation of traffic streams.
3. design of an *early warning device.*

Notes:

Goals 1., 2., and 3. are not mutually exclusive;

Goals 2. and 3. are *infrastructural* solutions, requiring resolution by statutory authorities.

From the above we could now specify the following *goal-directed* technical problem statement:

DESIGN GOAL:

Design an *early warning device* for level crossings.

TECHNICAL OBJECTIVES:

- suitable for crossings serviced by Gates/ Booms;
- cheap;
- reliable;
- easy to maintain;
- fail-safe.

CASE EXAMPLE 6.2: PORTABLE WATER HEATER

This is an already highly evolved, *goal-directed* problem, stated in specific terms.

DESIGN GOAL:

A new type of portable electric water heater is required: to be used for hot water supply at family outings, and picnics.

TECHNICAL OBJECTIVES:

- small enough to fit into the luggage space of a family automobile;
- it should operate from a 12 volt D.C. supply.

INFORMATION NEEDED:

The designer needs to gather information, to make it possible to intelligently negotiate the enformulation process, from the *technical objectives* to specific *design parameters*. These parameters will eventually allow us (the designer) to evaluate competing alternative designs.

SIZE:

- luggage space size?
- how much water do I need at a picnic?
- what are currently available portable hot-water storage (thermos) sizes?

POWER SUPPLY:

- what types of heating elements are available now?
- should we use conventional or exotic materials?
- is there a standard size?
- can they work with 12 volt? Is there some alternative form of power supply we couild adapt to 12 volt D.C.? Does this suit all automobile battery supply voltages?
- power? current drain?

OPERATIONAL MATTERS:

- how will it operate? controls?
- water in / water out?
- allowable skin temperature?
- safety cut-outs?

The above issues need to be evaluated before we can settle on the technical specification for this design problem.

(a) Facing the problem description

(b) Generating design goals

(c) Gathering data for problem evolution

(d) Negotiating with client

(e) Negotiating with Research & Development

(f) Negotiating with manufacturing

(g) Employing heuristics

(h) Innovative R&D

(j) Field test and review design

(k) Presentation to management

Figure 6.13 A lighthearted view of the design process[6.4]

6.4 These wonderfully perceptive sketches are due to A. E. Beard , reproduced by permission from Clarke, R.L. and Beard A.E. (1973) Lighter Engineering, London: The Institution of Mechanical Engineers.

6.3.6 Problem intensity parameters

We have explored generic problem evolution, as well as two case examples, representing the two ends of the problem description spectrum. The level-crossing problem was an *open-ended, ill-defined, need-directed* problem, while the portable water heater problem was a relatively, *well-defined*, specific *goal-directed* problem. The most common questions raised by young designers when exploring such case examples are: *"How will this help me with my problems?"*; *"What aspect of this, or that, case can I apply directly to other problems, that I am likely to face in my design career?"*

As we have already noted earlier, design is a life-long learning experience. No engineer designs always and no engineer designs never. As Rittel suggests (Rittel, 1987):

"All designers intend to intervene into the expected course of events by premeditated action. All of them want to avoid mistakes through ignorance and spontaneity. <u>They want to think before they act</u>. Instead of immediately and directly manipulating their surroundings by trial and error until these assume the desired shape, designers want to think up a course of action thoroughly before they commit themselves to its execution."

The descriptive form of the *design process* that emerges from these and other case examples only serves to illuminate the most general aspects of the problem solving processes involved in design. Experienced designers will probably target solutions rather quicker than inexperienced designers. These *design experts* make use of their developed, domain-specific, *wisdom* in reducing their search for a solution. The formal description for this process is called *making use of heuristics*. The Oxford English dictionary offers the following definition:

"heuristic (adjective): serving to find out or discover; using trial and error".

The experienced designer will make best use of received wisdom to select *trials*, that might lead to a solution with least likelihood of error. A parallel can be found in the game of chess. The rules of the game are defined by the permissible moves of the several pieces. It would seem reasonable that anyone could play like an expert if we had the mental capacity to evaluate the consequences of all legal moves remaining after every move during the game. The vast number of available possibilities that would need to be explored makes this process prohibitive, even for electronic computers. It has been estimated that the number of possible legal combinations of moves, to a depth of three moves, is of the order of 10^{14}. Chess playing has been extensively studied as a generic, complex, problem-solving process. Chase and Simon (1973) report on the perception of solutions in chess by novices and masters. Based on their studies they note that:

"Masters [of chess playing] *search through about the same number of possibilities as weaker players — perhaps even fewer, certainly not more — but they are very good at coming up with the* 'right' *moves for further consideration, whereas weaker players spend considerable time analyzing the consequences of bad moves."*

Chase and Simon go on to suggest that chess masters develop heuristic strategies that help them find the *right* moves more quickly than novices.

We can regard highly experienced, wise, design problem-solvers, as the *masters* of their game. They will have developed problem-solving tactics in their knowledge domain to enable them to target rapidly on substantially useful solution alternatives. Experience with problem-solving also allows expert designers to develop heuristic strategies for tackling the problem evolution and enformulation process. In what follows, we list some guidelines to be adopted when experience fails to offer better heuristics.

First of all, we identify some generic measures associated with design problems. We call these measures *problem intensity* parameters (Samuel and Weir, 1997). Secondly, we offer a general strategy for employing these parameters in negotiating our way through to problem enformulation.

Problem-boundary

We need to be aware of the limits within which the problem is being explored. When dealing with specific products, we need to identify the sphere of application for which our designed product might be used. For example a light fitting might be intended to serve as a *street light*, a *freeway light*, a *household light*, a *design-office light* or a *mood-light* for a discotheque. All of these applications have special constraints and design requirements. Each set of constraints and design requirements in turn represents the *problem-boundary* which delimits our solution space.

Novelty

This measure is associated with the technical novelty of the problem. Typically problems may be *evolutionary*, where previous technical solutions exist, or *innovative*, where radically new information needs to be generated for their solution. Of the products shown in Figure 6.6, the light fitting and waste bin are evolutionary problems, while the anamorphic printing and the Bishop steering gear are highly innovative, and technically novel.

Novelty is a subjective measure, depending on who is addressing the problem. If the designer involved needs to explore a new or infrequently experienced knowledge domain, the novelty, and the consequent cognitive load imposed by the problem, will be high.

Complexity

Problem complexity is a measure associated with:

- the number of knowledge domains spanned by the problem;
- the number of concepts (for example, structural integrity rules, thermo-fluids rules, electro-mechanical rules) involved;
- the degree of interconnectedness and number of links between the various concepts involved in the problem.

In total, this is a quantity that measures the cognitive load on the designer. A useful way of evaluating complexity is to draw a *concept-map* of the problem (see for example Buzan, 1974; Klein and Cooper, 1984; Samuel and Weir, 1995)

Discordance

Discordance measures the conflict between constraint values and physical reality in a problem (such as material property limitations —*"required yield strength = 1 GPa; required operating temperature = 2000°C"*), or between key objectives (*competing objectives*). As a degenerate example of discordance, the perpetual-motion machine is a very (intractably) difficult problem, because of the irreconcilable discordance between its constraints (that the system is closed and energy is to be continuously extracted from it), and physical reality (the first and second laws of thermodynamics; Ord-Hume, 1977). Locating a power-generating station is a socially discordant problem due to the high level of conflict between competing objectives. Everyone wants access to electrical power, cheaply and reliably, at the plug in the home. Yet, nobody is prepared to accept a power-generating station *in my backyard*.

Modelability

In general, all design problems involve some modelling, or abstraction. The scale of resources needed to generate an acceptable model is a measure of modelability. The power distribution system (Assignment 1.6) is relatively inexpensive to model, since a believable explicit model may be generated for this problem. In contrast, the level crossing problem, or the bicycle parking problem are intractably difficult to odel.

Seriousness

This measure identifies the social, economic and environmental impact of the problem. Finding an alternative fuel for automobiles that might result in reduced air pollution, or finding the appropriate site for a power generating station, are problems with a *high level of seriousness*. The level crossing problem had a high social impact, but relatively small economic impact. In its time, it might have been regarded as a very seious problem. Clearly, that is no longer the case, even though a technical solution has not been found. The portable water heater is a problem of relatively *low level of seriousness*, even if its economic impact on the manufacturer might be substantial.

A strategy for problem enformulation

Novelty, *complexity*, *discordance*, and *modelability* are *cost dimensions* which collectively represent the drain on resources or obstacles to solving a design problem. *Seriousness*, on the other hand, represents the *driving force* to develop a solution to the problem. It is the measure of the value of the opportunity to influence the social, economic and environmental status quo. The *problem boundary* serves to deliit our problem to a manageable sphere of influence. In general, designers need to be clear about the bounds on their problem to permit *enformulation* within reasonable time and economic scales. It would be unreasonable to expect a single design team to address such problems as world famine or global warming.

Developing a pollution-free alternative fuel (for example nuclear fusion) has high levels of complexity, discordance, novelty and modeling costs. Yet it is worthwhile to expend considerable resources world wide on this problem, due to the large benefit to be gained from its high degree of seriousness.

We can apply these qualitative *problem intensity parameters* to a diverse range of problems. Typically, a problem of high problem intensity has a high risk associated with seeking a solution. Hence such a problem needs to be one of correspondingly high seriousness measure to make this high solution risk worthwhile.

Ex. 6.3

In exercise 6.2 you were asked to carry out an initial *design audit* for the product design examples shown in Figure 6.6. In each case, arry out a *problem intensity analysis*. Estimate the problem intensity parameters for each example and draw up a table of comparative problem intensities for the six product items shown in Figure 6.6.

Clearly state all assumptions, particularly those needed to identify the *problem boundary* for each product.

6.4 The design process

In this section we present a formal procedure for engineering design. We have already alluded to the fact that engineering design can proceed informally, where experience and heuristics point the way to a more rapid isolation of solution alternatives. However, in general, we need to regard the design process as a formal, contractual arrangement between client and designer. In what follows we identify the operational rules of this arrangement. The rationale for employing this formal procedure is based on these assertions:

- enginering problem solving generally engages client and designer in a complex negotiation where it is easy to lose sight of the original objectives;
- client and designer act in both cooperative and adversarial roles at various times throughout the problem solving process;
- we need to keep a clear record of our trajectory through the problem-solving process. This can safeguard against back-tracking over previously covered ground and the inadvertant overlooking of some important design issue;
- there can be, and often is, significant discordance between a client's and designer's perception of the problem to be tackled;
- there can be, and often is, significant discordance between a client's and designer's perception of what can be delivered in response to the client's stated and perceived needs. We need to be certain that the problem solving process delivers exactly what has been agreed to.

6.4.1 The components of the process

1. *Problem recognition*: This is usually achieved through a recognised or articulated need.

Problems can come to the designer from many sources. They can be the result of a larger problem solving programme (for example, developing a new photocopying process), part of the day-to-day operation of a product manufacturer (for example, planning and design of new model automobile), or even self-generated (for example, *"I can desig a better mouse-trap"*). As a cautionary note, beware the colleague or friend who approaches you with:*"Have I got a great design problem for you!"*. Unless there is a clear statement of the problem, its boundaries and its restrictions fully agreed to by client and designer, preferably in writing, the problem doesn't really exist at all.

2. *Problem evolution* and *enformulation*: Evolution establishes the problem boundaries and the nature of the problem. Enformulation develops the several design parameters that allow the designers to make comparative judgement about competing design proposals.

Pahl and Beitz (1995) refer to this aspect of the design process as *problem clarification*. While this may be a useful, simplified, description of what takes place during prolem evolution, there s considerably more value added to the problem during this phase than the notion of mere clarification implies.

3. *Solution generation*: This is a divergent phase of the problem-solving process, closely coupled to problem evolution. This phase of the design relies on many skills and talents embodied in te designer (or the design team). Experienced designers use a great variety of aids to promote idea generation (see for example French, 1985, 1988; deBono, 1976; McKim, 1972).

4. *Prediction of outcomes:* This phase of the design relies most onmathematical modelling and estimation. The outcomes from alternative design proposals need to be predicted in terms of operating parameters developed during the enformulation phase.

5. *Evaluation of feasible alternatives:* Selection of those proposals that meet the agreed design objectives and restrictions within the problem boundaries.

Steps 4 and 5 constitute the covergent design phase of *decision making*: during this phase the designer (or team) develops the most favoured design proposal to match all the problem requirements restrictions and constraints.

6. *Specification of the solution*: This aspect of the deign process is the most crucial in terms of delivering the promised/hoped-for performance of the chosen solution alternative.

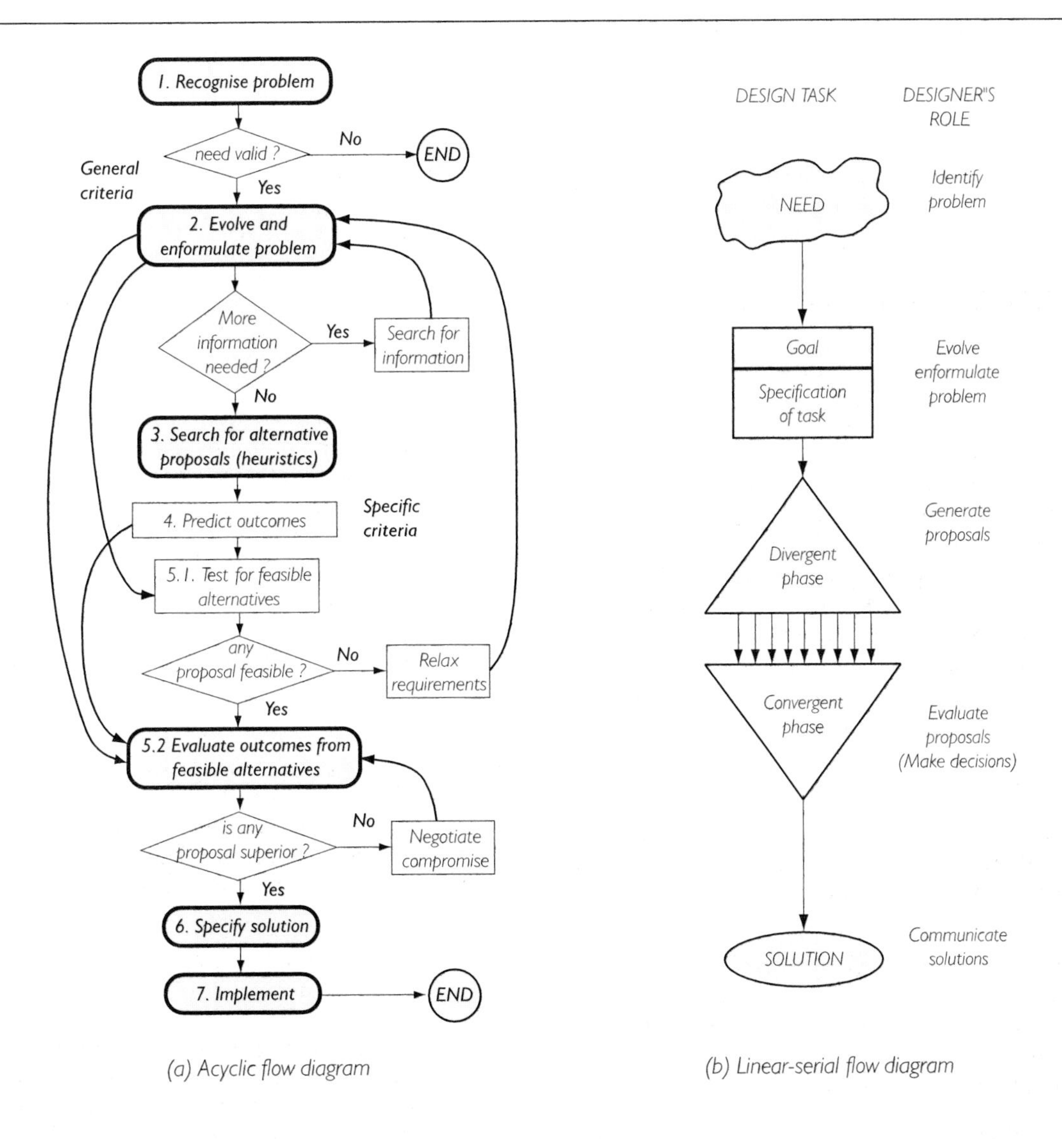

Figure 6.14 Flow diagram for the generic formal design process

The specification communicates, in unambiguous, precise detail, the elements of the solution that determine successful function of the solution. In practice, during this phase, further serious negotiation takes place between the designer (team), and the client

7. *Implementation*: This phase includes prototype testing and evaluation.

In the past much of the implementation aspect of design was devolved to manufacturing. Current ideas of quality planning and management (*concurrent* or *simultaneous* engineering) make it mandatory to include the designer in this phase.

Of course, after implementation comes maintenance, servicing in use, and, eventually, product retirement. Product retirement is urrently receiving considerable attention, due to its impact on environmental issues in product design (see for example Sivaloganathan and Shahin, 1998). All these *downstream* issues of design invariably esult in the ientification of further needs and the initiation of another cycle of the design process. Figure 6.14 shows two typical flow diagrams for the formal design process.

Figure 6.14(a) indicates te acyclic interconnected nature of the process, while figure 6.14(b) shows a simplified linear-serial version of the same process. Both types of diagram are useful for describing the several stages of design. The acyclic flow diagram is useful for keeping track of all the necessary steps and interconnections when preparing a design review. The linear-serial diagram is a useful simple abstraction of the whole process. It draws particular attention to the divergent-convergent nature of problem solving. Moreover, it clearly identifies the need for separating the *idea-generation* phase of the process from the *evaluative-judgemental* phase. This separation, referred to as *deferred-judgement*, is especially important when generating ideas, since judging *on the fly* tends to inhibit the free flow of ideas, particularly when working in a group.

6.4.2 The Initial Appreciation

To continue our discourse on formal design methods we need to introduce and define some key terms which allow us to formally describe the design problem.

- *Design goal* is the overall intent of the design: It needs to be stated in one sentence, without reference to the means of achieving it.
- *Design boundary* delimits the design search and investigation. The design boundary usually focuses on some specific application or locale, that the proposed design seeks to address.
- *Design objectives* (sometimes referred to as *technical objectives*) are the *desired features* of the design. Design ojectives are the expressions of major aspects of performance we seek to deliver with the design.
- *Design criteria* are the scales for easuring the success of a design proposal in meeting specific design objectives. Each objective must be associated with at least one criterion (otherwise the objective is unmeasurableand therefore pointless). Note that criteria are *not specific values* of perfrmance – they are merely the scales on which such values can be measured. Criteria usually have *units* of measurement.
- *Design parameters* are frmalised expressions of the operating performance characteristics. They allow comparative evaluation of alternative designs. They are special kinds of criteria that the designer will commonly try to optimise to achieve the most favourable design outcome. All design parameters are criteria, but not all criteria are design parameters. For example, a criterion for a low cost design objective is the unit of cost ($), while an associated design parameter might be cost per unit mass or cost per unit power.
- *Design requirements* are relaxable constraints.
- *Design constraints* are the mandatory requirements of the design. A constraint may be expressed as either a required maximum or minimum value on some criterion-scale, or possibly as an allowable range of such values. For example to say that *"my india-rubber ball must rebound to at least 50% of its drop-height"*, is a constraint based on the criterion of rebound-height to drop-height ratio

These seven formal problem descriptors need to be cmpiled by the designer and client cooperatively during the initial negotiation through the problem evolution phase of the design. The resulting information is set out in a document we call the *Initial Appreciation* of the design problem.

Pahl and Beitz (1995) use a *requirements list* to formally articulate the design problem. However, their list is less formal than the Initial Appreciation, and it is intended only as *internal communication* between members of the design team.

The *Initial Appreciation*:

- attempts to formalise the definition of the problem;
- records the designer's first responses to a problem;
- helps the designer make a start;
- is the first formal communication about the design problem between designer and client.

The Initial Appreciation contains:

- design objectives, and corresponding criteria;
- identification of the design-boundary;
- design requirements and constraints;
- ratings o priorities to be given to the several technical objectives;
- information required by the designer;
- a list of foreseen subproblems;
- a proposed design strategy.

To illustrate the application of these ideas, and how engineers plan their approach to a desin problem, we look at a particular example where some decisions have alredy been taken about the nature of he solution.

Design of wheelchair

Suppose then, it hasbeen decided to design a self propelled heelchair for people who are unable to walk because of their injury or disease, ut who retain almost full se of their mental faculties and their upper bodies, including their arms.

Table 6.5 shows the Initial Appreciation for this design problem. The design has been delimited to involve self-propelled chairs suited to 95% of traumatic quadriplegics and some cerebral palsy victims. The only mandatory requirement is conformance to an International Standards Office (*ISO*) Draft International Standard (*DIS*) on wheelchairs No. 7176. The *ISO* have been working on wheelchair standards for a number of years, and adherence to even a draf standard is a useful way of ensuring consistent design performance.

Defining an appropriate design parameter to allow ranking of competing alternative wheelchair designs is not an easy task. This process involves defining some utility parameter that can be minimised (or maximised) to achieve an *optimal* design choice. In the wheelchair design we seek to have a versatile, easily adjustable design. This is necessary to allow matching the chair to a wide range of possible users. We could argue that this implies a large number o adjustable components in the chair. If the number of available useful adjustments is N (we want this to be large), the total cost is C (we want this to be low) the weight is W (we want this to be low fo portability), then, all other things being equal, we would want N/CW to be as large as possible.

Clearly, the utility parameter N/CW is only one of several such utility parameters we could define for the wheelchair. Defining the utility parameters in a design is part of the enformulation process and it needs consistent negotiation with the design client. The above example is only offered as a demonstration of the type of reasoning that may be applied when developing these performance parameters. We recommend the use of a convenient document template for the *Initial Appreciation*, called Engineering Design - Project Analysis Document (*ED-PAD*). A typical ED-PAD is shown in Figure 6.15.

Ex 6.4

- Write an Initial Appreciation for one of the products shown in Figure 6.6.

[Note that you will need to imagine what might have been in the designer's mind when the original problem was enformulated, since the Initial Appreciation is usually generated for problems that have yet to be solved. Hence in essence this is a *Design Audit*.]

- Prepare an Initial Appreciatin for a can opener.

6.5 Generating ideas (solutions)

Perhaps we have given the impression so far that engineering design is a terribly serious activity and rich in procedure, bound up with structure, rules and definitions. This emphasis derives from the nature of the enformulation task. As the designer moves on from enformulation to the divergent, idea-generation phase of design evolution, we find a more creative – even playful – dimension emerging in the design process. This section deals with the third major activity in the Design Process,

Table 6.5 Initial Appreciation for the wheelchair design problem

OBJECTIVES	CRITERIA
1. EASE OF OPERATION BY INVALID	
• ease of propulsion	• forces/torques exerted (N; N.m)
• manoeuvrability	• turning circle (m)
• ease of control	• forces/torques exerted (N; N.m)
• steering, accelerating, braking, stability	• response time (sec.)
• chair to ascend/descend ramps steps, gutters	• slope of ramp; size of steps, gutters (deg.)
• ease of entering/leaving chair comfort	• forces exerted; time taken; subjective rating; number/size of bedsores (N; sec.;–)
• range of operation	• distance travelled without attention required to power source (km)
2. RELIABILITY AND MAINTENANCE	
• operation	• mean time to breakdown (days)
• cleaning and repair	• mean time to clean and repair (min.)
3. DURABLE CONSTRUCTION	
• long life, no parts easily broken due to manoeuvring	• forecast life of chair and components (years)
4. WEIGHT	
• chair to be light to assist portability and propulsion	• weight (kg)
5. SIZE	
• passage of chair through doorways transport of chair in automobiles and public transport	• width, overall dimension, weight (mm; kg)
6. SAFETY	
• emergency braking	• braking distance, slope (m; deg.)
• protection of invalid in crash overturn	• forecast injuries (# per annum)
7. PROPER USE OF RESOURCES	
• ease of production	• number and complexity of components (#; –)
• low cost	• manufacturing cost ($)
8. AESTHETICS	
• appearance	• subjective rating (–)
PRIORITIES : difficult to set priorities until criteria are expressed quantitatively. Objectives 1 to 6 seem to be the most important ones.	
REQUIREMENTS/CONSTRAINTS	
• chair must have adjustments to allow tailoring to 95% of traumatic quadriplegics (C4 to C7 range); • chair design should suit some cerebral palsy victims also; • chair design and testing procedure must conform to ISO/DIS 7176	

Table 6.5 (continued) Initial Appreciation for the wheelchair design problem

INFORMATION REQUIRED
— number of invalids (population)
— anthropometry (shape and size)
— mental/physical capability, disability and the effect of these on the control and operation of chair
— physical environment in which the chair is to be used; indoors/outdoors, home, hospital, street, transport
— geometry of the environment and surface properties
— range of actions to be performed; are urinating, excreting, eating included
— existing power sources and construction materials
SUBPROBLEMS FORSEEN
— difficulty of matching chair to existing environment; steps,doorways, toilets, vehicles
— diversity of people to be catered for
— difficulty of keeping weight of chair within acceptable limits
DESIGN STRATEGY
1. collect information
2. design chair around invalid
3. check design
• does it match the physical environment
• is it light/portable?
• is it likely to be cheap to produce?
• does it satisfy the other criteria?
• if the answer to any of these questions is no, iterate until a compromise solution is found

as typefied in Figure 6.14. During this divergent phase several (possibly many) alternative proposals are generated for solving the design problem. In this phase the designer must initially defer judgement and let the *mind go* and perhaps *dream of things that never were and ask:* "why not?".

6.5.1 Mind games

Some authors in design recommend various procedures for *freeing the mind* from the inhibitions of our past experience. McKim (1972) suggests, among other mind-freeing experiments, a game called *Barnyard*. In this game, members of the design team take on the role of various animals (sheep, goat, chicken etc.) and, while facing other members of the design team, loudly emit the appropriate animal sounds. If this experience makes the designers feel slightly ridiculous, then it has achieved its aim. One of the overwhelming barriers to offering *new* ideas is the *fear of ridicule*. *Barnyard* attempts to overcome this fear. McKim even goes so far as to suggest standing at a busy intersection, or in some public transport, and emitting the sound of some barnyard animal, as a daring form of training in feeling ridiculous. Another mind game, called *breathing*, referred to by Adams (1987), asks designers to conjure up strong mental images of familiar things in unfamiliar circumstances. Adams writes

"Let us imagine we have a goldfish in front of us. Have the fish swim around. /Have the fish swim into your mouth. / Take deep breath and have the fish go down into your lungs, into your chest. / Have the fish swim around in there. / Let out your breath and have the fish swim out into the room again. /"

The *breathing* game goes on to include further items, such as rose petals, and sand. Adams suggests, that the mental imagery is good training for making *the familiar strange*,and the *strange familiar*, as further mental joggers for free flow of ideas.

Concept building and creative combinatorial design have received substantial study in artificial

ENGINEERING DESIGN PROJECT ANALYSIS DOCUMENT (ED-PAD)		Date
Project Title:		Sheet ...of...
Project Goal:		Designer:
OBJECTIVES	CRITERIA	RESTRICTIONS (CONSTRAINTS)

Figure 6.15 ED-PAD template to be used for Initial Appreciation

intelligence (*AI*). Still, in 1993 Marvin Minsky wrote[6.5] :

"The mind ... is a tractor-trailer, rolling on many wheels, but AI workers keep designing unicycles."

Both McKim and Adams have exerted significant influence on creative design thinking. In 1972 McKim noted:

"Education that develops skill in thinking by the vehicle of sensory imagery has several advantages. First, it revitalises the sensory and imaginative abilities that are often allowed to atrophy by contemporary education. Secondly, it provides vehicles that are frequently more appropriate to the thinker than are language symbols."

While McKim's proposition may well be correct, thinking processes used to generate novel ideas are the result of many sensory inputs, not merely visual thinking. Moreover, as suggested by many studies of thinking, including Adams and de Bono, people communicate their thoughts most effectively by processes that are comfortable to them. Some do this in abstract mathematical form, while others might use sketches or words. As Adams remarked in 1987:

"... there is no complete and scientifically verified explanation of thinking which can result in universal rules for conceptualizing more productively, nor is there likely to be until a much more complete understanding of the mind is available."

We are still unable to point to any generic ways of thinking that might be most appropriate to solving design problems. Instead, our plan is to tackle the problem by exploring the known major barriers to creative thinking. In this way we hope to develop a *self-awareness* that may open the *floodgate of ideas* necessary in design problem solving.

6.5.2 Deferred judgement

The human mind is certainly capable of suspending judgement under certain circumstances. We are regularly prepared to *make-believe* in the entertainment industry, with such fanciful ideas as *superheroes* (Superman, Batman, Hercules), and *science-fiction* (Star Trek, Jurassic Park). We are even prepared to suspend judgement when several of our senses are involved in combination. It is precisely this suspension of judgemet that makes the whole notion of *virtual-reality* possible (see for example Rheingold, 1991; Thalmann and Thalmann, 1994; Clarke, 1996).

Yet for problem solving, in almost any technical sphere, fantasy and imagination are actively discouraged as wasteful, undisciplined forms of thinking.

THE WHEELBARROW

The following exercise is due to de Bono (1986). Consider a *new design* for the familiar wheelbarrow, shown in Figure 6.16.

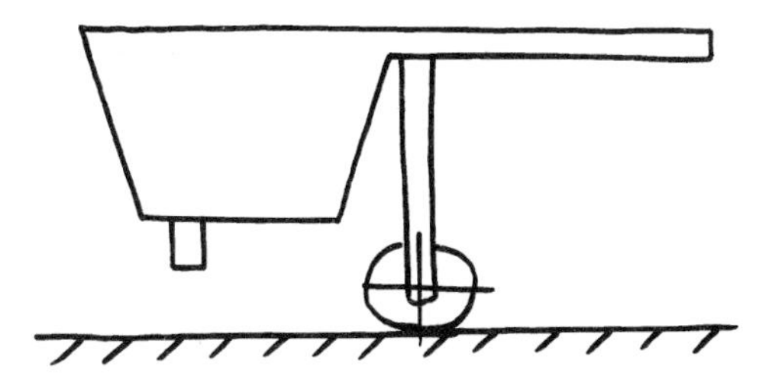

Figure 6.16 'New design idea' for wheelbarrow

When this *design* is shown to experienced engineers, or engineering students, the responses are almost invariably negative. Typical examples are :

> *"it will be hard on your back, the leverage is unfavourable"*;
>
> *"it's too high off the ground"* [we don't show any scale with the sketch];
>
> *"it will be hard to balance"*;
>
> *"handle is too thin, it might break easily"*.

In contrast to the above *conservative/negative* responses, the following were the responses of a group of *naive* designers:

> *"easy to kick mud off wheel"*;
>
> *"easy to go round corners"*;
>
> *"pushing down easier on back"*;
>
> *"easier to tip into hole, over wall"*;
>
> *"can add spring on wheel strut for automatic weighing"*.

Figure 6.17 shows some proposed improvements to the design offered by the *naive* designers (they were a group of *kindergarten children*).

The weheelbarrow design experience is an example of *deferred judgement*. In *"Six Thinking Hats"* de Bono (1986) identifies *negative thinking* with a

6.5 Marvin L. Minsky, in PROFILE, Scientific American, 269(3), November 1993

black hat and positive, *non-judgemental* thinking with a *yellow hat*. In the wheelbarrow example we see both examples of conservative/negative *black hat* and youthfully affirmative, *non-judgemental, yellow hat* thinking.

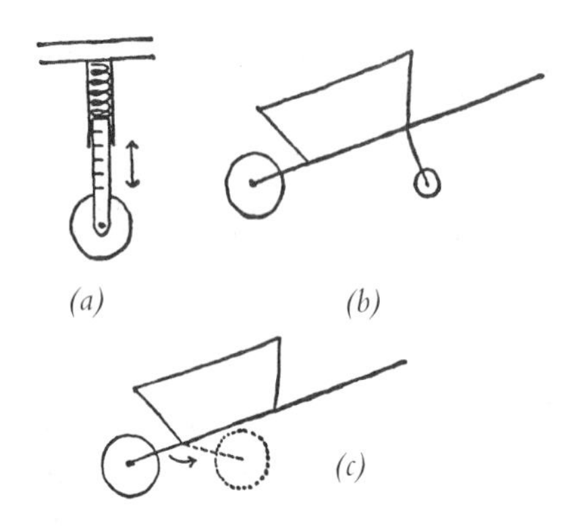

Figure 6.17 Some proposed improvements to the wheelbarrow design

Young children learn about the world by experimentingwith sensory images. No baby is initially afraid to touch or feel fire. Soon however, guided by adults/parents, children larn to create sensory limitation on the world around them. In addition, they soon learn to *"look before you leap"*. In scence, judgement is the very core of most doctrines. Even Karl Popper's (paraphrased) admonishment *unrefutable conjectures are not in the best interest of science*, is a clear signpost to the early acquisition of judgement. In design, under the weight of student or professional peer scrutiny, this tendency towards *"kill-the-idea"* thinking becomes almost paralysing.

Personal awareness of the need for clear differentiation and separation of *yellow hat* and *black hat* thinking, and of our own inner leaning towards negativity, can help in the free flow of ideas that is so essential in generating design solutions.

Ex 6.5

Consider the following *new* inventions :

1. *Pool ideator*: a special writing pad and pen to be used for recording ideas while in a swimming pool;
2. *Reflectaroads*: addition of a specially prepared glass to bitumen for creating a highly reflective road surfac;
3. *Paperfashon*: paper as a replacement for cloth in the fashion garment industry.

Prepare a list of five positive, *yellow hat* evaluations of each invention. Exchange these evaluations with a design partner and, where possible, expand on the ideas and evaluations proposed by your partner. Be sure to retain these lists. They will be useful in exercises on evaluation and decision making.

A cautionary tale about corporate judgement

A cautionary tale of how agressive *black hat* thinking cost one corporation substantially, is the story of the *'Alto'* personal computer (Smith and Alexander, 1988). In 1970, Xerox, a successful plain paper copying machine manufacturer, established the Palo Alto Research Center (PARC). The intention of Xerox for PARC was to *"invent systems that could support executives, secretaries, salesmen, and production managers, in what was to become known as* 'the office of the future'". One outcome of this plan was the *'Alto'*, a new personal computer, developed by the Information Systems Group at PARC. In 1979, predating both Apple and IBM, Xerox PARC comissioned a television commercial, based on the Alto. In the words of Smith and Alexander:

"The commercial highlights many parts of the Xerox system including the graphics-rich Alto screen, the mouse, the word processing program, the laser printer, and most prominently, the systems communication capabilities. ... Other than the Xerox name, nothing about it would surprise a television audience even if it were shown today."

However, after some financial difficulties experienced by Xerox, in 1975 they chose to shelve the Alto. One of the several reasons offered for this decision by management was :

"The strategic relevance of this product line is less than that of other programs for which we also need funding."

Xerox management might well have defended such a decision at the time, but in the light of subsequent events, the decision must rank among the *biggies* in *black hat* corporate errors of judgement.

The moral of this as with any early judgement *black hat* decision error is that, in generating novel ideas, it is far cheaper to *waste* consideration on ten *crazy, way-out* proposals, than to allow one unexpectedly clever possibility to slip away. A common technical cliché for this moral is *"don't let's throw out the baby with the bathwater!"*.

6.5.3 Flexibility and Fluency

Idea generation is a skill that can be developed and improved with practice. Fluency is the generation of ideas rapidly, while flexibility is the capacity to generate many distinctly different ideas. These two aspects of idea generation are almost, though not completely, independent of one another. Our experience indicates that fluent idea generators tend to be capable of generating many distinctly different ideas, while inflexible idea generators tend to be less fluent.

'Uses for X': a test for fluency nd flexibility

The following exercise is due t Adams (1987). The *Red Brick* Manufacturing Co. has engaged you as a technical consultant for expanding its market for red house bricks. Your task is to generate as many as possible new uses to which the brick might be put. As a precursor to discussing this exercise and its influence on developing fluency and flexibility, it is useful to ask designers to write down idea on a single sheet of paper in about five minutes.

If you came up with ten or more ideas in the allotted time (our experience with this exercise is in the range of between 10 and 15 ideas) you can regard yourself as a *fluent* generator of ideas. If these ideas were of the type :

> build a fence/ build a house/ build a garage/ build a playhouse for children/ build a barbecue;

then you must regard yourself as a somewhat *inflexible* idea generator. In any case, the Red Brick Company has already developed markets for red bricks in the building industry.

A considerably more productive way of approaching such problems is to focus on the product's various distinct attributes, as an alternative to considering the product itself.

Attribute list for a *red house brick*:

- size;
- weight, mass;
- colour;
- rectangularity;
- geometry;
- porosity;
- strength;
- surface texture;
- thermal capacity;
- insulating properties;
- hardness.

Once these attributes are identified, we can associate possible uses which employ one or more of them. Adams notes that once we have recognised weight as an attribute, it's an easy step to think up new, non-conventional, uses such as: anchor, ballast, doorstop, counterweight, holding down tarpaulins, or waste newspapers, projectiles in wars, riots. Other creative uses are: storing water/ bed warmers/ book shelf supports/ colour gauges/ the element of a new sporting event — *brick putting*.

Ex 6.6

Shopping at supermarkets for food, or in department stores for household goods, or clothes, always results in some disposable packaging. Fast-food outlets also tend to generate disposable containers and packaging. Also, at sporting venues or during holiday periods, the city's waste baskets overflow with empty plastic bottles, polystyrene containers and empty beverage cans. What particularly *bugs* us is the terrible waste of the following:

plastic shopping bags/ empty beverage containers/ plastic bottles/ bubble packaging/ used automobile tyres/ used batteries.

Prepare an attribute list for one of these items and generate a list of possible unconventional uses for your chosen item.

Further aids to fluency and flexibility: 'shape doodles'

Shape doodling is yet another exercise in self-awareness of creative blocks due to inflexible or restricted thinking styles. Figure 6.18 shows a set of 30 shapes (10 each of circles, squares, triangles). Each *shape* may be thought of as an uncompleted sketch of some familiar object. These shapes should be about 30mm across, with 10mm spacing between them.

On a clean sheet of A4 size paper, sketch a selection from these shapes, and complte each to a clearly recognisable form. Do this for as many shapes as you are able in five minutes. Figure 6.28 in Section 6.6.5 shows a selection of *finished shapes*. If you used all the circles to make different faces, for example, then you are a little less than flexible in idea generation. Because of its apparent child-like simplicity, this exercise is often seen by engineering students as wasteful, and of little relevance to *hard* technical matters.

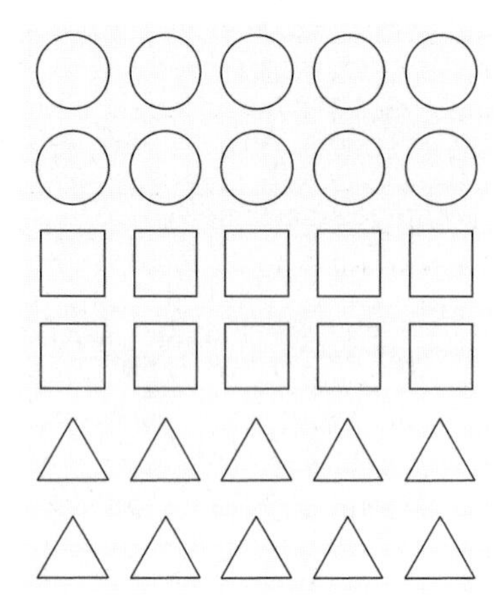

Figure 6.18 'Shape doodle': starting figures

It is useful to present this exercise as yet another opportunity to apply the *yellow-hat* deferred judgement to the outcome.

6.6 Generic barriers to idea generation

The free flow of ideas is the essence of creative problem solving. We have already alluded to fear of ridicule, *black-hat* early judgement and ingrained suspicion of fantasy and *wasteful* doodling as serious barrier to this free flow of ideas. Adams (1987) lists four major barriers to fluent and flexible idea generation. He calls the process of becoming self-aware of these barriers, and the personal attempts to overcome them, *conceptual blockbusting*.

While the notion of self-analysis may appear foreign to formal problem solving, there is considerable value in being fully aware of our bedevilment with *black-hat* avoidance of fantasy and imagination. The following is a brief review of the barriers to ideation explored by Adams.

6.6.1 Perceptual barriers

Inability to *see the whole picture* or to view problems in ways that present a clear vision of the information needed to solve them represent perceptual barriers. The often quoted cliché describing this type of barrier is: *not seeing the wood for the trees*. Some examples will serve to elaborate these perceptual barriers to problem solving.

Information overload or underload

Problems can be embedded in substantial information, much of which may be unnecessary to the problem evolution process. The problem solver may become confused with too much information (*cognitive overload*), or insufficient information (*not seeing the whole picture*). In problems of structural integrity we use generic models of structural elements such as *beams*, *columns*, and *shafts*, to deconstruct complex problems. In *need-directed* problem solving, we must find similar deconstructions to get to the very essence of the problem.

EXAMPLE 1. THE MEDITATING MONK (THIS PROBLEM IS DUE TO ADAMS)

A Buddhist monk is known to pend some days each year on a small mountain near his monastery. On the first day he starts early in the morning and climbs the mountain, stopping to rest several times along the way. By nightfall he reaches the mountain top, where he spnds several days in meditation. Finally, the monk commences his descent along the same route he took to ascend the mountain. Again, he starts in the morning and spends the whole day in his journey to the bottom of the mountain. Naturally, his descent is slightly faster than his ascent, since he finds it easier to walk downhill. By nightfall he reaches his monastery. It is conjectured that there is one specific location on his path up and down the mountain, where he would be at precisely the same time of the day during both ascent and descent. Can you prove or refute the conjecture?

EXAMPLE 2. TWO UNI-PARAMETRIC PROBLEMS

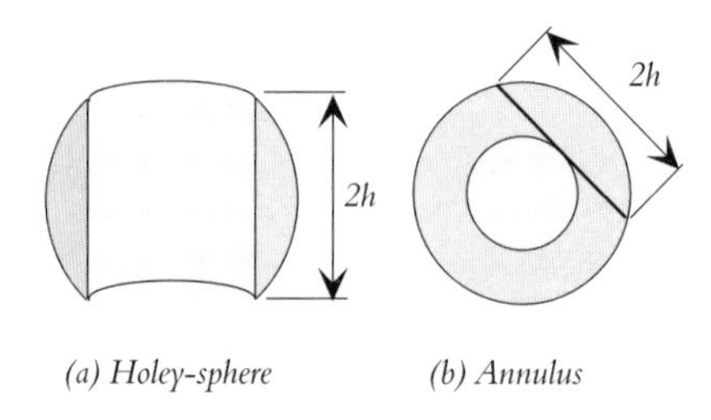

Figure 6.19 The sphere and the annulus

Figure 6.19(a) is a section through a sphere which has a cylindrical hole drilled through it, along a diameter. The remaining ring has a height of $2h$ as indicated. What is the volume of the ring?

Figure 6.19(b) is a regular annulus. The line segment shown touches the outer circle at two points and is tangent to the inner circle. What is the area of the annulus?

Delimitation

Perceiving boundaries and restrictions to problem solving, represents a common form of perceptual barrier known as *delimitation*. Two well known problems will exemplify this type of barrier.

EXAMPLE 3. THE NINE-DOTS PROBLEM

Figure 6.20 shows nine dots arranged in a rectangular shape. Can you draw four straight lines, that pass through each of the nine dots only once, without lifting the pen from the paper?

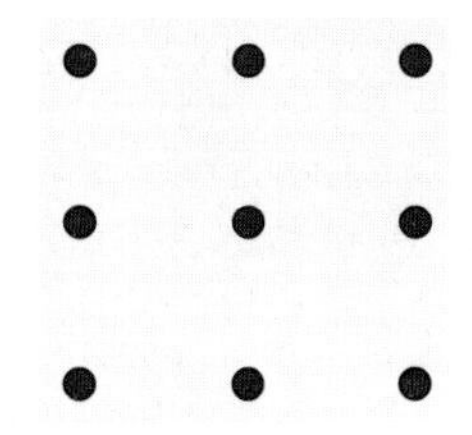

Figure 6.20 Nine dots problem

In solving this problem, you must assume thet the nine dots are mathematically infinitessimal points.

EXAMPLE 4. HE SIX MATCHES PROBLEM

Figure 6.21shows six matches arranged in a hexagon. Can you use these matches, without breaking them, to form four equilateral triangles?

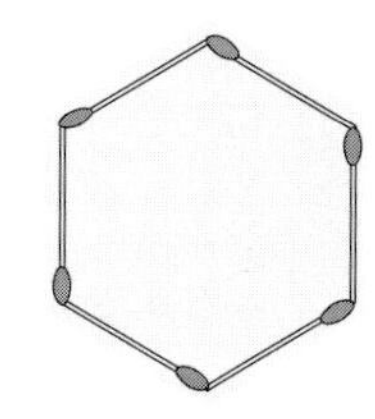

Figure 6.21 The six matches problem

Saturation

Our senses get dulled to information we see or feel continuously. If we press our fingertip to a moderately sharp object for any length of time, after some time, our haptic senses fail to register the object. We see familiar things in our everyday life which don't *register* in our consciousness. These are images such as the fine details of our front door, the precise features on the front of a bus or a train. This time-dependent dulling of senses is called *saturation*. Given all the information to which we are regularly exposed, we simply filter out some we regard as inessential. In a tenuous way, saturation is the result of information overload. Some of it simply *slips through the cracks* of our consciousness. In dealing with technical problems, we can not afford to let any information slip us by. As a useful exercise, cover the image in figure 6.22 and then show it for five seconds to someone, who has not seen it previously. Ask your subject to recall the text. Our general experience is, that the majority of first-time viewers will fail to see the repeated word.

Figure 6.22 What does this sign say?

Ex 6.7

Sketch the detail of your front door/ your television set/ the dashboard of your automobile/ the keyboard of your mobile phone.

Stereotyping

When we fail to utilise information from all of our senses, we fall into the trap of *stereotyping*, or *seeing what we have been conditioned to see*. This notion was astutely exploited by several graphic artists including Escher and Vasarely. Figure 6.23 shows the well known *Waterfall* image of the Dutch

Figure 6.23 Waterfall, by M. C. Escher

graphic artist Maurits Escher (1898-1972). This image, as many others of Escher, exploits improper use of perspective. The result is that our visual senses are *fooled* into accepting the image of the self-driven waterwheel.

ictor Vasarely (Vásárhelyi) is a Hungaro-French graphic artist, who is regarded as the father of *op-art*. He uses graphic imagery that exploits errors in visual perception. The resulting images again deceive the visual senses into *seeing* a degree of three-dimensionality that is not really there in the image.

Figure 6.24 Op-art by Victor Vasarely

Figure 6.24 is a typical example of Vasarely's art. Other well-known exploiters of improper perspective are the *engineer's security key* and the *mystic triangle*. Both of these *impossible objects* are shown in Figure 6.25.

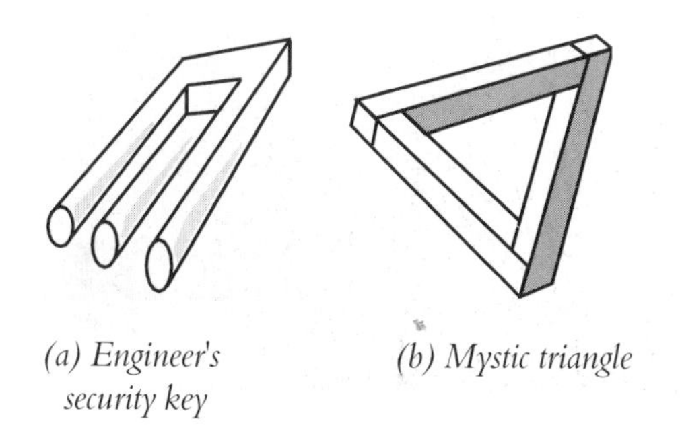

(a) Engineer's security key *(b) Mystic triangle*

Figure 6.25 Impossible geometric objects

6.6.2 Cultural and environmental barriers to idea generation

Cultural barriers are the result of developed forms of thinking, particularly influenced by one's upbringing and environment. Typically, children explore the world around them with all their senses. However, once in formal schooling, they soon learn that (in Adams' words):

"Fantasy and reflection are a waste of time, lazy, even crazy;

Playfulness is for children only;

Reason, logic, numbers, utility, practicality are good; feeling, intuiton, qualitative judgements, pleasure are bad."

Examples of cultural barriers are:

- authoritarian behaviour: *"do what I tell you"*;
- reluctance to share knowledge, resources, or information: *"this is my turf"*;
- corporate monoculture: *"this is the way we do things here"*;
- organisational hierarchy: *"management must not fraternise with technical staff"*;
- xenophobia: *"that idea was not invented here — it can't be used"* (the dreaded *NIH* complex).

Tunnel vision

Tunnel vision is the conceptual block caused by approaching a problem from *only one point of view*. The person suffering from tunnel vision may be an enthusiast for a particular technology, or skilled in a narrow sub-discipline of engineering. Perhaps they are subconsciously inclined to make their own responsibilities easier at the expense of other aspects of the problem – an attitude exacerbated by organisations that split designers into spcialist groups. In any case the result is an unbalanced focus on the elements of the problem, and a tendency to concentrate on one's own interests and sphere of expertise.

Consider the concept sketches shown in Figure 6.26, allegedly deriving from several engineeirng groups working on the design of a 1942 military aircraft.

Notes:

– the production engineering group is responsible for manufacturing the product efficiently;

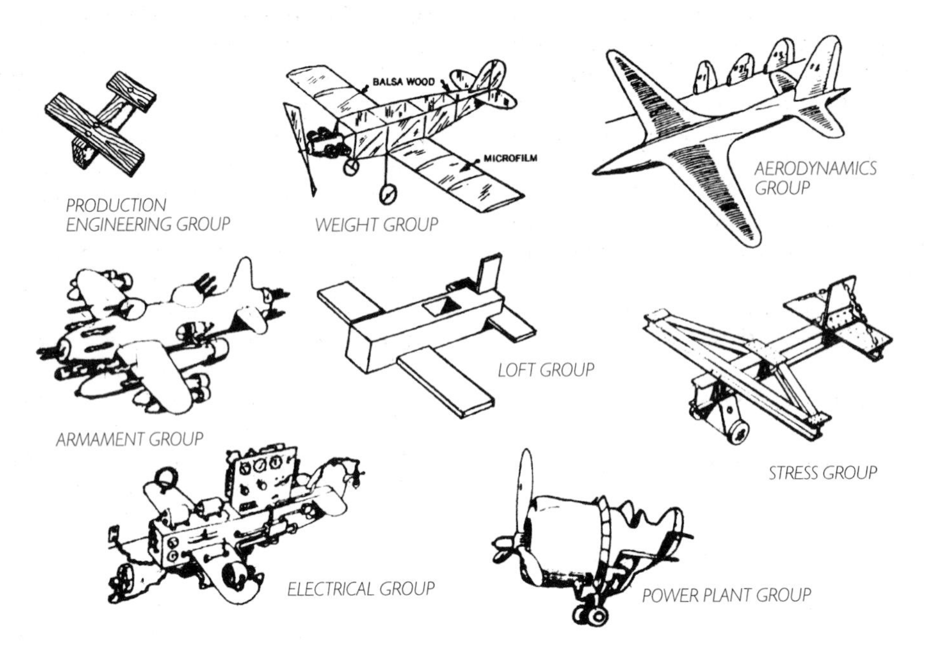

Figure 6.26 'Tunnel vision' in aircraft design

– the *loft group* specifies all the complex curved surfaces of the aircraft;

– the *power plant group* is clearly full of mechanical engineers (escapees from the racing car industry).

Environmental barriers

Environmental barriers are, unfortunately, all too common in engineering. Creative idea generation requires a supportive environment. We need the support of colleagues, design team members, as well as management.

McGowan's 'laws'[6.6]

Peter McGowan is a senior mechanical engineer in a large multi-national chemical company. The majority of his work involves *trouble-shooting* (recall that this is a form of focused problem solving). In his experience with major engineering projects, he has been able to formulate 'laws' (working hypotheses) about two significant environmental barriers to creative problem solving.

Law 1: Employee trust of management is directly proportional to management's trust of employees. This rule is easily extended to the work of design teams, their members and their tea leaders;

Law 2: The person nearest to the problem always knows more than they are given credit for. Therefore, ask the worker, the installer, the operator, the maintenance technician.

Closing the barn door after the horses have bolted

Potential Failure Mode and Effects Analysis (*PFMEA*) is a formalised design evaluation tool used by many large consumer product manufacturers. The procedure ranks the parts of the design requiring special attention to ensure its eventual safe and reliable operation. Although inportant, the procedure can be tedious and time consuming. Consequently, some organisations invoke *PFMEA*, not prior to releasing a design, but when *"time is available"*.

6.6.3 Emotional barriers

e have already referred to the *Barnyard* game of Bob McKim, as a means of overcoming fear of ridicule. Within any group, fear of rejection, and that of *making a fool of oneself*, are strong barriers to creative idea generation. As a simple exercise in any problem solving situation, each member of a team

6.6 McGowan, P. (1998) Design Tensions, Paper presented at the 10th Annual Conference on Engineering Education, Gladstone, September

writes down on a piece of paper up to five ideas for solving the problem, clearly numbering each suggestion. These lists of ideas are then taped onto a display board or the wall of the design office, so that all members of the team can see them. At this point each team member considers the proposed ideas with a *yellow hat* on, and writes down up to five positive or favourable outcomes that might result from each proposal. In this way, the encouragement to think positively heightens the team's awareness of its own emotional response to the problem under consideration.

6.6.4 Intellectual and expressive barriers

Problem solving and idea generation are highly personal procedures. Each individual will use different forms of communication of the problem and its solutions. Idea logs represent one form of communication, particularly useful for generating graphic *snap-shots* of our thinking processes. Yet, some people find it easier to think in verbal or symbolic languages.

THE 'CAT-IN-THE-BAG' EXERCISE

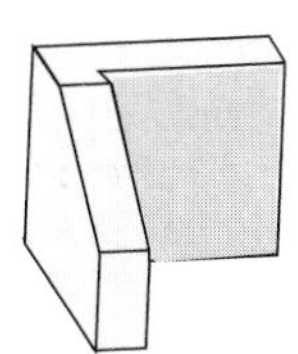

Figure 6.27 The 'cat-in-the-bag'

In this exercise, we take a relatively simple object, shaped as shown in Figure 6.27. This simple shape is placed in a brown paper bag. A student is asked to *"feel"* the object in the bag and describe its shape in word only. No arm or hand movements are permitted. Also excluded are references to trade names or non-generic shapes. The rest of the student group are then asked to sketch the object on a sheet of paper, based on the verbal description.

In general, people find this exercise quite challenging, and the resulting sketches can be quite amusing when compared to the *real* object.

THE CYCLISTS AND THE FLY

Two cyclists A and B start off from two towns twenty kilometers apart and head towards each other in a straight line at an average speed of 5 km/hr. At the instant when cyclist A starts out, a fly takes off from the front of his bicycle and flies to wards the other cyclist at an average speed of 7 km/hr. When the fly reaches cyclist B it immediately reverses its path and flys on at the same 7 km/hr back towards cyclist A. The fly continues to switch back and forth between cyclists until they meet, somewhere between towns A and B. How far did the fly travel?

This problem can be easily enformulated as an infinite series, which can be summed for the solution. But there is a much simpler approach to the problem, which most of us would pursue first. In their anthology of science stories, Weber and Mendosa (1973) recount how a young scientist colleague posed the problem to John von Neumann (1903-1957), the Hungarian born mathematician, popularly regarded as one of the originators of programmed computers. Von Neumann was supposed to have incredible skills in mental computation. When faced with the above problem, he almost immediately gave the correct answer (14 km). The young scientist's face fell, *"you have heard this problem already"*, he remarked. *"No"*, said von Neumann, *"I simply summed the infinite series."*

Was von Neumann unwise for adopting such an unwieldy procedure to solve this otherwise easy problem (the cyclists travel for 2 hours before they meet, in which time the fly has flown 2 hrs. @ 7 km/hr = 14 km)? Not necessarily – he was merely using the enformulation that suited his thinking style best. As a general rule it is useful to express and communicate problems in a way most suited to productive solution opportunities.

6.6.5 Answers to problems on conceptual barriers to creative problem solving

THE MEDITATING MONK:

In this problem it is useful to consider the monk and his *alter-ego* travelling the same path up and down the mountain. Where the *two* meet is the location correponding to the precise time of day when the original monk was at the same place and time on each trip.

UNI-PARAMETRIC GEOMETRY PROBLEMS:

Since in each case the geometry is described in terms of a single parameter (h), we must assume the problems to be independent of the size of the hole in the sphere, or the inner circle in the annulus. Consequently, we can imagine the *cutting* holes in each problem to be of zero radius, resulting in a sphere and a circle respectively, both of diameter 2h. Hence, the volume of the ring is $4\pi h^3/3$ and the area of the annulus is πh^2.

SHAPE DOODLE EXERTCISE

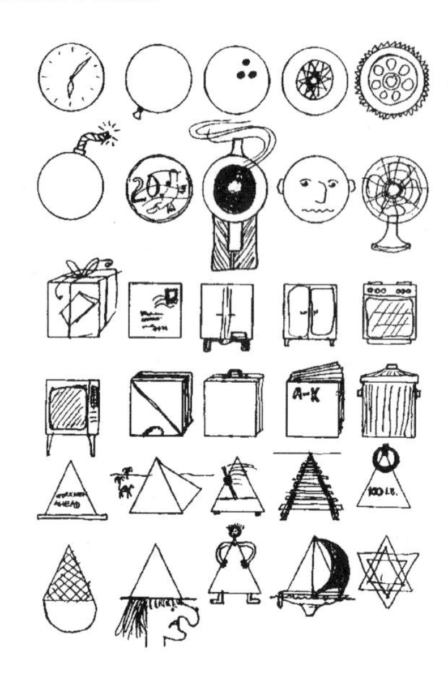

Figure 6.29 'Shape doodle': some completed examples

THE NINE DOTS PROBLEM

The solutions to this problem is indicated in Figure 6.28.

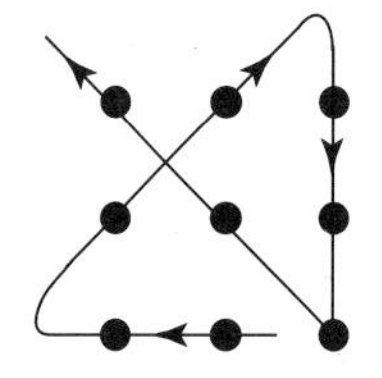

Figure 6.28 Solution to the nine dots problem

This is only one of many possible solutions, but all others are dependent on the dots having finite size. In this case, it was good to heed the admonition of lateral thinkers to *think outside the square*.

THE SIX MATCHES PROBLEM

This is a very entertaining problem, particularly among groups who have not seen the problem previously. This is a three-dimensional analogue of the nine dots problem. Most people delimit this exercise to the plane, when the simple solution is a *regular tetrahedron*.

Ex 6.8 Spanning a chasm

We continue to refer to the *thought experiment* as a powerful problem solving tool, especially in problems where we intentionally introduce what appears to be an artificial delimitation to the solution space. As a final exercise in this type of artificially delimited problem, imagine a *chasm* to be spanned by a *kind of bridge*. The *supports* of the bridge are two drinking glasses, placed on two desks, separated by the *chasm*. Your task is to take a single A4 sheet of paper and *construct* your bridge from it, such that it spans the chasm between the supports. Your single sheet of paper is the only permitted material of construction and no other material may be used. The *best design* is the bridge of the longest span. The bridge need only carry its own weight, and its midspan sag must not exceed 1/300 of the span.

Ex 6.9 Sharps disposal container

Hospitals make use of injection needles, scalpel blades and other sharp objects in their day-to-day operation. Disposal of these *sharps* is a serious waste management problem, particularly in the current climate of blood related transmission of diseases such as AIDS and Hepatitis-B.

Your client is a manufacturer of disposable plastic products and wants to explore the possibility of developing a new market in *sharps disposal* containers. There are severe constraints and restrictions on this design. The container must be capable of withstanding a specified drop test without leaking. The top cover should accept a wide range of sharp implements, needles and disposable syringes of various types. It should not be necessary to touch the container, and the lid must be designed to prevent the entry of fingers – this is to avoid *needle-stick* injuries.

There are International Standards relating to safe handling and hazard evaluation that must be adhered to.

The following pages show some extracts from an idea log of one group of professional designers faced with the above problem. The idea log is presented as an indication of the degree of detail to be included in generating ideas for product design. We are particularly grateful to Leong Weng Yue for permission to reproduce his wonderful sketches.

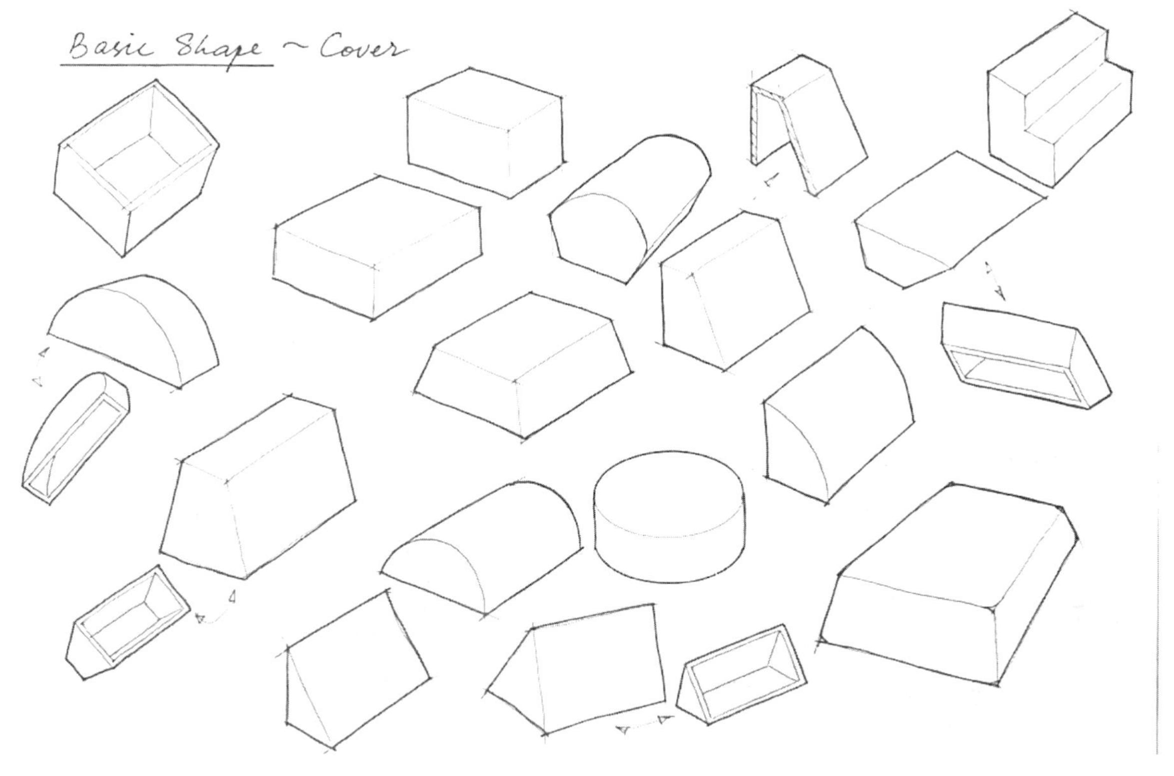
Basic Shape ~ Cover

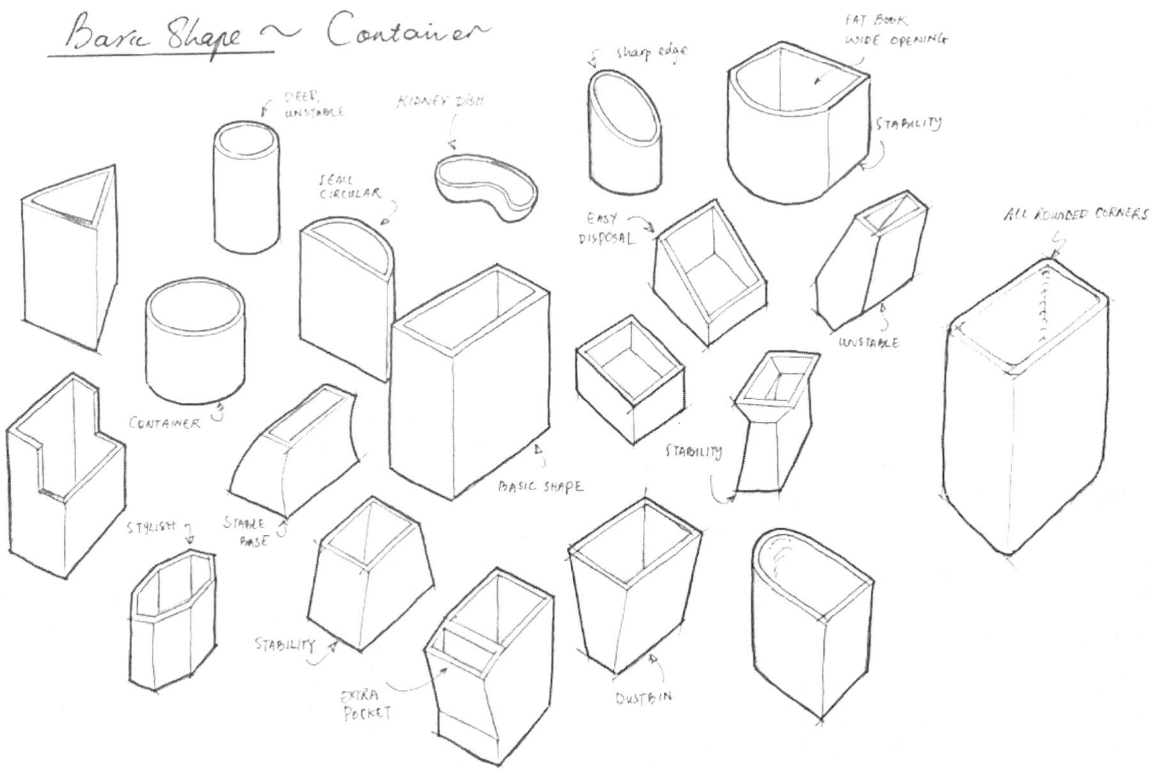
Basic Shape ~ Container
FAT BOOK
WIDE OPENING
sharp edge
DEEP, UNSTABLE
KIDNEY DISH
STABILITY
SEMI CIRCULAR
EASY DISPOSAL
ALL ROUNDED CORNERS
UNSTABLE
CONTAINER
STABILITY
BASIC SHAPE
STYLISH
STABLE BASE
STABILITY
EXTRA POCKET
DUSTBIN

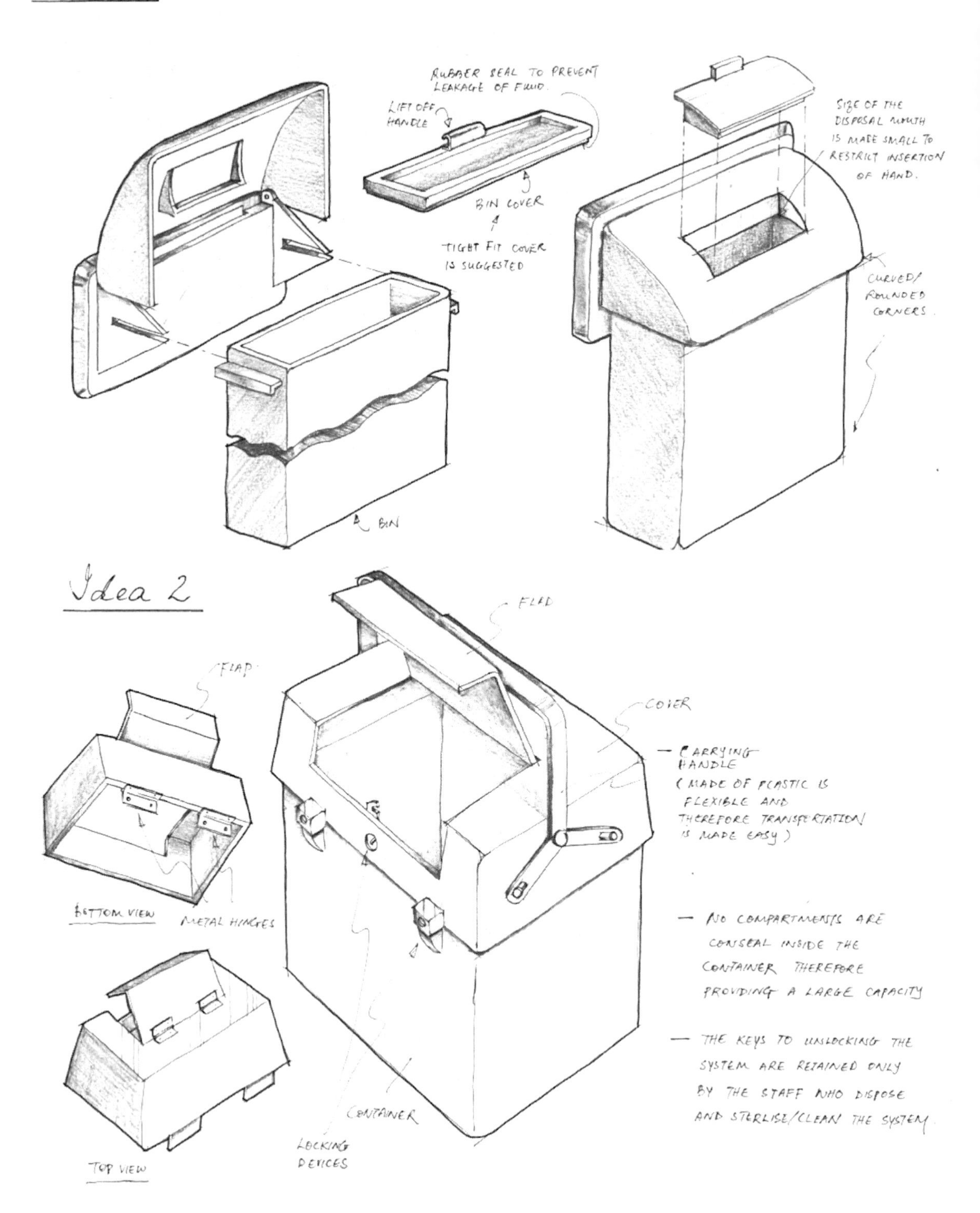
Idea 1
RUBBER SEAL TO PREVENT LEAKAGE OF FLUID.
LIFT OFF HANDLE
BIN COVER
TIGHT FIT COVER IS SUGGESTED
SIZE OF THE DISPOSAL MOUTH IS MADE SMALL TO RESTRICT INSERTION OF HAND.
CURVED/ ROUNDED CORNERS.
BIN
Idea 2
FLAP
FLAP
COVER
— CARRYING HANDLE (MADE OF PLASTIC IS FLEXIBLE AND THEREFORE TRANSPORTATION IS MADE EASY)
BOTTOM VIEW
METAL HINGES
— NO COMPARTMENTS ARE CONSEAL INSIDE THE CONTAINER THEREFORE PROVIDING A LARGE CAPACITY
— THE KEYS TO UNLOCKING THE SYSTEM ARE RETAINED ONLY BY THE STAFF WHO DISPOSE AND STERLISE/CLEAN THE SYSTEM
CONTAINER
LOCKING DEVICES
TOP VIEW

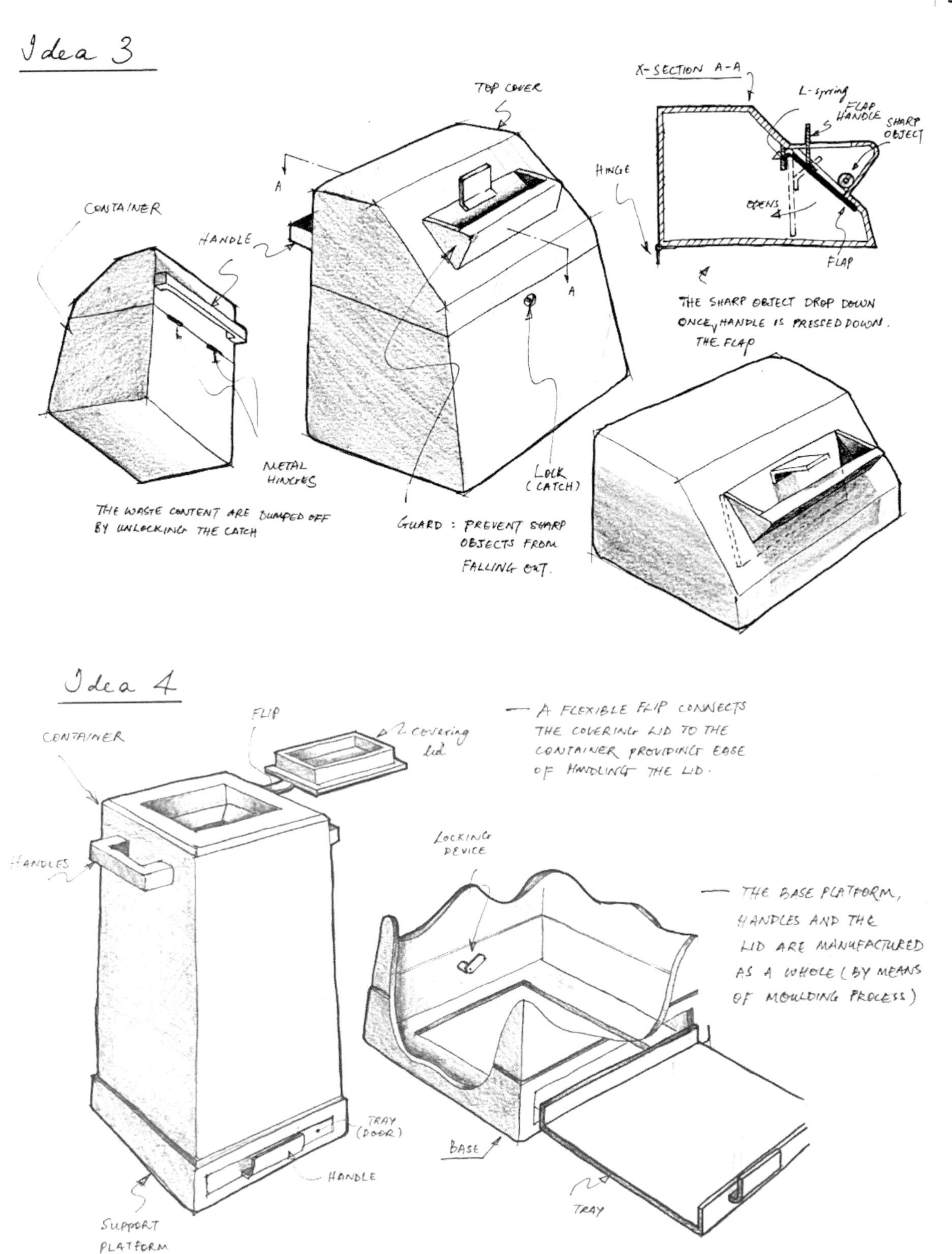
Idea 3
TOP COVER
X-SECTION A-A
L-spring
FLAP HANDLE
SHARP OBJECT
HINGE
OPENS
FLAP
CONTAINER
HANDLE
A
A
THE SHARP OBJECT DROP DOWN ONCE HANDLE IS PRESSED DOWN. THE FLAP
METAL HINGES
LOCK (CATCH)
THE WASTE CONTENT ARE DUMPED OFF BY UNLOCKING THE CATCH
GUARD : PREVENT SHARP OBJECTS FROM FALLING OUT.
Idea 4
FLIP
covering lid
CONTAINER
— A FLEXIBLE FLIP CONNECTS THE COVERING LID TO THE CONTAINER PROVIDING EASE OF HANDLING THE LID.
HANDLES
LOCKING DEVICE
— THE BASE PLATFORM, HANDLES AND THE LID ARE MANUFACTURED AS A WHOLE (BY MEANS OF MOULDING PROCESS)
TRAY (DOOR)
BASE
HANDLE
TRAY
SUPPORT PLATFORM

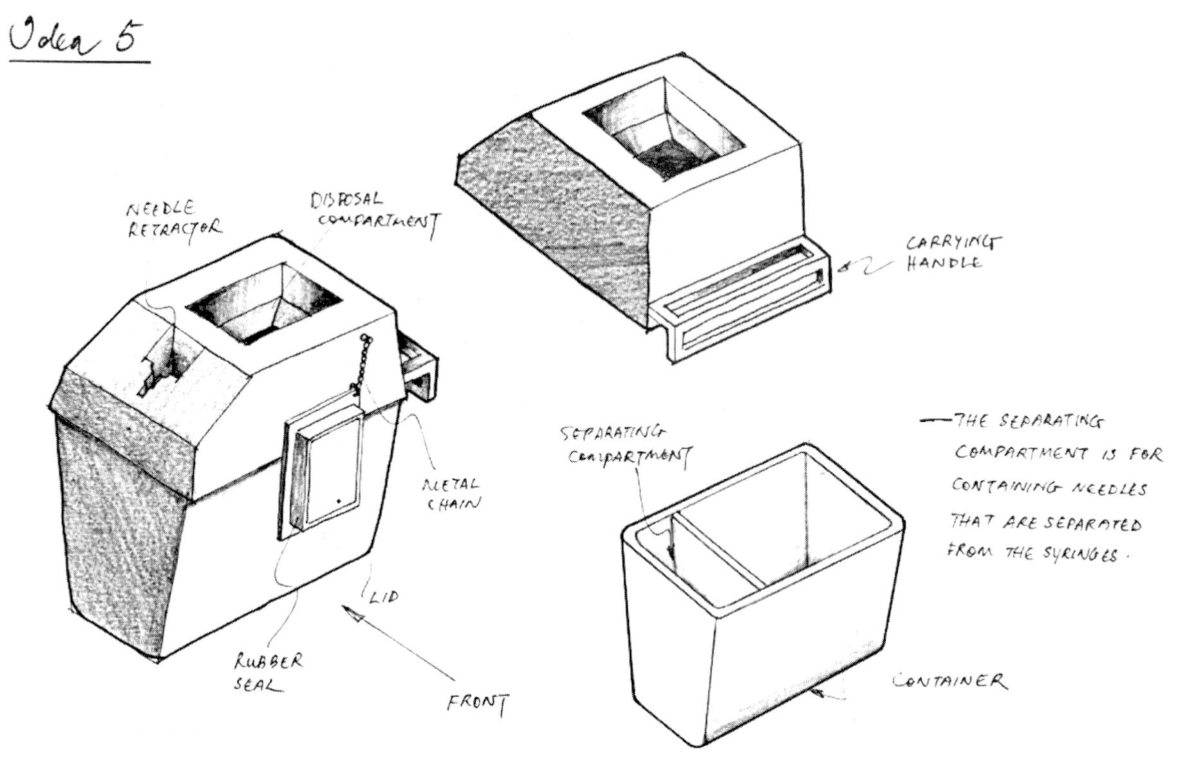
Idea 5
NEEDLE RETRACTOR
DISPOSAL COMPARTMENT
METAL CHAIN
LID
RUBBER SEAL
FRONT
CARRYING HANDLE
SEPARATING COMPARTMENT
THE SEPARATING COMPARTMENT IS FOR CONTAINING NEEDLES THAT ARE SEPARATED FROM THE SYRINGES.
CONTAINER

Development Of Chosen Idea ~ Container

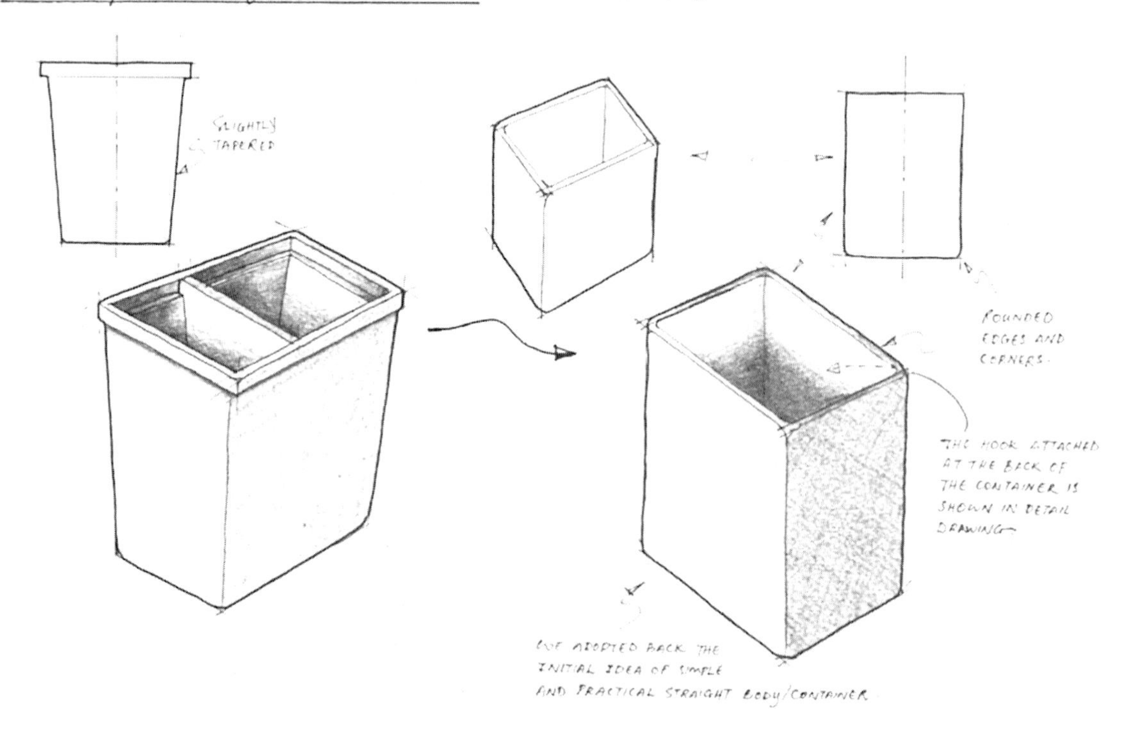
SLIGHTLY TAPERED
ROUNDED EDGES AND CORNERS.
THE HOOK ATTACHED AT THE BACK OF THE CONTAINER IS SHOWN IN DETAIL DRAWING
WE ADOPTED BACK THE INITIAL IDEA OF SIMPLE AND PRACTICAL STRAIGHT BODY/CONTAINER

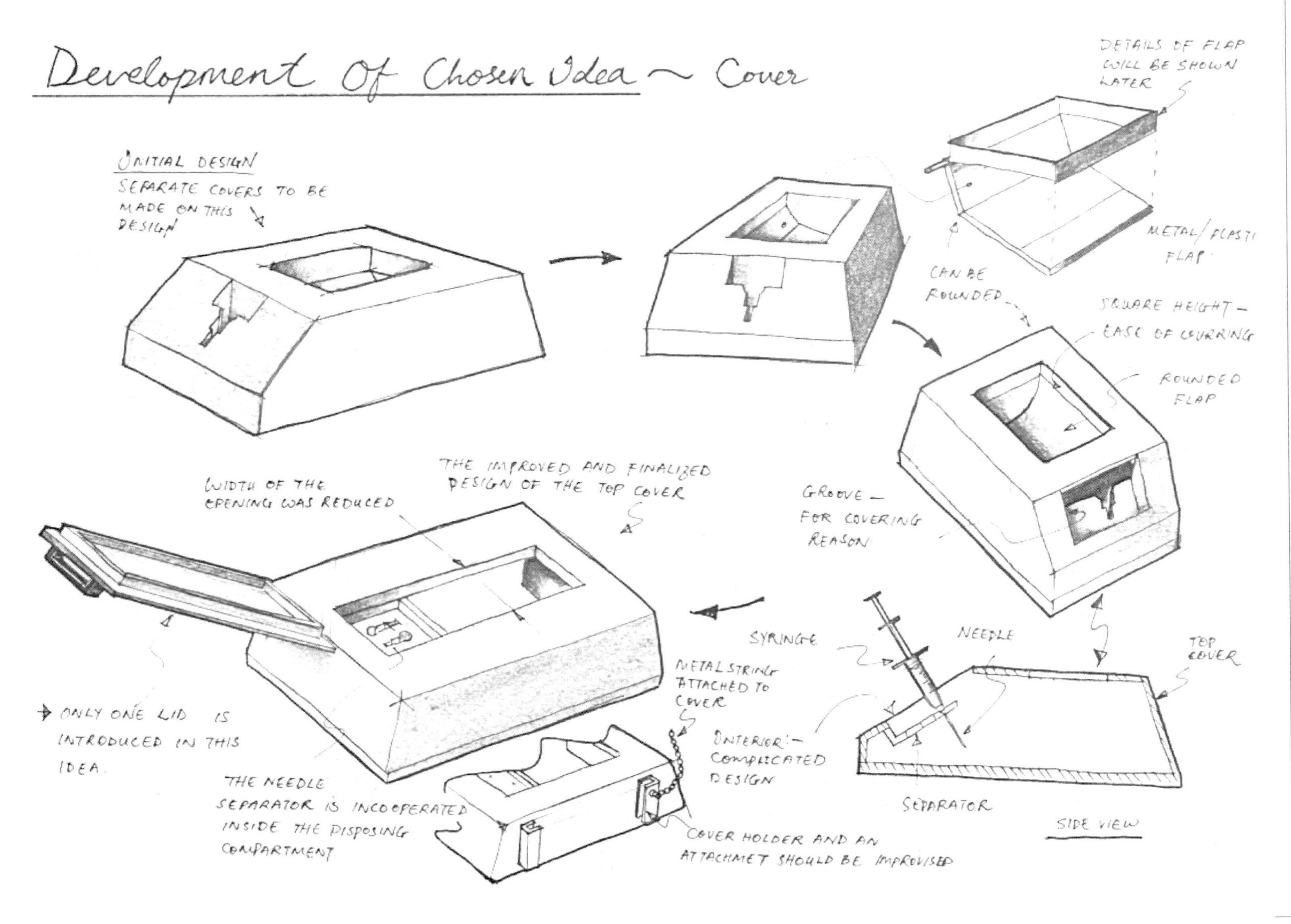
Development Of Chosen Idea ~ Cover
INITIAL DESIGN
SEPARATE COVERS TO BE MADE ON THIS DESIGN
DETAILS OF FLAP WILL BE SHOWN LATER
METAL/PLASTI FLAP
CAN BE ROUNDED
SQUARE HEIGHT – EASE OF COVERING
ROUNDED FLAP
GROOVE – FOR COVERING REASON
THE IMPROVED AND FINALIZED DESIGN OF THE TOP COVER
WIDTH OF THE OPENING WAS REDUCED
ONLY ONE LID IS INTRODUCED IN THIS IDEA.
SYRINGE
NEEDLE
TOP COVER
METAL STRING ATTACHED TO COVER
INTERIOR: – COMPLICATED DESIGN
THE NEEDLE SEPARATOR IS INCOOPERATED INSIDE THE DISPOSING COMPARTMENT
SEPARATOR
SIDE VIEW
COVER HOLDER AND AN ATTACHMET SHOULD BE IMPROVISED

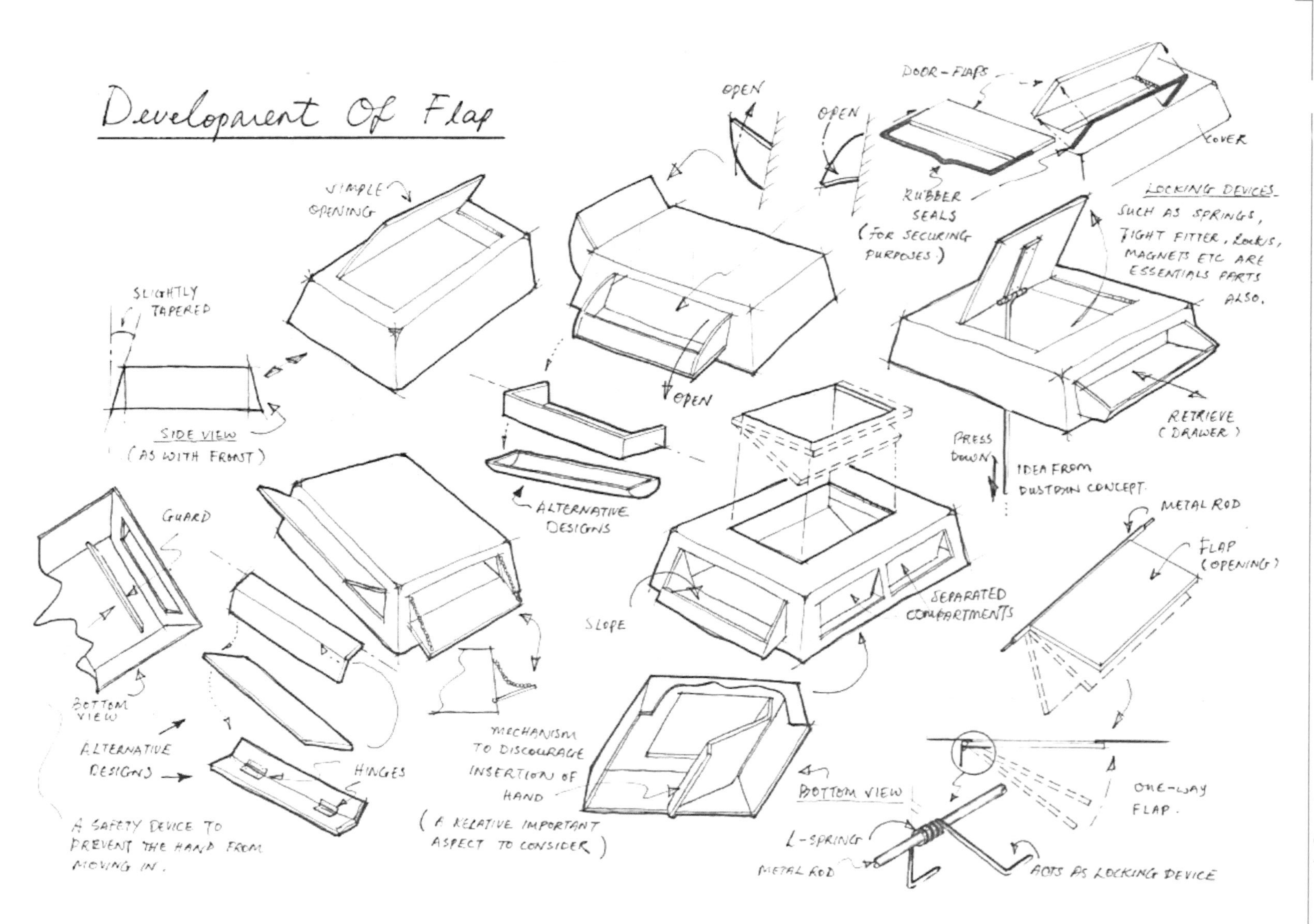
Development Of Flap
SIMPLE OPENING
SLIGHTLY TAPERED
SIDE VIEW (AS WITH FRONT)
OPEN
OPEN
OPEN
DOOR-FLAPS
COVER
RUBBER SEALS (FOR SECURING PURPOSES)
LOCKING DEVICES SUCH AS SPRINGS, TIGHT FITTER, LOCKS, MAGNETS ETC ARE ESSENTIALS PARTS ALSO.
RETRIEVE (DRAWER)
PRESS DOWN
IDEA FROM DUSTBIN CONCEPT.
ALTERNATIVE DESIGNS
METAL ROD
FLAP (OPENING)
GUARD
SLOPE
SEPARATED COMPARTMENTS
BOTTOM VIEW
ALTERNATIVE DESIGNS
HINGES
A SAFETY DEVICE TO PREVENT THE HAND FROM MOVING IN.
MECHANISM TO DISCOURAGE INSERTION OF HAND
(A RELATIVE IMPORTANT ASPECT TO CONSIDER)
BOTTOM VIEW
ONE-WAY FLAP.
L-SPRING
METAL ROD
ACTS AS LOCKING DEVICE

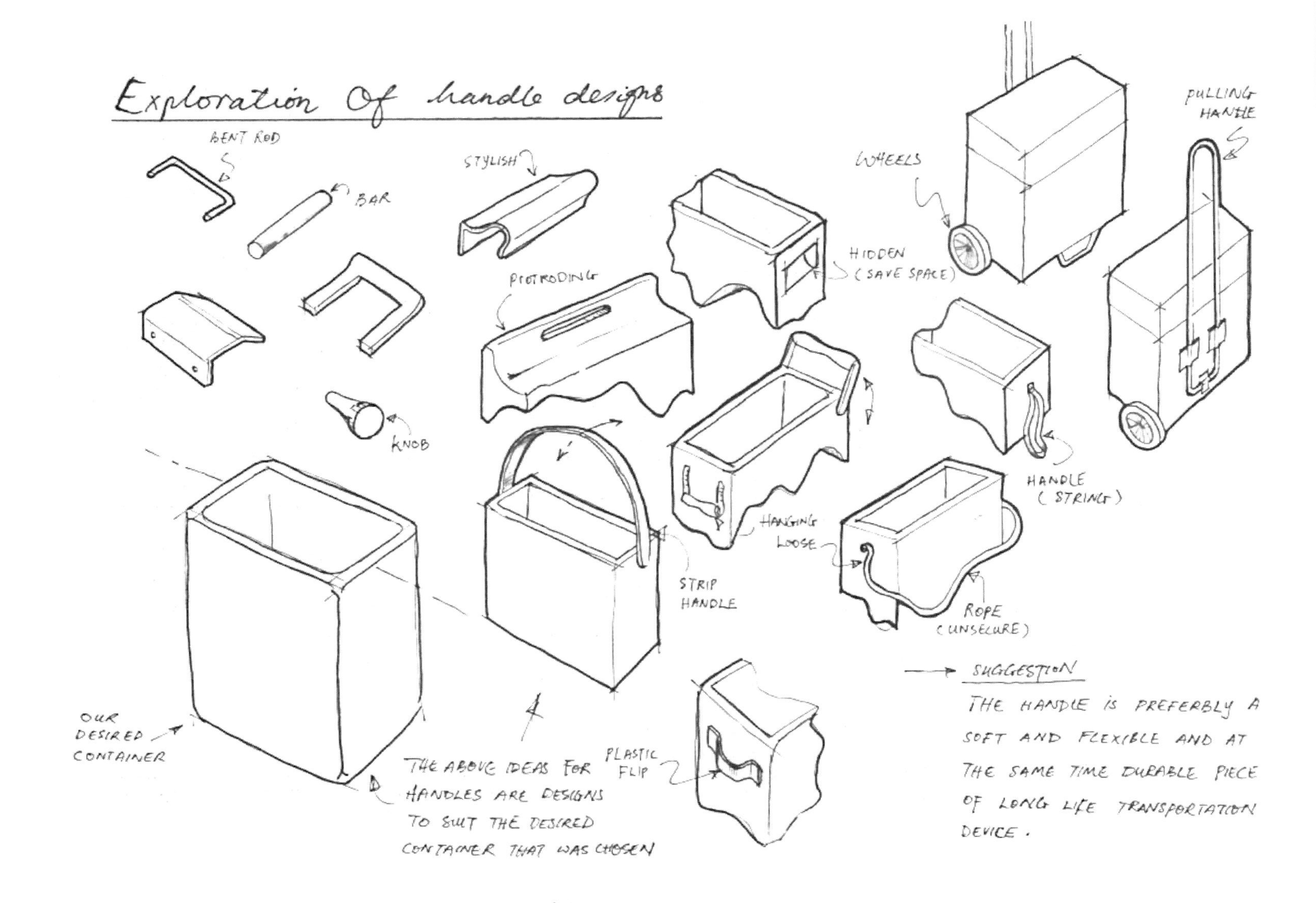
Exploration Of handle designs
BENT ROD
BAR
STYLISH
PROTRODING
KNOB
HIDDEN (SAVE SPACE)
WHEELS
PULLING HANDLE
HANDLE (STRING)
HANGING LOOSE
STRIP HANDLE
ROPE (UNSECURE)
OUR DESIRED CONTAINER
THE ABOVE IDEAS FOR HANDLES ARE DESIGNS TO SUIT THE DESIRED CONTAINER THAT WAS CHOSEN
PLASTIC FLIP
SUGGESTION
THE HANDLE IS PREFERBLY A SOFT AND FLEXIBLE AND AT THE SAME TIME DURABLE PIECE OF LONG LIFE TRANSPORTATION DEVICE.

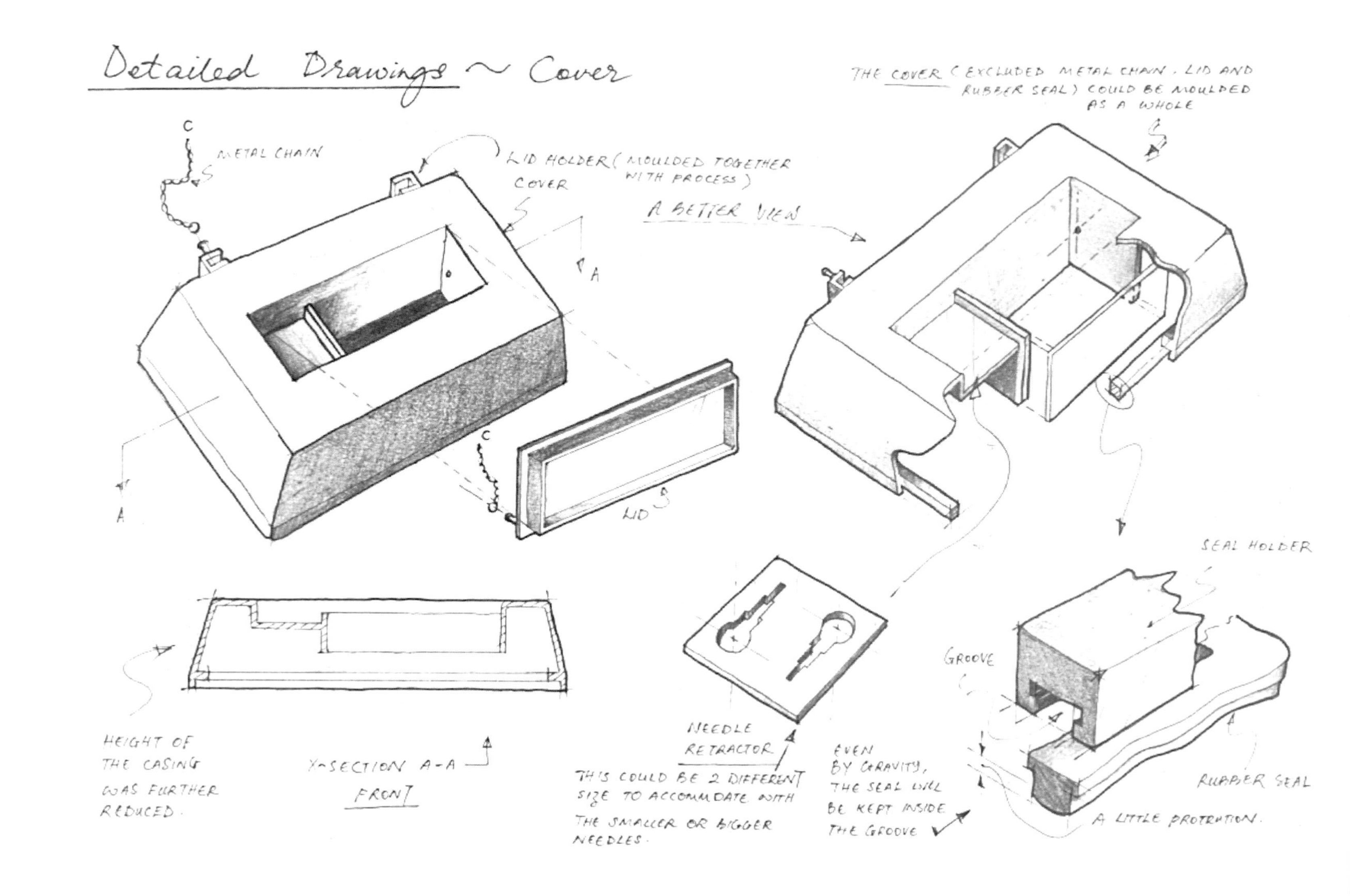
Detailed Drawings ~ Cover
C
METAL CHAIN
LID HOLDER (MOULDED TOGETHER WITH PROCESS)
COVER
A BETTER VIEW
THE COVER (EXCLUDED METAL CHAIN, LID AND RUBBER SEAL) COULD BE MOULDED AS A WHOLE
A
A
C
LID
SEAL HOLDER
GROOVE
HEIGHT OF THE CASING WAS FURTHER REDUCED.
X-SECTION A-A
FRONT
NEEDLE RETRACTOR
THIS COULD BE 2 DIFFERENT SIZE TO ACCOMMDATE WITH THE SMALLER OR BIGGER NEEDLES.
EVEN BY GRAVITY, THE SEAL WILL BE KEPT INSIDE THE GROOVE
RUBBER SEAL
A LITTLE PROTRUTION.

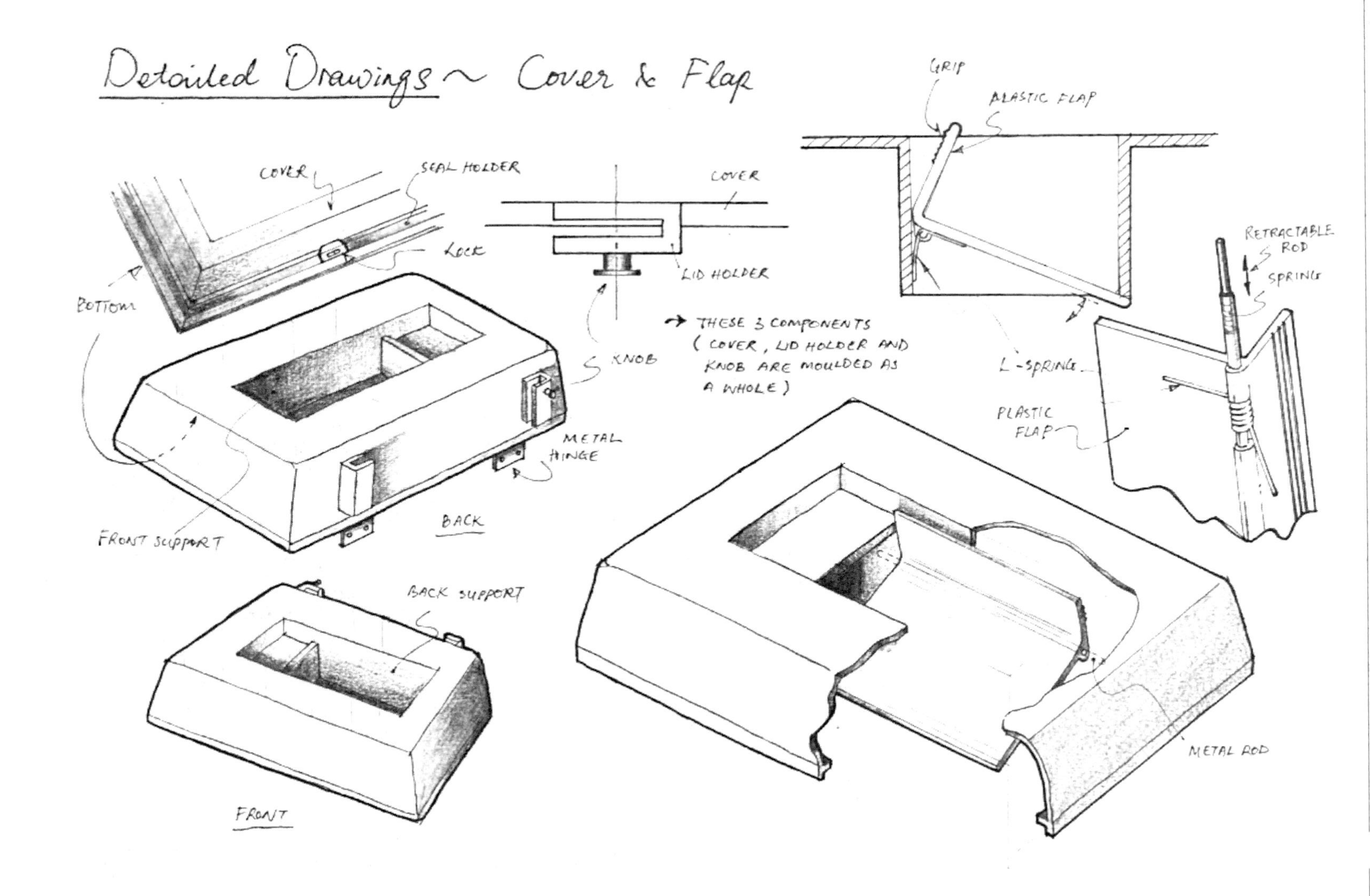
Detailed Drawings ~ Cover & Flap
COVER
SEAL HOLDER
LOCK
BOTTOM
COVER
LID HOLDER
KNOB
→ THESE 3 COMPONENTS (COVER , LID HOLDER AND KNOB ARE MOULDED AS A WHOLE)
METAL HINGE
FRONT SUPPORT
BACK
BACK SUPPORT
FRONT
GRIP
PLASTIC FLAP
L-SPRING
PLASTIC FLAP
RETRACTABLE ROD
SPRING
METAL ROD

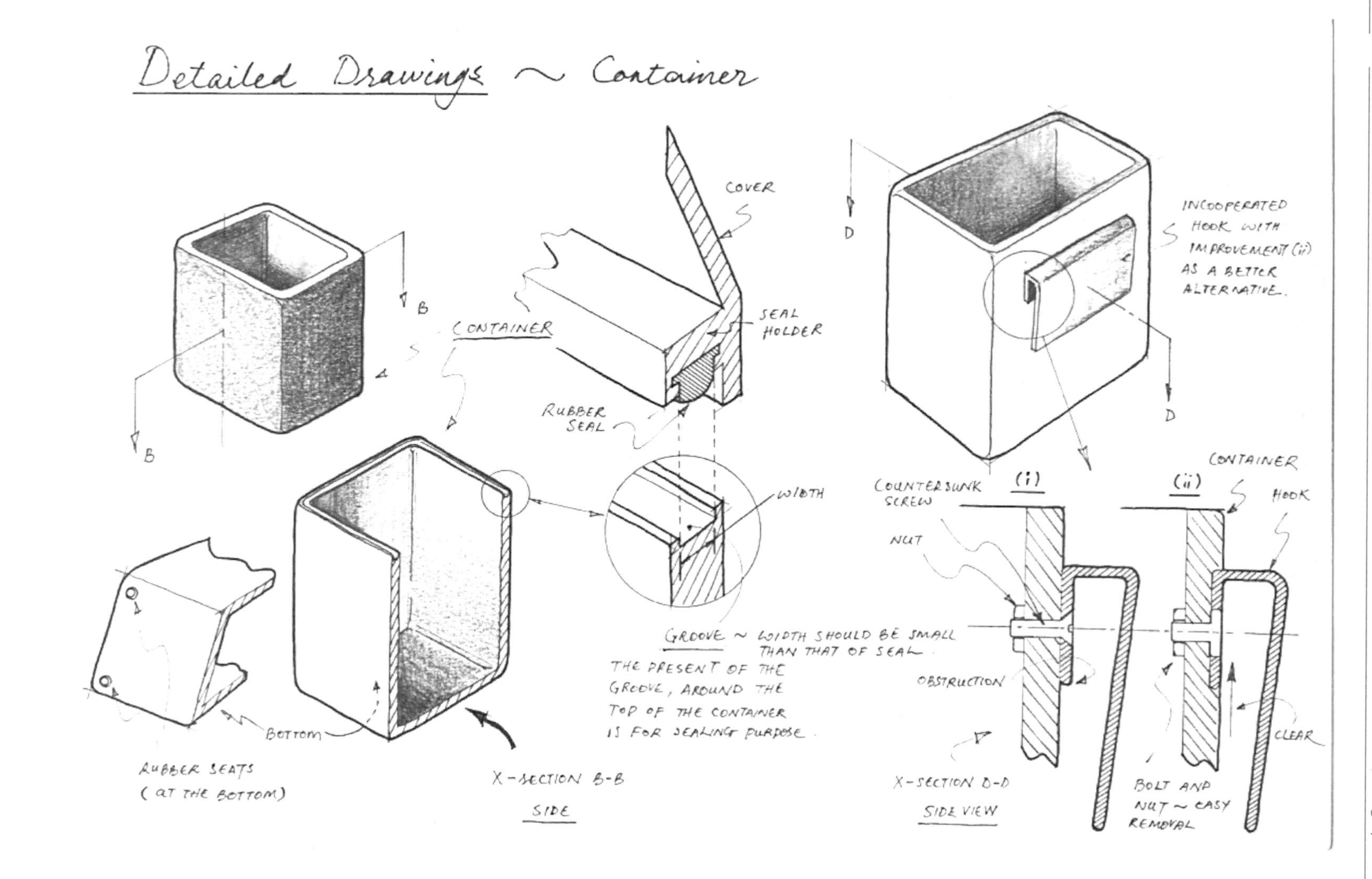
Detailed Drawings ~ Container
B
B
CONTAINER
COVER
SEAL HOLDER
RUBBER SEAL
WIDTH
GROOVE ~ WIDTH SHOULD BE SMALL THAN THAT OF SEAL.
THE PRESENT OF THE GROOVE, AROUND THE TOP OF THE CONTAINER IS FOR SEALING PURPOSE.
BOTTOM
RUBBER SEATS (AT THE BOTTOM)
X-SECTION B-B
SIDE
D
D
INCOOPERATED HOOK WITH IMPROVEMENT (ii) AS A BETTER ALTERNATIVE.
CONTAINER
HOOK
(i)
(ii)
COUNTERSUNK SCREW
NUT
OBSTRUCTION
CLEAR
BOLT AND NUT ~ EASY REMOVAL
X-SECTION D-D
SIDE VIEW

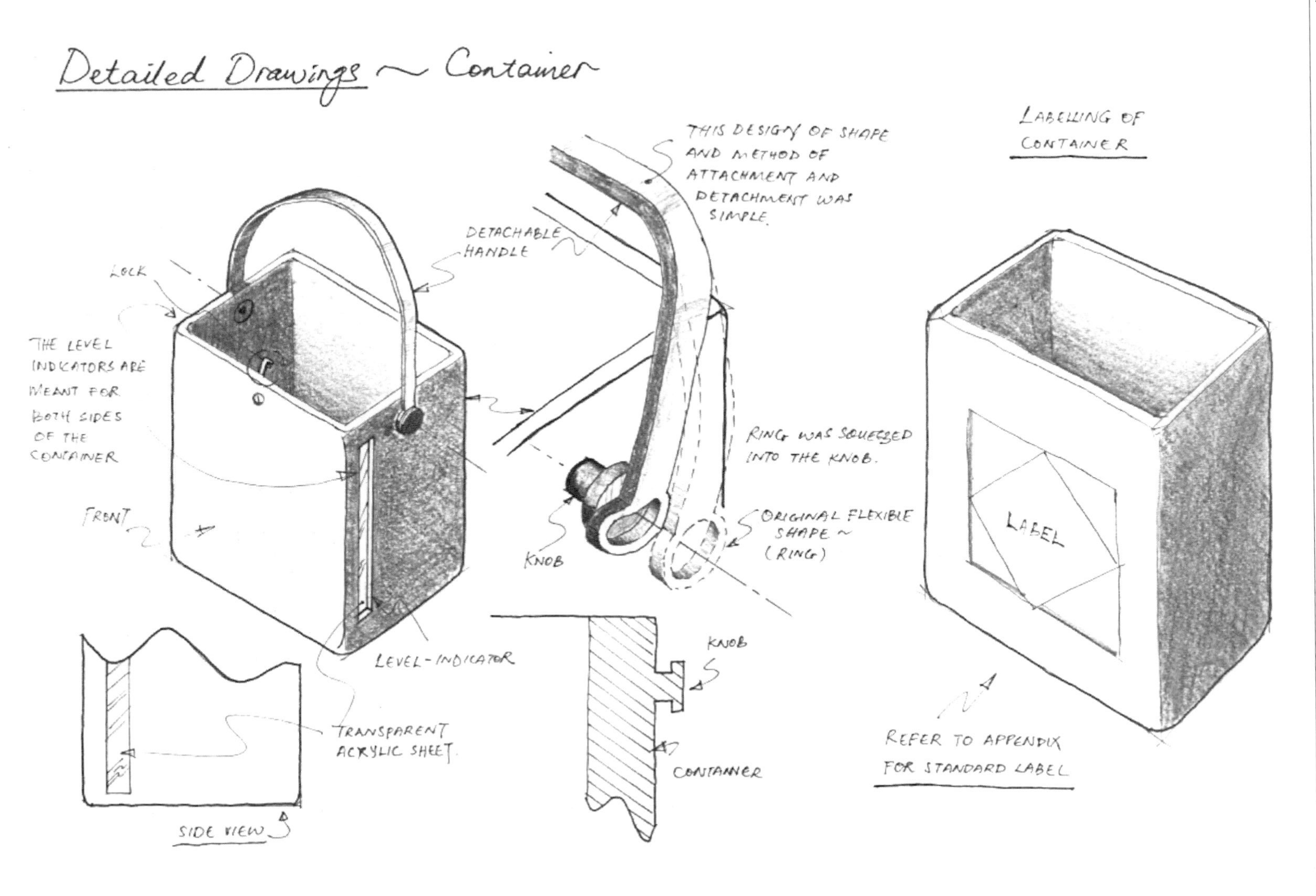
Detailed Drawings ~ Container
LABELLING OF CONTAINER
THIS DESIGN OF SHAPE AND METHOD OF ATTACHMENT AND DETACHMENT WAS SIMPLE.
DETACHABLE HANDLE
LOCK
THE LEVEL INDICATORS ARE MEANT FOR BOTH SIDES OF THE CONTAINER
FRONT
RING WAS SQUEEZED INTO THE KNOB.
ORIGINAL FLEXIBLE SHAPE ~ (RING)
KNOB
LABEL
LEVEL-INDICATOR
KNOB
TRANSPARENT ACRYLIC SHEET
CONTAINER
REFER TO APPENDIX FOR STANDARD LABEL
SIDE VIEW

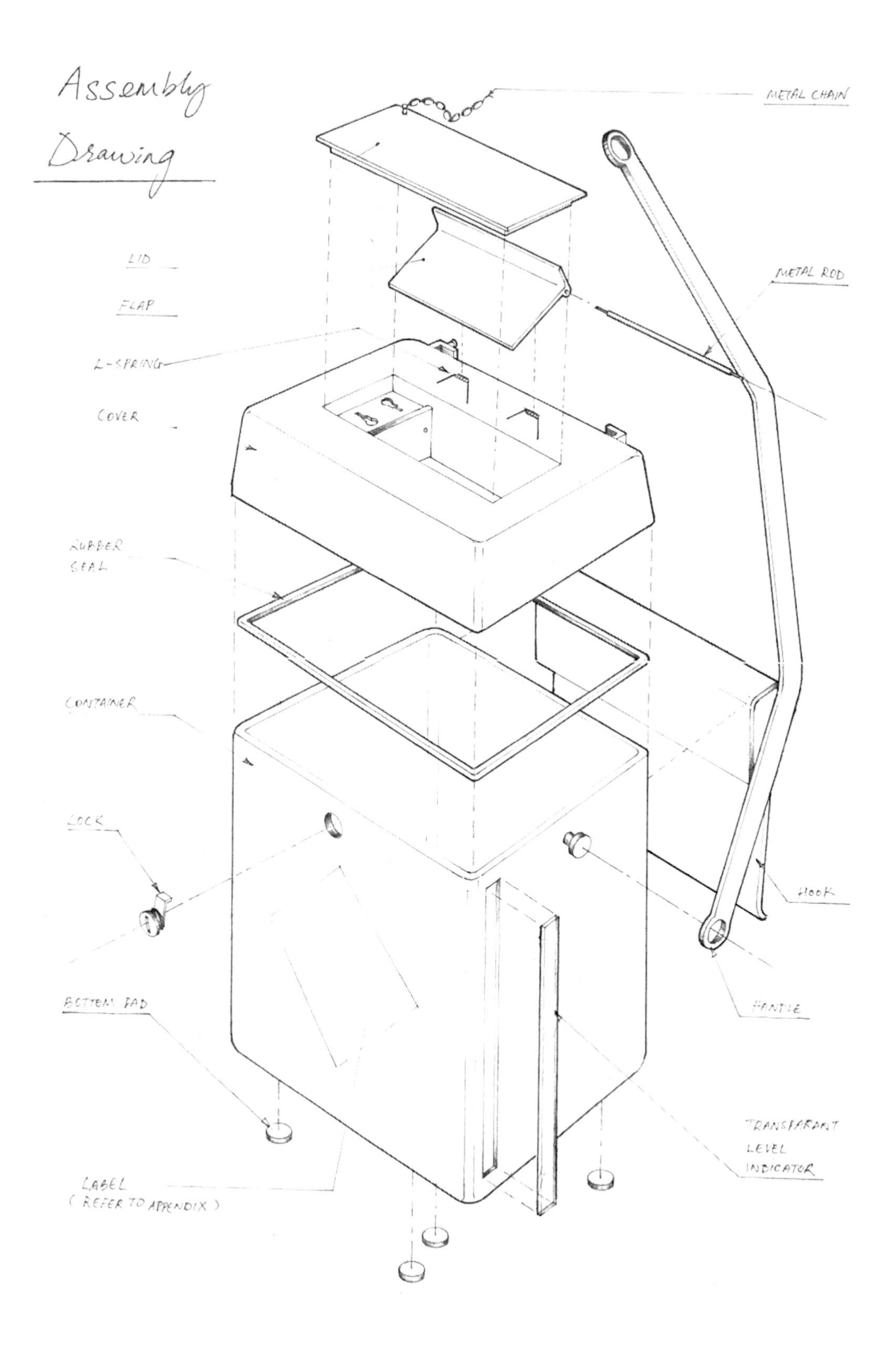
Assembly Drawing
METAL CHAIN
LID
FLAP
METAL ROD
L-SPRING
COVER
RUBBER SEAL
CONTAINER
LOCK
HOOK
HANDLE
BOTTOM PAD
TRANSPARANT LEVEL INDICATOR
LABEL (REFER TO APPENDIX)

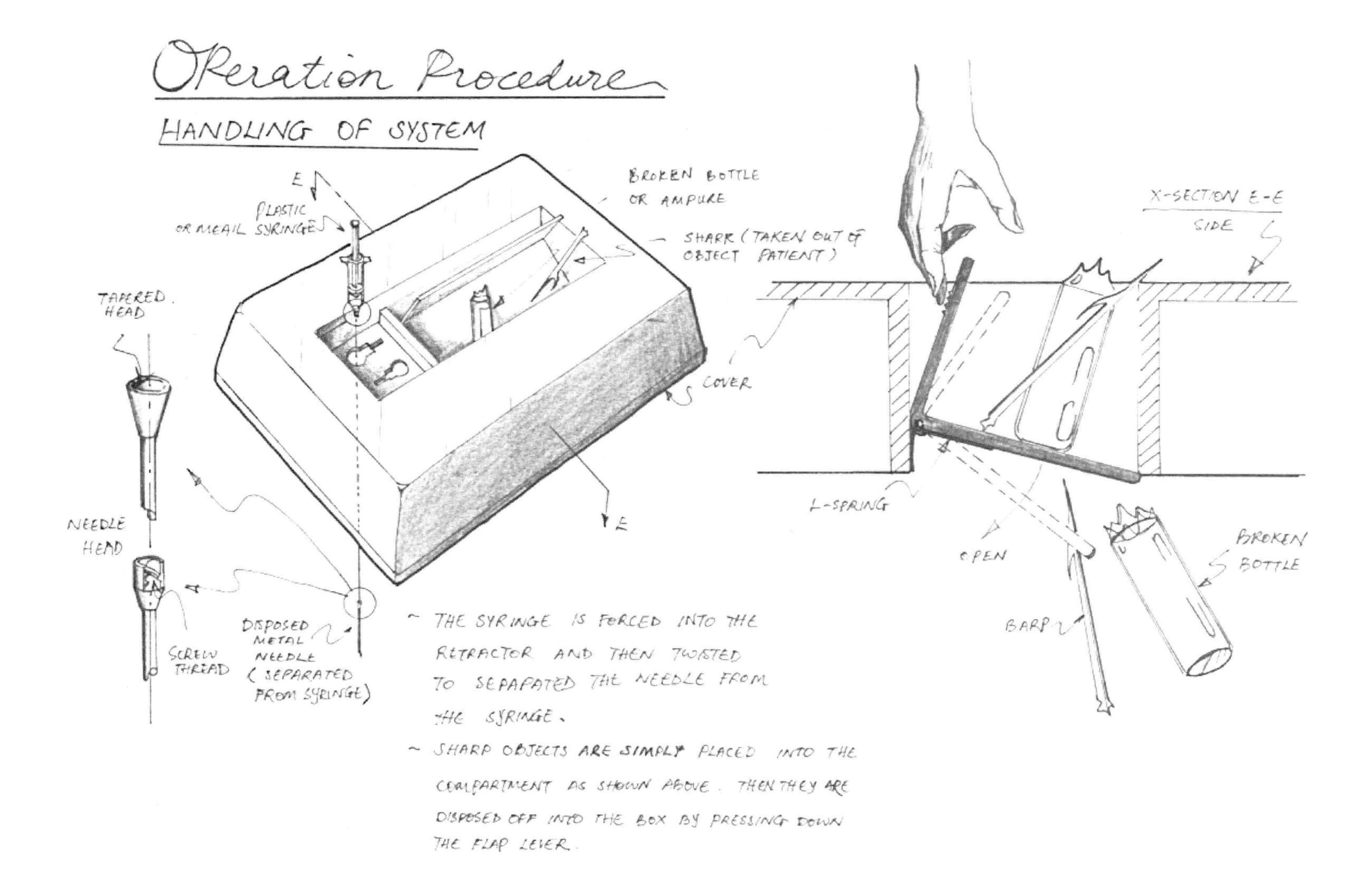
OPERATION PROCEDURE
HANDLING OF SYSTEM
E
PLASTIC OR MEAIL SYRINGE
BROKEN BOTTLE OR AMPURE
SHARK (TAKEN OUT OF OBJECT PATIENT)
TAPERED. HEAD
COVER
E
NEEDLE HEAD
SCREW THREAD
DISPOSED METAL NEEDLE (SEPARATED FROM SYRINGE)
~ THE SYRINGE IS FORCED INTO THE RETRACTOR AND THEN TWISTED TO SEPAPATED THE NEEDLE FROM THE SYRINGE.
~ SHARP OBJECTS ARE SIMPLY PLACED INTO THE COMPARTMENT AS SHOWN ABOVE. THEN THEY ARE DISPOSED OFF INTO THE BOX BY PRESSING DOWN THE FLAP LEVER.
X-SECTION E-E
SIDE
L-SPRING
OPEN
BARP
BROKEN BOTTLE

Operation Procedure

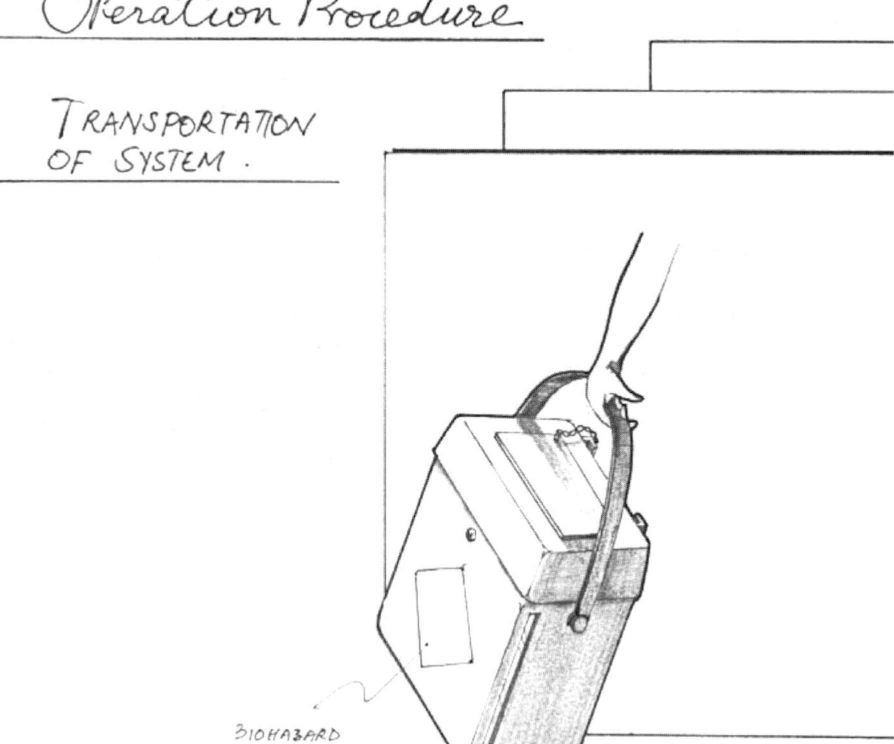

Presentation Drawing

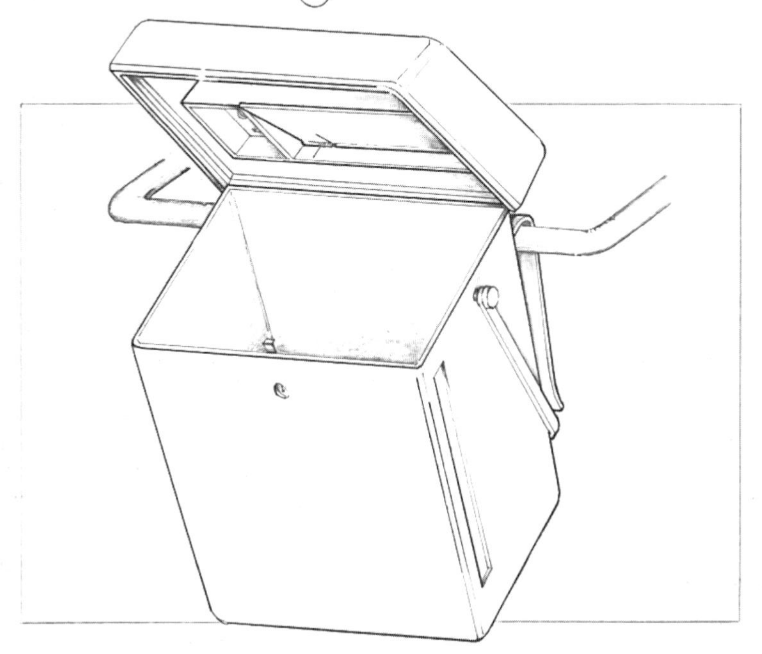

6.7 Evaluating designs and decision support systems

"Now for the evidence", said the King, "and then the sentence."
"No!" said the Queen, "first the sentence, and then the evidence!"
Lewis Carroll, Alice's adventures in Wonderland, 1865

Decision-making in the design process is often difficult because problems have *conflicting objectives*: objectives A and B are in conflict (*competition*) if an improvement towards A tends to cause a worsening of B. Common examples could include:

— *"high strength"* and *"low weight"*;

— *"low cost"* and *"long life"*.

When faced with conflicting objectives, the designer must exercise judgement to achieve a *compromise*. Conflicting objectives are *traded-off* against each other to achieve a balance

We begin thinking about decision-making in engineering design by considering a case example.

CASE EXAMPLE 6.3:
SITE FOR A POWER STATION.

FACTORS INFLUENCING THE SITING OF ELECTRIC POWER GENERATING STATIONS

- access to consumers
- access to fuel (15,000 tons/day)
- access to cooling water (250,000 litres/hr)
- disposal of ash (3,500 tons/day)
- space for stockpiling coal
- stability of site:
  — building foundations
  — coastal erosion
- construction:
  — transport of heavy equipment
  — heavy construction vehicles
  — cost of construction
- community:
  — attitudes of people
  — local trade; employment
  — supply of skilled operators
  — local community services
- environmental:
  — pollution via exhaust gases
  — dust; noise
  — coping with *warm water (thermal* load)
  — loss of habitat
- aesthetics:
  — buildings
  — power lines

This problem is notable since there are a large number of factors to be considered, and so many of them are in conflict (typically, access to cooling water and large scale ash disposal versus environmental impact; access to consumers and cheap distribution system versus aesthetic impact). Moreover, the range of alternatives (power station sites) is likely to be small.

DECISION FACTORS IN LARGE SCALE INFRASTRUCTURE PROJECTS

— load prediction
— effectiveness
— design life
— cost
— environmental issues
— political feasibility

These must be considered in relation to:

— construction phase
— operation during life of project
— decommissioning

It is interesting that engineering emphasis has traditionally been directed towards the first two of these, whereas the third issue (decommissioning) has been largely ignored at the design stage. One of the reasons for regarding nuclear energy in the 1950s and 60s such an economical option for the future, was that the as yet inestimably large cost of decommissioning a nuclear fission plant was not considered.

The power station example is typical of a generic decision making problem that we characterise as a *flow system*. Flow systems, in general, accept purposeful inputs, and transform them into desired outputs. Along the way the transformation process generates some uncontrolled, often undesirable, outputs. Some typical examples of flow systems are:

- electric power generating stations;
- urban transport systems;
- economic systems;

- chemical processing industries and any processing plant;
- schools and universities, libraries, museums, movie theatres, supermarkets;
- coffee grinders, washing machines, lawn mowers.

When facing a problem of designing, locating, or simply evaluating such a *flow system*, the procedure we use is an *input/output analysis*. The system is represented on an input/output diagram, as shown in Figure 6.29. Input/output analysis, corresponds to the problem evolution and enformulation phase of the design. As the name implies, the procedure focuses particular attention on the flow of materials in and out of the plant, without reference to the processing that occurs in the transformation of material from input to output. In general, with flow systems, the *processing embodiment* has already been determined.

In establishing performance parameters for our flow system, we often focus on the measure of *efficiency*. The term is commonly misused to describe another measure, *effectiveness*. While the confusion is a result of both words originating from the same root (*to have an effect*), we must be absolutely clear about the credible use of both terms in the engineering context.

Efficiency, η, is a ratio between useful outputs available from a system and the inputs required to produce the outputs. Typically

$$\eta = \frac{\textit{useful output}}{\textit{input required to produce output}}.$$

Generally, but not always, efficiency is a dimensionless ratio. In contrast to efficiency, which is a precise technical measure, *effectiveness* is a subjective measure of *fitness for purpose*. The automobile is a very *effective* personal transport

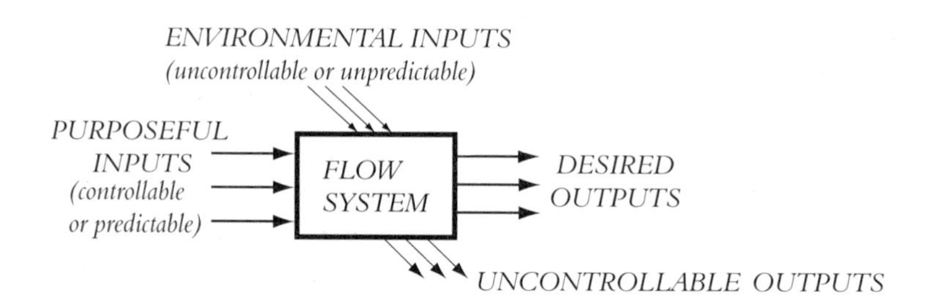

(a) The generic input-output diagram

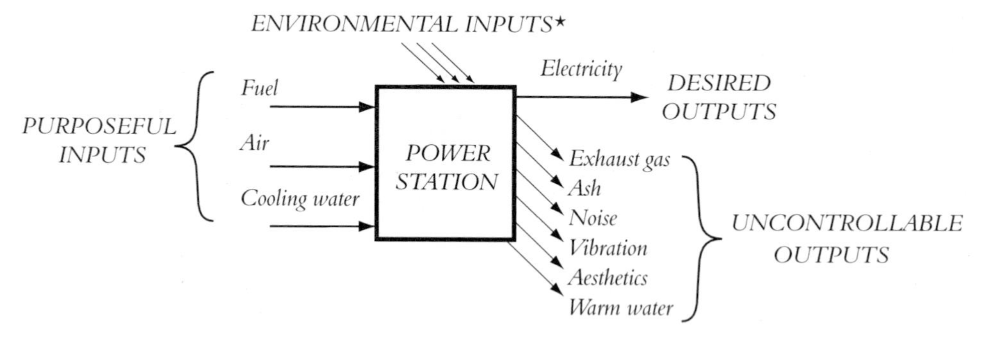

(b) Input-output diagram for power station example

Figure 6.29 Input-output diagrams (* for the power station example, environmental inputs may be unseasonable weather conditions, forcing possible power restrictions; unpredictable fluctuations in demand; equipment breakdowns; variations in cooling water temperature or availability)

system, but it is a *highly inefficient* people mover when compared to public transport systems. By contrast, a coal powered power station is *most effective* for delivering base-load electric power. Yet, it is a *quite inefficient converter* of thermal energy into electrical energy. In fact, the measure of effectiveness in a design is a direct function of the design parameters (*criteria*) used for comparing alternative design options. In the case of the automobile, if our criterion of performance were *convenience* (a somewhat subjective measure, but we could use for it *average time to reach destination*), then, in general, an automobile would be a more effective personal transport system than a train. By contrast, if we used *number of people per unit trip* as a criterion of performance, the train would be a far more effective transporter than the automobile.

Ex 6.10

Public transport systems use trains, buses, trams and trolley buses, as people movers. Enformulate efficiency measures for those people movers familiar to you. Compare the efficiency measure with similar measures for an automobile, a bicycle, a pogo stick, a horse, and a motor bike.

Draw up a table of comparative energy efficiencies for all of these people movers.

6.8 Decision making strategies

Having explored the *divergent*, creative, entirely *yellow-hat*, aspects of the design process, we now focus attention on the *convergent*, evaluative, *black-hat* aspects. The generic term for evaluation procedures in design is *decision making*. While we advocated the use of fantasy and playfulness in idea generation, in evaluating competing design proposals, we must become entirely ruthless in our comparative value judgement.

There is a legend about a farmer whose hired labourer was required to dig up a field of potatoes and then sort them into three categories of size, small, medium, and large. At the end of the day the farmer visited the potato patch to find that the labourer had dug up the whole crop. However, the potato sorting was not going so well. There were two small mounds of potatoes in the large and small categories. The remainder of the crop was in a huge mound, with the labourer standing next to it, scratching his head. *"What's the problem?"* asked the farmer. *"Ah, decisions, decisions"*, replied the unhappy labourer.

Designers use a range of evaluative tools for decision making. These decision support tools allow comparative evaluation of competing alternative designs. In general, all the evaluative procedures presuppose the existence of the design embodiment. There are no known, formal, evaluative tools for conceptual design.

In what follows we briefly review some of these decision support tools, and their spheres of application.

6.8.1 Input-output analysis (I/O)

I/O analysis aids in the identification of:

- constraints, restrictions;
- performance criteria;
- design objectives.

We have already seen a case example of a flow system in selecting a site for a power station. In that example we explored the several sources of information to be consulted. In the following, smaller scale, example we specify hardware requirements.

CASE EXAMPLE 6.4 :
DESIGN OF A WASHING MACHINE.

- Input: Soiled fabrics
- Input variables:
  - — size of load ($\leq$ 0.5 m^3)
  - — mass of load ($\leq$ 6 kg)
  - — type of fabric (all)
  - — type of dirt (all)
  - — mass of dirt ($\leq$ 0.3 kg)
- Output: Clean fabrics
- Output variables:
  - — amount of dirt ($\leq$ 1%)
  - — shrinkage ($\leq$ 0.05%)
- System: Washing machine
- System variables:
  - — size
  - — shape
  - — method of freeing dirt
  - — source of power
  - — materials of construction
- Requirements:

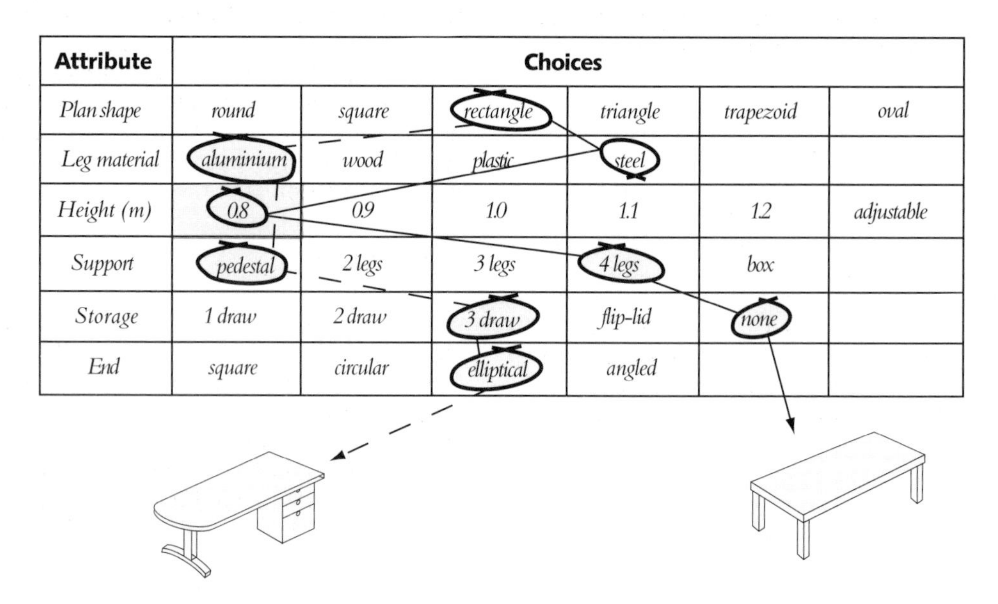

Attribute	Choices					
Plan shape	*round*	*square*	*rectangle*	*triangle*	*trapezoid*	*oval*
Leg material	*aluminium*	*wood*	*plastic*	*steel*		
Height (m)	*0.8*	*0.9*	*1.0*	*1.1*	*1.2*	*adjustable*
Support	*pedestal*	*2 legs*	*3 legs*	*4 legs*	*box*	
Storage	*1 draw*	*2 draw*	*3 draw*	*flip-lid*	*none*	
End	*square*	*circular*	*elliptical*	*angled*		

Figure 6.30 Morphological matrix for an office desk

— size must be ≤ 1m x 1m x 1m
— cost must be ≤ $250
— must be foolproof in operation

- Constraints:
  — must be approved by statutory Health and Safety Authority

6.8.2 Benefit-Cost-analysis

This type of analysis deals with comparative evaluation of various utility values associated with the design. A Benefit-Cost analysis:

- identifies all benefits to be gained from proposed solution (*yellow-hat*)
- assigns values to benefits on specific value scales (identifies criteria, usually $)
- identifies all costs and drains on resources generated by proposed solution (*black-hat*)
- assigns values to costs on specific value scales (identifies criteria, usually $)
- draws attention to the total value of the proposal in comparison to alternative ways of assigning resources: this is the ultimate outcome of the evaluation and it often involves some subjective evaluation. While this evaluation is probably the domain of business analysts, it is always useful to recall the experience of Xerox-PARC and the *'Alto'* (refer back to 6.5.1: our *cautionary tale* about deferred judgement). This is the part of the evaluation most subject to the risk of *throwing the baby out with the bathwater.*

Benefit-Cost analysis is further explored in Section 7.2.

6.8.3 Morphological analysis

Morphology is the *study of form and structure* of things. We use morphological analysis when there is already an established form of embodiment for our design problem. In general, morphological analysis allows us to generate a wide range of alternative embodiments. The process focuses on the attributes of the solution that already match the problem's requirements: recall the *"uses for-X"* exercise in idea generation; see also Marples 1961.

Morphological analysis requires that:

- the mechanism of the product's operation is already understood, and
- the major attributes, or the major functional parts, of the solution can be listed.

Figure 6.30 shows a partial morphological matrix for an office desk, with two of the many alternative embodiments indicated.

6.8.4 Fishbone diagrams

Designers involved in trouble-shooting or fault/failure analysis make use of formalised procedures for depicting *potential cause-effect* relationships. The fishbone diagram, also called *"Ishikawa diagram"*, after its developer, Kaoru Ishikawa (Ishikawa, 1985), provides a useful form of representation for such relationships. The main value of the fishbone diagram is that it draws attention to all those elements in the system that might contribute to some specific outcome. It is a first step in preparing a more detailed study of cause-effect investigations.

This type of diagram also provides a useful medium for representing product attributes, and their relationship to design requirements. Figure 6.31 is a type of *fishbone diagram* relating product attributes to customer requirements. Although the attribute list is based on our accepted notion of what constitutes a *"can opener"*, many possible embodiments might result from this description.

A can may be cut, sheared, *rolled* open (sardine cans), pierced or *ring pulled* (beverage cans), or *levered* open (paint cans). In each case, there is a specific embodiment that springs to mind during problem evolution.

6.8.5 The design tree

Earlier we referred to forward and backward chaining during problem evolution. The *design tree* provides a medium for forward chaining from technical objective to embodiment. Design trees are particularly useful when there are many alternative forms of embodiment to be considered.

In many parts of rural Australia (and no doubt in other parts of the world) electrified fences are used to control livestock in paddocks. The fences keep livestock in, and keep threatening wildlife out. Farmers make use of wooden fence-posts to string the bare electric conductors. Specially designed insulators are used to attach the wire to the fence-

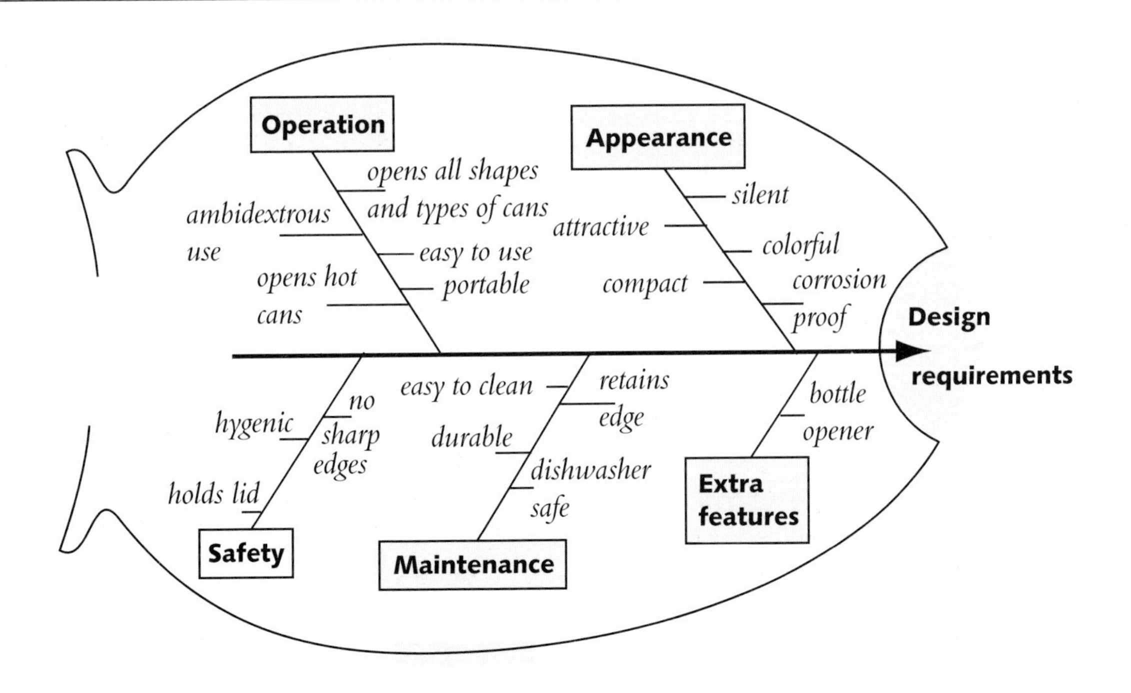

Figure 6.31 Fishbone diagram for a 'can opener'

posts. Some of these insulators are of moulded polymer and others of a ceramic/metal construction. There are many possible embodiments. Figure 6.32(a) shows a partial design tree for the electric fence insulators. The photo in Figure 6.32(b) shows a small selection from the alternative embodiments available for this design.

6.8.6 Decision tables and other scaled check lists

Project evaluation often calls for comparative evaluation of two or more alternative proposals. The designer needs to predict the outcomes of these alternative proposals, in order to make a reasoned evaluation. Some outcomes may be predicted in terms of cost ($), but often these *predictions* are estimates only. Moreover, some estimated outcomes can only be expressed on a subjective scale such as *preferred* or *not preferred*. For these types of evaluation, decision tables provide a useful comparative procedure.

CASE EXAMPLE 6.5:
PROPOSAL FOR A PROCESSING PLANT.

In this example we are considering a new capital project to build a processing plant. Other embodiments already exist and we use a decision table to compare the two alternative proposals, by relating the predicted outcomes to those from existing plants. This is a first order qualitative decision process, where we only consider scaled positive (+1 = better; to +5 = best), negative (-1 = worse to -5 = the worst) or neutral (similar) performances.

In the scaled decision table, Table 6.6 , we have compared two proposals for the new processing plant. Taking *Alternative A*, we note that it is slightly more expensive to construct than the old plant (inflation and increased cost of labour might be the reason for this assessment). Hence this *factor* is regarded as worse (denoted by a *-1*) than the old plant. However, the new plant is estimated to outperform the old plant in almost all other areas. Similarly, the engineering staff have predicted performance estimates for *Alternative B*. Interesting to note that, based on *raw* performance estimates, *Alternative A* is a better risk than *Alternative B*. However, when the weightings are taken into consideration, *Alternative B* is estimated to outperform *Alternative A*.

Scaled decision tables are helpful in formalising the evaluation of design alternatives, and in focusing attention on the factors to be considered in the evaluation. However, we caution against placing too much stock in the aggregate results from such an analysis. Recall, we started with the objective of performing a *qualitative* evaluation only. While the performance estimates might be near enough to the mark, the weightings are, at best, speculative only.

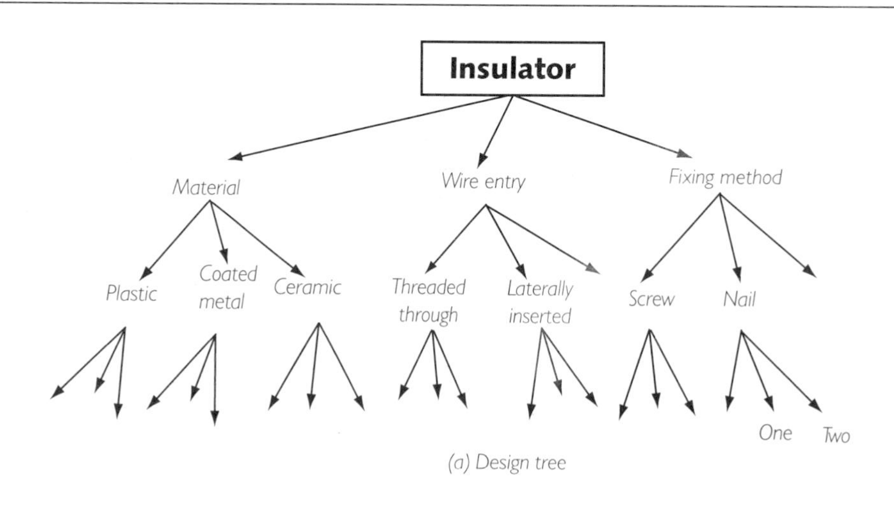

(a) Design tree

(b) Sample embodiments

Figure 6.32 Partial design tree for electric fence insulators

Table 6.6 : Scaled decision table (After Holick, 1993)

Factor	*Importance (weight)*	**Alternative A**		**Alternative B**	
		Raw evaluation	*Weighted*	*Raw evaluation*	*Weighted*
Construction cost	4	-1	-4	-2	-8
Return on investment	4	+2	+8	+6	+24
Noise	2	+1	+2	-3	-6
Emission	5	+2	+10	+1	+5
Soil erosion	2	-2	-4	-1	-2
Aggregate		+2	+12	+1	+15

Yet another form of qualitative decision table is the *"Pugh matrix"*, attributed to Stuart Pugh, who was Professor of Engineering Design at University of Strathclyde. Pugh evolved a simple, but useful, qualitative decision matrix as a means of evaluating competing design concepts.

To make use of the Pugh matrix approach, we need to have access to a *datum* embodiment of the proposed design. All the competing concepts are then compared to this *datum* embodiment. The comparison invokes several performance factors and assigns only positive (*better*), negative (*worse*), or zero (the *same*) values to each performance factor, relative to the *datum*, in the matrix.

CASE EXAMPLE 6.6: AUTOMOBILE HORN.

Figure 6.33 shows eight alternative embodiments for an automobile horn. There are many other possibilities, but these few will suffice for a demonstration applying the Pugh matrix.

The *datum* concept is number 1 in the figure, and all other concepts are evaluated relative to this

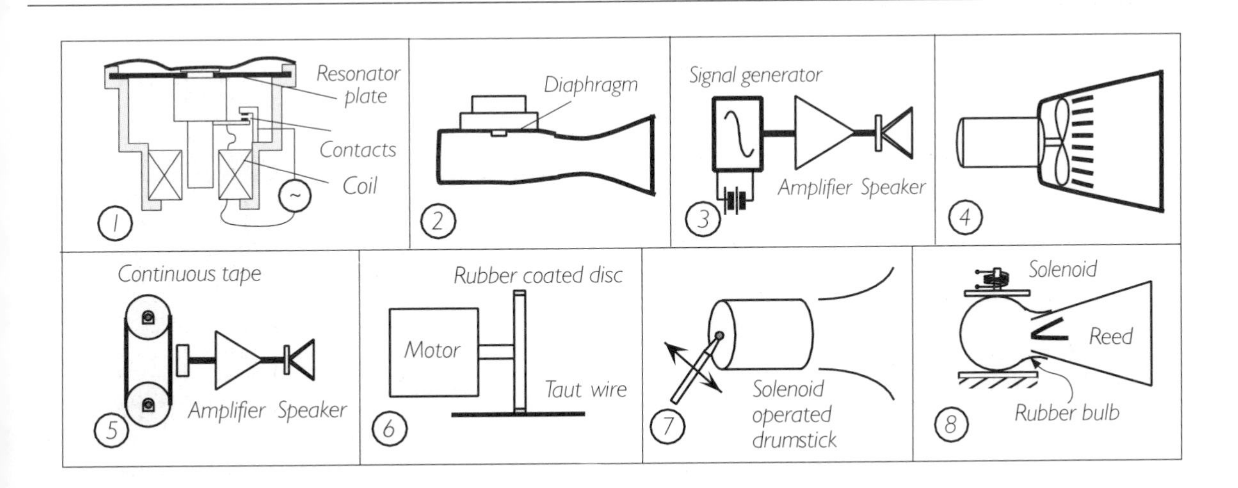

Figure 6.33 Some alternative embodiments for an automobile horn

Table 6.7: Pugh matrix for automobile horn concept evaluation (After Pugh, 1981)

	Concept							
Technical objective	1	2	3	4	5	6	7	8
Ease of producing 105-125 dB	D A T U M	s	+	-	+	-	-	+
Ease of producing 2-5 kHz		s	+	s	+	s	-	s
Corrosion, erosion		-	s	-	s	-	s	-
Shock, vibration		s	s	-	-	-	s	s
Temperature sensitivity		s	+	-	-	-	s	-
Aggregate score		-1	+3	-4	0	-4	-2	-1

conventional, well known, embodiment of the automobile horn.

The technical objectives for the horn are :

- sound level = 105 to 125 dB (sound pressure level is measured on a logarithmic scale of bels, relating sound energy to a datum energy level. The usual sound level scale is calibrated in tenth of bels or decibels, denoted dB);
- frequency = 2 to 5 kHz;
- corrosion and erosion proof;
- impervious to shock and vibration;
- operation in wide range of temperatures.

Table 6.7 shows the evaluation matrix for the eight design concepts considered. A useful way of comparing competing design concepts is to compare aggregate results from such a table. The various concepts may be developed using a morphological analysis of the design in the first place.

6.8.7 Mathematical modelling

In chapters 1 to 5 we have explored mathematical modelling for evaluating structural integrity of engineering components. In essence, the substance of those chapters was to make reasoned decisions about the likelihood of failure. Our reasoning was based on mathematical modelling of the behaviour of engineering components. Complex engineering systems can be the subject of mathematical modelling. The formal procedures used in such decision making are multi-objective optimisation and search methods.

Multi-objective techniques attempt to reconcile, *trade-off*, conflicting requirements, by developing suitable *utility functions* that place proper priorities on the design objectives. When the number of design objectives and the factors influencing them are large, computer search procedures are used for evaluating *best fitness for purpose*. Here we only draw attention to the existence of these advanced decision making tools, as they are outside the scope of a general design text. For an introduction to these specialist search procedures see, for example, Wilde (1978), Goldberg (1989) and Parmee (1997).

CASE EXAMPLE 6.7:

ELECTRIC POWER DISTRIBUTION.

In this example we consider the choice of a suitable conductor in an electric power distribution system. The two available alternatives are copper and aluminium.

- Copper (Cu) : resistivity = 1.7241 microhms per cm^2 per cm; specific gravity = 8.99;
- Aluminium (Al) : resistivity = 2.828 microhms per cm^2 per cm; specific gravity = 2.70.

The choice of conductor will determine the distribution losses and the structural character of supporting pylons. Cu is a 64% better conductor, but weighs 3.3 times as much as Al.

This is a simple deterministic problem with a single parameter utility function (total cost over the life of the installation). We can easily optimise for each material as shown on Figure 6.34.

Resistive losses are a function of conductor area and length. Hence, we can plot distribution losses as a function of conductor cross section. The conductor size will also determine the cost of installation, in terms of support structures and conductor material cost. This too can be plotted on the same diagram. The aggregate of the two cost curves is the total installation cost, and this curve identifies the lowest cost. As seen on Figure 6.34, the minimum cost is not very sensitive to conductor diameter. Hence we are able to choose a suitable 'preferred size' without the need to manufacture a special batch of conductor sizes. A similar figure can be drawn for the aluminium conductor and the minimum costs compared for a decision between the two alternatives.

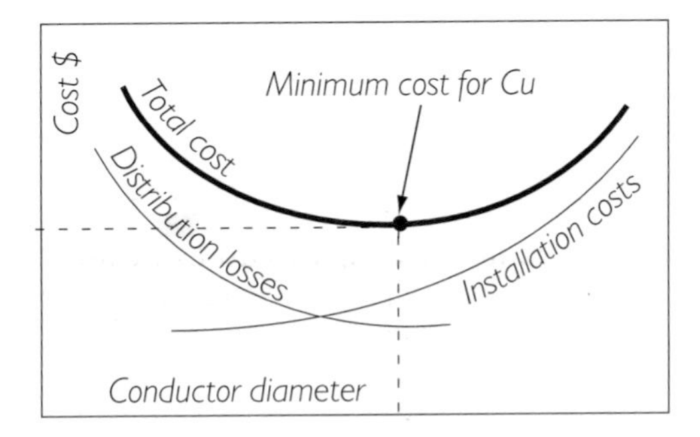

Figure 6.34 Choice of optimum conductor diameter

CASE EXAMPLE 6.8

WIND LOAD ON CHIMNEY.

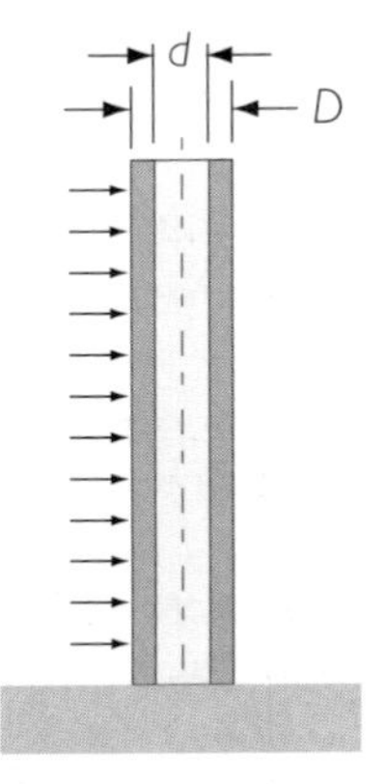

Figure 6.35 Incinerator chimney under wind loading (schematic)

In some process industries incinerators are used to burn hazardous substances. The resulting discharge must be released and dispersed into the free-stream air, well above ground level. Consequently, the chimneys of these incinerators can be quite high. Wind loading on the chimney becomes a significant design consideration. Figure 6.35 is a schematic sketch of such a chimney installation.

It is easy to show that the two governing equations for safe structural behaviour of the chimney are

$$\text{allowable stress: } \frac{k_1 D^2}{D^4 - d^4} \le 1, \text{ and}$$

$$\text{allowable deflection: } \frac{k_2 D}{D^4 - d^4} \le 1,$$

where k_1 and k_2 are constants for a given installation. We can plot these two inequalities on the same diagram, as seen in Figure 6.36. Note, that the stress inequality increases more rapidly (governed by D^2) than the deflection inequality (governed by D).

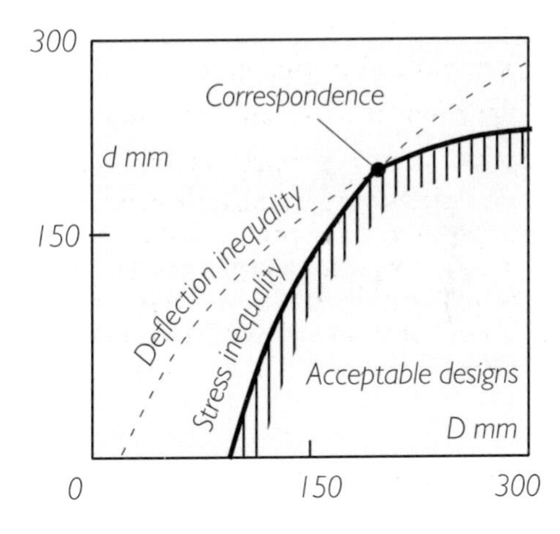

Figure 6.36 Design plot for chimney problem

At first sight, it may seem reasonable to select the dimensions of the chimney so that the two design inequalities correspond. This would appear to make the most effective use of the construction material. However, in practice, there are many other aspects of the design that will determine the best choice of dimensions.

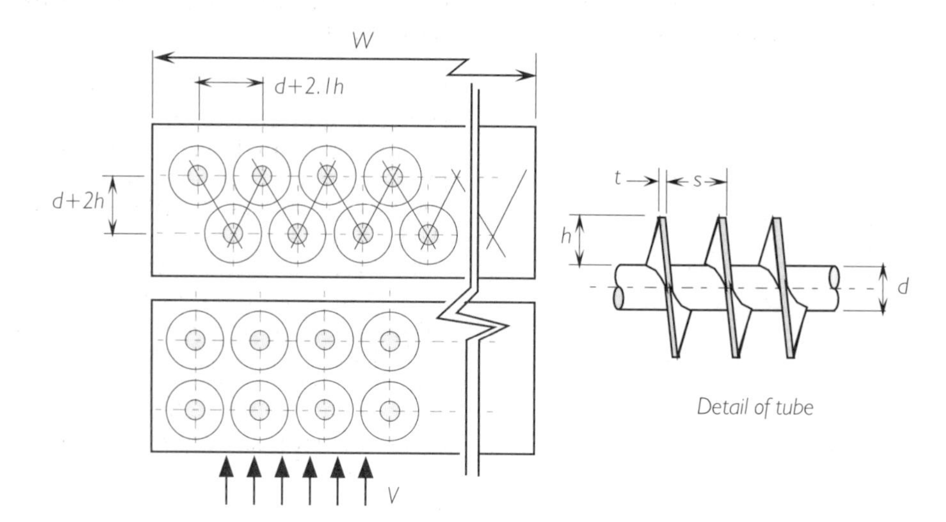

Figure 6.37 Heat exchanger detail, indicating design variables and two alternative tube configurations (Field, 1970)

Ex 6.11

Assuming that total installed cost of the chimney is proportional to its mass, determine the form of a curve of constant cost on the (*D*, *d*) plane of Figure 6.36. Hence propose a method of selecting a safe design for minimum cost.

CASE EXAMPLE 6.9:

HEAT EXCHANGER DESIGN.

We have just looked at two small-scale problems with relatively few design variables to be determined for the *best*, or at least the most favourable, design choice. Most design problems have many more variables than we saw in these problems. It is specifically for large-scale multi-variable problems that search and response surface methods are designed to provide decision support (see for example Box, Hunter and Hunter, 1978).

In this case example we consider a relatively common cross-flow heat exchanger. A schematic cross section and tube detail for such a heat exchanger are shown in Figure 6.37. On the face of it, the problem has only five variables: tube diameter (*d*), fin height (*h*), fin pitch (*s*), fin thickness (*t*), and overall face width (*W*). However, the configuration of tubes within the exchanger has considerable influence on performance. Only two of the many possible alternative arrangements are shown in Figure 6.37.

In this type of problem, the appropriate decision support system is a response surface approach. We first determine the overall sensitivity of the heat exchanger's performance to changes in variables. Output variables are pumping energy (E_p - we want this to be low), fluid temperature change (DT - we want this to be large), and total mass of copper (m_{Cu} - we want this to be small). Hence, a suitable technical objective might be to

$$\text{maximise } P_{HE} = \frac{\Delta T}{E_p\, m_{Cu}}.$$

Setting design variable limits, we examine the changes in P_{HE} as a function of each variable alone, keeping the others fixed at one or other limit. Following this evaluation, we choose the two variables that exert most influence on P_{HE} and plot these as a surface (the *response surface*) to find the *best* combination of values for design variables.

Figure 6.38(a) shows one of the many typical sensitivity analyses used in this problem, and Figure 6.38(b) shows the chosen response surface. The approach can certainly be used for virtually any number of design variables, but the resulting response *hyper-surface* needs to be searched for global maxima or minima, using one of the many computer based *search engines* available.

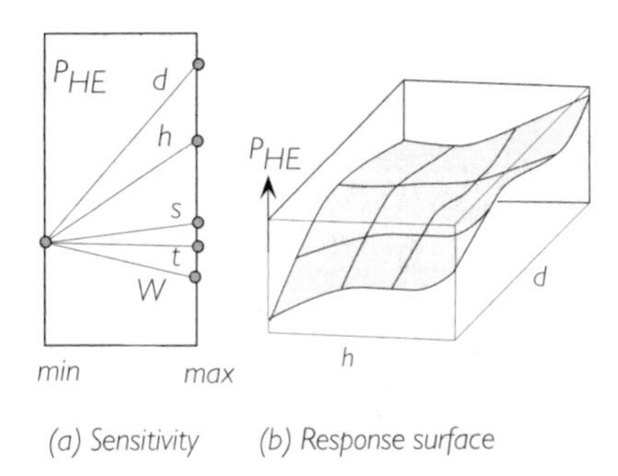

(a) Sensitivity (b) Response surface

Figure 6.38 Sensitivity analysis and response surface for the heat exchanger problem (not to scale)

6.8.8 A comparative evaluation of case examples

We have examined a series of case examples in engineering problem solving. These case problems are now classified according to our problem-intensity parameters identified in Section 6.3.6. Table 6.8 shows a brief qualitative classification of the case examples we presented in this chapter. It is extremely useful to prepare a *problem intensity* evaluation of design problems as part of the *Initial Appreciation*. This will alert both client and designer to the degree of effort to be spent in later stages of problem development. Well-defined, clearly articulated problems of high seriousness, deserve focused attention. Ill-defined, poorly bounded, problems of low seriousness should not be allocated valuable design resources.

Table 6.8 Qualitative classification of the case examples

CASE	Level crossing	Portable water heater	Power station site	Washing machine	Desk	Insulator	Automobile horn	Can opener	Cu v. Al conductor	Chimney	Heat exchanger
Problem-boundary	H	VL	VH	M	L	VL	M	M	VL	VL	M
Complexity	H	L	VH	M	L	L	M	L	VL	VL	H
Novelty	L	L	M	H	M	M	H	M	L	L	M
Discordance	M	L	M	L	L	L	L	L	L	M	VL
Modelability	VL	M	VL	M	VH	H	VH	M	VH	VH	H
Seriousness	H	L	H	L	L	L	L	L	H	H	H
Solution type	Ev	N	Ev	Ev	Ex	Ex	N	N	Opt	Opt	Is

Legend

(a) Problem intensity parameters: VL= very low; L=low;M=medium; H=high; VH=very high;
[Note that for a problem boundary to be 'high' (H) implies a broadly based problem with wide implications, while a 'low' (L) problem boundary indicates a narrowly based problem.]

(b) Solution type: N = novel; Ex = exploratory; Ev = evolutionary; Opt = mathematical optimisation; Is = iterative search.

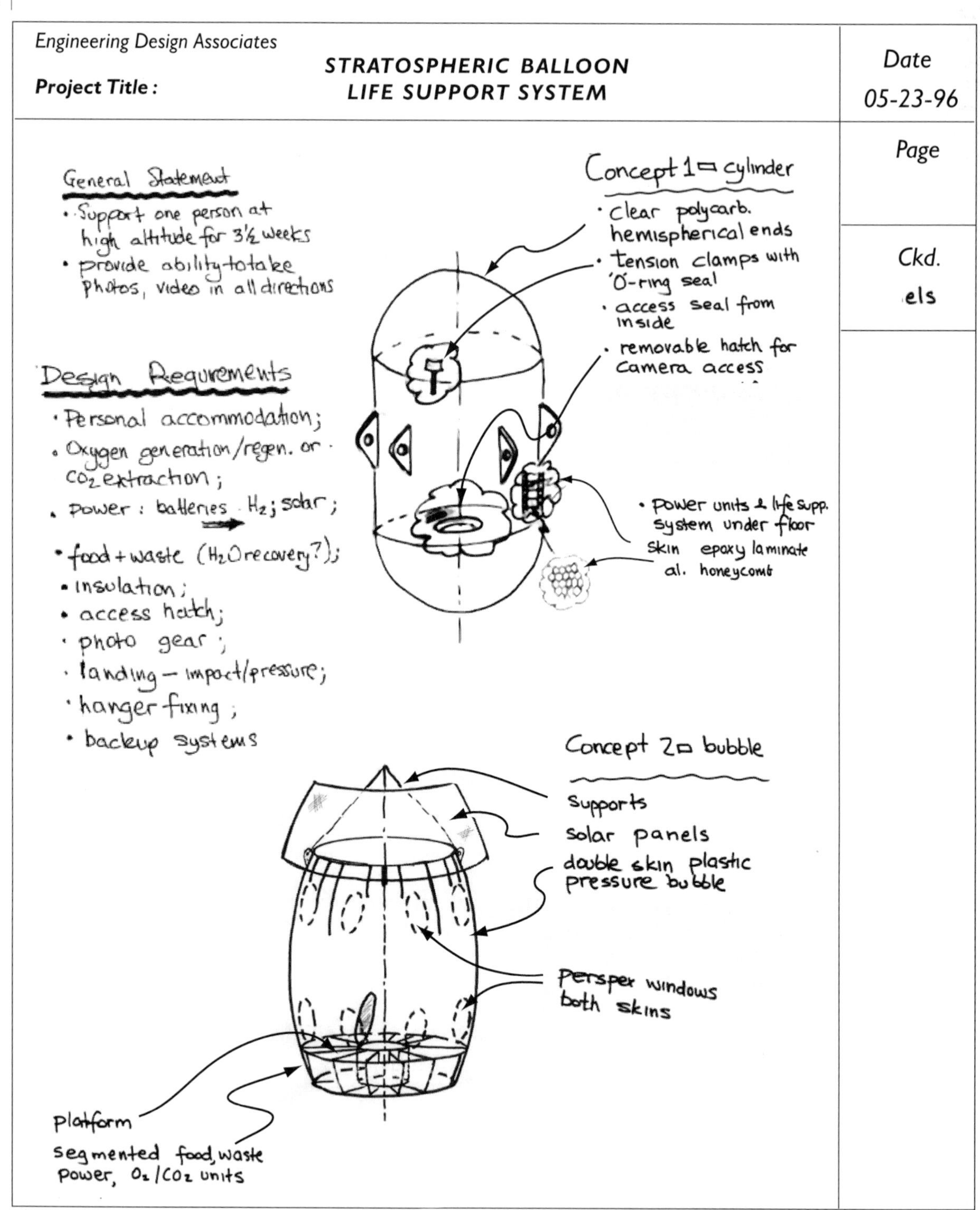

Engineering Design Associates
Project Title :
STRATOSPHERIC BALLOON
LIFE SUPPORT SYSTEM
Date
05-23-96
Page
Ckd.
els
General Statement
• Support one person at high altitude for 3½ weeks
• provide ability to take photos, video in all directions
Design Requirements
• Personal accommodation;
• Oxygen generation/regen. or CO2 extraction;
• Power : batteries H2; solar;
• food + waste (H2O recovery?);
• insulation;
• access hatch;
• photo gear;
• landing – impact/pressure;
• hanger fixing;
• backup systems
Concept 1 cylinder
• Clear polycarb. hemispherical ends
• tension clamps with 'O'-ring seal
• access seal from inside
• removable hatch for camera access
• Power units & life supp. system under floor
skin epoxy laminate
al. honeycomb
Concept 2 bubble
Supports
Solar panels
double skin plastic pressure bubble
Persper windows both skins
platform
segmented food, waste power, O2/CO2 units

Figure 6.39 A page from a designer's workbook

Ex 6.12

A wealthy explorer wishes to make a stratospheric solo balloon flight around the world. He has sold the rights to his experiences to National Geographic and he has contracted to take considerable numbers of photographs of his journey. It has been estimated that the total journey should take approximately 25 days. Figure 6.39 is a page from a designer's workbook, showing the notes taken during preliminary discussion with the client. Write a formal initial appreciation for this problem, and identify the technical objectives.

6.9 Assignment on bicycle security

(Suitable for teams of 4 to 6 designers)

SCENARIO

[This scenario reflects the situation in 1994. Since that time, several commendable measures to improve campus bicycle security have been implemented, largely arising from a design study conducted along the lines of this exercise]

The University of Melbourne has a serious problem with the theft of bicycles from campus, and with the congestion of current bicycle parking facilities. The existing facilities for bicycle parking are well below modern standards. The University only provides facilities that enable the front wheel of the bicycle to be locked, whereas modern bicycles require that the frame and both wheels are to be locked securely. By introducing parking facilities with a higher level of security the number of bicycles stolen from the university will be reduced, and by situating these facilities at appropriate locations the level of convenience provided to both riders and pedestrians will be improved.

In an effort to encourage a greater proportion of students to travel to university by bicycle, the administration has decided to spend funds on upgrading the bicycle parking infrastructure at the Parkville campus. Your design team has been given the task of writing a detailed recommendation for the proposed facilities. This recommendation is to include the type, number and location of the new bicycle parking facilities, particularly as they relate to the Faculty of Engineering.

BACKGROUND DATA

- Recent data (the 1993 student enrolment survey) indicates that 11% of people enrolled at the university nominated cycling as their main mode of transport to the university. Currently the number of student enrolment is about 25,000;
- The average value of a bicycle that is being ridden to the university is over $500;
- Bicycle theft on campus approaches 70 cycles each year (1992 data);
- A survey of university bicycle riders on 31st March 1994 (114 responses to questionnaires tied to 280 parked bicycles) indicated that 75% of respondents rode to university every day, and parked their bicycles for more than 5 hours;
- Security is the major concern of these riders — most wanted parking facilities where they can lock both the frame and the wheels of the bicycle. Many believe that the introduction of even higher security facilities is most important, and just over 50% of the riders are prepared to pay for the privilege of using a locker-type facility (in the vicinity of $15 – $20 per semester);
- A significant number of these cyclists stated that they are not prepared to walk further than 20m from their bike to their destination. When using high security facilities cyclists were prepared to walk further, indicating a limiting distance of about 200m;
- A site survey on 26th April 1994 (12:00 – 2:00pm) counted 590 bicycles on campus, of which 373 were parked at existing facilities, and 217 were locked to other facilities such as fences;
- The Australian Standard on Bicycle Parking Facilities (AS 2890.3–1993) nominates acceptable varieties of parking facility, and divides them into three defined levels of security, as follows:

  — Class 1 — *bicycle lockers* (see Figure 6.40);

  — Class 2 — *lockup cages* and *no-go compounds* (see Figure 6.41);

  — Class 3 — *inverted U-bars* and *leaning rails* (see Figure 6.42). The best method of attaching bicycles to Class 3 facilities is to use a commercially available *D-lock* (see Figure 6.43);

- Only class 1,2 or 3 type of security are to be considered here.

PRELIMINARY COST ESTIMATES

Capital cost of *bicycle lockers* is $1,100 for a double locker (not including site preparation). Capital cost of *inverted U-bars* is $90 each (not including site preparation).

Site preparation (paving/installation) will cost around $150 per square metre. The cost of a *D-lock* is approximately $30, paid by the user. The major issues to be considered in selecting suitable locations for facilities are:

— proximity to cyclists' destination; accessibility;

— security (Note: visibility by passing pedestrians is good; well lit areas are also desirable);

— weather protection (especially a roof);

— pedestrian safety (avoiding protrusion onto footpaths);

— protection from other vehicles.

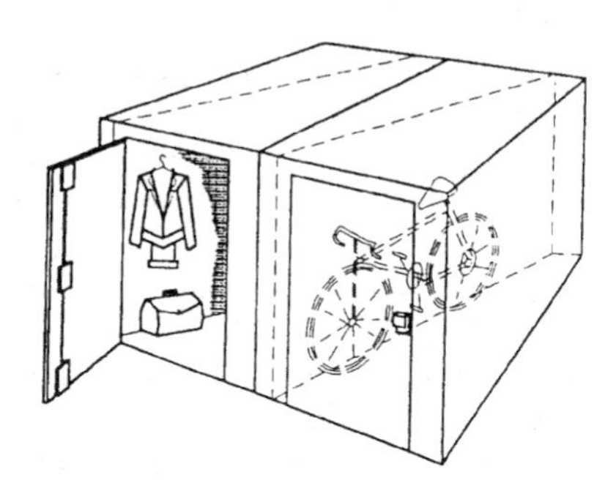

Figure 6.40 'Bicycle locker' for two bikes

DESIGN TASK

1. Discuss the advantages and disadvantages of each type of facility listed above. Summarise your evaluations in a *Decision Table* (scaled check list).
2. Make a decision about the type of facility which should be adopted for the University of Melbourne. Justify your choice with a brief statement.
3. Make a decision concerning the number of units of bicycle parking which should be installed around the campus. Again, briefly document the reasons for your choice.
4. List and estimate the *costs and benefits* of your proposal.

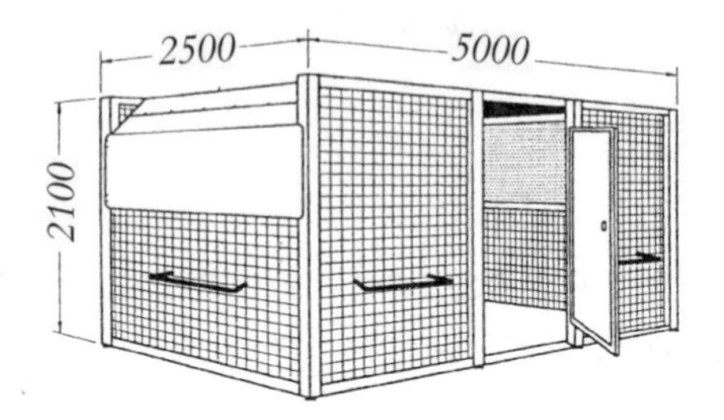

(a) Lockup cage

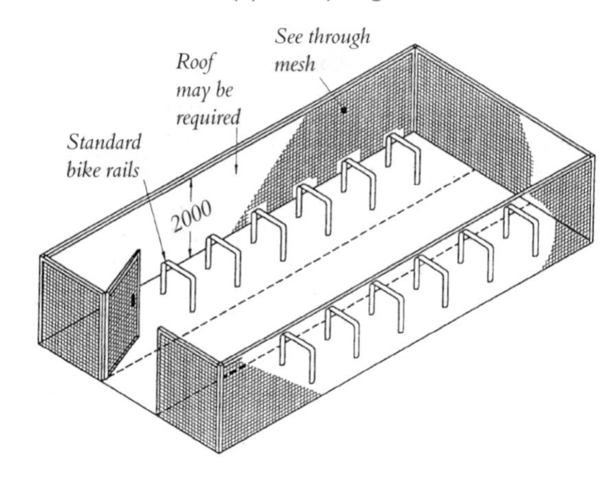

(b) 'No-go compound

Figure 6.41 'Lockup cages' and 'no-go' compounds

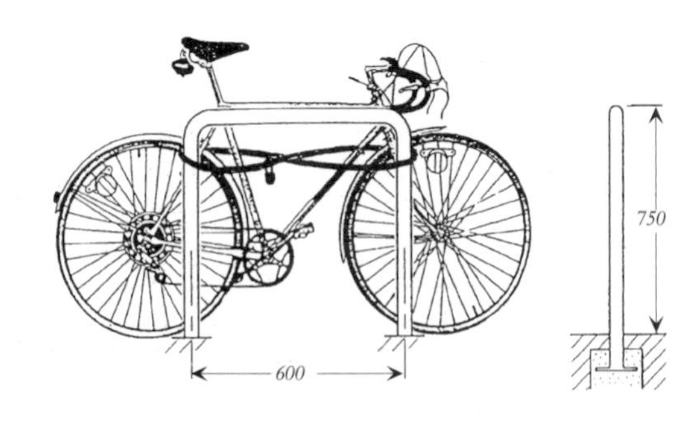

Figure 6.42 'Lean-to' rails

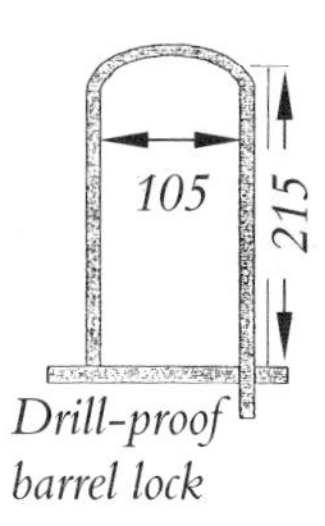

Figure 6.43 D-lock

6.10 A brief note about technical reporting

Note: The Lord's Prayer comprises 56 words, the 23rd Psalm 118 words, the Gettysburg address 226 words and the Ten Commandments 297 words. By comparison, the U.S. Department of Agriculture's order on the recommended price of cabbages has 15,629 words.

Design projects may result in a working model, a set of drawings and sketches, an equation, or whatever is acceptable to the client. Since technical reports are the main form of communication used by professional engineers, it is essential that a clear and accessible presentation style is adopted. The design report is the culmination of many hours of work, and represents the sum total of current knowledge of a particular topic, leading to plans for alternative courses of action and selection of the best plan and recommendations for its implementation. Reporting is to engineering as language is to communication, and competence in report writing is mandatory for the successful designer. The skilful selection and management of words and phrases is essential for conveying appropriate messages to colleagues, clients and the general public. Even more importantly, the substance of technical reports often represents the detailed specification for design action. Consequently, the report must contain not only the desired criteria, but also the acceptance limits of performance. Useful guidance on the preparation of design specifications is provided by *BS7373:1998 "Guide to the Preparation of Specifications"*, published by the British Standards Institution.

When planning a report perhaps the most important aspect is the way we organise the information contained in it. A good approach to achieving a well organised information structure is to map out the report's contents list – generally to two levels of headings and subheadings – at an early stage of the report writing process. The following is a list of items to be included, arranged in order:

(a) *Title and workers* – the title must reflect the main mission of the project and the workers list must identify all participants in the project, together with their respective roles and contributions;

(b) *Contents list* – must clearly identify contents and pagination;

(c) *Executive summary* – design reports invariably commence with a brief summary, clearly enunciating the major findings and recommendations of the report. In all reports, the executive summary should *set the scene and command the attention of the reader*;

(d) *Acknowledgment* – any associates who may have helped with the project must be properly acknowledged;

(e) *Objectives* – reports must begin with a clear statement of objectives and end with an assessment of how well these objectives have been met;

(f) *The body* – the main substance of the report should be presented in a clear, concise section including a statement of major results of technical analyses used. This section must also include well prepared, neat sketches or drawings or plots, all sized to fit into the report. This part of the report must be sufficiently self-contained to provide a good understanding of the work in the project. The details of the analyses and any supporting calculations should be included in an Appendix;

(g) *References* – References may be cited in any of the standard formats recognised by archival journals. They should be referred to in the text according to one of the following styles.

- Name/Date system (sometimes called the *Cambridge style*):

" *..when the spinnaker rattled free, Clancy lower'd the boom (Clancy, 1987). ...*" Complete titles of references must be supplied in the reference list, listed alphabetically, attached to the report. For example :

Clancy, Seamus H. (1987), Boom lowering under stress, Proceedings of the IEEE 4th International Marine Computing Conference on Microfiche and Silicon Chips, *V3(1):226-234*

- Numbered references as end notes:

"...The prod fusillator was bent experimentally through an angle of 90 degrees [52]...": where the number in brackets identifies the fifty-second reference listed. Complete titles of references must be supplied in the reference list listed in numerical order, corresponding to the order in which they were cited in the text.

- Numbered references as footnotes:

"...When the fuel consumption reached 37 litres per kilometre, Clancy[5] lowered the boom ...": where the superscript identifies a footnote at the bottom of the same page. This style is often used when a reference is used many times throughout a long report. The reference list at the end of the report should still contain the complete reference.

The same method of citation should be used consistently for all references in a report.

Some important issues about report writing

Presentation and style

An organised report reflects a well planned project. Formalising the structure of the report will aid thinking about the project and the nature of its achievements. Point form summaries are often acceptable in technical reporting. Remember; *"Omit needless words"* (Strunk and White, 1979).

The misconception that quality and quantity are transmutable

On an occasion when admonished for verbosity, Sir Winston Churchill is credited with the comment, *" I don't seem to find time to write short reports."* Short reports are a distillation of thought and skill with a dash of experience and flair for choosing the telling components of a story.

Credibility is our business

In the progress towards a solution to engineering problems the most effective weapon in the designer's armoury is credibility.

The printed word commands authority. Its very existence derives from the fact that somebody has gone to the trouble of preparing an article. Inevitably the reader assumes that all the work has not been in vain, that the words so presented have intrinsic significance.

There is a spectrum of credibility ranging from well-established facts (the speed of light is 300,000 km per second), through believable anecdotes based on personal interviews (*"I slipped on the soap while having a cold shower this morning"*), to unsupported generalizations (usually prefaced with hopeful phrases like *"as generally agreed"*). Designers, of necessity, plan for uncertain futures. It is their responsibility to delineate clearly the credence to be given to all statements in their reports. The reader must be given every opportunity to follow the thread of an argument, indeed to retrace the steps in the designer's thinking and reveal the foundations on which it is constructed.

7

ECONOMIC, SOCIAL AND ENVIRONMENTAL ISSUES

Rational decisions need not be based upon quantitative evaluations– indeed, some forms of rationality can not be quantified. When individuals and groups disagree on the interpretation of evaluation data, this is usually not a sign of irrationality, but rather an indication of differing values and objectives.
Malcolm Hollick, An Introduction to Project Evaluation, 1993, Longman Cheshire

7.1 Economic imperatives in design

Engineers propose; managers dispose. This little aphorism sums up the process of decision making in the planning of engineering projects. Often, the *project manager* will be an engineer *wearing a different hat*. This notion of wearing different hats, also described by de Bono (1986), suggests the different views project engineers take of project proposals. The two extreme views are the optimistic, creative, *yellow hat* view, and the pragmatic, *bottom-line* oriented, *black hat* view. In the context of making economic decisions, project managers are faced with the unenviable task of choosing to support those projects that are more likely to yield better returns on investment, than those that are rejected. The process is unenviable because, often, the rejected projects may have received significant creative engineering design input.

In this section we review some simple economic evaluations of project investment opportunities. Clearly, by investing in a specific project, one loses the opportunity of using the investment in some alternative, perhaps more profitable, or less financially risky way. Consequently, all the early economic decision processes involved in project engineering rely on the criteria of *greatest utility* or *least loss of opportunity*.

Greatest utility of a project is often determined by the contribution of the project outcome to the whole operation of a specific organisation. For example, investing in an expensive foundry, as opposed to *out-sourcing* casting requirements, may seem a costly decision for an automobile engine manufacturer. However, the utility of maintaining control on the workflow through the foundry may compensate for the investment, in terms of improved production scheduling. These utility issues are organisation-specific and we don't deal with them, beyond cautioning decision makers about over-commitment to the least loss of opportunity view. The cautionary tale offered in section 6.5.1, about the Xerox-PARC *Alto*, should serve to further underscore this warning.

7.1.1 The components of projects cost.

Figure 7.1 shows a simple example of evaluating the cost of a project.

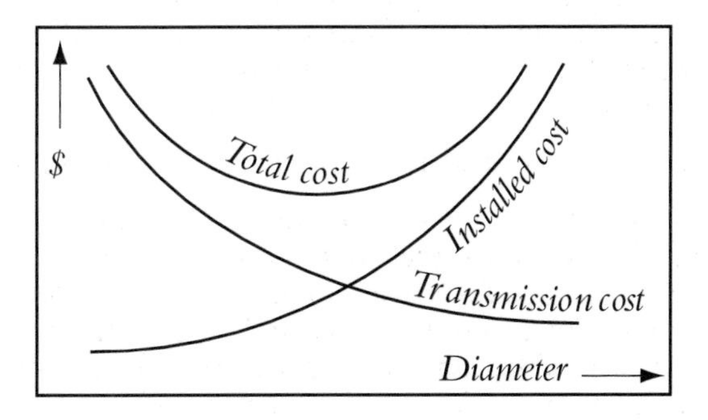

Figure 7.1 Cost of a natural gas pipeline

Here the total cost is a function of one single parameter, pipe diameter. In most practical situations, the total cost of a project will be a complex function of many factors.

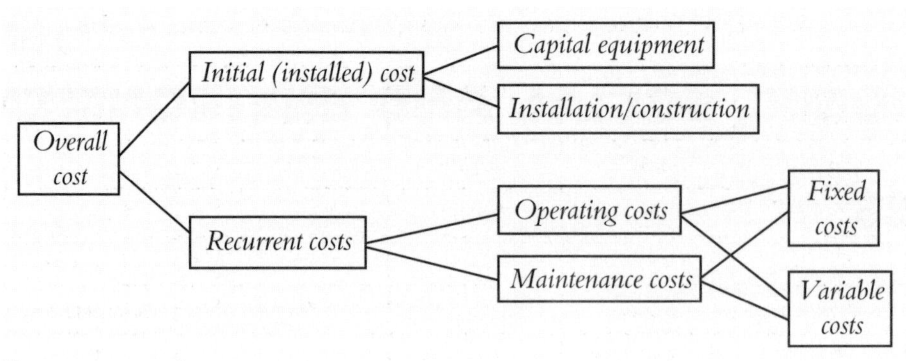

Figure 7.2 The components of project cost

The following is a list of generic project cost components:

- *overall cost* of a plant: must include the initial cost and the recurrent costs;
- *initial cost* of a plant consists of capital equipment cost and the installation costs;
- *recurrent (annual) costs* consist of operating costs and maintenance costs;
- *operating (running) costs* can be further subdivided into:
  - *fixed costs* (for example rent, labour), and
  - *variable costs* (for example raw materials.

Note:

- *fixed costs* are those operating costs which are incurred regardless of how much product is made;
- *variable costs* are those operating costs which are incurred roughly in proportion to the amount of product made.

Figure 7.2 shows the schematic relationships between these various cost components.

7.1.2 Opportunity loss: a simple approach to evaluating project costs

Evaluating least loss of opportunity is based on the general idea that money invested in a capital project represents an annual cost to the investor. There are two factors which influence this cost. The first is the cost of borrowing money from lending institutions. Banks and other money lending institutions use the terms *interest-rate*, or occasionally *lending-rate*, to describe this cost. The second factor is the rate at which money loses its value over time. This rate of loss, called *inflation,* effects the cost of goods and services over time. In times of war, revolution or other major national upheavals, inflation rates tend to become extreme. In the instability following the end of World War II, some European countries experienced inflation rates of several hundred percent. Currently, many underdeveloped countries still experience double digit inflation rates. In the developed countries, the ten year average inflation rate has been estimated at around 5%. The *real* cost of money, when considering its investment in an engineering project, is determined by the difference between *lending rate* and *inflation rate*. We call this difference the *real interest rate*, denoted by the symbol r. Table 7.1 shows the way interest rate affects the value of money. To simplify our discussion, we have assumed a *real interest rate* of 10%.

Interest rate, r is the real cost of borrowing money for capital projects. As indicated in Table 7.1, due to the influence of interest rate, the value of a sum of money S in the future is worth somewhat less in the present. We refer to value P of a sum of money S, n years in the future, as the *present worth* of S. The following equations indicate the relationship.

$$P = S \Big/ \left(1+r\right)^n .$$

$$p = \frac{P}{S} = 1 \Big/ \left(1+r\right)^n .$$

Table 7.1 Effect of interest rate on the value of sums of money (assume r=10%)

Value in Year 0	Value in Year 1	Value in Year 2	Value in Year 3	•	Value in Year n
\$ 1	\$ 0.909	\$ 0.826	\$ 0.751	•	$\$1/(1.1)^n$
\$ 1.1	\$ 1	\$ 0.909	\$ 0.826	•	$\$1.1/(1.1)^n$
\$ 1.21	\$ 1.1	\$ 1	\$ 0.909	•	$\$1.21/(1.1)^n$
\$ 1.331	\$ 1.21	\$ 1.1	\$ 1	•	$\$1.331/(1.1)^n$
$P = \$ S/(1+r)^n$	$P = \$ S/(1+r)^{n-1}$	$P = \$ S/(1+r)^{n-2}$	$P = \$ S/(1+r)^{n-3}$	•	\$ S

The ratio p is referred to as the *present worth factor*. This is the factor that is used to multiply S to get its present worth.

In accounting terminology, we refer to the *real cost* of a capital project as its *recovery*. This is the sum of money we need to *recover*, in order to *break even* on our initial investment. To measure the cost of a capital investment over some period of time, say n years, we consider the *cash flow* during the life of the project. As an example we consider the capital recovery for a project with an estimated life of n years into the future, and an initial investment $\$I$. The cash flow during the life of this project will be taken as n equal payments of $\$R$ at the end of each year. Figure 7.3 shows the procedure schematically.

Based on the assumed regular annual payment, we can calculate the present worth of the total stream (or series) of payments over the life of the project.

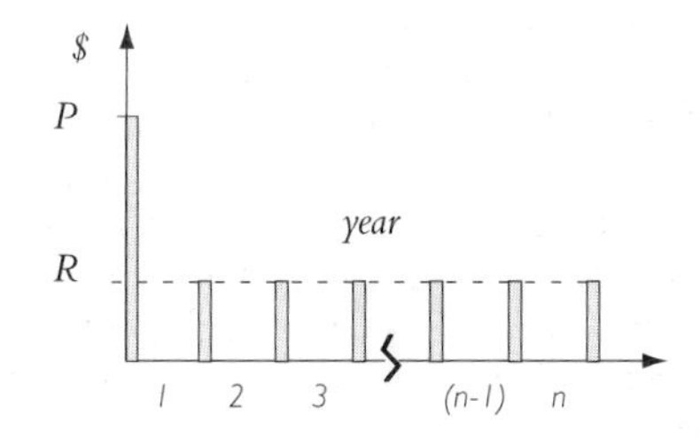

Figure 7.3 Cash flow during a project (schematic only)

From Figure 7.3

$$P = \frac{R}{(1+r)^1} + \frac{R}{(1+r)^2} \cdots + \frac{R}{(1+r)^{n-1}} + \frac{R}{(1+r)^n},$$

$$\therefore\ P(1+r)^n = R\left[1 + (1+r) \cdots + (1+r)^{n-1}\right].$$

Summing the geometric series on the right hand side of this equation we get

$$P(1+r)^n = R\,\frac{(1+r)^n - 1}{(1+r) - 1} = R\,\frac{(1+r)^n - 1}{r}.$$

We define the *capital recovery factor*, c as

$$c = \left(\frac{R}{P}\right) = \frac{r(1+r)^n}{(1+r)^n - 1}.$$

The *capital recovery factor*, c, expresses the proportion of the invested sum I that needs to be recovered annually as benefits, to break even in the project (so that P is at least equal to I). The following example will clarify the terminology used.

CASE EXAMPLE 7.1:

PROPOSED WIDGET PLANT

A news widget plant is contemplated by management. The following *islands of certainty* exist in relation to the proposed plant:

- capital works: \$100,000 (estimated);
- projected life of plant: 10 years;
- prime lending rate (for this size capital): 10%;

- expected inflation rate over the life of the plant: 4%.

The *real interest rate* for the life of the project is (10% - 4% =) 6%. Hence the capital recovery factor, c, is found to be

$$c = \frac{0.06\left(1+0.06\right)^{10}}{\left(1+0.06\right)^{10}-1} = 0.136.$$

The interpretation of this result is that in each year of he widget plant's life it needs to return at least 0.136 × \$100,000, or \$13,600 in order to break even on the investment within the life of the project. Put another way, an investment of \$100,000 in a 10-year project represents an *opportunity loss* (under the assumed *real interest rate* of 6%) equivalent to \$13,600 per year, and this must be *at least recovered* from the project to make it worthwhile.

7.1.3 The annual cost equivalent

A common way of expressing the cost of a project is in terms of its annualised cost to the investor. This cost is called the *annual cost equivalent*, denoted A. The annual cost equivalent is only slightly different from the annual capital recovery found in the previous section. For a plant installed cost of I, where the eventual salvage value of the plant is V, and the *capital recovery factor* is c,

$$A = c(I - pV),$$

where p is the present worth factor. In general, it is difficult to estimate the salvage value for a plant unless it is similar to already existing plants. In that case, an estimate of its worth after some years of use may be available.

As an example consider a \$1 million capital investment. Let us assume that the salvage value of the plant is \$ 200,000 after an operating life of 25 years. The real interest rate (ie. lending - inflation rate) is 6%.

Capital recovery factor,

$$c = \frac{0.06\left(1+0.06\right)^{25}}{\left(1+0.06\right)^{25}-1} = 0.0782.$$

Present worth factor,

$$p = \frac{1}{1.06^{25}} = 0.233.$$

Annual cost equivalent,

$$A = \left(10^{6} - 0.233\times 2\times 10^{5}\right)\times 0.0782$$

$$= \$\ 74{,}556.$$

As before, we interpret this last figure as the benefits to be returned from the investment in order to break even. Alternatively we can also interpret the result as the total lifetime cost of our investment:

$$25\times \$\ 74{,}556 = \$\ 1.864\times 10^{6}.$$

This method of evaluating the total cost of a project, allowing for *"time-value"* of money, leads to a measure of the economic effectiveness of a project known as the *discounted payback period*. With a total cost of $T=nA$ and annual return, or benefit, R, the discounted payback period is defined as $DPBP=T/R$. Contrast this with the *simple payback period*, $SPBP=I/R$, which *takes no account of opportunity loss*.

CASE EXAMPLE 7.2 :

RELAMPING A FACTORY

We consider two alternative options.

(a) *Incandescent globes*:
— \$1 per globe;
— 100 W (1200 Lumen);
— L_{10} life = 1000 hours.

(b) *Fluorescent globes* (fit into same bayonet socket):
— \$30 per globe;
— 20 W (1200 Lumens);
— L_{10} life = 6000 hours.

Factory requires 200 lights:
— average usage = 8 hours/day;
— 250 working days/year;
— ie. 2000 hours/year;
— electricity costs (say) 6 cents/ kW hour (for domestic use it costs 12 cents/ kW hour), and is paid quarterly;
— real interest rate r = 8%.

Note: L_{10} life is a measure of reliability. This is the length of time after which, on average, 10% of the globes have *expired*.

OPTION A

Cost to light factory for three years with incandescent globes (i.e. six generations of globes).

Installation:

— 200 globes x $1 x 6 (globes are bought annually in lots of 400);

Payments:

— 400 x $1 (now) = $400;

— 400 x $1 (after 1 year) = $400/1.08 = $370;

— 400 x $1 (after 2 years(= $400/(1.08)^2 = $343;

Present worth of capital cost = $400 + $370 + $343 = $1,113.

Running costs:

$$200\ \ globes \times \left(\frac{2000\ \ hr\ /\ year}{4\ \ quarters\ /\ year} \times \frac{100 \times 0.06}{1000} \right)$$

= $600/quarter.

For this example we need to calculate the *capital recovery factor* over three years: i.e. 12 quarters, with the interest rate @ 8% pa = 2% /quarter.

$$c\ = \frac{0.02\left(1+0.02\right)^{12}}{\left(1+0.02\right)^{12} - 1} = 0.00946\ .$$

The present worth of the running cost over three years is $600/*c* = $6,345.

Total cost for option A

= $1,113 + $6,343
= $7,458.

OPTION B

Cost to light factory with fluorescent globes (i.e. one generation of globes).

Installation:

— 200 globes @ $30 = $6,000 now;

Running costs:

$$200\ \ globes \times \left(\frac{2000\ \ hr\ /\ year}{4\ \ quarters\ /\ year} \times \frac{20 \times 0.06}{1000} \right)$$

= $120/quarter.

The *capital recovery factor* is the same for both options. Hence, the present worth of the running cost over three years is $120/*c* = $1,269.

Total cost for option B

= $6,000 + $1,269
= $7,269.

Clearly, there is little to choose between these two alternatives. However, with different power costs the results might well turn out to be different.

Ex 7.1

Work out the comparisons in case example 7.2, with a power cost of 12 cents/kW hour.

Ex 7.2

Your employer is considering the installation of a new widget machine. The cost of installation is expected to be $50,000 with an estimated operating life of ten years. A ten-year old machine costs approximately $10,000. Widget sales estimate is $10,000 in the first year with inflation-indexed income increasing in each subsequent year at a rate of 15% per annum. Current bank lending rate for investment is 12% p.a., and inflation rate is estimated at 5% p.a. Estimate the *discounted payback period* (*DPBP*)(the time needed to break even on the investment, taking account of the *time-value* of money).

Ex 7.3

Expansions to a petrochemical plant require the purchase of a special 200 kW electric motor to drive a new gas compressor, the rated motor output being 200 kW.

Three motor manufacturers have submitted tenders as shown in Table 7.2:

Table 7.2 Motor manufacturer's costing

Manufacturer	A	B	C
Cost of motor delivered to site ($,000)	28.875	34.870	36.520
Guaranteed minimum motor efficiency %	0.91	0.92	0.93

The plant will be operated for 100 hours a week, 50 weeks a year, for ten years. At the start of the 10 year period it is predicted that electrical power will be available at a cost of 5.0 cents per kWh, increasing by equal annual increments to 6.8 cents per kWh at the start of the tenth year. Electricity bills are paid quarterly.

(a) On the basis of the information given which motor would you recommend be purchased?

(b) Is your answer to (a) sensitive to variations in the cost of power, for example, if power costs were to double over the ten year period to 10 cents per kWh, would this affect your recommendation?

(c) What additional information would you seek in order to reach a final decision in this matter?

Ex 7.4

An extract is given below from a consultant's report in which an economic evaluation is made of a proposal to reduce energy and maintenance costs in a university building by replacing existing hot and cold water pipelines. The estimated cost of the installation is $85,000.

Derive the following equation which is quoted without proof in the extract.

$$P = AS\frac{1+i}{r-i}\left[1-\left(\frac{1+i}{1+r}\right)^n\right],$$

where

P = present worth equivalent to current annual savings of AS dollars;

n = number of years;

i = annual real rate of increase of energy and maintenance costs;

r = real interest rate.

Show that the discounted payback period for the capital investment of new pipe installation is 9.4 years, as stated in the extract, for the case where i= 0.1, and r= 0.1.[Hint: the series summation simplifiesappreciably in this case]

Extract:

"The present worth of these savings over 15 years provides an indication of whether replacement of existing lines is an economic proposition compared with investing the capital amount at a real interest rate of ten percent.

Thus the present worth of the $9,000 p.a. savings is $135,000 compared with the capital cost of $85,000 to replace pipelines, which suggests that replacement of the piping would be an economic proposition. The following table sets out the present worth of the anticipated savings for different values of ci and n and confirms that replacement of the pipelines would be an economic proposition for lower rates of increases in energy and maintenance costs over a 15 year period (see shaded area in table) Values are in whole dollars."

Table 7.3 Cost savings generated by proposal

i %	n years			
	5	10	15	20
0	$34,400	$55,300	$68,500	$76,600
4	$38,100	$67,000	$88,700	$105,210
8	$40,320	$73,800	$101,700	$124,800
10	$45,000	$90,000	$135,000	$180,000

7.2 Project evaluation and benefit-cost analysis

Evaluating the economic impact of several alternative project options is only the first step in project evaluation. While we can express some aspects of a project in terms of dollars, evaluating its potential environmental and social impact may present us with substantially greater challenges. For example, freeways are clearly seen as economic and environmental benefits in terms of fuel efficiency. Yet their negative social impact can be considerable. Electric power generating stations and power transmission lines provide an effective and economic energy distribution system. Yet, they too represent significant negative environmental and social impact.

As Hollick (1993) suggests:

"Different people often have different objectives and, even if they have similar ones, are likely to give them different priorities. Thus two equally rational decision-makers may evaluate the same data and come to quite different conclusions. ....often when two groups disagree over a proposed project, they accuse each other of being irrational when really they have different views and objectives."

One thing is certain: when facing a project evaluation programme, emotion and sentiment should be left out of the decision making process. Typically, we could ask ourselves questions like:

"can we put a dollar value on human life?"

"how can we compare the social value of a tree-lined streetscape with a barren freeway alternative?"

"while I want easy access to electric power, would I be happy living next to a power station, or a power distribution pylon?"

Benefit-cost analysis (often *cost-benefit analysis*) does not answer questions like these. However, it does provide a rational procedure for weighing up competing alternatives. As with structural integrity, we tend to learn considerably more from failed applications of decision tools than from successful ones. The following two case examples illustrate both of these extremes of applying cost-benefit analysis in making technical project evaluations.

CASE EXAMPLE 7.3:

THE FORD PINTO CASE

The Ford Pinto case is mentioned in most Business Ethics texts as an example of *Cost-Benefit* analysis (see for example Birsch and Fielder, 1994).

The following case story is based on an article compiled by Mark Dowie (general manager, Mother Jones' business operations), and Alexandria Woods[7.1]. It relates to events that took place in the 1970s.

Facing strong competition from Volkswagen for the lucrative small-car market, the Ford Motor Company rushed the Pinto into production in much less than the usual time.

Lee Iacocca, the new president of Ford, argued forcefully that European and Japanese small-car manufacturers were going to capture the entire American subcompact market unless Ford put out its own alternative, and almost immediately began a rush programme to produce the *Pinto*. Iacocca wanted that little car in the showrooms of America with the 1971 models. So he ordered his engineering vice president, Bob Alexander, to oversee what was probably the shortest production planning period in modern automotive history. The normal time span from conception to production of a new car model (*cradle-to-launch*) is about 43 months. The Pinto schedule was set at just under 25.

Design, styling, product planning, advance engineering and quality assurance all have flexible time frames, and engineers can generally work at these simultaneously. Tooling, on the other hand, has a lead time of about 18 months. Normally, an automobile manufacturer doesn't begin tooling until the other processes are almost complete. Tooling involves the commitment of substantial costs and management normally doesn't make such commitments until the production prototypes have been thoroughly tested. But Ford's proposed rush programme meant that Pinto tooling went on in parallel with product development. Consequently, when crash tests revealed a serious defect in the fuel tank, the tooling was already well under way.

Ford engineers discovered in pre-production crash tests that rear-end collisions would rupture the Pinto's fuel tank extremely easily. Because of the already substantial tooling commitment, when engineers found this defect, Ford management chose to manufacture the car with the known defect, even though Ford owned the patent on a much safer fuel tank. For more than eight years afterwards, Ford successfully lobbied, with extraordinary vigour, against a key government safety standard that would have forced the company to change the Pinto's fire-prone fuel tank.

In 1972 a woman, SG[7.2], was driving along the Minneapolis highway in her new Ford Pinto. Riding with her was a young boy, RC. As she entered a merge lane, SG's car stalled, and another car *rear-ended* hers at an impact speed of 28 miles per hour. The Pinto's fuel tank ruptured, and fuel vapours from it mixed with the air in the passenger compartment. A spark ignited the mixture and the car exploded in a ball of fire. As a result of this accident SG was killed and her passenger, 13-year-old RC, who is still alive, has suffered substantial burns to the majority of his body.

By conservative estimates, Pinto crashes have caused between 500 to 1000 burn deaths to people who would not have been seriously injured if the car had not burst into flames. We might well be concerned as to why SG's Ford Pinto caught fire so easily, seven years after Government action that brought more safety improvements to cars than any other period in automotive history? (Nader, 1972)

Ford was permitted to wait these years because they managed to reach an informal agreement with the major public servants who would be making automobile safety decisions. This was an agreement that *cost-benefit* would be an acceptable mode of analysis for evaluating safety improvements in automobile manufacture. As a result of this agreement, *cost-benefit* analysis quickly became the basis of Ford's argument against safer car design.

Mother Jones has studied hundreds of reports and documents on rear-end collisions involving Pintos. These reports conclusively show that rear-ending a Pinto at about 30 miles per hour would result in the whole rear end collapsing up to the back seat. In addition, the tube leading to the fuel-tank cap would be torn away from the tank itself,

7.1 Mother Jones is a consumer magazine, published by The Foundation for National Progress, 731 Market Street San Francisco CA 94103. URL: http//ww.mothwerjones.com

7.2 See also Grimshaw v. Ford Motor Co., 174 Cal. Rptr. 348

and fuel would immediately begin pouring onto the road around the car. The buckled fuel tank would be jammed up against the differential housing, which contains four sharp, protruding bolts likely to rip holes in the tank and exacerbate the fuel spill. A spark under these conditions would result in both cars exploding in flames. If the rear-ending accident took place at about 40 miles per hour, the doors of the Pinto would most likely jam and the passengers would be condemned to burn in the car.

Internal Ford company documents show that the Pinto has been crash-tested more than 40 times and that every test made at over 25 mph without special structural alteration of the car has resulted in a ruptured fuel tank. Despite this knowledge the company chose to manufacture and release the Pinto, based on their curiously *amoral cost-benefit* analysis. To quote the Mother Jones article:

"When it was discovered the fuel tank was unsafe, did anyone go to Iacocca and tell him? 'Hell no', replied an engineer who worked on the Pinto, a senior company official for many years, who, unlike several others at Ford, maintains a necessarily clandestine concern for safety. 'That person would have been fired. Safety wasn't a popular subject around Ford in those days. With Lee it was taboo. Whenever a problem was raised that meant a delay on the Pinto, Lee would chomp on his cigar, look out the window and say: 'Read the product objectives and get back to work'. ... As Lee Iacocca was fond of saying, 'Safety doesn't sell'."

It is interesting to speculate on how Ford was able to successfully lobby for eight years against Federal Motor Vehicle Safety Standard 301, the rear-end provisions of which would have forced Ford to redesign the Pinto. Ford's success is partly due to their capacity to *"educate"* the new federal automobile safety bureaucrats. Their *"education"* was able to implant the official industry ideology in the minds of the new officials regulating automobile safety. Briefly summarized, that ideology states that automobile accidents are caused not by cars, but by 1) people and 2) highway conditions. Mother Jones notes:

"This philosophy is rather like blaming a robbery on the victim. 'Well, what did you expect? You were carrying money, weren't you?' It is an extraordinary experience to hear automotive 'safety engineers' talk for hours without ever mentioning cars. They will advocate spending billions educating youngsters, punishing drunks and redesigning street signs. Listening to them, you can momentarily begin to think that it is easier to control 100 million drivers than a handful of manufacturers. They show movies about guardrail design and advocate the clear-cutting of trees 100 feet back from every highway in the nation. If a car is unsafe, they argue, it is because its owner doesn't properly drive it. Or, perhaps, maintain it. "

Cost-benefit analysis was used only occasionally in government until President Kennedy appointed Ford Motor Company president Robert McNamara to be Secretary of Defense. McNamara, originally an accountant, preached cost-benefit with all the force of a Biblical zealot. As a management tool in a business where profit is the major driving force, cost-benefit analysis makes a certain amount of sense. Serious problems arise, however, when public officials, who ought to have more than corporate profits at heart, apply cost-benefit analysis to every conceivable decision. The inevitable result is that they must place a dollar value on human life.

Once Ford was able to convince federal regulators to measure automobile safety in terms of *cost-benefit* analysis, Ford needed to have a dollar value figure for the *benefit*. Rather than be so vulgar as to suggest a price tag itself, the automobile industry pressured the National Highway Traffic Safety Administration to do so. In a 1972 report the agency decided a human life was worth $200,725. Inflationary forces have recently pushed the figure up to $278,000. Furnished with this useful tool, Ford immediately went to work using it to prove why various safety improvements were too expensive to make.

Nowhere did the company argue harder that it should make no changes than in the area of rupture-prone fuel tanks. Not long after the government arrived at the $200,725-per-life figure, it surfaced, rounded off to a cleaner $200,000, in an internal Ford memorandum. This cost-benefit analysis argued that Ford should not make an $11-per-car improvement that would prevent 180 deaths by incineration per year. (This minor change would have prevented fuel tanks from breaking so easily both in rear-end collisions, like SG's, and in rollover accidents, where the same thing tends to happen.)

Ford's cost-benefit table is presented in a seven-page company memorandum entitled *"Fatalities Associated with Crash-Induced Fuel Leakage and Fires"*. The memo argues that there is *no financial benefit* in complying with proposed safety standards that would admittedly result in fewer automobile fires, fewer burn deaths and fewer burn injuries. Faced with this terrible decision of justifying expenditures to make the Pinto safer, it is worthwhile to examine the *real costs* that might have been incurred.

One Ford document showed that crash fires could be largely prevented for considerably less than $11 a car. The cheapest method involves placing a heavy rubber bladder inside the gas tank to keep the fuel from spilling if the tank ruptures. Goodyear had developed the bladder and had demonstrated it to the automotive industry. Crash-test reports showed that the Goodyear bladder worked well. The total purchase and installation cost of the bladder would have been $5.08 per car. That $5.08 could have saved the lives of SG and several hundred others.

Unfortunately, neither is the Pinto an isolated case of corporate malpractice in the automobile industry, nor is Ford a lone offender. There probably isn't a car on the road without a safety hazard known to its manufacturer. Furthermore, cost-valuing human life is not used by Ford alone. Ford was just the only company careless enough to let such an embarrassing calculation slip into public records. The process of wilfully trading lives for profits is built into corporate capitalism. Commodore Vanderbilt publicly scorned George Westinghouse and his *"foolish"* air brakes while people died by the hundreds in accidents on Vanderbilt's railroads.

The original draft of the Motor Vehicle Safety Act provided for criminal sanction against a manufacturer who wilfully placed an unsafe car on the market. Early in the proceedings the automobile industry lobbied the provision out of the bill. Since then, there have been those damage settlements, of course, but the only government punishment meted out to auto companies for noncompliance to standards has been a minuscule fine, usually $5,000 to $10,000. One wonders how long automobile manufacturers would continue to market lethal cars if the fines were significantly more severe, including gaol terms.

CASE EXAMPLE 7.4:

INJECTABLE CONTRAST AGENTS

This case example is taken from the Australian Broadcasting Corporation's *'Health Report'*, hosted by Dr. Norman Swan.

Injectable contrast agents are radioactive dyes used in computerised axial tomography (*CAT*) scans, to provide contrast in specific organs or skeletal structure being scanned. There has been some concern in medical circles about adverse reactions, sometimes deaths, caused in some patients by the injection of these contrasting agents. A new injectable contrasting agent, that can potentially eliminate these adverse reactions, has recently come onto the market. However, the new agent costs significantly more than the older agent. The problem is to evaluate the two alternatives in some rational manner.

It is *not true* that *"money is no object when lives are at stake"* since resources are limited, and society must make choices about which life-saving or public health activities to invest in. We need to ask whether we are committing funds which could be more effectively used elsewhere. We must make comparisons between various life-saving or public health procedures to see which gives the lowest *cost* for a similar level of *benefit*.

The following is a simplified summary of the comparison between *old* and *new* injectable contrast agents:

- *new* agent costs 5 times as much (*relative cost*)
- *old* agent causes 5 times the number of deaths (*relative risk*)

Comparative Risk (using worst-case results from performance studies)

— *old* agent: 1 death/40,000 injections, could be as low as 1 death/100,000 injections.

As a comparison, we note that of the 9.5 million cars in Australia 3,200, or 1/3,000 are involved in fatal accidents each year. Non steroidal anti-inflammatory drugs used by 20% of the population over 50 years of age cause 1 death/4,000 from haemorrhaging ulcers. Hence, in absolute terms, the risk in using the *old* contrasting agent is comparatively low.

— *new* agent: 1 death/200,000 injections. Based on the worst case results for the *old* agent, the *new* agent is one fifth as risky.

We need to develop a suitable evaluation scale on which we can compare the absolute performance of these two agents in terms of a *cost/benefit* ratio. In developing this ratio we need to recognise that the cost of the *old* agent needs to be augmented by the expense of treatment needed in cases of adverse reactions.

It has been estimated that the annual extra cost of switching to the *new* agent over the *old* agent is:

$ 2.7 million/death prevented (based on the low risk group);

$ 1.5 million/death prevented (based on the average risk group);

$ 800,000/death prevented (based on high risk group).

The ultimate measure of the *effectiveness* of changing to the *new* agent is the cost/years of life saved by the change. If we assume that the life expectancy of these *CAT*-scan patients is 20 years, we find the following comparative *cost/benefit* ratios:

$40,000/life-year saved (high risk group);

$140,000/life-year saved (low risk group).

These figures are significantly worse than those for other life-saving procedures, such as cardiac surgery, heart transplants, kidney transplants and hospital-based kidney dialysis.

Clearly, there are significantly more effective ways of spending medical funds than on the *new* injection agent. Naturally, there are moderating factors that will impact on this decision. Screening (selectively filtering out) *CAT*-scan patients, who may be at high risk of adverse reaction to the *old* agent, will certainly reduce the costs per life year saved.

Ex 7.5

Develop comparative cost/benefit ratios for the following:

(a) changing from gas to solar powered heating of a home swimming pool;

(b) travelling by car or by aeroplane between San Francisco and Los Angeles (approximately 800 km), based on 2, 3 or 4 passengers travelling together.

Ex 7.6

In aircraft design the hydraulic control lines to the air control surfaces (*ailerons*) are critical to flight safety. Most aircraft are designed with threefold redundancy of these control lines (i.e. there are three separate sets of lines to each control surface, so if any one of them fails, one or other of the remaining two can take over).

Develop a suitable cost/benefit ratio for comparing threefold redundancy of control lines with fourfold redundancy (the Boeing 747 has fourfold redundancy).

Ex 7.7

Your employer is a progressive company intent on improving working conditions for office staff. It has been proposed that the administrative offices should be air-conditioned. The offices house 20 full-time employees and a further 20 *casuals*, who spend about 30% of their time in these offices situated on the top floor of the company building. It has been estimated that work improvement, following air-conditioning, might offset running and installation costs. Develop a suitable *cost/benefit* ratio for this application. You may assume that severe variations from the temperature norm can reduce human work capacity and effectiveness by approximately 20%.

7.3 Design for human users: introductory ergonomics

Behaviourism is indeed a kind of flat-earth view of the mind .... it has substituted for the erstwhile anthropomorphic view of the rat, a ratomorphic view of man. Arthur Koestler, The Ghost in the Machine, 1967

The origins of human measurement can be traced back to the journeys of Marco Polo (1273-1295), during which the measurement of the humans were conducted mainly for the purposes of racial or national comparisons (Roebuck, Kroemer and Thompson, 1975). With increasing industrial mechanisation, the collection and recording of human measurements became an integral part of workspace design. Time and motion study, introduced by F. W. Taylor in the late 19th century and the work of the Gilbreths in scientific motion management (1912) led to quantification of both human size as well as motion dynamics (Giedion, 1948; Barnes, 1949). A major motivation for large scale studies of human body size was provided by the need to design and build military equipment (Randall, 1948). So, it seems that modern workplace design in peaceful applications owes its origins to war. Roebuck et al. (1975) note :

"The need for an integration of life science disciplines for engineering applications was brought into focus by World War II, which created a whole new series of man-machine environment problems. In addition to such problems as defining clothing dimensions for Army troops, great numbers of accidents in training and operational aircraft pointed the need for study of basic causes. Psychologists, who were asked to study the actions of men under stress of flying, found that the complexity of modern military equipment was outstripping the abilities of men to operate them."

7.3.1 Anthropometry

The term anthropometry is derived from the Greek $\alpha\nu\theta\rho\omega\pi o\varsigma$ (man) and $\mu\varepsilon\tau\rho\iota\alpha$ (measuring). Originally, the task of anthropometrists was to measure human body size. The current science of anthropometry measures much more than merely size. Human behaviour is extremely complex. Designers use a number of different conceptual models to help order their thinking about the different aspects of human behaviour which affect the equipment they are designing. These are :

(a) Occupant of a workspace;

(b) Source of power;

(c) Sensor or transducer;

(c) Processor of information;

(d) Tracker and controller;

(e) Person with motives, emotions, habits, preferences, maybe even some prejudices.

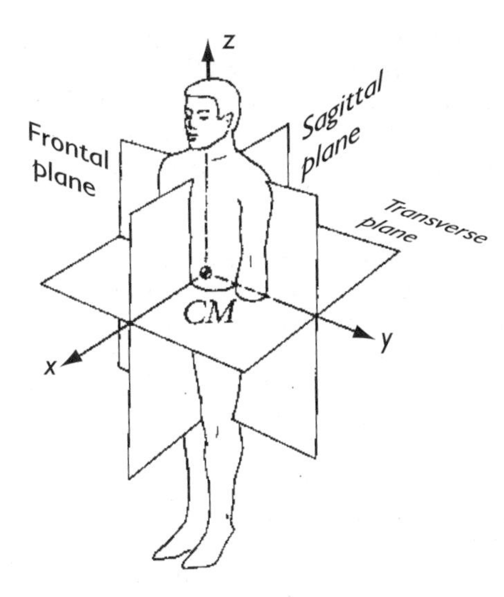

Figure 7.4 Standard reference planes used for human scale measurement (After Roebuck et al., 1975)

We also differentiate between static and dynamic anthropometry.

Static anthropometry refers to body sizes and limb dimensions of the population for whom we are designing.

Dynamic anthropometry is the measure of body and limb movements and space requirements for them. This includes preferred types of control movement, magnitudes of forces and torques which can be applied, arrangements of instruments and control devices such as pedals, levers and handwheels.

Workspace design is only one aspect of the general need to collect human measurements. To design successful environments for human-machine interactions we require a deep understanding of all the operational needs of the human in these complex environments. Our first task then is to examine both the nature of human-machine interactions and the range of measurements that we might reliably collect about human scale and behaviour.

Typically, body size measurements are expressed relative to some reference planes in the body and Figure 7.4 shows the standard reference planes used for human scale measurement. Added to this set of reference planes are the notions of relative measurements taken from origins associated with limb joints. Typically, relative to the shoulder joint the *proximal* arm joint is the elbow and the *distal* joint is the wrist. Motion away from the centreline outwards is referred to as *abduction*, and the reverse move is called *adduction*. *Flexion* and *extension* are the folding and straightening of limbs respectively. Roebuck et al. provide a useful glossary of terms for all motion measurements available for workspace design. Figure 7.5 is a sketch of a typical portable measuring frame used in the Royal Airforce (see also Morant, 1947).

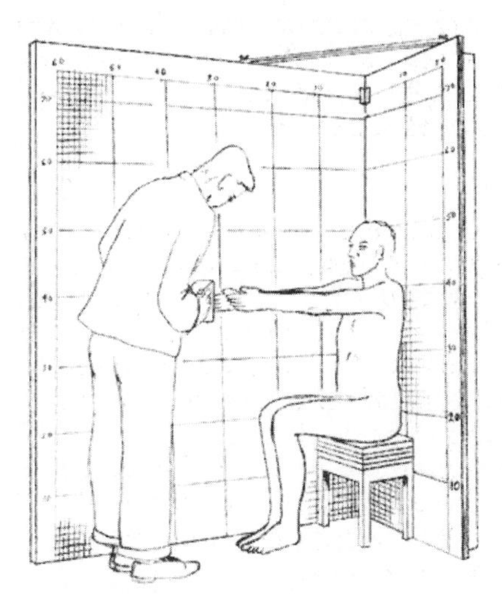

Figure 7.5 Portable measuring frame for body size and reach envelope assessment (After Roebuck et al., 1975)

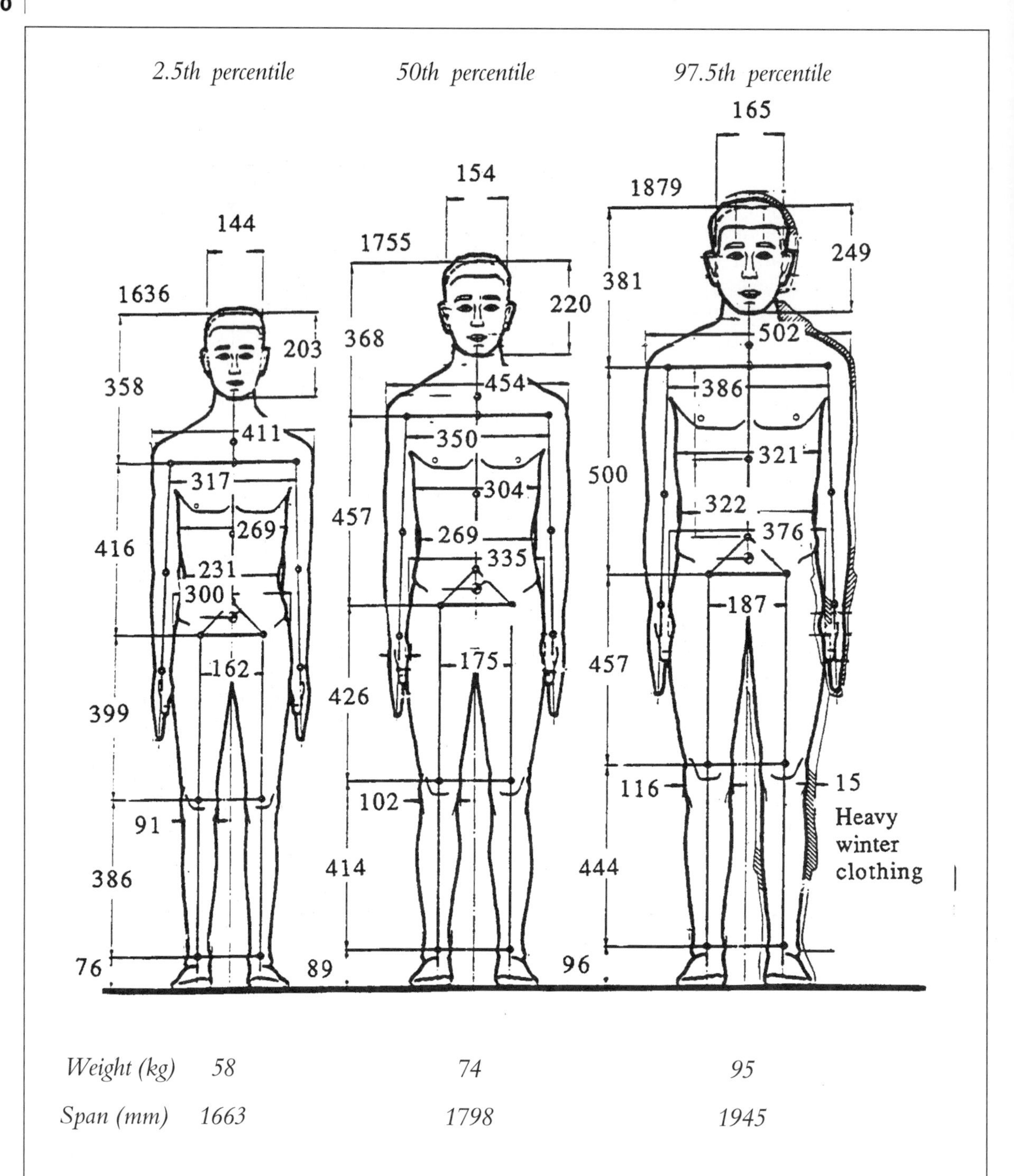

Figure 7.6 (a) Front vew of standing adult male - including 95% of population (dimensions mm) based on the charts of Dreyfuss (1967)

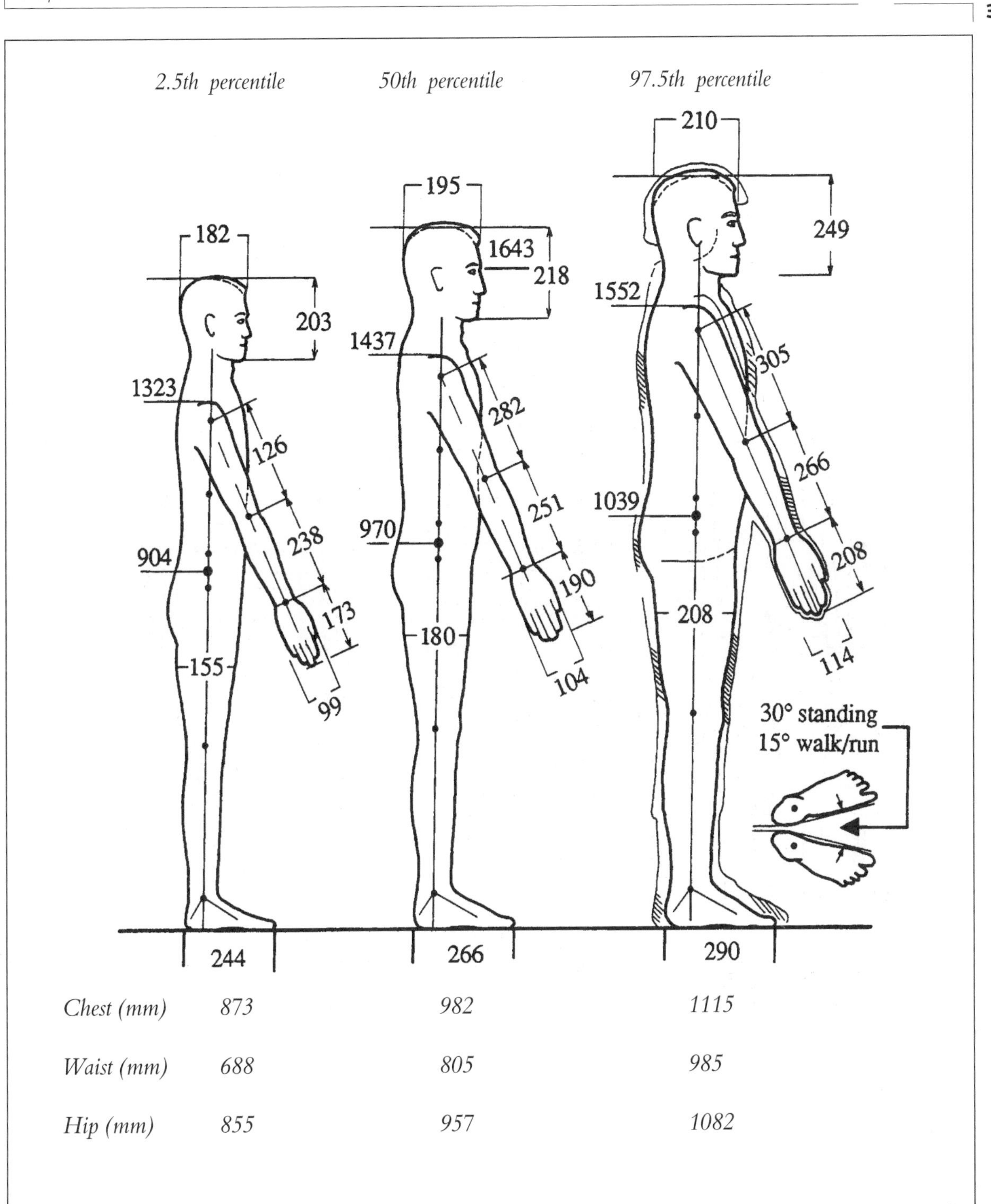

	2.5th percentile	50th percentile	97.5th percentile
Chest (mm)	873	982	1115
Waist (mm)	688	805	985
Hip (mm)	855	957	1082

Figure 7.6 (b) Side vew of standing adult male - including 95% of population (dimensions mm) based on the charts of Dreyfuss (1967)

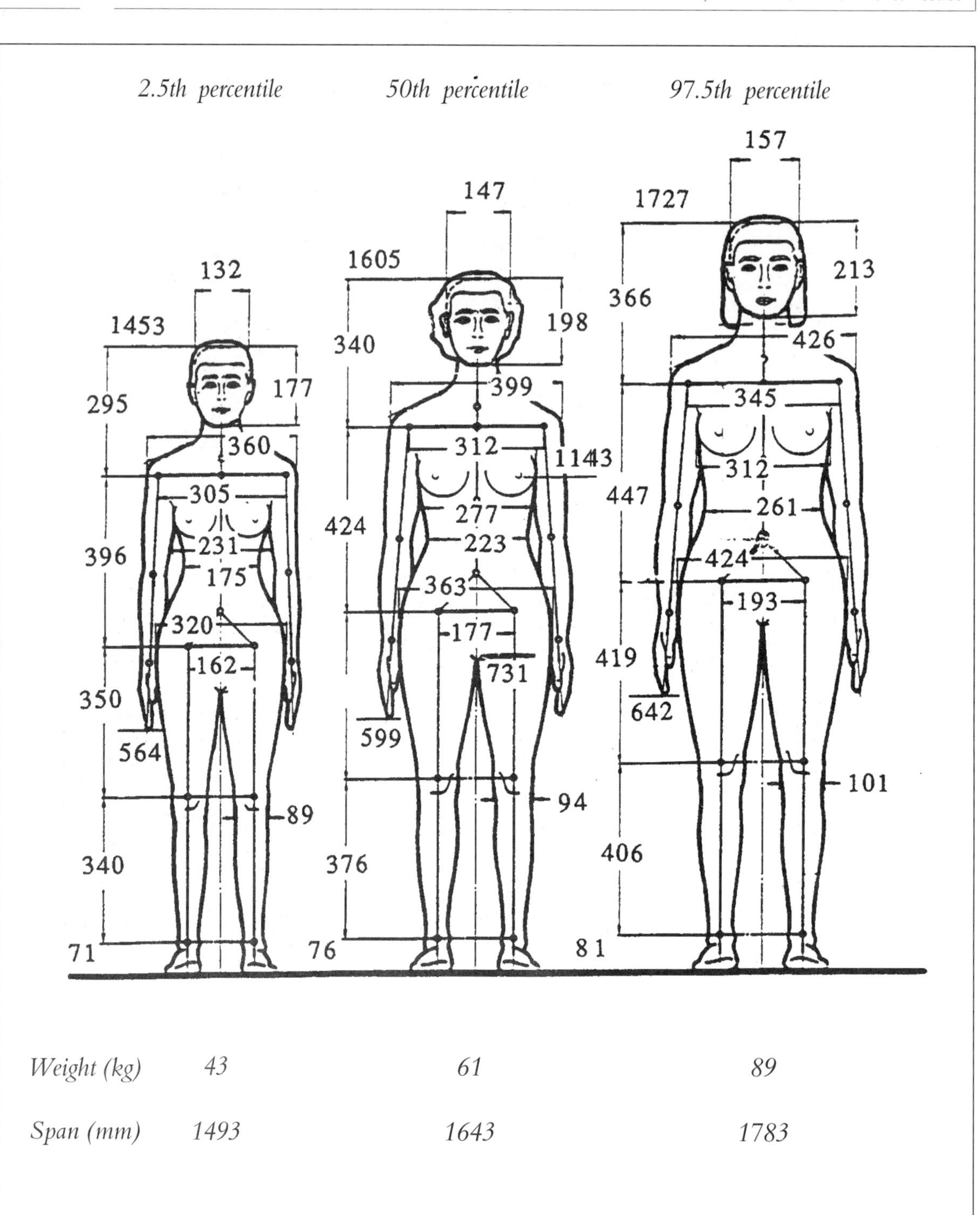

Figure 7.6 (c) Frontal vew of standing adult female - including 95% of population (dimensions mm) based on the charts of Dreyfuss (1967)

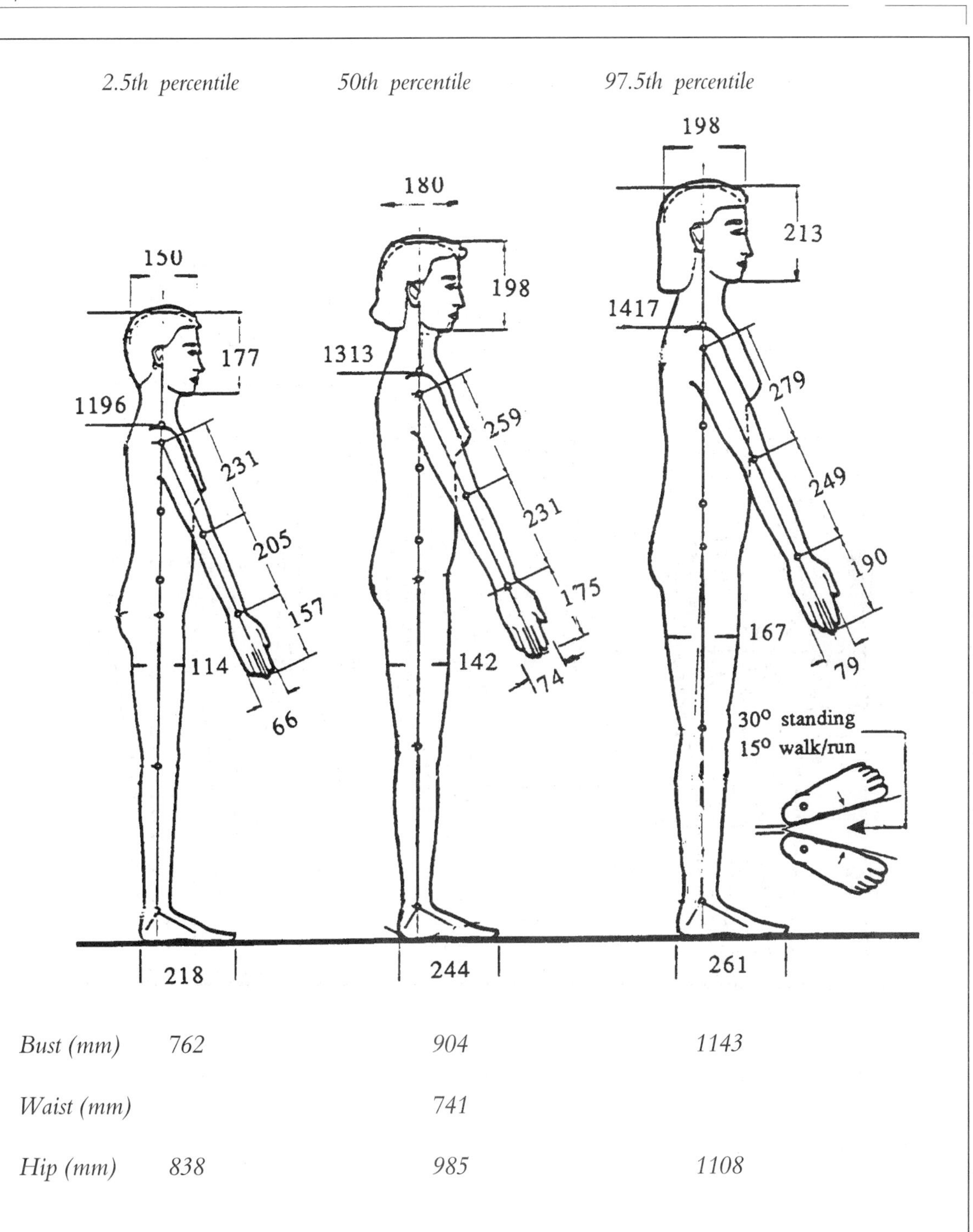

Figure 7.6 (d) Side vew of standing adult female - including 95% of population (dimensions mm) based on the charts of Dreyfuss (1967)

7.3.2 People in engineering systems

Occupant of workspace

This is the anthropometric model: data is presented in Figures 7.6(a) through 7.6(d) based on the charts of Dreyfuss (1967), collected mainly from surveys of the U.S. defence forces. While similar data for other human groups may also exist, the U.S. data may be used with slight corrections for special racial types where necessary. In Australia, for example, we need to correct for greater girth and hip breadth (Roebuck et al., 1975).

An important principle is to design for a range of users, usually from the 2.5th to 97.5th percentile, that is, all but the smallest 2.5% and the largest 2.5%. As the data follow a normal curve (*Gaussian distribution*), the 2.5th percentile point is approximately equal to the mean minus 2.25 standard deviations and the 97.5th percentile is correspondingly the mean plus 2.25 standard deviations. To accommodate 95% of potential users is considered a reasonable design objective, bearing in mind the trouble and expense required to accommodate the extremes. However, we use common sense in applying these limits. An escape hatch or doorway needs to accommodate the 97.5th percentile (largest) male, and a console for a control task needs to have everything within reach of the 2.5th percentile (smallest) female. Figure 7.7 shows a comparison of the extreme heights for 90% of the male population.

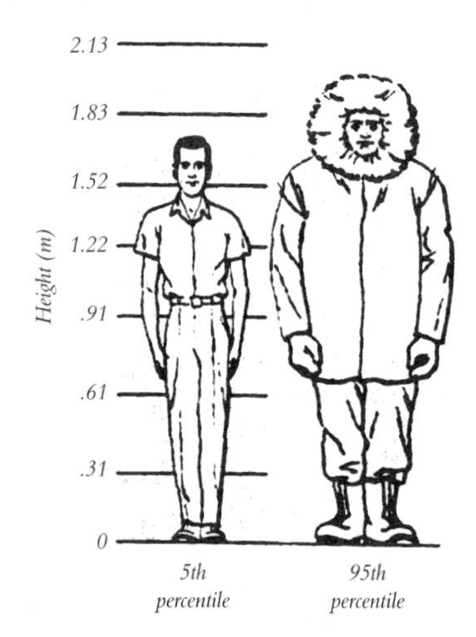

Figure 7.7 Comparative sizes in males (After Damon, Stoudt and McFarland, 1966)

Source of power

When a person is used to produce power it is usually because the power has to be applied in a variety of different ways at different places, for example, stacking goods on shelves. The maximum continuous power output a person is capable of, depends on which muscles are called into play but in general, it is of the order of 250W. A good rule of thumb for workplace design is, that a person should not be required to make a continuous effort greater than about 50% of the maximum of which he/she is capable. If the limit is kept down to this level, work can be performed reasonably steadily all day: exertion is kept within the aerobic level, the level at which normal metabolic processes replace the energy consumed without an oxygen deficit.

Sensor or transducer

If we define the function of a transducer as the detection of any change in the physical environment, then we can recognise a corresponding human function. However, in this capacity a person may function in two different ways:

(*a*) *detector* of a signal and initiator of some process in response to that signal;

(*b*) *monitor* of an on-going process where he/she is expected to take some action if a signal indicates a change in the process.

The two tasks are different. In the first case the sensory load tends to be high, and the function of the human perceptual system is to discriminate, and identify the stimuli that are relevant from those that are not. Almost any industrial task serves as an example: assembly work, machining and press operation.

On the other hand, in the watch-keeping task, there is, typically, a low level of stimulation. The task is to determine when (or *if*) a signal has occurred. Usually, the task is monotonous: for example, inspecting dimensions of industrial components, operating control rooms of power stations, or chemical plants, and watching radar screens in aircraft control towers.

Errors may occur in both types of task but for different reasons. In the first case, they are the result of overload, where discriminations between stimuli are too difficult to make, or are required too rapidly. In the second case, errors may be the result of underloading, when there is insufficient activity to keep the operator alert. The latter situation was described as a *vigilance* task by the American psychologist, N. H. Mackworth (1950). The classic vigilance effect is a sharp drop in the probability that a signal will be detected after a person has been on watch for half an hour or so.

Processor of information

First of all, we must quantify the information in incoming stimuli and in human responses to them. To do this we use the special mathematical theory of communication due to Shannon and Weaver (1949).

Information implies a gain of knowledge in some manner. In order for information to be conveyed there must initially be some uncertainty. The amount of information potentially available increases with the amount of uncertainty in the situation. In this context, information is associated with reduction in uncertainty. The basic notion of *information theory* is to measure the amount of information in a series of equally likely outcomes (for example, successive tosses of a coin) by measuring the total number of binary *yes-no* decisions that must be made in order to obtain a precise identification. The unit of measurement is a single binary decision, called a *bit*, a contraction of binary digit. The *bit* of information is the amount required to reduce uncertainty by a half. Thus if there are 32 equally likely alternatives, the number of binary decisions required to identify one alternative is five, that is, $\log_2 (32)$.

In general, the amount of information H contained in a set of equally likely alternatives is equal to the logarithm to the base 2 of the number of alternatives n from which the choice is made, or

$$H = \log_2 (n).$$

The rate at which people can process information depends on how that information is coded (see Miller, 1957). In one experimental task based on readings of ordinary English prose rates of 45 bits per second were observed. If a person is engaged in a sensorimotor task, such as driving a car, then the amount of alphanumeric information he/she can process (for example, from road signs) is greatly reduced.

It takes people time to react to a signal. Reaction times can vary enormously according to individual differences, level of alertness of the person concerned, level of fatigue, information load experienced, and whether or not a signal is expected. The correct detail design of information displays is often crucial to the successful performance of engineering systems. This is highlighted by an Air Force study of the types of error most frequently made in responding to instruments and signals. Misinterpretation of multipointer instruments (for example, some types of altimeter) was commonest - 18% of all errors made. Misinterpretation of the direction of indicator movement was next - 17%. Failure to respond to warning lights or sounds - 14%. Errors associated with poor legibility - 14%. The likelihood of such errors being made can be significantly reduced by attention to detail design.

Murrell (1965) gives a comprehensive set of recommendations for information to be displayed and for selecting the appropriate design for displays:

(a) *Quantitative reading*—The exact value in some conventional unit has to be available: for example, aircraft altimeter, wattmeter, digital clock.

(b) *Qualitative reading* — An approximate indication of the state of the system is required: for example, high, normal, or low, or within agreed tolerance limits for normal performance.

(c) *Check or dichotomous reading* — Is the state of the system O.K. or not? Is the warning light on or off ?

(d) *Tracking* — A rather special case in which a desired level of performance of a system has to be achieved and maintained by the active control of the operator. It is necessary to provide rate information: for example, a moving pointer on a dial which indicates closure onto the desired level.

Tracker and controller

The general definition of a tracking task is any situation in which a changing input has to be matched in some way by a system output, which is under the control of the human operator. As tracker and controller, a person functions as an element of a dynamic system. The system may be stable or unstable. A stable system tends towards a steady state if, when it is perturbed by a change in input, it tends afterwards to settle down again, either in a new steady state appropriate to the input it has received, or in its former state. If an unstable system is perturbed from its setting by a change in input, it tends to diverge from its previous state instead of settling down. Most systems employing human controllers are neither completely stable nor completely unstable: they are stable within a limited range, and tend towards instability as these limits are approached. An aircraft or motor car is stable within normal operating limits. Ships, helicopters, or submarines are unstable. Without constant attention to its controls, any one of these vehicles will not maintain a steady course.

Human use of human beings

Norbert Wiener (1894-1964), who is credited with establishing the science of cybernetics, towards the end of his life became concerned about the relationship between humans and machines. In his book, *Human use of human beings* (Wiener, 1954) he argues strongly for the separation of mindless and monotonous tasks that machines are capable of performing from the more associative thought processes possessed by humans. We have already made reference earlier to *vigilance tasks*. In this context we interpret Wiener's thesis to suggest that errors in human performance don't only result from insufficient levels of signal stimuli, but also from insufficiently engaging task content.

In addition, it is easy to forget that people have attitudes, habits, preferences, emotions and ambitions. These factors may complicate a design problem, but they must be considered nonetheless. There is no point in designing a switch or control to be operated in one way when everyone has a strong preference for operating it in some other way. Some expectations are almost universal and very strong: for example, we almost universally expect that clockwise movement on a dial indicates an increase. These are known as *population stereotypes*, and a system which runs counter to them is sure to fail through incorrect operation, as soon as the operator becomes flustered or overloaded. Naturally, this would be most likely to occur in an emergency, when correct operation is more important than ever.

Motivation

Why do people behave in the way they do ? What are their motives ? Attempts by psychologists to come to grips with the complexities of human motivation have given rise to a variety of theories. One of the best known is that due to Maslow (1970), who postulated a hierarchy of needs which human beings may strive to satisfy. Once a need toward the lower end of the hierarchy is satisfied, the one above it becomes the most important to an individual.

(a) Physiological needs - food, shelter, sleep, sex;

(b) Safety needs - security, freedom from fears of want, danger, unemployment;

(c) Social needs - need for love, affection, a sense of belonging;

(d) Esteem - need for self-respect and the respect of others;

(e) Need for *self-fulfilment, self-actualization*.

Maslow distinguishes between behaviour motivated by deficiency needs, *(a)* to *(d)*, and that motivated by growth needs, *(e)*. To put the matter briefly, he argues that the satisfaction of growth motivation leads to growing health of the individual, whereas the satisfaction of deficiency needs only prevents illness.

Another influential social scientist, Herzberg, has developed a similar theory and applied it to industrial conditions (Herzberg et al., 1959). He argues that there is one set of factors that lead to positive job satisfaction (*growth needs*), and another set that, while not involved in positive satisfaction, is sufficient to arouse job dissatisfaction (*hygiene needs*). The factors which prevent or end job dissatisfaction are the major environmental aspects of work. Those which produce positive attitudes to work are achievement, recognition, responsibility, and advancement. According to Herzberg's view of their role, managers have a responsibility not only to arrange for the effective use of hygiene factors (for example, by providing good working conditions) but also to aid the psychological growth of employees and encourage them to become self-actualisers. Cotgrove, Dunham and Vamplew (1971) describe a case where these ideas were applied to improve the operations of a nylon-spinning plant.

Aesthetics

Whereas renaissance nudes tended to be bulkier and more lumpy than today's slim fashion models, renaissance architecture is still much admired. Our conception of beauty is very subjective and may vary over a period of time. In the 1950s, as we emerged from the war years, aesthetic elegance was associated with streamlined contours on almost all household equipment, from table lamps to refrigerators. None of these items was intended to *fly* anywhere, but by association of human imagination with what, at that time, seemed the height of engineering achievement, streamlined design may have been an appropriate extension to household items. Current design of consumer goods pays much more attention to function than appearance. This is precisely the approach suggested by the American architect Louis Henry Sullivan when he wrote *"form follows function"* in his 1896 book, *The Tall Office Building Artistically Considered*.

It is difficult for the engineer to find a firm foundation in the shifting sands of fashion and taste. A lack of response to aesthetics sometimes induces hostile attitudes in community leaders. In one controversy, in England, regarding the construction of electricity transmission lines across a stretch of beautiful countryside, the editor of a leading newspaper, *The Times*, addressed the following question to the head of the electricity authority, *"When electricity pylons are being evolved, are designers concerned with aesthetics brought in or is this work left to the engineers?"* The implications of this question are obvious: aesthetics ought to be considered at the early design stage, and not during construction.

One of the prime uses of aesthetics, in the design of an engineering product, is to indicate function and purpose. Simple, easy-to-perceive forms, which recognise and reflect function and purpose, automatically do this. A suburban train travels forwards: it should look as though it does in fact do this. Long horizontal lines should be emphasised in the appearance of its carriages instead of arrays of short vertical lines corresponding to the openings for doors and windows.

Important concepts relevant to discussions on aesthetics include:

form and proportion;

visual balance - achieved by a subconscious association with the principles of physical balance, for example, weights and levers;

clarity of form and expression - lack of clutter, a product looking as though it will do what it does;

colour;

rhythm - an effect gained by repetition, for example, a balustrade;

surface texture;

unity - a coherence of parts forming the whole, achieved through some fundamental similarity.

For further reading see Ashford (1969). An interesting study of the adaptation of form to purpose in living creatures and plants is that by Thompson (1961). More recent texts dealing with aesthetics in architecture, art and culture are by Holgate (1992), Crowther (1993), Dipert (1993) and Matravers (1998).

Environmental factors

Important environmental factors are: air temperature and humidity, air velocity, illumination, noise and vibration. Any discussion of these matters must be based on a firm foundation of specialised branches of engineering science, which is outside the scope of this book. Murrell's treatment (1965) is particularly helpful. See also Kryter (1970).

CASE EXAMPLE 7.5: A CEMENT MIXER

The sequence of photographs shown in Figure 7.8 are due to Dr Errol Hoffmann, an internationally respected human factors specialist. The pictures show the construction of the premixer section of a motorised cement mixer, and its operation by a 97.5th percentile person. The photos shown in 7.8(h) and 7.8(g) elicit particularly horrified responses from a first-time audience. While this is only a *cautionary* example, suggesting *"please don't design this way"*, it is salutary to realise that someone has gone to great lengths in designing and constructing the machine shown in Figure 7.8.

Design for people with disabilities

Little systematic research has been carried out into the capacities and requirements of people handicapped by physical disability or disease. The one precept that can be laid down with certainty is that the design of equipment to be used by people with disabilities must closely involve the end user. The essence of our precept is that without this involvement the result may be seriously misdirected.

Goldsmith (1967) has considered in great detail *architectural design criteria for people with physical disabilities who are handicapped by buildings*. Although mainly concerned with people in wheelchairs, his book is an essential starting point for any designer in this field.

(a) Premixer parts

(b) Assembly in progress

(c) Fully assembled

(d) Loading premixer

(e) Ready to deliver to mixer

(f) Delivery complete

(g) Loading pedal returning

(h) Loading pedal fully returned

Figure 7.8 Assembly and operation of a cement premixer

Ex 7.8

You have been contracted to design a student carrel (private work area) in a university library. The carrel is in a progressive library with computing and video facilities. The design will be the prototype for all the libraries of your university and possibly a model for other learning institutions. Prepare an initial appreciation for this design problem and develop some preliminary sketch plans for the carrel showing major dimensions.

Ex 7.9

A large legal business has asked your help in planning their mail sorting and distribution system at their company headquarters, installed on the top three floors of a 20 storey building. The company has several hundred employees, and receives mail and parcels from national and international clients. Mail is delivered to the mail room, situated in the basement of the building, each day in the morning and it needs to be distributed to the several floors on which the legal offices are situated. A key component of the mail distribution system is the mail trolley used to deliver mail to each floor.

Prepare an initial appreciation of this design problem. Also prepare a preliminary design for the mail trolley.

Ex 7.10

People in wheelchairs have access to classrooms in university departments. There is a need for a special fitting to be used by wheelchair users to allow them to take notes, and to write reports, while still in their chair. Prepare an initial appreciation for this special fitting.

Ex 7.11

Consider the photographs in Figure 7.8. Identify and list as many of the potential operational problems with the premixer shown as you are able. Based on your list of potential operational problems, write a set of design specifications for a consultant, to be briefed to redesign the premixer.

In preparing your design specification, it is not intended to repair problems with the premixer shown in Figure 7.8. You should specify the redesign based on the dimensions and operating characteristics of a standard cement mixer. The purpose of the premixer is to permit dry mixing of a batch of the ingredients while the mixer is mixing the previous batch.

7.4 Managing risk and hazards: fault, failure and hazard analysis in engineering systems

When things fail, never trust the guy with the smile: he has found someone to blame it on. Anonymous

The goal of every engineer is to retire without getting blamed for a major catastrophe. Scott Adams, "The Dilbert Principle"

In any design problem there is a risk element to be considered. Designers continually face the challenge of risking failure. In structural integrity problems we deal with this risk by making use of factors of *safety*, to reduce the risk to a *credible* level. In the design of engineering products and systems we need to develop procedures for assessing the nature and level of risk involved.

We are primarily concerned with designing and planning for product quality, which in this context is distinguished by two factors:

PRESENCE – the presence of product performance characteristics which conform to quality targets and give value for money; and

ABSENCE – the absence of defects due to failures of components and assemblies to perform their intended functions.

An unacceptably large deviation from a quality target constitutes a failure. Designers and planners are charged with the responsibility of preventing all failures thought by them to be potentially possible, or at least of reducing the incidence of such failures to an acceptably low level.

The arrays of events leading up to and deriving from a failure are usually sufficiently complex to warrant the development of special aids to thinking and analysis. Aerospace, chemical and automotive industries have taken the lead in doing this. Important aids are :

- *Causal Networks* - these identify clear cause/effect relationships between preceding and subsequent events: Fishbone diagrams, discussed in the section on decision making in Chapter 6 may be used as causal networks;
- *Fault Tree Analysis* (*FTA*) - these show the causal relationships between some failure event and the preceding causes that may have lead to the failure event, by backward chaining;

- *Event diagrams* - these allow forward chaining from a specific initiating fault to some ultimate hazard event.

We now briefly examine these tools in connection with some simplified case examples.

7.4.1 Causal networks

Causal networks display large and complex arrays of information, and thus enhance our ability to manage and solve quality problems. They are used as aids for :

(a) diagnosing the cause(s) of failures which have already occurred (*diagnosis*);

(b) planning the prevention of undesired failures (*prognosis*).

Imagination and judgment are required in both (a) and (b) to construct causal networks which encompass all credible paths to and from a failure, and also in (b) to decide what constitutes a *credible failure* in the first place.

CASE EXAMPLE 7.6:

FAILURE OF MAIN BEARING ON TUNNELLING MACHINE

This is an example of causal construction.

A large and expensive bearing has failed in a tunnel boring machine. The machine was used for constructing the Melbourne underground rail-loop tunnel, commencing at the Jolimont rail yards, and proceeding North-West towards the Treasury building. Figure 7.9 shows the style of operation with a tunnel boring machine.

Figure 7.10 The failed main bearing with the works supervisor (approximately 1.7 m tall) indicating the scale. The bearing is situated in the vicinity of the "muck ring" indicated in Figure 7.9.

The machine cost approximately $2 million and was expected to complete the works in about 3000 hours. After approximately 900 hours the machine reached the Treasury building site and the cutting head of the machine, weighing nearly 50 tonnes, fell off. This cutting head carried 40 disc cutters and was held on the drive mechanism by a large, 2m diameter, double row tapered roller bearing,

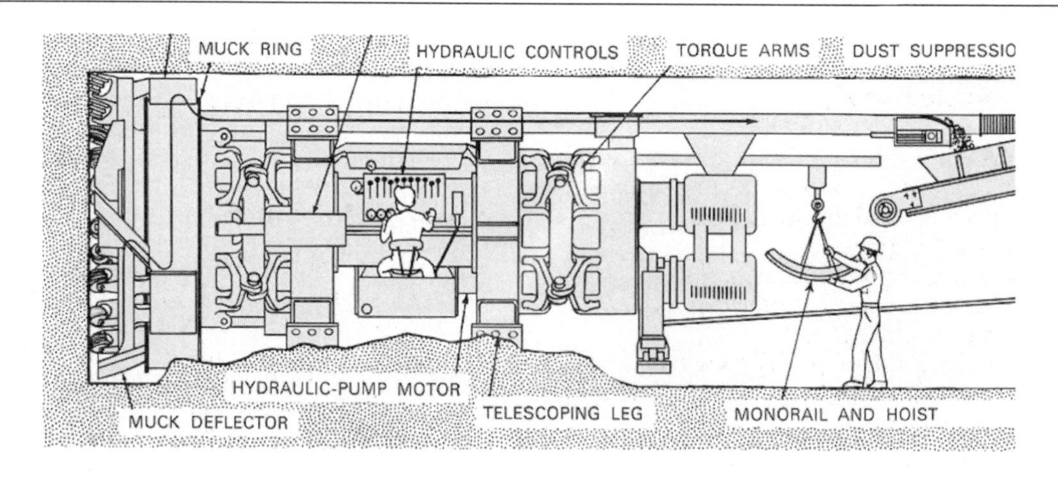

Figure 7.9 Schematic diagram of tunnelling with a tunnel borer

part of which broke and this was thought to be the main cause of the failure (refer to Figure 7.10).

The cost of repairs and delays to the works was estimated at $0.5 million. A consultant was asked to examine the design to determine the cause and to advise if the planned repair could guarantee the return of the tunnelling machine to a state capable of completing the project safely.

MODES OF FAILURE FOR BEARINGS:

(a) surface fatigue - pitting and deep surface damage;

(b) part of the bearing outer race fails (parts break off);

(c) part of the inner race fails (parts break off).

POSSIBLE CAUSES:

(a) overloading;

(b) inadequate lubrication;

(c) misalignment;

(d) material failure (inadequate heat treatment or wrong material).

POSSIBLE CONSEQUENCES:

(a) bearing makes low rumbling noise, machine overhaul is planned and eventually bearing is replaced at a convenient time;

(b) bearing breaks at inconvenient time; machine stops, it is overhauled with associated delays.

7.4.2 Fault tree analysis

A *Fault Tree* is a network representing causal relations for a system or component. The relationship is described in terms of a logic diagram using *'and'* and *'or'* gates.

- The failure appears as the *top event* in the tree and is linked to more basic *underlying causes* by event statements and logic gates.
- *Fault Tree Analysis* (*FTA*) uses backward-chaining logic based on successive posing of questions of the form:

"What prior event could have led to failure event X ?"

Figure 7.11 shows a partial *causal net* for the tunnelling machine bearing failure and Figure 7.12 shows a generic form of a fault tree.

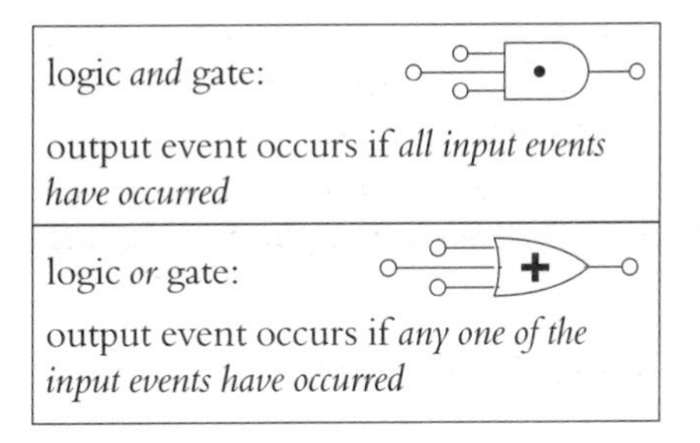

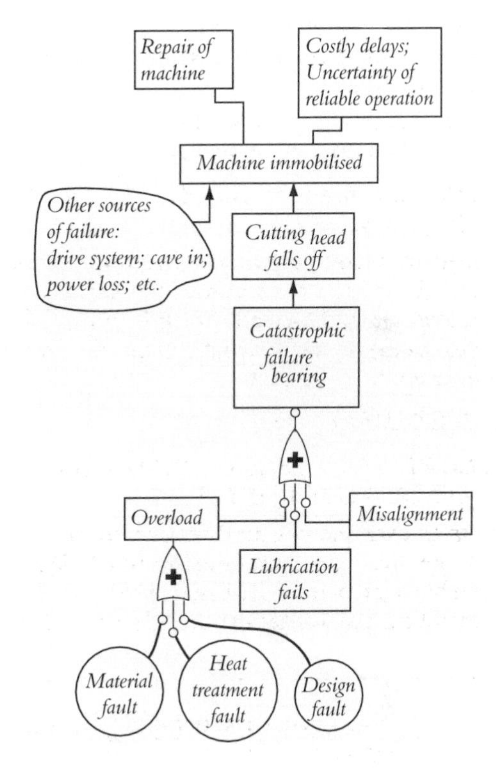

Figure 7.11 Logic gates and partial fault tree for bearing failure

Structure of a fault tree

The undesired failure appears as the top event, and this is linked to more basic fault events by event statements and logic gates. The analysis is focused on possible combinations of components and events that lead to the ultimate failure.

The method depends on a backward chaining logic and identifies sets of causal relations leading to the particular quality failure under investigation.

Note that a *command fault* is defined as a component being in a failed condition due to improper control signals or noise.

Table 7.4 Symbols used in Fault Tree Analysis

Primary fault events resulting from combinations of more basic faults acting through logic gates; a primary failure for which the component is held accountable.	Rectangle	▭
Secondary failure; a failure for which the component is not held accountable; for example excessive stress caused by neighbouring components, such as out-of-tolerance conditions of amplitude, frequency, residence-time and energy inputs from external sources.	Diamond	◇
Basic component failure; the limit of resolution of a fault tree, usually an event for which quality and reliability data are available.	Circle	○

Table 7.4 shows the type of symbols used in Fault Tree analysis, together with their interpretations.

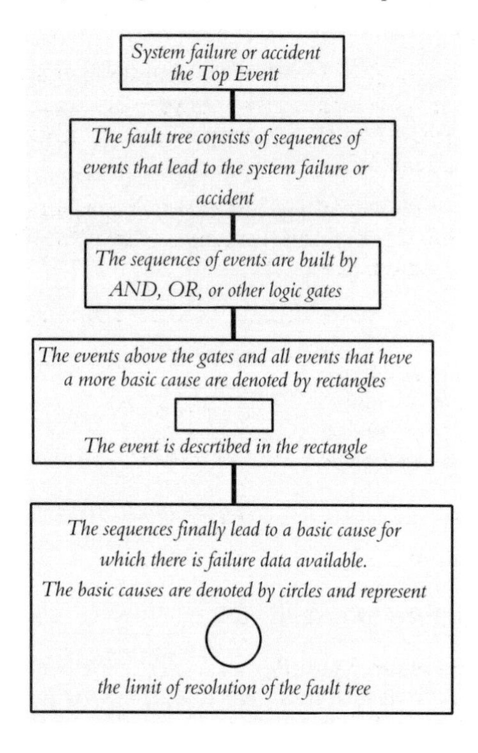

Figure 7.12 A generic fault tree

Advantages and disadvantages of fault trees:

Advantages :

- direct the attention of the designer to possible sequences of events leading to quality failure;
- provide a concise but information-rich graphic aid to the thinking of designers and managers;
- provide insights, not easily obtained by other means, into the behaviour of complex systems (e.g. automotive systems) and into the interactions between the components of such systems;
- provide an effective and convenient way of communicating ideas between different *Design and Planning Teams*.

Disadvantages:

- based on binary decisions, either something has happened or it hasn't, shades of grey are excluded;
- can become excessively complicated, unless a modicum of practical judgment is exercised by *Design and Planning Team*.

Some guidelines for construction of fault trees :

- Replace an abstract event by a less abstract event;

Example: *"Current to motor on too long"* instead of *"Motor operates too long"*.

- Classify an event into more elementary events;

Example: *"Explosion by overfilling "*, or *"Explosion by runaway reaction"*, instead of *"Explosion of tank"* .

- Identify distinct causes of an event;

Example: *"Excessive feed"* and *"loss of cooling"* instead of *"Runaway reaction"*.

- Couple *"trigger event"* with *"No protective action"*;

Example: *"Loss of cooling"* coupled with *"No system shutdown"*, instead of *"Overheating"*.

- Find co-operative causes for an event;

Example: *"Leak of flammable fluid"* and *"Relay sparks"*, instead of *"Fire"*.

- Pinpoint a component failure event;

Example: *"No current in wire"*, instead of *"No current to motor"*; *"Main valve is closed"* coupled with *"Bypass valve is not opened"*, instead of *"No cooling water"*.

7.4.3 Event Tree Analysis

Event Tree Analysis (*ETA*) makes use of forward-chaining logic by asking questions such as

"What happens if this component fails in some particular failure mode ?"

or, in general, what consequences follow from a postulated initiating event.

ETA is thus closely related to *Potential Failure Modes and Effects Analysis* (*PFMEA*), a quality management tool used by many manufacturers to improve their products (see for example Hawkins and Woollons, 1998; Perkins, 1996; Astridge, 1996; Clark and Paasch, 1996; Hatty and Owens, 1995). *ETA* is often applied to chains of events hypothesized to describe accidents having severe or catastrophic consequences.

Figure 7.13, shows a partial event tree developed during a study of the safety of nuclear power reactors. The critical part of the reactor is its cooling system, and the chosen initiating event (the *worst credible accident*) is breakage or fracture of the coolant pipe, *event A*, with a probability of occurrence of P_A.

Note (1): Here we have made the approximation

$$P_A' = P_A(1-P_B)(1-P_C)(1-P_D)(1-P_E)$$
$$\approx P_A,$$

since all the various probabilities of failure are very small.

- At the first branch of the event tree, the status of the electric power supply is considered, (*event B*): P_{B1} is the probability of electric power failing.
- If it is available, the next-in-line system, the emergency core cooling system (*ECCS*) is studied, (*event C*).

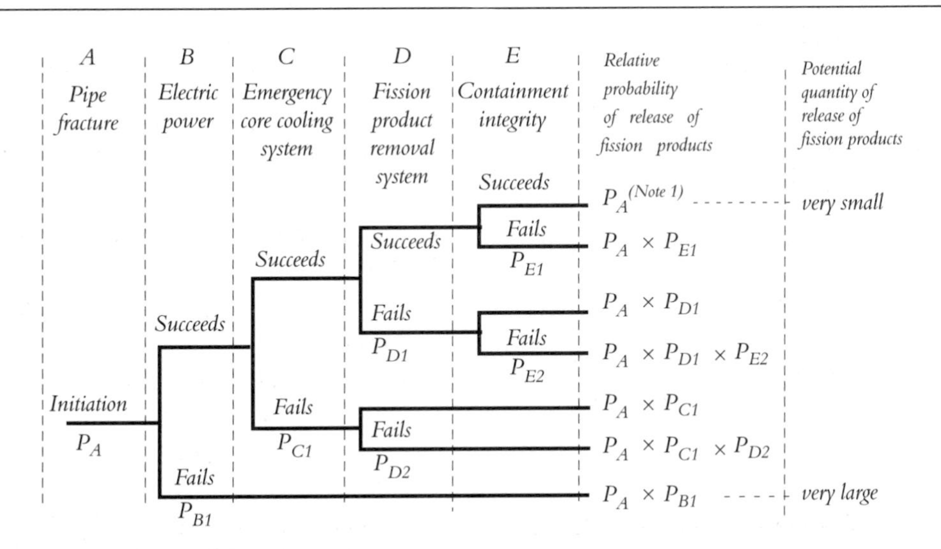

Figure 7.13 Partial event tree for nuclear power plant

- Failure of the *ECCS* results in fuel meltdown and varying amounts of fission product released (*event D*), depending on the integrity of the reactor containment system, (*event E*).
- If there is no electric power, the emergency core cooling pumps and sprays do not function, in fact none of the post-accident corrective functions can be performed. Potentially a very large quantityof fission products may be released, the probability of this combination of events being $P_A \cdot P_{B1}$.
- If electric power is available, the next choice for study is the availability of the *ECCS*. If it is unavailable (probability P_{C1}), then we follow the appropriate path in the event tree, and so on for the radioactivity removal system (probability of failure P_{D1}) and containment integrity (probability of failure P_{E1}).

By working through the entire event tree, we produce a spectrum of release magnitudes and their probabilities for the various accident sequences.

In practice, nuclear design is predicated on the possibility of coolant pipe fracture, but each of the safety systems is assumed to operate successfully (i.e. it is assumed that the various probabilities of failure are very small, leading to the approximation identified in *Note 1*).

Ex 7.12

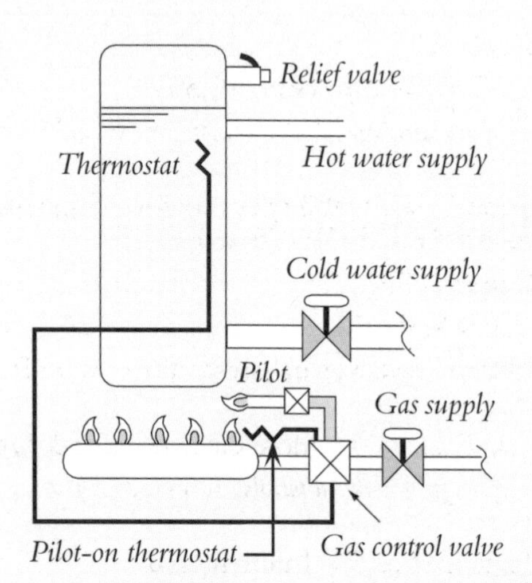

Figure 7.14 Schematic layout of a domestic hot water supply system

Figure 7.14 shows the main components of a domestic gas hot water supply system.

(a) Draw up a fault tree, considering *tank explosion* as the *top event*.

(b) Consider the condition that the pilot light is blown out as an *initiating event*. Draw up an event tree to indicate the way in which one could estimate the probability of a tank explosion resulting from this event.

Clearly state all assumptions.

Ex 7.13

In the pumping system shown in Figure 7.15, the tank is filled in 10 minutes and empties in 50 minutes: thus the cycle time is one hour.

After the *switch* is closed, the *timer* is set to open the *contacts* in 10 minutes. If the mechanisms fail, then the *alarm horn* sounds and the *operator* opens the switch to prevent a *tank rupture* due to overfilling.

(a) Prepare a fault tree for this system.

(b) Prepare an event tree with the timer failure as the initiating event.

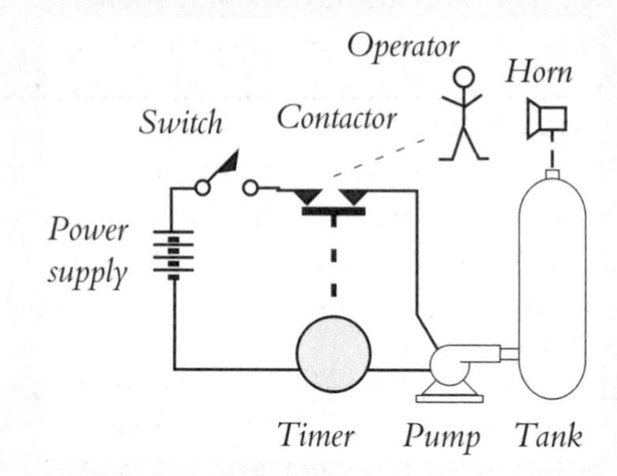

Figure 7.15 Tank pressure cycling system

REFERENCES

Adams, J. L. (1987) *Conceptual blockbusting*, Harmondsworth,Middlesex: Penguin

Ashby, M.F. and Jones, D.R.H.(1980) *Engineering materials*, Oxford:Pergamon

Ashby, M. F. (1992), *Material selection in mechanical design*, Oxford:Pergamon (ISBN 0-08-041907-0)

Ashford, F. (1969) *The aesthetics of engineering design*, London: Business Books

ASM[1]: *Engineering materials reference book* / editor, Bauccio, M. (ed., 1994) Materials Park, OH: ASM International, (ISBN 0871705028)

ASM International. Handbook Committee (1995): *Engineered materials handbook*; prepared under the direction of the ASM International Handbook Committee ; volume chair, Michelle M. Gauthier. (ISBN 0871702835)

Astridge, D. G. (1996) Design safety analysis of helicopter rotor and transmission systems, *Proceedings of the Institution of Mechanical Engineers*, Part G: Journal of Aerospace Engineering 210 (4): 345-355

Barnes, R. M. (1949) *Motion and Time Study*, New York: John Wiley

Bevan, T. (1955) *Theory of Machines*, London: Longmans, Green and Co.

Birsch, D. and Fielder, J. H.(eds., 1994) *The Ford Pinto case: A Study in applied ethics, business, and technology*, New York: State University of New York Press

Bloom, B.S. (ed., 1956) *Taxonomy of educational objectives– Handbook I: Cognitive domain*, New York: David McKay

Box, G. E. P., Hunter, W. G. and Hunter, J. S. (1978) *Statistics for experimenters*, New York: Wiley

Braun, H. (1987) *Allianz handbook of loss prevention*, Berlin:Allianz Versicherungs A.G.

Brown, D.C., Waldron, M.B. and Yoshikawa, H. (eds., 1992) *Intelligent computer aided design*, North Holland, Amsterdam

Buzan, A. (1974) *Use your head*, London:BBC

Buzan, A. and Buzan, B. (1993) *The mind map book*, London:BBC

Chase, W. G. and Simon, H. A. (1973) Perception in chess, *Cognitive Psychology*, (4):55-81

Chawla, S.L. and Gupta, R.K. (1994) *Materials selection for corrosion control*, Materials Park, OH : ASM International

Clark, G. E. and Paasch, R. K. (1996) Diagnostic modeling and diagnosability evaluation of mechanical systems, *Journal of Mechanical Design*, Transactions of the ASME 118 (3): 425-431

Clarke, Arthur C. (1996) *3001 : The final odyssey*, London: HarperCollins

Collins, J. A. (1993) *Failure of materials in mechanical design*, New York: Wiley

Constantin, C. G. (ed., 1979) *Romanian inventions and prioritites in aviation*, Bucharest: Albatros (in Romanian)

Cotgrove, S., Dunham, J. and Vamplew, C. (1971) *The Nylon Spinners*, London: George Allen & Unwin

Crane, F. A. A. and Charles, J. A.(1984) *Selection and use of engineering materials*, London: Butterworth (ISBN 0-408-10859-2)

Craig, B. D.(ed., 1989) *Handbook of corrosion data*, Metals Park, OH : ASM International

Crowther, P. (1993). *Art and embodiment : from aesthetics to self-consciousness*, Oxford: Clarendon Press

CRC materials science and engineering handbook, Shackelford, J. F., Alexander, W. and Park, J. S. (eds., 1994), 2nd ed. Boca Raton : CRC Press, (ISBN 0849342503)

Damon, A., Stoudt, H. W. and McFarland, R. A.(1966) *The human body in equipment design*, Cambridge, Mass.: The Harvard University Press

1. American Society of Metals

de Bono, E. (1986) *Six thinking hats*, London: Little Brown & Co.

Dipert, R. R. (1993) *Artifacts, art works, and agency*, Philadelphia : Temple University Press,

Dreyfuss, H. (1967) *The measure of man*: Human Factors in Design, New York: Whitley Library of Design

Eddy, P., Potter, E. and Page, B. (1976) *Destination disaster*, London: Granada

Einstein, A. (1936) Relativity-the special and general theory, London:Methuen

Einstein, A. and Infeld, L. (1938) *The evolution of physics*, Cambridge: The University Press

Farquharson, F. B. (1949) *Aerodynamic stability of suspension bridges, with special reference to the Tacoma Narrows Bridge*, Part I: Investigations Prior to October 1941. Report. Structural Research Laboratory, University of Washington.

Field, B. W. (1970) *Some computer aided design strategies*, Unpublished Master of Engineering Science Thesis, University of Melbourne

French, M. J. (1988) Research in enginering design, aimed at better practice and better teaching: some proposals, Chapter 5, *Engineering Design and Manufacturing Management*, Samuel, A. E. (ed.), Proceedings of the International Workshop on Engineering Design held at the University of Melbourne, April (ISBN 0-444-88132-9)

Frost, N. E., Marsh, K. J., Pook L. P. (1974) *Metal fatigue*, Oxford: Clarendon Press

Fuchs, H. O. and Stephens, R. I. (1980) *Metal fatigue in engineering*, New York:Wiley

Gale, W. K. V. (1952) *Boulton, Watt and the Soho undertakings*; with an appendix by W.A. Seaby. Birmingham, England: City of Birmingham Museum and Art Gallery, Department of Science and Industry.

Giedion, S. (1948) *Mechanisation takes command*, New York: Oxford University Press

Goldberg, D. E. (1989) *Genetic algorithms in search, optimisation and machine learning*, Reading, Mass.: Addison Wesley

Goldsmith, S. (1967) *Designing for the disabled*, London: Royal Institute of British Architects

Gordon, J. E. (1976) *The new science of strong materials or why you don't fall through the floor*, Harmondsworth: Penguin

Gordon, J. E. (1979) *Structures or why things don't fall down*, Harmondsworth: Penguin

Gregory, S. A. (ed., 1966) *The design method*, London:Butterworth

Hatty, M. and Owens, N. (1995) Potential failure modes and effects analysis: A business perspective, *Quality Engineering* 7(1):169

Hawkins, P. G. and Woollons, D.J. (1998)Failure modes and effects analysis of complex engineering systems using functional models, *Artificial Intelligence in Engineering* 12 (4): 375-397

Herzberg, F., Mausner, B. and Snydermann, B. B. (1959) *The motivation to work* (2nd. edn.), New York: Wiley

Hewat, T. and Waterton, W.A. (1956) *The Comet riddle*, London: Frederick Muller

Holgate, A. (1992) *Aesthetics of built form*, Oxford : Oxford University Press

Holick, M. (1993) *An Introduction to project evaluation*, Melbourne: Longman Cheshire

Ishibashi, A. and Yoshino, H. (1985) Power transmission efficiencies and friction coefficients at teeth of Novikov-Wildhaber and involute gears, *ASME J. Mechanical Transmissions and Automation Design*, 1071(1):74-81

Ishikawa, Kaoru (1985) *What is total quality control?: The Japanese way*, Englewood Cliffs, N.J.: Prentice Hall

Juvinall, R. C. (1967) *Stress, strain and strength*, New York: McGraw Hill

Jahnke, E. and Emde, F. (1945) *Tables of functions with formulae and curves*, New York:Dover

Kelly, F. C. (ed. 1996) *Miracle at Kitty Hawk: The letters of Wilbur and Orville Wright*, New York: Da Capo

Klein, J.H. and Cooper, D.F. (1982) Cognitive maps of decision makers in a complex game, *J. Opl. Res. Soc.* 33:63-71

Kratwohl, D.R., Bloom, B.S. and Masia, B.B. (1964) *Taxonomy of educational objectives– Handbook II: Affective domain*, London: Longmans

Kryter, K. D. (1970) *The Effects of noise on man*, New York: Academic Press

Kusiak, A. (ed., 1992) *Intelligent design and manufacturing*: A collection of articles on concurrent design and manufacturing, New York: Wiley

Leeman, F. (1975) *Hidden images: games of perception, anamorphic art, illusion*, New York: Henry Abrahams, Inc.

Lipson, C. and Juvinall, C.R. (1961) Application of stress analysis to design and metallurgy, *The University of Michigan Summer Conference*, Ann Arbor, Michigan

Mackworth, N. H. (1950) Researches on the measurement of human performance, *Special Report Series, No. 268, Medical Research Council,* London

Mann, J. Y.(1970) *Bibliography on the fatigue of materials, components and structures*, 1838-1950. Oxford:Pergamon

Manson, S.S.(1965) Fatigue:A complex subject–Some simple approximations, *Experimental Mechanics*, 3(7)193-226

McKim, R. H. (1972) *Experiences in visual thinking*, Monterey: Brooks/Cole Publishing Co.

McLuhan, M. and Fiore, Q. (1967) *The medium is the massage*, Harmodsworth, Middlesex: Penguin

McPherson, N. (1994) *Machines and economic growth : the implications for growth theory of the history of the industrial revolution*. Westport, Conn. : Greewood Press

Marples, D. L. (1961) The decisions of engineering design, *IEEE Transactions on Engineering Management*, (EM-8):55-71

Maslow, A.H. (1970) *Motivation and personality* (2nd. edn), New York: Harper and Row

Matravers, D. (1998). *Art and emotion*, Oxford : Clarendon Press

Miller, G. A. (1956) The magical number seven, plus or minus two: Some limits on our capacity for processing Information, *Psychological Review*, (63): 81-97

Milner, P. (1977) *Steam power and the industrial revolution*, Parkville, Vic.: University of Melbourne, Department of Mechanical and Industrial Engineering.

Miner, M. A. (1946) Cumulative damage in fatigue, *J. appl. Mech*.12(3):A159-A164

Mitchell, M. R. and Landgraf, R. W. (1991) *Advances in fatigue lifetime predictive techniques, Philadelphia*:ASTM

Monk, R. (1996) *Bertrand Russell: The spirit of solitude*, London: Jonathan Cape

Morant, G. M. (1947) Anthropometric problems in the Royal Airforce, *British Medical Bulletin*, (5):25-32

Morgan, M. H.(1960) *Vitruvius' The ten books on architecture, traslated from De Architectura* (1st century BCE), New York:Dover (Originally published in 1914)

Murrell, K. F. H. (1965) *Ergonomics*, London: Chapman and Hall

Nader, R. (1972) *Unsafe at any speed; the designed-in dangers of the American automobile*, New York: Grossman

Niebel, B.W. and Draper, A.B.(1974) *Product design and process engineering*. New York: McGraw-Hill(ISBN 0-07-046535-5)

Ord-Hume, A. W. (1977) *Perpetual motion: The history of an obsession*, New York: St. Martins Press

Osgood, C. C. (1979) *Fatigue design*, 2nd. Ed. New York:Pergamon

Pahl, G. and Beitz, W. (1996) *Engineering design: A systematic approach,* 2nd. edition, Translated into English by Wallace, K. (ed.), Blessing, L. and Bauert, F., London: Springer

Palmgren, A. (1924) *Die lebensdauyer von Kugellagern* [The life of ball bearings] Z. V. D. I. 68(14):339-341

Palmgren, A. (1959) *Ball and roller bearing engineering*, Philadelphia: SKF Co. publishing

Parmee, I. C. (1997) Strategies for the integration of evolutionary/adaptive search with the engineering design process, in: Dasgupta, D. and Michelewitz, Z. (eds. 1997) *Evolutionary Algorithms in Engineering Applications*, Berlin: Springer

Perkins, D. (1996) FMEA for a real (resource limited) design community, Annual Quality Congress Transactions, *Proceedings of the 1996 ASQC's 50th Annual Quality Congress* May 13-15 1996 Chicago, IL, USA, pp 435-442

Peterson, R.E.(1974) *Stress concentration factors*, New York:John Wiley

Petroski, H. (1994) *Design paradigms:Case histories of error and judgement in engineering*, Cambridge University Press

Petroski, H. (1996) *Invention by design : How engineers get from thought to thing*, Cambridge Mass.: Harvard University Press

Popper, K. R. (1972) *Conjectures and refutations:The growth of scientific knowledge*, London:Rutledge and Kegan Paul

Pugh, S. (1981) Concept selection - A Method That Works, *Proceedings of International Conference on Engineering Design*, ICED 81, Rome: M3:16.1-16.10

Randall, F. E. (1948) Applications of anthropometry to the determinationof size in clothing, *Environment Protection Division Report* No. 133, Natick, Mass. Quartermaster's Research and Development Command

Rheingold, H. (1991) *Virtual reality*, London: Mandarin

Rittel, H. (1987) The reasoning of designers, unpublished article presented at the *International Conference on Engineering Design, ICED 87*, Boston, August

Roebuck, J. A. Jr., Kroemer, K. H. E. and Thompson, W. G. (1975) *Engineering anthropometry methods*, New York: Wiley Interscience

Rodenacker, W.G. (1970) *Methodisches Konstruieren, Konstructionsbücher*, Bd 27, Berlin: Springer

SAE[2]:Fatigue Design Handbook (1997) 3rd Edition, Warrendale, Pa.:SAE International

Samuel, A. E. (1984) Educational objectives in engineering design, *Instructional Science*, 13, 243-273

Samuel, A. E. (ed., 1988) *Engineering design and manufacturing management*, The 1988 International Workshop on Engineering Design held at the University of Melbourne (ISBN 0-444-88132-9)

Samuel, A. E. and Horrigan, D. (1995) A comparative study of two finite element packages,*J. Eng. Design* 6(1):57-68

Samuel, A. E. and Weir, J. G. (1991) The acquisition of wisdom in engineering design, *Instructional Science*, 20: 419-442

Samuel A. E. and Weir J.G.(1995), Concept learning in engineering design using cognitive maps, *Proceedings of International Conference on Engineering Design, ICED 95* Prague, August 22-24, p 303-311

Samuel, A.E. and Weir J.G.(1997) A parametric approach to problem enformulation in engineering design, Proceedings of *Proceedings of International Conference on Engineering Design, ICED* 97, Tampere Finland, August, (3): 83-90

Shannon, C. E. and Weaver, W. (1949) *The mathematical theory of communication*, Urbana, Ill.: University of Illinois Press

Sivaloganathan, S. and Shanin, T.M.M. (eds., 1998) Design reuse, Collected papers from *Engineering Design Conference'98*, Brunel University, UK, 23-25 June, London: Professional Engineering Publishers

Shackelford, J. F., Alexander,W. and Park, J. S. (eds., 1995). *CRC practical handbook of materials selection*, Boca Raton : CRC Press Companion vol. to: CRC materials science and engineering handbook, 2nd ed. (ISBN 0849337097)

Shigley, J. E. (1986) *Mechanical engineering design*, New York: McGraw Hill

Schweitzer, P. A (ed., 1983) *Corrosion and corrosion protection handbook*, New York: M. Dekker

Shreir, L. L., Jarman, R.A. and Burstein G. T. (eds., 1994) *Corrosion*, Oxford ; Boston: Butterworth-Heinemann

Smith, D.K. and Alexander, R. C. (1988) *Fumbling the future*, New York: William Morrow & Co.

Soderberg, C. R. (1935) Working stresses, *J. Appl. Mech.* (57):A-106

Stewart, D. (1965) A platform with six degrees of freedom, *Proc. I. Mech. E.* vol. 180, part 1(15): 371-386

Strunk, W. Jr. and White, E. B. (1979) *The elements of style*, New York: McMillan

Thalmann, N. M. and Thalmann, D. (eds., 1994) *Artificial life and virtual reality*, New York: Wiley

Thompson, W. D'A (1961) *On growth and form* (Abridged version edited by J. T. Bonner), Cambridge, U.K.: Cambridge University Press

Timoshenko, S. P. (1953) *History of strength of materials*, New York:McGraw Hill

Timoshenko, S. P. (1955) *Strength of materials-Part I: Elementary theory and problems*, New York:Van Nostrand

Timoshenko, S. P. (1956) *Strength of materials-Part II: Advanced theory and Problems*, Princeton, New Jersey: Van Nostrand (Third edition)

Timoshenko, S. P. and Goodier, J. N. (1983) *Theory of elasticity*, London:McGraw Hill

van Vlack, L.H.(1985) *Elements of materials science and engineering*, Reading Mass.: Addison Wesley

Weber, R. L. and Mendosa, E. (1973) *A random walk in science*, London:The Institute of Physics

West, J. M.(1980) *Basic corrosion and oxidation*, New York : Halsted Press

Wiener,N. (1954) *Human use of human beings*, New York: Penguin

Wilde, D. J. (1978) *Globally optimal design*, New York: Wiley

Woebcken, W. (ed., 1995) *Saechtling international plastics handbook for the technologist, engineer and user*, 3rd. Edition, Cincinnati:Hanser/Gardner

Wöhler,A.(1858) Bericht über die Versuche, welche auf der Königl. Niederschleesisch-Märkischen Eisenbahn mit Apparaten zum Messen der Biegung und Verdrehung von Eisenbahnwagen-Achsen während de Fahrt, angestellt wurden. [Report on tests of the Königl. Niederschleesisch-Märkischen Eisenbahn made with apparatus for the measurement of the bending and torsion of railway axles in service] *Z. Bauw.*, (8):642-651; reported in Mann(1970)

Young, W.C.(1989) *Roark's formulas for stress and strain* (6th edn). New York: McGraw-Hill (ISBN 0-07-072541-1)

Annotated bibliography of engineering design

Adams, J. L. (1974) *Conceptual blockbusting*, San Francisco:Freeman

Easy-to-absorb text dealing with creative thinking. The basic objective of the text is to free the human mind from cognitive boundaries (mind-sets). The popular terminology used for the style of thinking advocated throughout the text is *thinking outside the square*, or *lateral thinking*[3].

Alger, J. R. M. and Hays, C. V. (1964) *Creative synthesis in design*, Englewood Cliffs N.J.: Prentice-Hall

Series of introductory, descriptive, essays on the design process, by two experienced practicing engineers.

Andreasen, M.M., Kähler, S. and Lund, T. (1983) *Design for assembly*, Berlin:Springer

Introductory text on design for ease of assembly, with well researched case examples.

Archer, L. B. (1974) *Design awareness and planned creativity*, London: The Design Council

2. Society of Automotive Engineers

Asimow, M. (1962) *Introduction to design*, Englewood Cliffs N.J.: Prentice-Hall

Excellent, concise, introductory overview on the general procedural approach to design.

British Standards Institution, (1989) BS7000: *Guide to managing product design*, London: BSI

Creamer, R. H. (1984) *Machine design*, Reading, Mass: Addison-Wesley

Third edition (originally published in 1968) of a standard text on machine element design.

Cross, N.(1994) *Engineering design methods: Strategies for product design*, Chichester: John Wiley

Introductory level text about strategies for synthesis in product design.

de Bono, E. (1971) *The use of lateral thinking*, Harmondsworth: Penguin

Articulate presentation of the author's ideas on *lateral thinking*, as an aid to creative problem solving.

Dieter, G. (1983) *Engineering design: A materials and processing approach*, New York: McGraw-Hill

This text is addressed to designers with a specific interest in materials selection and process management. It also contains a brief introductory section on general procedural design methods.

Deutschman, A. D., Michaels, W. J. and Wilson, C. E. (1975) *Machine design*, London: Macmillan

A comprehensive, practical text on the analysis and selection of specific machine elements.

French, M. J. (1985) *Conceptual design for engineers*, London/Berlin: The Design Council/Springer

Challenging, thoughtfully written text advocating combinatorial and parametric evaluation of designs at the conceptual level. The examples offered are thoroughly grounded in engineering science.

French, M. J. (1988) *Invention and evolution: Design in nature and engineering*, Cambridge: University Press

While this text claims to be an introduction to design for function, with several examples drawn from nature, the author's deep understanding of the underlying engineering principles makes this a scholarly presentation. Insightful parallels are drawn between design in nature and products designed by humans.

French, M. J. (1992) *Form, structure and mechanism*, London: Macmillan

Further case examples of embodiment design based on the author's substantial design experience.

Haugen, E. B. (1968) *Probabilistic approaches to design*, New York: Wiley

A specialist text devoted to reliability analysis in the context of product design.

Hindhede, U., Zimmerman, J. R., Hopkins, B. R., Erisman, R. J., Hull, W. C., Lang, J. D. (1983) *Machine design fundamentals*, New York: John Wiley

A comprehensive practical text on machine element selection for function and structural integrity.

Hubka, V., Andreasen, M. M. and Eder, E. (1988) *Practical studies in systematic design*, London: Butterworth

Well presented case examples on the development of embodiment design.

Johnson, R. C. (1980) *Optimum design of machine elements*, 2nd. edition, New York: John Wiley

Systematic approach to setting up objective functions in the design of machine elements.

Jones, J. C. (1980) *Design methods*, New York: John Wiley

Seminal text on design methods, originally published in 1970. This text is the forerunner of many ideas eventually articulated in specialist design texts, including removal of mental blocks (see also Adams: *Conceptual blockbusting*).

3. 'lateral thinking' is a term coined by Edward de Bono

Juvinall, R. C. and Marshek, K. M. (1991) *Fundamentals of machine component design*, 2nd. edition, New York: John Wiley

A comprehensive text dealing with the selection and analysis of machine components for structural integrity.

Krick, E. V. (1969) *An introduction to engineering and engineering design*, New York: John Wiley

Well written text for an introductory course on decision support tools in engineeering design, with many excellent examples.

Krick, E. V. (1976) *An introduction to engineering and engineering: Methods, concepts and issues*, New York: John Wiley

Excellent introduction to the broad spectrum of problems that can face engineers. There are many engaging examples for design scholars.

Lawson, B. (1990) *How designers think: The design process demystified*, London:Butterworth

Levinson, I. J. (1978) *Machine design*, Reston, Virginia: Reston Publishing Co.

Introductory text on machine element design.

Matousek, R. (1963) *Engineering design: a Systematic approach*, London: Blackie & Son, (translated from German by A. H. Burton and D. C. Johnson)

An Introductory text on the systematic design of embodiment for machine elements, based on function.

Nevins, J. L. and Whitney, D. E. (eds., 1989) *Concurrent design of products and processes*, New York: McGraw-Hill

A collection of articles on specialist topics with the focus on manufacturing and its relation to design.

Niebel, B.W. and Draper, A.B.(1974) *Product design and process engineering*, New York: McGraw-Hill

Although somewehat dated, this is a valuable text for design related matters in material selection and process engineering.

Pahl, G. and Beitz, W. (1996) *Engineering design: A systematic approach*, 2nd. edition, Translated into English by Wallace, K. (ed.), Blessing, L. and Bauert, F., London: Springer

Currently the most often quoted text in engineering design. The objective of the authors is to present a systematic approach to machine element embodiment design. This is a purely functional evaluation of the embodiment, and structural integrity is not considered in this context.

Petroski, H. (1994) *Design paradigms: Case histories of error and judgement in engineering*, Cambridge: University Press

An excellent anecdotal text about engineering judgement and its role in design. The majority of examples are drawn from civil engineering, due to the author's special interest in that discipline.

Petroski, H. (1996) *Invention by design : How engineers get from thought to thing*, Cambridge Mass.: Harvard University Press

Well researched case examples on the evolution of inventions, ranging from paper clips to elevators.

Sharpe, J. (ed., 1996) AI System support for conceptual design: *Proceedings of the 1995 Lancaster International Workshop on Engineering Design*, London: Springer

A mixed collection of articles about the evolution of computer based autonomy in generating design concepts.

Shigley, J. E. (1986) *Mechanical engineering design*, 5th edition (orginally published in 1963), New York: McGraw-Hill

A seminal text on the design of generic machine elements for structural integrity. The clarity and scholarly presentation makes this a most popular text on machine element design.

Spotts, M. F. (1978) *Design of machine elements*, 5th edition (orginally published in 1948), Englewood Cliffs N. J. : Prentice-Hall

A substantial text dealing with the design of a selection of generic machine elements for structural integrity.

Wilde, D. J. (1978) *Globally optimal design*, New York: Wiley

A general approach to setting up objective functions for guaranteed arrival at a global optimum.

Conferences and Journals on Engineering Design

The Design Theory and Methodology Conference (DTM, since 1987), The American Society of Mechanical Engineers, New York

International Conference on Engineering Design (ICED, since 1981) WDK series, Heurista, Zürich

Journal of Engineering Design (since 1990) Oxford: Carfax

Research in Engineering Design (since 1988) London: Springer

APPENDIX A: CONVERSION TABLES

INDEX OF CONVERSION TABLES:

Table A.1 — LENGTH
Table A.2 — AREA
Table A.3 — VOLUME
Table A.4 — VOLUME RATE OF FLOW (volume/time)
Table A.5 — MASS
Table A.6 — MASS RATE OF FLOW (mass/time)
Table A.7 — VELOCITY
Table A.8 — DENSITY (mass/volume)
Table A.9 — SECOND MOMENT OF AREA
Table A.10 — MOMENT OF INERTIA
Table A.11 — FORCE
Table A.12 — MOMENT OF FORCE (torque)
Table A.13 — PRESSURE
Table A.14 — STRESS
Table A.15 — ENERGY, WORK, HEAT
Table A.16 — POWER, HEAT FLOW RATE
Table A.17 — DYNAMIC VISCOSITY (μ)
Table A.18 — KINEMATIC VISCOSITY (μ/ρ)
Table A.19 — DENSITY OF HEAT FLOW RATE (heat /area /time)
Table A.20 — HEAT TRANSFER COEFFICIENT (heat /area /time /temperature_interval)
Table A.21 — THERMAL CONDUCTIVITY (heat x length /area /time /temperature_interval)
Table A.22 — SPECIFIC HEAT CAPACITY (heat /mass /temperature_interval)
Table A.23 — SPECIFIC ENERGY (heat /mass)
Table A.24 — VOLUMETRIC ENERGY (heat /volume)

GENERAL NOTES:

- these tables are adapted from a format used by Orica Engineering (formerly ICI Australia)
- table format used by permisssion, Orica Engineering
- the units within each table are listed in order of increasing unit magnitude (e.g mm < cm < ft)
- equivalent values are read horizontally across the table, with units stated in the top row
- preferred SI units are highlighted in the top row
- values in the tables are quoted to a maximum of 6 significant figures (5 if in exponential notation)
- values with fewer significant figures are generally exact
- subscripts and exponents in unit names are printed in-line for clarity

Table A.1 LENGTH

mm	*cm*	*in*	*ft*	*yd*	*m*	*km*	*mile*
1	0.1	3.9370E-02	3.2808E-03	1.0936E-03	1E-03	1E-06	6.2137E-07
10	1	0.39370	3.2808E-02	1.0936E-02	0.01	1E-05	6.2137E-06
25.4	2.54	1	0.08333	2.7778E-02	0.0254	2.54E-05	1.5783E-05
304.8	30.48	12	1	0.33333	0.3048	3.048E-04	1.8939E-04
914.4	91.44	36	3.00	1	0.9144	9.144E-04	5.6818E-04
1E+03	100	39.3701	3.28084	1.09361	1	1E-03	6.2137E-04
1E+06	1E+05	39370.1	3280.84	1093.61	1E+03	1	0.62137
1609344	160934	63360	5280	1760	1609.34	1.60934	1

Notes:

1 m (micron) =	1E-06 m =	3.937E-05 in
1 thou =	0.0254 mm =	0.001 in
1 (ngstrm) =	1E-10 m	
1 UK nautical mile =	6080 ft =	1853.2 m
1 international nautical mile =	6076.1 ft =	1852 m

Table A.2 AREA

mm2	*cm2*	*in2*	*ft2*	*yd2*	*m2*	*acre*	*hectare*	*km2*	*mile2*
1	0.01	1.5500E-03	1.0764E-05	1.1960E-06	1E-06	2.4711E-10	1E-10	1E-12	3.8610E-13
100	1	0.155000	1.0764E-03	1.1960E-04	1E-04	2.4711E-08	1E-08	1E-10	3.8610E-11
645.16	6.45160	1	6.9444E-03	7.7160E-04	6.4516E-04	1.5942E-07	6.4516E-08	6.4516E-10	2.4910E-10
92903.0	929.030	144	1	0.111111	9.2903E-02	2.2957E-05	9.2903E-06	9.2903E-08	3.5870E-08
836127	8361.27	1296	9.00	1	0.836127	2.0661E-04	8.3613E-05	8.3613E-07	3.2283E-07
1E+06	1E+04	1550.00	10.7639	1.19599	1	2.4711E-04	1E-04	1.0000E-06	3.8610E-07
4.0469E+09	4.0469E+07	6272640	43560	4840	4046.86	1	0.404686	4.0469E-03	1.5625E-03
1E+10	1E+08	1.5500E+07	107639	11959.9	1E+04	2.47105	1	1E-02	3.8610E-03
1E+12	1E+10	1.5500E+09	1.0764E+07	1195990	1E+06	247.105	100	1	0.386102
2.5900E+12	2.5900E+10	4.014E+09	2.7878E+07	3.0976E+06	2.5900E+06	640	258.999	2.58999	1

Notes: 1 are = 100 m2

Table A.3 VOLUME

mm3	*ml*	*in3*	*l*	*US gal*	*UK gal*	*ft3*	*yd3*	*m3*
1	1E-03	6.1024E-05	1E-06	2.6417E-07	2.1997E-07	3.5315E-08	1.3080E-09	1E-09
1E+03	1	6.1024E-02	1E-03	2.6417E-04	2.1997E-04	3.5315E-05	1.3080E-06	1E-06
16387.1	16.3871	1	1.6387E-02	4.3290E-03	3.6047E-03	5.7870E-04	2.1433E-05	1.6387E-05
1E+06	1E+03	61.0237	1	0.264172	0.219969	3.5315E-02	1.3080E-03	1E-03
3.7854E+06	3785	231.000	3.785	1	0.832674	0.133681	4.9511E-03	3.7854E-03
4.5461E+06	4546.09	277.419	4.54609	1.20095	1	0.160544	5.9461E-03	4.5461E-03
2.8317E+07	28316.8	1728	28.3168	7.48052	6.22884	1	3.7037E-02	2.8317E-02
7.6455E+08	764555	46656	764.555	201.974	168.179	27	1	0.764555
1E+09	1E+06	61023.7	1E+03	264.172	219.969	35.3147	1.30795	1

Notes:

1 l = 1.00 dm3 (according to the 1974 definition of the litre)
1 l = 1.000028 dm3 (according to the 1901 definition of the litre)
1 ml = 1.000028 cm3 (according to the 1901 definition of the litre)
1 US barrel = 42 US gal = 34.97 UK gal
1 fluid oz = 28.41 ml
1 UK pint = 568.2 ml
1 litre = 1.760 UK pints

Table A.4 VOLUME RATE OF FLOW (volume/time)

litres /h	*ml /s*	*gal /h*	*m3 /d*	*l /min*	*gal /min*	*m3 /h*	*ft3 /min*	*l /s*	*ft3 /s*	*1E6 gal /d*	*m3 /s*
1	0.277778	0.219969	0.024	1.6667E-02	3.6662E-03	1E-03	5.8858E-04	2.7778E-04	9.8096E-06	5.2793E-06	2.7778E-07
3.6	1	0.791889	8.64E-02	0.060	1.3198E-02	3.6E-03	2.1189E-03	1E-03	3.5315E-05	1.9005E-05	1E-06
4.54609	1.26280	1	0.109106	7.5768E-02	1.6667E-02	4.5461E-03	2.6757E-03	1.2628E-03	4.4595E-05	2.4E-05	1.2628E-06
41.6667	11.5741	9.16539	1	0.694444	0.152756	4.1667E-02	2.4524E-02	1.1574E-02	4.0873E-04	2.1997E-04	1.1574E-05
60	16.6667	13.1982	1.44	1	0.219969	0.060	3.5315E-02	1.6667E-02	5.8858E-04	3.1676E-04	1.6667E-05
272.765	75.7682	60	6.54637	4.54609	1	0.272765	0.160544	7.5768E-02	2.6757E-03	1.44E-03	7.5768E-05
1000	277.778	219.969	24	16.6667	3.66615	1	0.588578	0.277778	9.8096E-03	5.2793E-03	2.7778E-04
1699.01	471.947	373.730	40.7763	28.3168	6.22884	1.69901	1	0.471947	1.6667E-02	8.9695E-03	4.7195E-04
3600	1000	791.889	86.4	60	13.1982	3.6	2.11888	1	3.5315E-02	1.9005E-02	1E-03
101941	28317	22423.8	2446.58	1699.01	373.730	101.941	60	28.3168	1	0.538171	2.8317E-02
189420	52617	41666.7	4546.09	3157.01	694.444	189.420	111.489	52.6168	1.85814	1	5.2617E-02
3.6E+06	1E+06	791889	86400	60000	13198.2	3600	2118.88	1000	35.3147	19.0053	1

Notes:

1 l = 0.001 m3 (according to the 1974 definition of the litre)
gal = UK or Imperial gallon
1 US gal = 0.833 UK gal

Table A.5 MASS

g	*oz*	*lb (lbm)*	*kg*	*slug*	*cwt*	*US ton*	*t (tonne)*	*UK ton*
1	3.5274E-02	2.2046E-03	1E-03	6.8522E-05	1.9684E-05	1.1023E-06	1E-06	9.8421E-07
28.3495	1	6.25E-02	2.8350E-02	1.9426E-03	5.5804E-04	3.125E-05	2.8350E-05	2.7902E-05
453.592	16	1	0.453592	3.1081E-02	8.9286E-03	5E-04	4.5359E-04	4.4643E-04
1E+03	35.2740	2.20462	1	6.8522E-02	1.9684E-02	1.1023E-03	1E-03	9.8421E-04
14593.9	514.785	32.1740	14.5939	1	0.287268	1.6087E-02	1.4594E-02	1.4363E-02
50802.3	1792	112	50.8023	3.48107	1	5.6E-02	5.0802E-02	0.05
907185	32000	2000	907.185	62.1619	17.8571	1	0.907185	0.892857
1E+06	35274.0	2204.62	1E+03	68.5218	19.6841	1.10231	1	0.984207
1016047	35840	2240	1016.05	69.6213	20	1.12	1.01605	1

Notes:
the tonne is often abbreviated "te" to distinguish it from the UK ton or the US ton
the unit of pound-mass is often abbreviated as "lbm" to distinguish it from the pound-force, "lbf"
the US Customary System preferred base units are the foot, pound-force, and second
the slug is the unit of mass in the foot-pound-second (FPS) system
1 slug = 1 lbf.s2/ft
1 quintal = 100 kg

Table A.6 MASS RATE OF FLOW (mass/time)

t /yr	*UK ton /yr*	*lb /h*	*kg /h*	*g /s*	*lb /min*	*t /d*	*UK ton /d*	*t /h*	*UK ton /h*	*lb /s*	*kg /s*
1	0.984207	0.251669	0.114155	3.1710E-02	4.1945E-03	2.7397E-03	2.6965E-03	1.1416E-04	1.1235E-04	6.9908E-05	3.1710E-05
1.01605	1	0.255708	0.115987	3.2219E-02	4.2618E-03	2.7837E-03	2.7397E-03	1.1599E-04	1.1416E-04	7.1030E-05	3.2219E-05
3.97347	3.91071	1	0.453592	0.125998	1.6667E-02	1.0886E-02	1.0714E-02	4.5359E-04	4.4643E-04	2.7778E-04	1.2600E-04
8.76	8.62165	2.20462	1	0.277778	3.6744E-02	0.024	2.3621E-02	1E-03	9.8421E-04	6.1240E-04	2.7778E-04
31.5360	31.0379	7.9366	3.6	1	0.132277	8.64E-02	8.5035E-02	3.6E-03	3.5431E-03	2.2046E-03	1E-03
238.408	234.643	60	27.2155	7.55987	1	0.653173	0.642857	2.7216E-02	2.6786E-02	1.6667E-02	7.5599E-03
365	359.235	91.8593	41.6667	11.5741	1.53099	1	0.984207	4.1667E-02	4.1009E-02	2.5516E-02	1.1574E-02
370.857	365	93.3333	42.3353	11.7598	1.55556	1.01605	1	4.2335E-02	4.1667E-02	2.5926E-02	1.1760E-02
8760	8621.65	2204.62	1E+03	277.778	36.7437	24	23.6210	1	0.984207	0.612395	0.277778
8900.57	8760	2240	1016.05	282.235	37.3333	24.3851	24	1.01605	1	0.622222	0.282235
14304.5	14078.6	3600	1632.93	453.592	60	39.1904	38.5714	1.63293	1.60714	1	0.453592
31536	31037.9	7936.64	3600	1E+03	132.277	86.4	85.0354	3.6	3.54314	2.20462	1

Notes:
1 yr taken as 365 days
1 US ton = 2000 lbm ("short ton")

Table A.7 VELOCITY

mm /s	*ft /min*	*cm /s*	*km /h*	*ft /s*	*mile /h*	*m /s*	*km /s*
1	0.196850	0.1	3.6E-03	3.2808E-03	2.2369E-03	1E-03	1E-06
5.08	1	0.508	1.8288E-02	1.6667E-02	1.1364E-02	5.08E-03	5.08E-06
10	1.96850	1	0.036	3.2808E-02	2.2369E-02	0.01	1E-05
277.778	54.6807	27.7778	1	0.911344	0.621371	0.277778	2.7778E-04
304.8	60	30.48	1.09728	1	0.681818	0.3048	3.048E-04
447.040	88	44.7040	1.60934	1.46667	1	0.44704	4.4704E-04
1E+03	196.850	100	3.6	3.28084	2.23694	1	1E-03
1E+06	196850	1E+05	3600	3280.84	2236.94	1E+03	1

Notes: 1 UK knot = 1.853 km /hr
1 international nautical knot (Kn) = 1.852 km /hr

Table A.8 DENSITY (mass/volume)

kg /m3	*lb /ft3*	*lb /UK gal*	*g /cm3*	*ton /yd3*	*lb /in3*
1	6.2428E-02	1.0022E-02	1E-03	7.5248E-04	3.6127E-05
16.0185	1	0.160544	1.6018E-02	1.2054E-02	5.7870E-04
99.7764	6.22884	1	9.9776E-02	7.5080E-02	3.6047E-03
1E+03	62.4280	10.0224	1	0.752480	3.6127E-02
1328.94	82.9630	13.3192	1.32894	1	4.8011E-02
27679.9	1728	277.419	27.6799	20.8286	1

Notes: 1 g/cm3 = 1 kg/dm3 = 1 t/m3 = 1 g/ml = 1 kg/litre
(based on the 1901 definition of the litre)

Table A.9 SECOND MOMENT OF AREA

mm4	*cm4*	*in4*	*ft4*	*m4*
1	1E-04	2.4025E-06	1.1586E-10	1E-12
1E+04	1	2.4025E-02	1.1586E-06	1E-08
416231	41.6231	1	4.8225E-05	4.1623E-07
8.6310E+09	863097	20736	1	8.6310E-03
1E+12	1E+08	2.4025E+06	115.862	1

Notes: this unit is often wrongly called the "moment of inertia"

Table A.10 MOMENT OF INERTIA

kg.mm2	*kg.cm2*	*lbm.in2*	*lbm.ft2*	*kg.m2*
1	0.01	3.4172E-03	2.3730E-05	1E-06
100	1	0.341717	2.3730E-03	1E-04
292.640	2.92640	1	6.9444E-03	2.9264E-04
42140.1	421.401	144	1	4.2140E-02
1E+06	1E+04	3417.17	23.7304	1

Notes: the unit of pound-mass is often abbreviated as "lbm"
this is to distinguish it from the pound-force, "lbf"

Table A.11 FORCE

pdl	*N*	*lbf*	*kgf*	*kN*	*tonf*
1	0.138255	3.1081E-02	1.4098E-02	1.3825E-04	1.3875E-05
7.23301	1	0.224809	0.101972	1E-03	1.0036E-04
32.1740	4.44822	1	0.453592	4.4482E-03	4.4643E-04
70.9316	9.80665	2.20462	1	9.8067E-03	9.8421E-04
7233.01	1E+03	224.809	101.972	1	0.100361
72069.9	9964.02	2240	1016.05	9.96402	1

Notes: the unit of pound-force is often abbreviated as "lbf"
this is to distinguish it from the pound-mass, "lbm"
the kgf is sometimes known as the kilopond, "kp"
the poundal, "pdl", is the force required to accelerate 1lbm at 1 ft/s2
1 pdl = 1 lbm.ft/s2
1 N = 1E+05 dyn

Table A.12 MOMENT OF FORCE (torque)

N.mm	*pdl.ft*	*lbf.in*	*N.m*	*lbf.ft*	*kgf.m*	*UK tonf.in*	*UK tonf.ft*
1	2.3730E-02	8.8507E-03	1E-03	7.3756E-04	1.0197E-04	3.9512E-06	3.2927E-07
42.1401	1	0.372971	4.2140E-02	3.1081E-02	4.2971E-03	1.6651E-04	1.3875E-05
112.985	2.68117	1	0.112985	8.3333E-02	1.1521E-02	4.4643E-04	3.7202E-05
1E+03	23.7304	8.85075	1	0.737562	0.101972	3.9512E-03	3.2927E-04
1355.82	32.1740	12	1.35582	1	0.138255	5.3571E-03	4.4643E-04
9806.65	232.715	86.7962	9.80665	7.23301	1	3.8748E-02	3.2290E-03
253086	6005.82	2240	253.086	186.667	25.8076	1	8.3333E-02
3037032	72069.9	26880	3037.03	2240	309.691	12	1

Notes: 1 N.m = 1E+07 dyn.cm

Table A.13 PRESSURE

dyn /cm2	*Pa*	*lbf /ft2*	*mbar*	*mm Hg*	*in H20*	*kPa*	*in Hg*	*lbf /in2*	*kgf /cm2*	*bar*	*atm*
1	0.1	2.0885E-03	1E-03	7.5006E-04	4.0146E-04	1E-04	2.9530E-05	1.4504E-05	1.0197E-06	1E-06	9.8692E-07
10	1	2.0885E-02	1E-02	7.5006E-03	4.0146E-03	1E-03	2.9530E-04	1.4504E-04	1.0197E-05	1E-05	9.8692E-06
478.803	47.8803	1	0.478803	0.359131	0.192222	4.7880E-02	1.4139E-02	6.9444E-03	4.8824E-04	4.7880E-04	4.7254E-04
1E+03	100	2.08854	1	0.750062	0.401463	0.1	2.9530E-02	1.4504E-02	1.0197E-03	1E-03	9.8692E-04
1333.22	133.322	2.78450	1.33322	1	0.535240	0.133322	3.9370E-02	1.9337E-02	1.3595E-03	1.3332E-03	1.3158E-03
2490.89	249.089	5.20233	2.49089	1.86832	1	0.249089	7.3556E-02	3.6127E-02	2.5400E-03	2.4909E-03	2.4583E-03
1E+04	1E+03	20.8854	10	7.50062	4.01463	1	0.295300	0.145038	1.0197E-02	1E-02	9.8692E-03
33863.9	3386.39	70.7262	33.8639	25.4	13.5951	3.38639	1	0.491154	3.4532E-02	3.3864E-02	3.3421E-02
68947.6	6894.76	144	68.9476	51.7149	27.6799	6.89476	2.03602	1	7.0307E-02	6.8948E-02	6.8046E-02
980665	98066.5	2048.16	980.665	735.559	393.701	98.0665	28.9590	14.2233	1	0.980665	0.967841
1E+06	1E+05	2088.54	1E+03	750.062	401.463	100	29.5300	14.5038	1.01972	1	0.986923
1013250	101325	2116.22	1013.25	760	406.782	101.325	29.9213	14.6959	1.03323	1.01325	1

Notes:
1 kgf/cm2 = 1 kp/cm2 = 1 technical atmosphere
1 torr = 1 mm Hg (to within 1 part in 7 million)
1 Pa = 1 N/m2

Table A.14 STRESS

dyn /cm2	*Pa*	*pdl /ft2*	*lbf /ft2*	*kPa*	*lbf /in2*	*kgf /cm2*	*MPa*	*kgf/mm2*	*hbar*	*UK tonf /in2*
1	0.1	6.7197E-02	2.0885E-03	1E-04	1.4504E-05	1.0197E-06	1E-07	1.0197E-08	1E-08	6.4749E-09
10	1	0.671969	2.0885E-02	1E-03	1.4504E-04	1.0197E-05	1E-06	1.0197E-07	1E-07	6.4749E-08
14.8816	1.48816	1	3.1081E-02	1.4882E-03	2.1584E-04	1.5175E-05	1.4882E-06	1.5175E-07	1.4882E-07	9.6357E-08
478.803	47.8803	32.1740	1	4.7880E-02	6.9444E-03	4.8824E-04	4.7880E-05	4.8824E-06	4.7880E-06	3.1002E-06
1E+04	1E+03	671.969	20.8854	1	0.145038	1.0197E-02	1E-03	1.0197E-04	1E-04	6.4749E-05
68947.6	6894.76	4633.06	144	6.89476	1	7.0307E-02	6.8948E-03	7.0307E-04	6.8948E-04	4.4643E-04
980665	98066.5	65897.6	2048.16	98.0665	14.2233	1	9.8067E-02	0.01	9.8067E-03	6.3497E-03
1E+07	1E+06	671969	20885.4	1E+03	145.038	10.1972	1	0.101972	0.1	6.4749E-02
9.8067E+07	9.8067E+06	6.5898E+06	204816	9806.65	1422.33	100	9.80665	1	0.980665	0.634971
1E+08	1E+07	6.7197E+06	208854	1E+04	1450.38	101.972	10	1.01972	1	0.647490
1.5444E+08	1.5444E+07	1.0378E+07	322560	15444.3	2240	157.488	15.4443	1.57488	1.54443	1

Notes:
the standard SI unit of pressure is the pascal (Pa)
however the megapascal (MPa) is frequently a more convenient unit in structural integrity design
the unit "pounds-force per square inch" is often abbreviated "psi"
1000 lbf/in2 = 1 kpsi
1 MN/mm2 = 1 MPa = 1 N/mm2

Table A.15 ENERGY, WORK, HEAT

J	ft.lbf	cal	kgf.m	kJ	Btu	Chu	kcal	MJ	hp.h	kW.h	therm
1	0.737562	0.238846	0.101972	1E-03	9.4782E-04	5.2657E-04	2.3885E-04	1E-06	3.7251E-07	2.7778E-07	9.4782E-09
1.35582	1	0.323832	0.138255	1.3558E-03	1.2851E-03	7.1393E-04	3.2383E-04	1.3558E-06	5.0505E-07	3.7662E-07	1.2851E-08
4.18680	3.08803	1	0.426935	4.1868E-03	3.9683E-03	2.2046E-03	1E-03	4.1868E-06	1.5596E-06	1.1630E-06	3.9683E-08
9.80665	7.23301	2.34228	1	9.8067E-03	9.2949E-03	5.1638E-03	2.3423E-03	9.8067E-06	3.6530E-06	2.7241E-06	9.2949E-08
1E+03	737.562	238.846	101.972	1	0.947817	0.526565	0.238846	1E-03	3.7251E-04	2.7778E-04	9.4782E-06
1055.06	778.169	251.996	107.586	1.05506	1	0.555556	0.251996	1.0551E-03	3.9301E-04	2.9307E-04	1E-05
1899.10	1400.70	453.592	193.654	1.89910	1.8	1	0.453592	1.8991E-03	7.0743E-04	5.2753E-04	1.8E-05
4186.80	3088.03	1E+03	426.935	4.18680	3.96832	2.20462	1	4.1868E-03	1.5596E-03	1.1630E-03	3.9683E-05
1E+06	737562	238846	101972	1E+03	947.817	526.565	238.846	1	0.372506	0.277778	9.4782E-03
2.6845E+06	1.9800E+06	641187	273745	2684.52	2544.43	1413.57	641.187	2.68452	1	0.745700	2.5444E-02
3.6E+06	2.6552E+06	859845	367098	3600	3412.14	1895.63	859.845	3.6	1.34102	1	3.4121E-02
1.0551E+08	7.7817E+07	2.5200E+07	10758576	105506	1E+05	55555.6	25199.6	105.506	39.3015	29.3071	1

Notes: the "cal" is the International Table calorie

1.014 hp.h (metric) = 1 hp.h = 745.7 W.h = 2.685 MJ

1 thermie = 1.163 kW.h = 4.186 MJ = 999.7 kcal

1 ft.pdl = 0.04214 J

1 erg = 1E-07 J

Table A.16 POWER, HEAT FLOW RATE

Btu /h	Chu /h	W	kcal /h	ft.lbf /s	kgf.m /s	metric hp	hp	kW	tcal /h	MW
1	0.555556	0.293071	0.251996	0.216158	2.9885E-02	3.9847E-04	3.9301E-04	2.9307E-04	2.5200E-04	2.9307E-07
1.8	1	0.527528	0.453592	0.389085	5.3793E-02	7.1724E-04	7.0743E-04	5.2753E-04	4.5359E-04	5.2753E-07
3.41214	1.89563	1	0.859845	0.737562	0.101972	1.3596E-03	1.3410E-03	1E-03	8.5985E-04	1E-06
3.96832	2.2046	1.16300	1	0.857785	0.118593	1.5812E-03	1.5596E-03	1.1630E-03	1E-03	1.1630E-06
4.62624	2.5701	1.35582	1.16579	1	0.138255	1.8434E-03	1.8182E-03	1.3558E-03	1.1658E-03	1.3558E-06
33.4617	18.5898	9.80665	8.43220	7.23301	1	1.3333E-02	1.3151E-02	9.8067E-03	8.4322E-03	9.8067E-06
2509.63	1394.237	735.499	632.415	542.476	75	1	0.986320	0.735499	0.632415	7.3550E-04
2544.43	1413.574	745.700	641.187	550.000	76.0402	1.01387	1	0.745700	0.641187	7.4570E-04
3412.14	1895.63	1E+03	859.845	737.562	101.972	1.35962	1.34102	1	0.859845	1E-03
3968.32	2204.62	1163.00	1E+03	857.785	118.593	1.58124	1.55961	1.163	1	1.1630E-03
3412142	1895634	1E+06	859845	737562	101972	1359.62	1341.02	1E+03	859.845	1

Notes: 1 W = 1 J/s

1 cal /s = 3.6 kcal /h

1 ton of refrigeration = 3517 W = 12000 Bth /h

1 erg /s = 1E-07 W

Table A.17 DYNAMIC VISCOSITY (μ)

N.s /m2	*kg /m.h*	*lbm /ft.h*	*cP*	*P*	*Pa.s*	*pdl.s /ft2*	*kgf.s /m2*	*lbf.s /ft2*	*lbf.h /ft2*
1	3.6E-03	2.4191E-03	1E-03	1E-05	1E-06	6.7197E-07	1.0197E-07	2.0885E-08	5.8015E-12
277.778	1	0.671969	0.277778	2.7778E-03	2.7778E-04	1.8666E-04	2.8325E-05	5.8015E-06	1.6115E-09
413.379	1.48816	1	0.413379	4.1338E-03	4.1338E-04	2.7778E-04	4.2153E-05	8.6336E-06	2.3982E-09
1E+03	3.6	2.41909	1	0.01	1E-03	6.7197E-04	1.0197E-04	2.0885E-05	5.8015E-09
1E+05	360	241.909	100	1	0.1	6.7197E-02	1.0197E-02	2.0885E-03	5.8015E-07
1E+06	3600	2419.09	1E+03	10	1	0.671969	0.101972	2.0885E-02	5.8015E-06
1488164	5357.39	3600	1488.16	14.8816	1.48816	1	0.151750	3.1081E-02	8.6336E-06
9806650	35303.9	23723.2	9806.65	98.0665	9.80665	6.58976	1	0.204816	5.6893E-05
4.7880E+07	172369	115827	47880.3	478.803	47.8803	32.1740	4.88243	1	2.7778E-04
1.7237E+11	6.2053E+08	4.1698E+08	1.7237E+08	1723689	172369	115827	17576.7	3600	1

Notes:

1 cP ("centipoise") = 1 mN.s /m2 = 1 g /m.s
1 P ("poise") = 1 g /cm.s = 1 dyn.s /cm2
1 Pa.s = 1 N.s /m2 = 1 kg /m.s
1 pdl.s /ft2 = 1 lbm /ft.s
1 lbf.s /ft2 = 1 slug /ft.s

Table A.18 KINEMATIC VISCOSITY (μ/ρ)

in2 /h	*mm2 /s (*)*	*ft2 /h*	*cm2 /s (*)*	*m2 /h*	*in2 /s*	*ft2 /s*	*m2 /s*
1	0.179211	6.9444E-03	1.7921E-03	6.4516E-04	2.7778E-04	1.9290E-06	1.7921E-07
5.58001	1	3.8750E-02	1E-02	3.6000E-03	1.5500E-03	1.0764E-05	1E-06
144	25.8064	1	0.25806	9.2903E-02	0.04	2.7778E-04	2.5806E-05
558.001	100	3.87501	1	0.36	0.15500	1.0764E-03	1E-04
1550.00	277.778	10.7639	2.77778	1	0.43056	2.9900E-03	2.7778E-04
3600	645.160	25	6.45160	2.32258	1	6.9444E-03	6.4516E-04
518400	92903.0	3600	929.030	334.451	144	1	9.2903E-02
5580011	1E+06	38750.1	1E+04	3600	1550.00	10.7639	1

Notes:

1 mm2 /s = 1 cSt (centistokes)
1 cm2 /s = 1 St (stokes)

Table A.19 DENSITY OF HEAT FLOW RATE (heat /area /time)

W /m2	*kcal /m2.h*	*Btu /ft2.h*	*Chu /ft2.h*	*kcal /ft2.h*	*kW /m2*
1	0.859845	0.316998	0.176110	7.9882E-02	1E-03
1.163	1	0.368669	0.204816	9.2903E-02	1.1630E-03
3.15459	2.71246	1	0.555556	0.251996	3.1546E-03
5.67826	4.88243	1.8	1	0.453592	5.6783E-03
12.5184	10.7639	3.96832	2.20462	1	1.2518E-02
1000	859.845	316.998	176.110	79.8822	1

Notes: this quantity is also called "ENERGY FLUX"

Table A.20 HEAT TRANSFER COEFFICIENT (heat /area /time /temperature_interval)

W /m2.K	*kcal /m2.h.°C*	*Btu /ft2.h.°F*	*kcal /ft2.h.°C*	*kW /m2.K*	*Btu /ft2.s.°F*	*kcal /cm2.s.°C*
1	0.859845	0.176110	7.9882E-02	1E-03	4.8919E-05	2.3885E-05
1.163	1	0.204816	9.2903E-02	1.1630E-03	5.6893E-05	2.7778E-05
5.67826	4.88243	1	0.453592	5.6783E-03	2.7778E-04	1.3562E-04
12.5184	10.7639	2.20462	1	1.2518E-02	6.1240E-04	2.9900E-04
1E+03	859.845	176.110	79.8822	1	4.8919E-02	2.3885E-02
20441.7	17576.7	3600	1632.93	20.4417	1	0.488243
41868.0	36000	7373.38	3344.51	41.8680	2.04816	1

Notes: this quantity is also called "THERMAL CONDUCTANCE"
a temperature interval of 1°C (degree Celsius) is the same as 1 K (kelvin)
a temperature interval of 1°C (degree Celsius) is the same as 1.8°F (degree Fahrenheit)
temperature conversion from Fahrenheit ("TF") to Celsius ("TC"): TC(°C) = [TF(°F)-32] / 1.8
temperature conversion from Celsius ("t") to kelvin ("T"): T(K) = t(°C) + 273.16
1 Btu /ft2.h.°F = 1 Chu /ft2.h.°C
1 W /m2.°C = 1E-04 W /cm2.°C

Table A.21 THERMAL CONDUCTIVITY (heat x length /area /time /temperature_interval)

Btu.in /ft2.h.°F	*kcal.in /ft2.h.°C*	*W /m.K*	*kcal /m.h.°C*	*Btu /ft.h.°F*	*cal /cm.s.°C*
1	0.453592	0.144228	0.124014	8.3333E-02	3.4448E-04
2.20462	1	0.317968	0.273403	0.183719	7.5945E-04
6.93347	3.14497	1	0.859845	0.577789	2.3885E-03
8.06363	3.65760	1.16300	1	0.671969	2.7778E-03
12	5.44311	1.73073	1.48816	1	4.1338E-03
2902.91	1316.74	418.680	360	241.909	1

Notes: 1 Btu.in /ft2.h.°F = 1.0000 Chu.in /ft2.h.°C
1 Btu /ft.h.°F = 1.0 Chu /ft.h.°C
1 W /m.°C = 1E-02 W /cm.°C = 1.00 kW.mm /m2.°C

Table A.22 SPECIFIC HEAT CAPACITY (heat /mass /temperature_interval)

J /kg.K	*ft.lbf /lbm.°F*	*kgf.m /kg.°C*	*kJ /kg.K*	*Btu /lbm.°F*	*kcal /kg.°C*
1	0.185863	0.101972	1E-03	2.3885E-04	2.3885E-04
5.38032	1	0.548640	5.3803E-03	1.2851E-03	1.2851E-03
9.80665	1.82269	1	9.8067E-03	2.3423E-03	2.3423E-03
1E+03	185.863	101.972	1	0.238846	0.238846
4186.80	778.169	426.935	4.18680	1	1.00000
4186.80	778.169	426.935	4.18680	1.00000	1

Notes: 1 Btu /lbm.°F = 1 Chu /lbm.°C

Table A.23 SPECIFIC ENERGY (heat /mass)

J /kg	*ft.lbf /lbm*	*kgf.m /kg*	*kJ /kg*	*Btu /lbm*	*kcal /kg*	*MJ /kg*
1	0.334553	0.101972	1E-03	4.2992E-04	2.3885E-04	1E-06
2.98907	1	0.304800	2.9891E-03	1.2851E-03	7.1393E-04	2.9891E-06
9.80665	3.28084	1	9.8067E-03	4.2161E-03	2.3423E-03	9.8067E-06
1E+03	334.553	101.972	1	0.429923	0.238846	1E-03
2326	778.169	237.186	2.326	1	0.555556	2.3260E-03
4186.80	1400.70	426.935	4.18680	1.8	1	4.1868E-03
1E+06	334553	101972	1E+03	429.923	238.846	1

Notes: examples of quantities with this unit include:
"CALORIFIC VALUE, MASS BASIS"
"SPECIFIC LATENT HEAT"
1 J /g = 1 kJ /kg
1 kcal /kg = 1 Chu /lbm

Table A.24 VOLUMETRIC ENERGY (heat /volume)

J /m3	*kJ /m3*	*kcal /m3*	*Btu /ft3*	*Chu /ft3*	*MJ /m3*
1	1E-03	2.3885E-04	2.6839E-05	1.4911E-05	1E-06
1E+03	1	0.238846	2.6839E-02	1.4911E-02	1E-03
4186.80	4.18680	1	0.112370	6.2428E-02	4.1868E-03
37258.9	37.2589	8.89915	1	0.555556	3.7259E-02
67066.1	67.0661	16.0185	1.8	1	6.7066E-02
1E+06	1E+03	238.846	26.8392	14.9107	1

Notes: examples of quantities with this unit include:
"CALORIFIC VALUE, VOLUME BASIS"
1 therm (100000 Btu) /UK gal = 23208 MJ /m3
1 thermie /litre = 4185 MJ /m3
1 MJ /m3 = 1 J /cm3

APPENDIX B: STANDARD SIZES AND PREFERRED NUMBER SERIES

Standard Sizes

Standardised sizes and modular components are widely used in modern engineering and design practice. The ongoing processes of standardisation and modularity are reflected in the standard sizing applied to the myriad range of consumer and industrial products. Historically this approach has evolved from the standardisation and mechanisation techniques developed at the US Armory at Springfield, Pennsylvania, prior to and during the United States Civil War (1861–1865). The modern application of such techniques is popularly attributed to Henry Ford and the Ford Motor Company and is a reflection of the ongoing need for engineers to achieve efficient, elegant and economical design solutions.

The selection of standard sizes follows two broad approaches: *dimensional coordination* and *preferred numbers*.

Dimensional Coordination

The application of dimensional coordination varies with the engineering discipline. It is a tool primarily for architects, builders and the industries supplying building components. Products should be based on a standard module or appropriate sub-modules and multi-modules. With the increased use of prefabricated and pre-assembled units in the building industry it is important that dimensional coordination is used to determine component sizes for the industry.

The International Standards Organisation and many national standards bodies have adopted a base module of 100 mm.

Preferred horizontal multi-modules are 300 mm, 600 mm, 900 mm, 1200 mm, 1500 mm, 3000 mm, and 6000 mm. Preferred vertical multi-modules are 100 mm, 300 mm and 600 mm. Increments of 100 mm are recommended up to 3000 mm maximum.

The system of dimensional coordination provides for controlling planes in both the vertical and horizontal directions. Controlling planes can represent the outer faces of load-bearing walls or columns, the floor level and the underside of ceilings. The distance between controlling planes across a wall column or floor is a *control zone* and between two adjacent *controlling planes* bounding control zones is the *controlling dimension*. The controlling dimension is made up of one or more *coordinating dimensions*. The dimensions of standard components should be based on coordinating dimensions.

The dimensions of components are derived from the modular coordination space they are to occupy. The modular coordinating dimension should be an appropriate multiple or sub-multiple of the base module. The manufacturing size of the component provides for manufacturing tolerances for the component, positioning tolerances and joint allowance which may be positive (i.e. an overlap) or negative (i.e. a clearance). The component size is usually designated by the nominal size derived from the coordinating dimensions.

Preferred Numbers

To promote efficiency and economy in design, designers tend to use convenient *round* numbers for the dimensions they choose. The notion of a *round number* is formalised in so-called *tables of preferred numbers*. Such tables of numbers give guidance to designers in choosing convenient values for dimensional sizes. The system of preferred sizes is based on a number of geometric series derived by Col. Charles Renard in the 1870s. They have been adopted and promulgated by various government and international standards organisations such as the ISO (International Organisation for Standardisation), ANSI (American National Standards Institute), SAE (Society of Automotive Engineers), BSI (British Standards Institution), and Standards Australia.

Preferred sizes based on a Renard series of preferred numbers have more significance in engineering than does dimensional coordination. They are used in equipment rating and sizing and the sizing of semi-finished products used generally by engineering industry. The common series and their ratios are given in the table below.

This table lists the preferred numbers in the *R5*, *R10*, *R20* and *R40* series and gives rounded values which may be preferable under some circumstances. The first rounding is indicated as *R'* and the more rounded series by *R"*.

The Renard series are given names such as *R 5*, *R 10*, *R 20*. This naming reflects the number of intervals per decade in the series:

- *R 5* gives 5 increments in each decade (i.e. 5 increments between 1 and 10; 5 increments between 10 and 100; 5 increments between 100 and 1000, etc);
- *R 10* gives 10 increments in each decade;
- *R 20* gives 20 increments in each decade, and so on.

The series may be used by manufacturers for the determination of product size ranges and by users in design selection to reduce the factory stores inventory. The decision required is how fine or coarse the size progression should be and the upper and lower limits of size.

APPENDIX B: COMMON RENARD SERIES

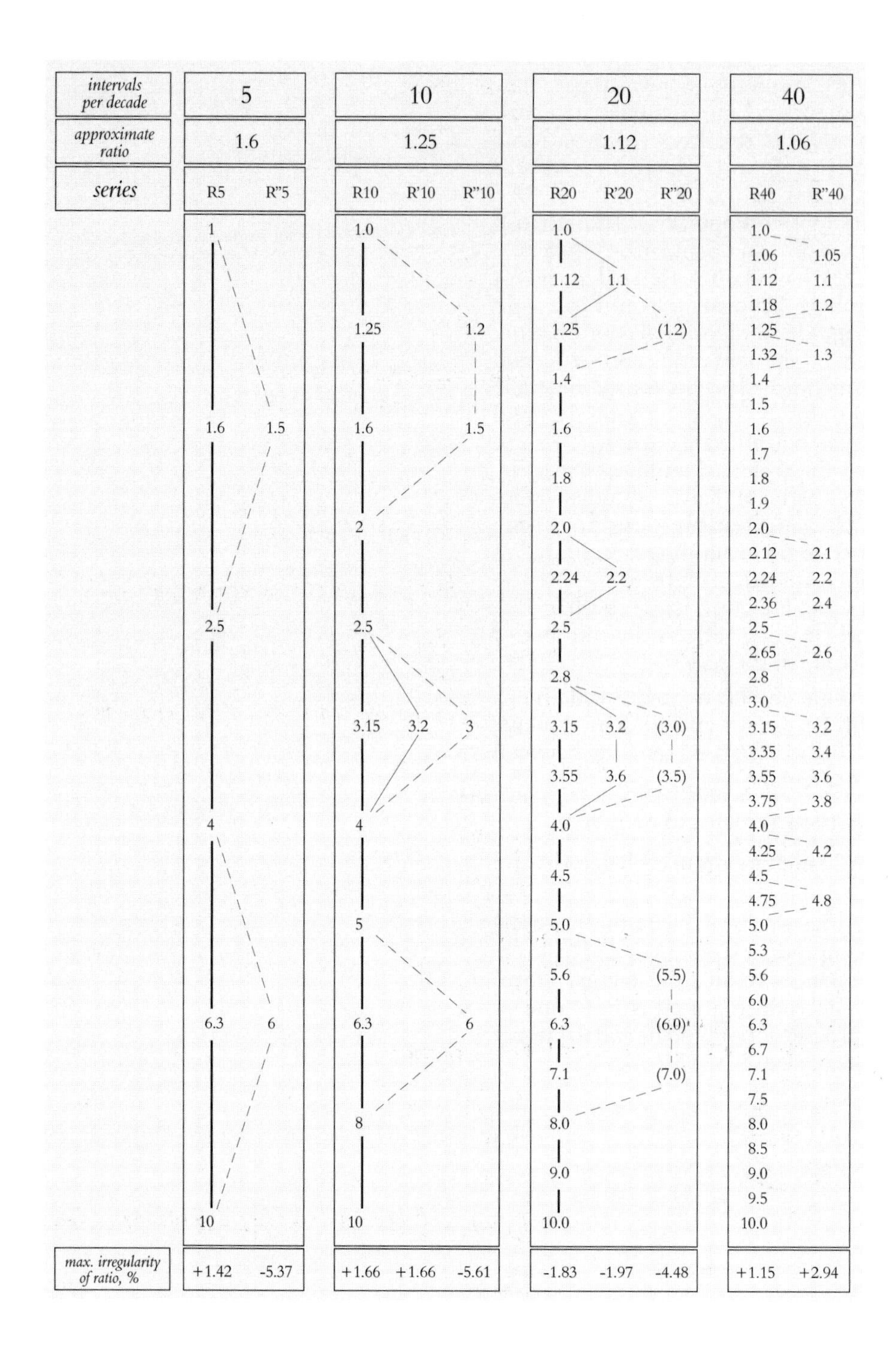

intervals per decade	5		10			20			40	
approximate ratio	1.6		1.25			1.12			1.06	
series	R5	R"5	R10	R'10	R"10	R20	R'20	R"20	R40	R"40
	1		1.0			1.0			1.0	
									1.06	1.05
						1.12	1.1		1.12	1.1
									1.18	1.2
			1.25		1.2	1.25		(1.2)	1.25	
									1.32	1.3
						1.4			1.4	
									1.5	
	1.6	1.5	1.6		1.5	1.6			1.6	
									1.7	
						1.8			1.8	
									1.9	
			2			2.0			2.0	
									2.12	2.1
						2.24	2.2		2.24	2.2
									2.36	2.4
	2.5		2.5			2.5			2.5	
									2.65	2.6
						2.8			2.8	
									3.0	
			3.15	3.2	3	3.15	3.2	(3.0)	3.15	3.2
									3.35	3.4
						3.55	3.6	(3.5)	3.55	3.6
									3.75	3.8
	4		4			4.0			4.0	
									4.25	4.2
						4.5			4.5	
									4.75	4.8
			5			5.0			5.0	
									5.3	
						5.6		(5.5)	5.6	
									6.0	
	6.3	6	6.3		6	6.3		(6.0)	6.3	
									6.7	
						7.1		(7.0)	7.1	
									7.5	
			8			8.0			8.0	
									8.5	
						9.0			9.0	
									9.5	
	10		10			10.0			10.0	
max. irregularity of ratio, %	+1.42	-5.37	+1.66	+1.66	-5.61	-1.83	-1.97	-4.48	+1.15	+2.94

APPENDIX C: PROPERTIES OF SECTIONS

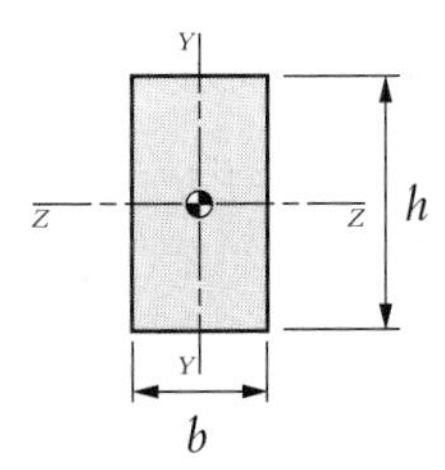

rectangular:

$A = bh$

$\bar{y} = 0$

$\bar{z} = 0$

$I_{yy} = \frac{hb^3}{12}$

$I_{zz} = \frac{bh^3}{12}$

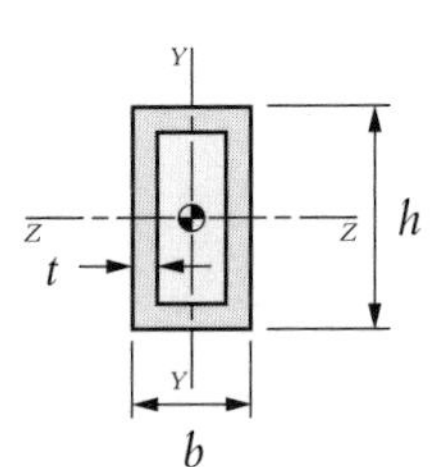

hollow rectangular:

$A = 2t(b + h - 2t)$

$\bar{y} = 0$

$\bar{z} = 0$

$$I_{yy} = \frac{1}{12}\left[hb^3 - (h-2t)(b-2t)^3\right] = \frac{t}{6}\left(3hb^2 - 6hbt + 4ht^2 + b^3 - 6b^2t + 12bt^2 - 8t^3\right)$$

$$I_{zz} = \frac{1}{12}\left[bh^3 - (b-2t)(h-2t)^3\right] = \frac{t}{6}\left(3bh^2 - 6bht + 4bt^2 + h^3 - 6h^2t + 12ht^2 - 8t^3\right)$$

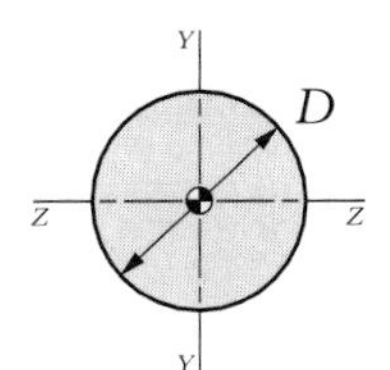

circular:

$A = \frac{\pi D^2}{4}$

$\bar{y} = 0$

$\bar{z} = 0$

$I_{yy} = I_{zz} = \frac{\pi D^4}{64}$

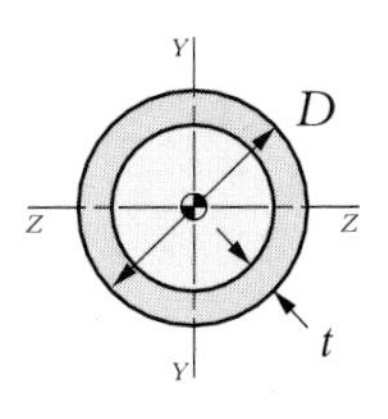

hollow circular:

$A = \frac{\pi}{4}\left(D^2 - d^2\right) = \pi(D - t)t$

$\bar{y} = 0$

$\bar{z} = 0$

$I_{yy} = I_{zz} = \frac{\pi}{64}\left(D^4 - d^4\right) = \frac{\pi t}{8}(D - t)\left(D^2 - 2Dt + 2t^2\right)$

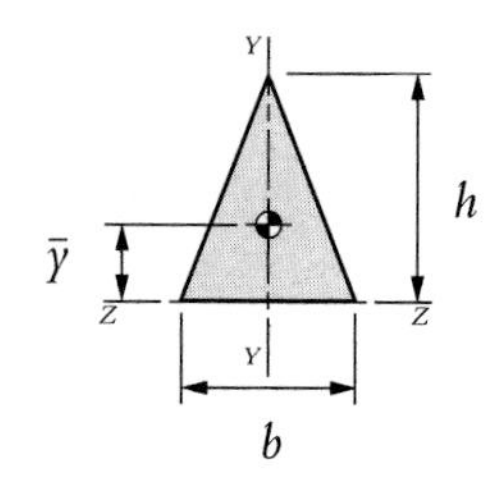

triangular:

$$A = \frac{bh}{2}$$

$$\bar{y} = \frac{h}{3} \quad ; \qquad \bar{z} = 0$$

$$I_{GG} = \frac{bh^3}{36}$$

$$I_{yy} = I_{GG} + A\bar{y}^2 = \frac{bh^3}{12} \quad ; \qquad I_{zz} = \frac{hb^3}{48}$$

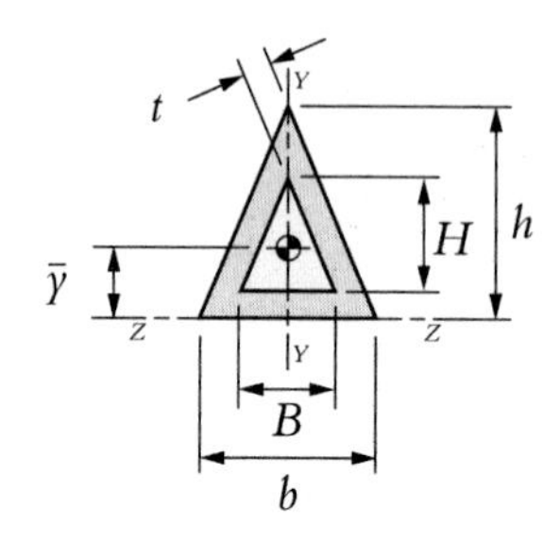

hollow triangular (constant thickness):

$$\lambda = 1 - (t/h) - \sqrt{(2t/w)^2 + (t/h)^2} \quad ; W = \lambda w \quad ; \quad H = \lambda h$$

$$A = \frac{wh}{2}\left(1 - \lambda^2\right)$$

$$\bar{y} = \frac{h}{3}\frac{1 - \lambda^2(\lambda + 3t/h)}{1 - \lambda^2} \quad ; \qquad \bar{z} = 0$$

$$I_{GG} = \frac{bh^3}{36}\left[3 - \lambda^4 - 2\lambda^2(\lambda + 3t/h)^2 - \frac{2}{1 - \lambda^2}\left[1 - \lambda^2(\lambda + 3t/h)\right]^2\right]$$

$$I_{yy} = \frac{hb^3}{48}\left(1 - \lambda^4\right) \quad ; \qquad I_{zz} = \frac{bh^3}{36}\left[3 - \lambda^4 - 2\lambda^2(\lambda + 3t/h)^2\right]$$

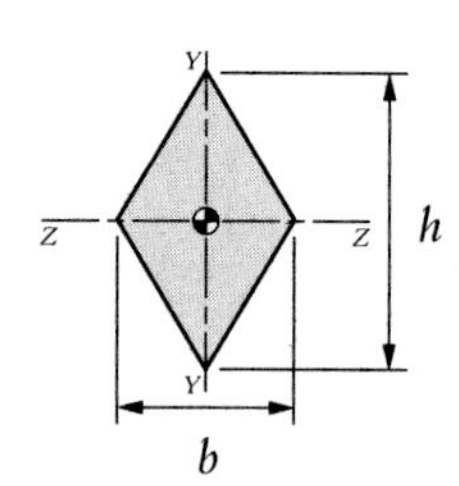

diamond:

$$A = \frac{bh}{2}$$

$$\bar{y} = 0 \quad ; \qquad \bar{z} = 0$$

$$I_{yy} = \frac{hb^3}{48} \quad ; \qquad I_{zz} = \frac{bh^3}{48}$$

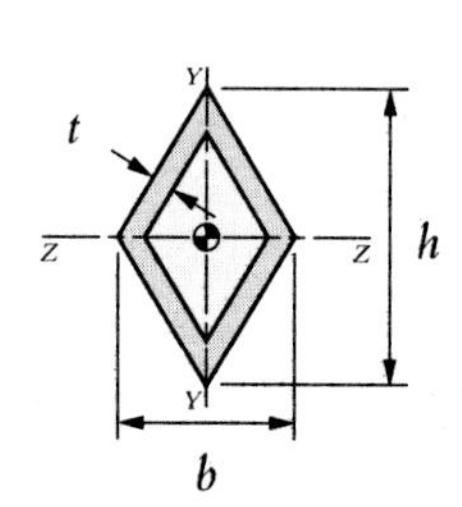

hollow diamond:

$$\zeta = 1 - \frac{2t}{w}\sqrt{1 + (w/h)^2}$$

$$A = \frac{bh}{2}\left(1 - \zeta^2\right)$$

$$\bar{y} = 0 \quad ; \qquad \bar{z} = 0$$

$$I_{yy} = \frac{hb^3}{48}\left(1 - \zeta^4\right) \quad ; \qquad I_{zz} = \frac{bh^3}{48}\left(1 - \zeta^4\right)$$

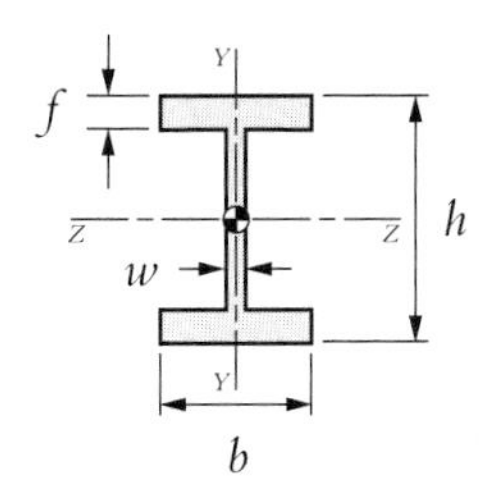

I - beam:

$$A = 2bf + h\left(w - 2f\right)$$

$$\bar{y} = 0 \quad ; \qquad \bar{z} = 0$$

$$I_{yy} = \frac{2fb^3 + \left(h - 2f\right)w^3}{12}$$

$$I_{zz} = \frac{wh^3 + \left(b - w\right)2f\left(3h^2 - 6hf + 4f^2\right)}{12}$$

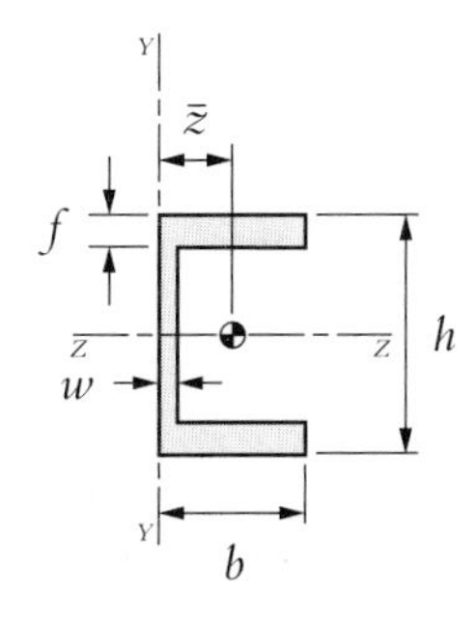

channel:

$$A = hw + 2f\left(b - w\right)$$

$$\bar{y} = 0 \quad ; \qquad \bar{z} = \frac{1}{2} \cdot \frac{hw^2 + 2f\left(b^2 - w^2\right)}{hw + 2f\left(b - w\right)}$$

$$I_{GG} = I_{yy} - A\bar{z}^2$$

$$I_{yy} = \frac{2fb^3 + \left(h - 2f\right)w^3}{3}$$

$$I_{zz} = \frac{bh^3 - \left(b - w\right)\left(h - 2f\right)^3}{12}$$

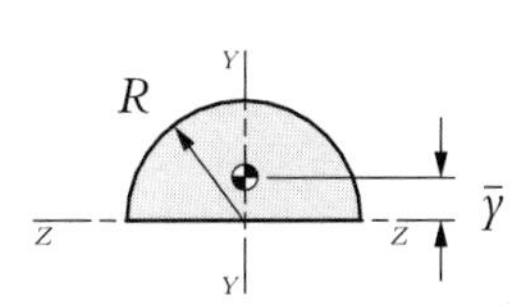

semi - circular:

$$A = \frac{\pi R^2}{2}$$

$$\bar{y} = \frac{4R}{3\pi} \quad ; \qquad \bar{z} = 0$$

$$I_{GG} = R^4 \frac{\left(3\pi - 8\right)\left(3\pi + 8\right)}{72\pi}$$

$$I_{yy} = I_{zz} = \frac{\pi R^4}{8}$$

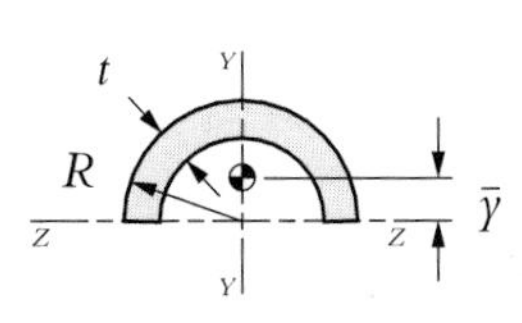

hollow semi - circular:

$$\chi = \left(t/R\right)$$

$$A = \frac{\pi R t}{2}\left(2 - \chi\right)$$

$$\bar{y} = \frac{4R}{3\pi} \cdot \frac{3 - 3\chi + \chi^2}{2 - \chi} \quad ; \qquad \bar{z} = 0$$

$$I_{GG} = R^3 t\left(2 - \chi\right)\left\{\frac{\pi}{8}\left[\left(1 - \chi\right)^2 + 1\right] - \frac{8}{9\pi}\left[\left(1 - \chi\right)^2 + 1 + \frac{\left(1 - \chi\right)^2}{\left(2 - \chi\right)^2}\right]\right\}$$

$$I_{yy} = I_{zz} = \frac{\pi R^3 t}{2}\left(1 - \frac{3\chi}{2} + \chi^2 - \frac{\chi^3}{4}\right)$$

APPENDIX D: BEAM FORMULAE

1. ***concentrated load cantilever beam***

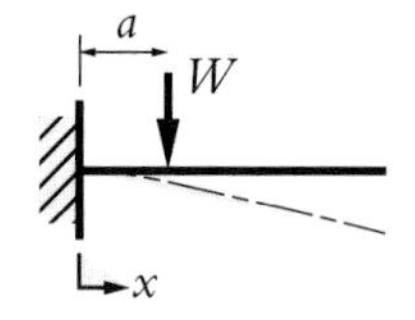

2. ***distributed load cantilever beam***

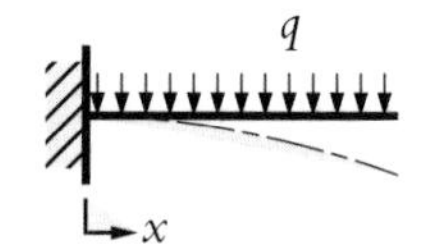

3. ***moment load cantilever beam***

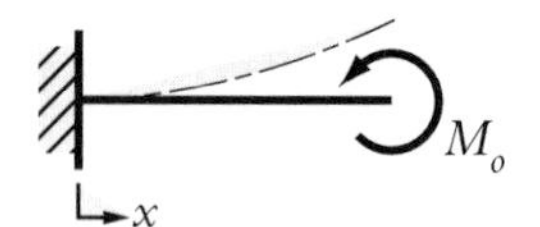

4a. ***concentrated load simple supports***

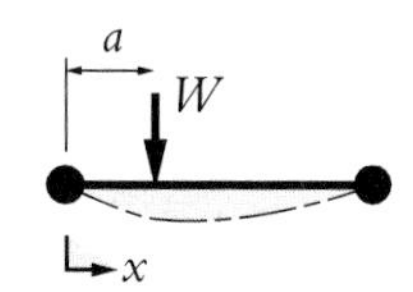

5. ***distributed load simple supports***

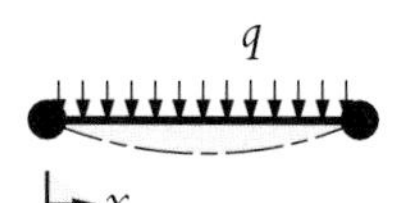

6. ***moment load simple supports***

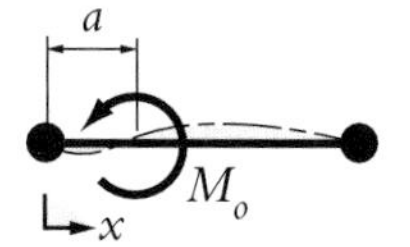

4b. ***concentrated load simple supports (overhung)***

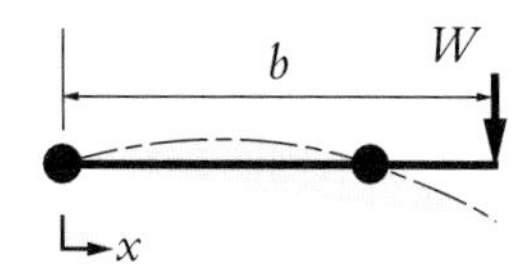

7. ***concentrated load fixed supports***

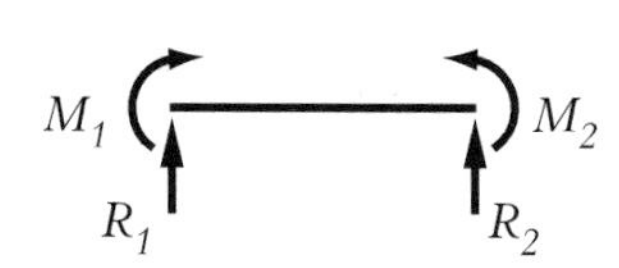

8. ***distributed load fixed supports***

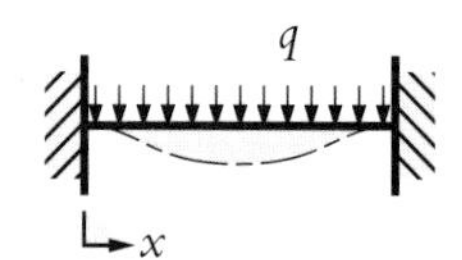

9. ***moment load fixed supports***

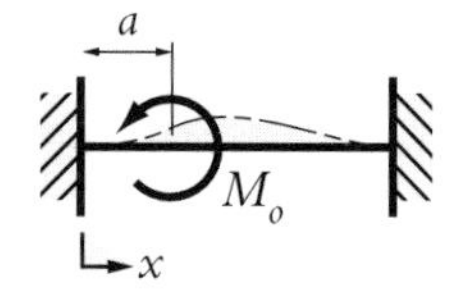

external load conventions:

variable conventions:

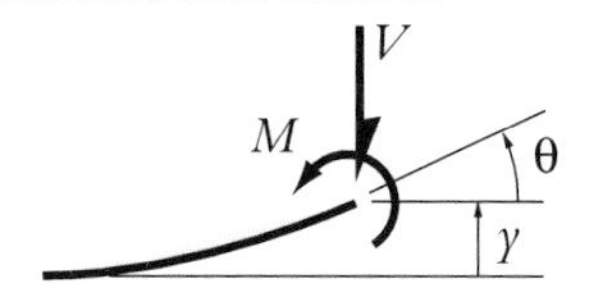

1. concentrated load cantilever beam

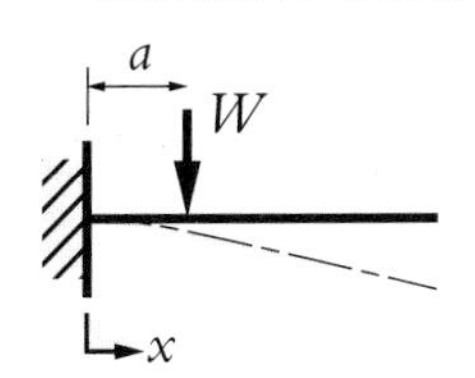

reactions: $R_1 = W$; $(R_2 = 0)$

$M_1 = -Wa$; $(M_2 = 0)$

$(0 \le x \le a)$:

$$V = W$$
$$M = W(x - a)$$
$$\theta = \frac{Wx}{2EI}(x - 2a)$$
$$y = \frac{Wx^2}{6EI}(x - 3a)$$

$(a < x \le L)$:

$$V = 0$$
$$M = 0$$
$$\theta = -\frac{Wa^2}{2EI}$$
$$y = \frac{Wa^2}{6EI}(a - 3x)$$
$$y_{\max} = \frac{Wa^2}{6EI}(a - 3L) \quad @x = L$$

2. distributed load cantilever beam

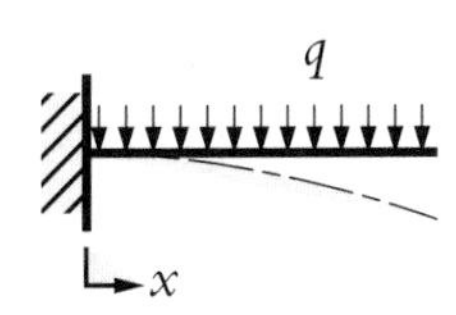

reactions: $R_1 = qL$; $(R_2 = 0)$

$M_1 = -\frac{qL^2}{2}$; $(M_2 = 0)$

$(0 \le x \le L)$:

$$V = q(L - x)$$
$$M = -\frac{q}{2}(L - x)^2$$
$$\theta = \frac{qx}{6EI}(3Lx - x^2 - 3L^2)$$
$$y = \frac{qx^2}{24EI}(4Lx - x^2 - 6L^2)$$
$$y_{\max} = -\frac{qL^4}{8EI} \quad @x = L$$

3. *moment load cantilever beam*

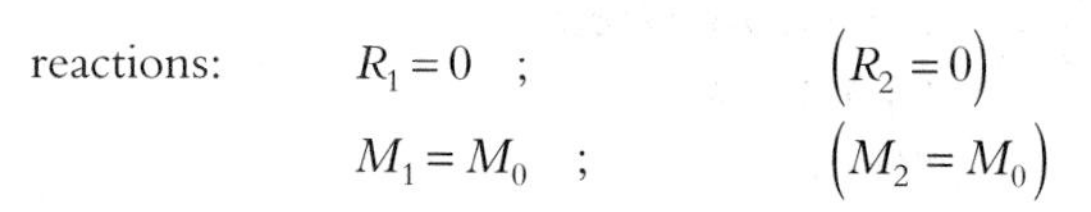

reactions: $R_1 = 0$; $(R_2 = 0)$

$M_1 = M_0$; $(M_2 = M_0)$

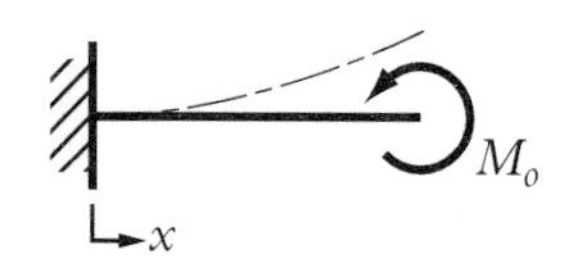

$(0 \le x \le L)$:

$$V = 0$$

$$M = M_0$$

$$\theta = \frac{M_0 x}{EI}$$

$$y = \frac{M_0 x^2}{2EI}$$

$$y_{\max} = \frac{M_0 L^2}{2EI} \quad @ x = L$$

4a. *concentrated load simple supports*

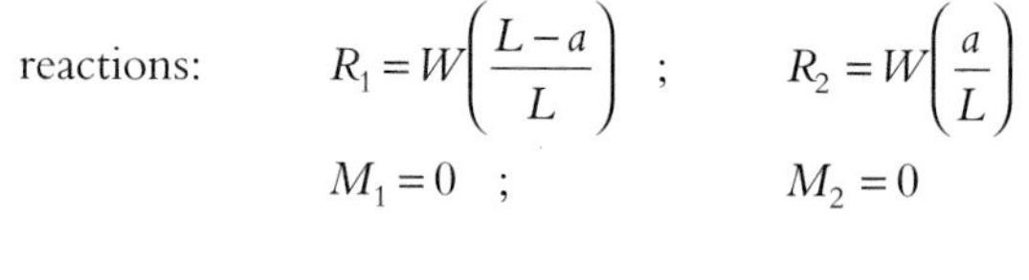

reactions: $R_1 = W\left(\frac{L-a}{L}\right)$; $R_2 = W\left(\frac{a}{L}\right)$

$M_1 = 0$; $M_2 = 0$

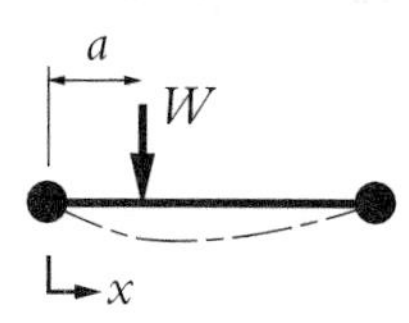

$(0 \le x \le a)$:

$$V = R_1$$

$$M = W\left(\frac{L-a}{L}\right)x$$

$$\theta = \frac{W(L-a)}{6EIL}\left[3x^2 + a(a-2L)\right]$$

$$y = \frac{W(L-a)x}{6EIL}\left[x^2 + a(a-2L)\right]$$

$(a < x \le L)$:

$$V = -R_2$$

$$M = W\left(\frac{a}{L}\right)(L-x)$$

$$\theta = \frac{Wa}{6EIL}\left[x(4L-3x) - a^2\right]$$

$$y = \frac{Wa(L-x)}{6EIL}\left[a^2 - x(2L-x)\right]$$

$$y_{\max} = -\frac{WL^3}{48EI} \quad @ x = \frac{L}{2}$$

4b. ***concentrated load simple supports (overhung)***

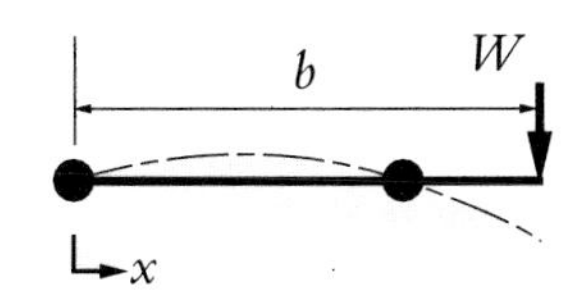

reactions: $R_1 = W\left(\frac{L-b}{L}\right)$; $R_2 = W\left(\frac{b}{L}\right)$

$M_1 = 0$; $M_2 = 0$

$(0 \le x \le L)$:

$$V = R_1$$

$$M = W\left(\frac{L-b}{L}\right)x$$

$$\theta = \frac{W(b-L)}{6EIL}\left(L^2 - 3x^2\right)$$

$$y = \frac{W(b-L)x}{6EIL}\left(L^2 - x^2\right)$$

$(L < x \le b)$:

$$V = W$$

$$M = W(x-b)$$

$$\theta = \frac{W}{6EI}\left[3x^2 - 6bx + L(4b-L)\right]$$

$$y = \frac{W(x-L)}{6EI}\left[x^2 + x(L-3b) + Lb\right]$$

$$y_{max} = -\frac{Wb(b-L)^2}{3EI} \quad @\, x = b$$

5. ***distributed load simple supports***

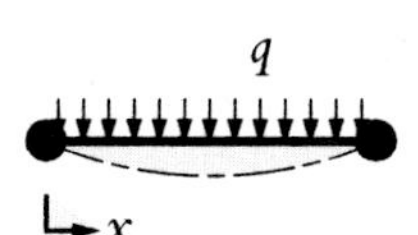

reactions: $R_1 = \frac{qL}{2}$; $R_2 = \frac{qL}{2}$

$M_1 = 0$; $M_2 = 0$

$(0 \le x \le L)$:

$$V = q\left(\frac{L}{2} - x\right)$$

$$M = \frac{qx}{2}(L-x)$$

$$\theta = \frac{q}{24EI}\left(6Lx^2 - 4x^3 - L^3\right)$$

$$y = \frac{qx}{24EI}\left(2Lx^2 - x^3 - L^3\right)$$

$$y_{max} = -\frac{5qL^4}{384EI} \quad @\, x = \frac{L}{2}$$

6. *moment load simple supports*

reactions: $R_1 = \dfrac{M_0}{L}$; $R_2 = -\dfrac{M_0}{L}$

$M_1 = 0$; $M_2 = 0$

$(0 \le x \le a)$:

$$V = \frac{M_0}{L}$$

$$M = M_0\left(\frac{x}{L}\right)$$

$$\theta = \frac{M_0}{6EIL}\left(3x^2 + 3a^2 - 6aL + 2L^2\right)$$

$$y = \frac{M_0 x}{6EIL}\left(x^2 + 3a^2 - 6aL + 2L^2\right)$$

$(a < x \le L)$:

$$V = \frac{M_0}{L}$$

$$M = M_0\left(\frac{x - L}{L}\right)$$

$$\theta = \frac{M_0}{6EIL}\left(3x^2 - 6Lx + 2L^2 + 3a^2\right)$$

$$y = \frac{M_0}{6EIL}\left[x^3 - 3Lx^2 + x\left(2L^2 + 3a^2\right) - 3a^2L\right]$$

7. *concentrated load fixed supports*

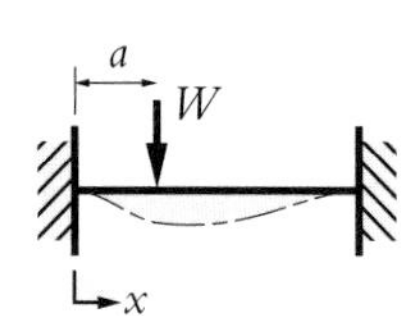

reactions: $R_1 = \dfrac{W(L-a)^2(2a+L)}{L^3}$; $R_2 = \dfrac{Wa^2(3L-2a)}{L^3}$

$M_1 = -\dfrac{Wa(L-a)^2}{L^2}$; $M_2 = \dfrac{Wa^2(L-a)}{L^2}$

$(0 \le x \le a)$:

$$V = R_1$$

$$M = \frac{W(L-a)^2}{L^3}\left[x(2a+L) - aL\right]$$

$$\theta = \frac{W(L-a)^2 x}{2EIL^3}\left[x(2a+L) - 2aL\right]$$

$$y = \frac{W(L-a)^2 x^2}{6EIL^3}\left[x(2a+L) - 3aL\right]$$

$(a < x \le L)$:

$$V = -R_2$$

$$M = W\left\{\frac{(L-a)^2}{L^3}\left[x(2a+L) - aL\right] - (x-a)\right\}$$

$$\theta = \frac{Wa^2}{2EIL^3}\left[2L(2L-a)x - (3L-2a)x^2 - L^3\right]$$

$$y = \frac{Wa^2(L-x)^2}{6EIL^3}\left[x(2a-3L) + aL\right]$$

8. ***distributed load fixed supports***

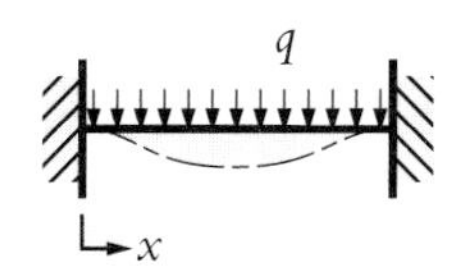

reactions: $R_1 = \frac{qL}{2}$; $R_2 = \frac{qL}{2}$

$M_1 = -\frac{qL^2}{12}$; $M_2 = -\frac{qL^2}{12}$

$(0 \le x \le L)$:

$$V = \frac{q}{2}(L - 2x)$$

$$M = \frac{q}{12}(6Lx - 6x^2 - L^2)$$

$$\theta = -\frac{qx}{12EI}(L - 2x)(L - x)$$

$$y = -\frac{qx^2}{24EI}(L - x)^2$$

$$y_{max} = -\frac{qL^4}{384EI} \quad @\, x = \frac{L}{2}$$

9. ***moment load fixed supports***

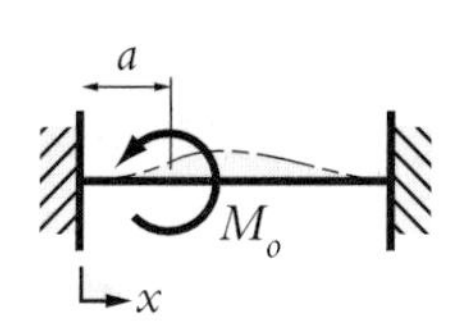

reactions: $R_1 = M_0 \frac{6a(L-a)}{L^3}$; $R_2 = -R_1$

$M_1 = M_0 \frac{(L-3a)(L-a)}{L^2}$; $M_2 = M_0 \frac{(2L-3a)a}{L^2}$

$(0 \le x \le a)$:

$$V = R_1$$

$$M = \frac{M_0(L-a)}{L^2}\left[\frac{6a}{L}x + (L-3a)\right]$$

$$\theta = \frac{M_0(L-a)}{EIL^2}\left[\frac{3a}{L}x^2 + (L-3a)x\right]$$

$$y = \frac{M_0(L-a)}{EIL^2}\left[\frac{a}{L}x^3 + \frac{(L-3a)}{2}x^2\right]$$

$(a < x \le L)$:

$$V = R_1$$

$$M = \frac{M_0 a}{L^2}\left[\frac{6a(L-a)}{L}x - (4L-3a)\right]$$

$$\theta = \frac{M_0 a}{EIL^2}\left[\frac{3(L-a)}{L}x^2 - (4L-3a)x + L^2\right]$$

$$y = \frac{M_0 a}{EIL^2}\left[\frac{(L-a)}{L}x^3 - \frac{(4L-3a)}{2}x^2 + L^2x + \frac{aL^2}{2}\right]$$

AUTHOR INDEX

Adams, J.L. 297, 299, 302, 305
Adams, S. 359
Agricola 112, 113
Alexander, R.C. 300
Ashford, F. 357
Astridge, D.G. 363
Barnes, R.M. 348
Bauer, G. 112
Beard, A.E. 289
Beitz, W. 292
Benham, P.P. 264
Bernoulli, D. 144
Birsch, D. 345
Bloom, B.S. 241
Box, G.E.P. 332
Buzan, A. 291
Carroll, L. 323
Chase, W. G. 290
Clapeyron, B.P.E. 166
Clark, G.E. 363
Clarke, A. C. 299
Clarke, R.L. 289
Collins, J.A. 82, 83, 85, 86, 89
Cooper, D.F. 291
Cotgrove, S. 356
Crowther, P. 357
Damon, A. 354
deBono, E. 292, 299, 339
Dinnik, A.N. 108
Dipert, R.R. 357
Dowie, M. 345
Dreyfuss, H. 354
Dunham, J. 356
Einstein, A. 242
Emde, F. 103
Emerson, R.W. 275
Euler, L. 144, 147
Farquharson, F.B. 4
Field, B.W. 332
Fielder, J.H. 345
Fiore, Q. 270, 271
FItzgerald, E. 3
Flügge, W. 144
French, M. J. 271, 272, 273, 292
Frost, N.E. 89
Fuchs, H.O. 85
Gale, W.K.V. 112
Giedion, S. 348
Goldberg, D.E. 330
Goldsmith, S. 357
Goodier, J.P. 87, 103, 108
Goodman, J. 40
Gordon, J.E. 265
Hatty, M. 363
Hawkins, P.G. 363
Hertz, J. 101, 107
Herzberg, F.M.B. 356
Hewat, T. 4
Holgate, A. 357
Holick, M. 329, 339, 344
Holmes, O.W. 240
Horne 270
Horrigan, D. 87
Huber, M.T. 108
Hunter, J.S. 332
Hunter, W.G. 332
Infeld, L. 242
Ishibashi, A. 108
Ishikawa, K. 327
Jahnke, E. 103

Juvinall, C.R. 86
Kelly, F.C. 239, 272
Klein, J.H. 291
Koestler, A. 348
Kratwohl, D.R. 241
Kroemer, K.H.E. 348
Kryter, K.D. 357
Kusiak, A. 269
Lagrange, J.L. 144
Lamé, G. 166
Landgraf, R.W. 82
Leeman, F. 278
Lipson, C. 86
Mann, J.Y. 82
Manson, S.S. 85
Marples, D.L. 326
Marsh, K.J. 89, 90
Masia, B.B. 241
Maslow, A.H. 356
Matravers, D. 357
McFarland, R.A. 354
McGowan, P. 306
McKim, R.H. 281, 292, 297, 299, 306
McLuhan, M. 270, 271
McPherson, N. 112
Mendosa, E. 14
Miller, G.A. 355
Milner, P. 112
Miner, M.A. 86
Minsky, M. 299
Minsky, M.L. 299
Mitchell, M.R. 82
Monk, R. 29
Morant, G.M. 349
Morgan, M.H. 49
Murrell, K.F.H. 355, 357
Ord-Hume, A. W. 291
Osgood, C.C. 82
Owens, N. 363
Paasch, R.K. 363
Pahl, G. 292
Palmgren, A. 86, 108
Parmee, I.C. 330
Perkins, D. 363
Peterson, R.E. 88, 260
Petroski, H. 4, 49, 241, 273
Pook, L.E. 89, 90
Popper, K. 300
Prandtl, L. 144
Pugh, S. 329
Rheingold, H. 299
Rittel, H. 239, 241, 290
Rodenacker, W.G. 278
Roebuck, J.A. Jr. 348, 349, 354
Samuel, A.E. 87, 290, 291
Shannon, C.E. 355
Shigley, J.E. 85, 86, 149
Simon, H. A. 290
Smith, D.K. 300
Soderberg, C.R. 90
Stephens, R.I. 85
Stoudt, H.W. 354
Strunk, W. 338
Sullivan, L.H. 357
Thalmann, D. 299
Thalmann, N.M. 299
Thompson, W.D'A. 357
Thompson, W.G. 348
Timoshenko, S.P.
25, 49, 50, 62, 63, 69, 82, 87, 101, 103, 108, 144
Vamplew, C. 356
von Mises, R. 144
Wahl, A.M. 139
Waterton, W.A. 4
Weaver, W. 355
Weber, R.L. 14
Weir, J. G. 290, 291
White, E.B. 338
Wiener, N. 356
Wilde, D. J. 330
Wöhler, A. 82, 83, 88, 90
Woods, A. 345
Woollons, D.J. 363
Yoshino, H. 108
Young, W.C. 25, 107, 264
Zimmerli, F.P. 139, 140

SUBJECT INDEX

A

A-M diagram 90–94
  constant life curves 90
  infinite life curve 90
  springs 140
aesthetics 356
annual cost equivalent 342, 342–348
anthropometry 349
  body size 349
  design for displays 355
  dynamic 349
  population stereotypes 356
  processor of information 355
  sensor or transducer 354
  source of power 354
  static 349
  tracker and controller 355
  workspace 354
audit
  design 295
  structural 241
axial loading
  columns 144
axial stress
  pressure vessels 170

B

backward-chaining 280
beam behaviour 25–26
beam columns 155
beams
  conventions 25
  hogging 26
  neutral layer 25
  sagging 26
behaviour
  motivation 356
bending stress
  welded joints 228
benefit-cost-analysis 326, 344–348
bolt and screw terminology
  bolted joints 211–212
bolted joints 195
  stiffness 200
bug-list 284
butt-welds
  pressure vessels 171

C

capital recovery factor 341, 342
cash flow 341
causal networks 360
codes of practice
  bolted joints 206
  pressure vessels 164
columns 144–155. *See also* generic engineering components
  beam columns 155
  buckling 144, 147
    long/slender columns 147
  critical buckling load 148
  design algorithms 152
  design method 152
  design rules 155
  eccentric loading 153
  eccentricity ratio 154
  effective length 149
  elastic curves 144
  end fixity 149

Euler buckling 144
fixed-pinned 149
General Method of Design 146
Johnson parabola 152
Johnson-Euler curve 152
radius of gyration 149, 150–151
secant formula 154
short 151
slenderness ratio 149
complexity 291
component 239
key load-bearing 239
concept-map 291
concurrent engineering 294
constraint
in design 294, 325
contact phenomena 101
contact stresses 101–110
brinelling 103
conforming contact 108
deformations 104
failure in bearing 103
general contact between ellipsoidal bodies 106–107
geometry 104
non-conforming contact 108
spheres in contact 103–106
stress fields generated 107
corrosion 57–59
corrosion allowance
pressure vessels 172
cost
fixed 340
initial 340
of project 339
annual cost equivalent 342, 342–348
capital recovery factor 341, 342
cash flow 341
discounted payback period 342
present worth factor 341, 342
simple payback period 342
operating 340
overall 340
recurrent 340
variable 340
criteria
in design 294, 325

D

decision support systems
benefit-cost-analysis 326, 344–348
design tree 327
fishbone diagrams 327
in design 323
input/output analysis 325
mathematical modelling 330
morphological analysis 326
optimisation 330
response surface 332
scaled check lists 328
decision tables 328
deferred judgement 299–300
DEFP
Distortion energy failure predictor 69–70
delimitation 303
design
enformulation 278
problem solving 272–273
design boundary 294
design constraint 294
design goal 294
design life 240
design method
pressure vessels 168
design parameter 294
design process 292–295
design requirement 294
design rules
axially loaded components 155
bolted joints 215
pinned joints 221
pressure vessels 171, 179
shafts 122
springs 143
welded joints 230
design tree 327–328
disabled
design for 357
discipline
in design 271
discordance 291
discounted payback period 342

E

ED-PAD 295
effective length

welded joint 226, 227
effectiveness 324
embodiment
in design 280
end fixity
columns 149
endurance limit 82
shafts 121
springs 139
enformulation
in design 278–292, 284–288
engineering component 3
design synthesis 111
engineering materials 52–57
aluminium 57
copper (brass, bronze) 57
free-cutting steel 54
hardening 54
high-strength low-alloy (HSLA) steels 55
nickel alloys 57
plain carbon steels 53
plastics and composites 59
selection 60–66
special steels 56
stainless steels 56
steel and cast iron 53
wood and concrete 60
ergonomics 348–359
anthropometry 349
estimation 15–21
rational 19
event diagrams 360

F

factor of safety 72–74
allowable-stress 72
design factor 72
load factor 72
safe-load/working-load 72
stress factor 72
working-stress 72
factors of safety 5
failure predictors
brittle materials 66
distortion energy 69–70, 118
shafts 122
maximum principal stress 66–67
shafts 123
maximum shear stress 67–69, 118, 146
multiaxial stress 67
static loading 62–66
von Mises-Hencky theory 69
failure
fatigue
crack-growth 82, 83
crack-initiation 82
fracture 82
intensity 5
operational 5
planes of failure
pinned joints 218
technical 5
unpredictable (acts of God) 5
worst credible accident 5
fatigue 82–94
beachmarks 83
bolted joints 208
cumulative damage 83, 86
dislocation theory 85
endurance limit
effect of geometry 87–89
effect of surface coating 89
effect of surface finish 89
notch radius 87
notch sensitivity 87
fluctuating load 83
fully-reversed load 83
Gerber assumption 90
high-cycle 83
ideal endurance limit 86
load cycles 83–84
low-cycle 83
Miner's rule 86
modified Goodman line 90, 92
repeated load 83
S-N diagram 85
stress field 85
fault tree analysis 359, 361–363
fault/failure analysis 327, 359–364
causal networks 359, 360
event diagrams 360
fault tree analysis 359
fillet welds
parallel loading 225. *See also* mechanical connections: welded joints
fishbone diagrams 327

fitness for purpose 324, 330
flexibility
  in design 301
fluency
  in design 301
forward-chaining 280

G

gedankenexperiment 242
generic engineering components 23–24, 239
  columns 144–155
  shafts 112–124

H

hard information 284
hazards 359–364
heuristic 15
heuristics 290
hoop stress
  pressure vessels 169

I

idea-log 281
information theory
  bit 355
initial appreciation 294–299
input/output analysis 324, 325–326
Ishikawa diagram 327
islands of certainty
  in design 284

L

Liberty ship 11
loss of opportunity 339, 340–342
Lueders lines 63

M

mathematical modelling 330
mechanical connections 187–195
  bolted joints 195
    anchor point 196
    Bolt and screw terminology 211–212
    bolt force 199, 204–205
    codes of practice 206
    connection regions 196
    contact surface 196
    design equations 203, 208
    design rules 215
    effective gasket seating width 206
    elements in parallel 202
    external force on 196
    failure criteria 205–206
    fatigue 208–209
    flange 196
    friction joints 206–207
    gasket contact force 199
    grades of SAE bolting 210
    joint force 205
    modes of failure 198
    nuts,dilation 212
    pre-loaded 200
    pre-loading torque 212
    proof strength of a bolt 209
    sealing force 196
    series of elements 201
    shank thinning 208
    simple mathematical model 198
    stiffness of bolt and joint 200–202
    structural failure of bolts 207
    thread terminology 210–211
    threaded fasteners 209–212
    types of threaded fasteners 212
  clamped joints 189
  classification 188–194
  fasteners 189, 195
  glued joints 194
  load-carrying 188
  pinned joints 193, 215–221
    design rules 221
    eye-plates 216
    key dimensions 217
    modes of failure 218
    planes of failure 218
    shear-connectors 215
    typical applications 216
  structural connection 187
  welded joints 221–230
    allowable stresses 225
    bending moments 228–229
    bending stress 228
    combined stresses 229
    design for strength 224–225
    design rules 230
    effective length 226, 227
    fillets-parallel loading 225–226

leg length 225
oblique loads 228
throat 225
transverse loading 226–228
types 223–224
mode of failure 51–52
axial compression
columns 147
axial tension 146
bolted joints 198
buckling 52
columns 144
corrosion 52
creep 52
excessive-deflection 52
fatigue 52
fracture (rupture) 52
pinned joints 218
pressure vessels 164
wear/erosion 52
yielding 52
modelability 291
Mohr's circle construction 119, 176
moment of resistance 25
morphological analysis 326–327
motivation 356
MPFP
Maximum principal stress failure predictor 66–67
MSFP
Maximum shear stress failure predictor 67–69
multiaxial stress
bending and torsion 112, 131
pressure vessels 167

N

necking down 64
novelty 290

O

objectives
in design 294, 325
One-Hoss Shay 240
optimisation 330

P

parameter
in design 294
permanent set 63
pinned joints 215
design rules 221
eye-plates 216
modes of failure 218
pin deflection 221
planes of failure 218
typical applications 216
pre-loading torque
bolted joints 212
present worth factor 341, 342
pressure vessels
axial stress 170
butt-welds 171
corrosion allowance 172
deformations 174–175
design 164–179
design method 168
design rules 171, 179
design stress intensity 168
external bending 175
full radiography 171
hoop stress 169
large compressive bending stress 177
large tensile bending stress 176
mode of failure 164
modified design rule 172
Mohr's circle construction 176
multiaxial stress 167
non-cylindrical 177
quench-annealing 171
radial stress 170
rupture 164
small bending stress 176
spot radiography 171
standard codes of practice 164
strakes 171
stress-relieving 171
thick walled cylindrical
principal stresses 172
thick-walled cylindrical 172–174
peak values of stress 173
thin-walled cylindrical 167–172
typical 165–167
welded joint efficiency 171
welding and inspection 171
yielding 164
principal stress 65

pressure vessels
thick walled cylindrical 172
problem
complexity 291
discordance 291
enformulation 278, 284–288
evolution 279
modelability 291
novelty 290
seriousness 291
statement
goal-directed 281
need-directed 281
problem clarification 292
problem intensity 290–292
problem solving
backward-chaining 280
barriers to idea generation 302
conceptual blockbusting 302
deferred judgement 299–300
delimitation 303
flexibility 301
fluency 301
forward-chaining 280
generating ideas 295–302
heuristics 290
in design 272–273
clarification 292
hard information 284
islands of certainty 284
soft information 284
perceptual barriers 302–305
tunnel vision 305
problem-boundary 290
proportional limit 63

R

radial stress
pressure vessels 170
radius of gyration
columns 150–151
real interest rate 340
received wisdom 271
requirement
in design 294
response surface 332
rotating beam test 82
rule of thumb. *See also* estimation

S

scaled check lists 328
secant formula
columns 154
second moment of area 25
seriousness 291
service load 3
shafts 112–124. *See also* generic engineering components
actual stress 120
deflection 123
bending 121
torsional 122
design codes 120
design rules 122
dynamic loading 120
fully reversed torque 121
gearbox 114
hollowness ratio 117
keyway 115
limiting stress 120
modes of failure 118–119
out-of-balance forces 116
output shaft 114
overhung 116
principal stresses 119. *See also* principal stress
pulsating torque 121
resolved loads 114–115
solids pump 116–117
steady torque 121
stress concentration 115
von Mises' stress 119
shear-connectors
pinned joints 215
simple payback period 342
simultaneous engineering 294
slenderness ratio 149
soft information 284
springs
A-M diagram 140
active coils 137, 141, 143
bending and torsion 131
binding 140
bottoming 140
buckling 140
common types 132
compression

static loading 136–138
design rules 143
elastic phenomena 132
materials 142
mechanical 131–143
shear stress endurance limit 139
special types 133
spring index 137
spring rate 138
stiffness 132
tension springs 140
torsion springs 140
vibration 142
Wahl correction 139
standard codes of practice
pressure vessels 164
structural audit 241
structural distillation 21–23

T

Tacoma Narrows Bridge 4
theories of failure. *See also* failure predictors: static loading
thought experiment 242
thread terminology
bolted joints 210
trade-off 323, 330
triaxial stress
pressure vessels 167

U

units
ampere 14
candela 14
Kelvin 14
kilogram 14
metre 14
mole 14
second 14
SI (International System of Units) 14
utility 339. *See also* loss of opportunity
utility function 330

V

von Mises' stress 70

W

welded joints. *See* mechanical connections: welded joints
bending moments 228–229
bending stress 228
combined stresses 229
design rules 230
effective length 226, 227
leg length 225
oblique loads 228
parallel loading 225–226
shear stress 229
throat 225
transverse loading 226–228
wisdom
in design 271
received 271
worst credible accident 5, 13, 239, 363